Climate Resilience and Molecular Adaptation in Alpine Medicinal Plants

Pankaj Kumar • Hemant Sood
Rajinder Singh Chauhan
Editors

Climate Resilience and Molecular Adaptation in Alpine Medicinal Plants

Ecophysiology, Metabolism and Plantomics

Springer

Editors
Pankaj Kumar
Department of Biotechnology
College of Horticulture
Dr. Yashwant Singh Parmar University of Horticulture and Forestry
Solan, Himachal Pradesh, India

Hemant Sood
Department of Biotechnology & Bioinformatics
Jaypee University of Information Technology
Solan, Himachal Pradesh, India

Rajinder Singh Chauhan
Center for Life Sciences
Mahindra University
Hyderabad, Telangana, India

ISBN 978-981-95-6922-9 ISBN 978-981-95-6923-6 (eBook)
https://doi.org/10.1007/978-981-95-6923-6

This Springer imprint is published by the registered company Springer Nature Singapore Pte Ltd.
The registered company address is: 152 Beach Road, #21-01/04 Gateway East, Singapore 189721, Singapore

Preface

In the context of global climate change, alpine ecosystems represent some of the most sensitive and ecologically complex environments on Earth. The high-altitude regions of the Western Himalayas, characterized by extreme fluctuations in temperature, intense ultraviolet radiation, and limited nutrient availability, present a unique natural laboratory for studying plant adaptation and resilience. Within these ecosystems, alpine medicinal plants exhibit extraordinary physiological and molecular strategies that enable survival under severe abiotic stress. The present volume, *Climate Resilience and Molecular Adaptation in Alpine Medicinal Plants: Ecophysiology, Metabolism and Plantomics*, provides an integrative and comprehensive overview of the mechanisms underpinning such resilience. It examines the intricate interactions between environmental pressures, plant physiology, and molecular networks that define the adaptability of these species to changing climatic conditions.

This compilation brings together multidisciplinary contributions from experts in plant ecophysiology, molecular biology, biotechnology, and environmental sciences. The initial chapters introduce the diversity of alpine medicinal flora and outline the ecological and evolutionary challenges imposed by high-altitude environments. Subsequent sections address resource use patterns, ecosystem services, and functional diversity along elevational gradients, providing insight into how ecological processes shape plant function and resilience. The book advances into molecular and omics-driven perspectives, detailing the applications of next-generation sequencing (NGS), transcriptomics, metabolomics, and co-expression network analyses in elucidating adaptive traits and secondary metabolite biosynthesis in alpine medicinal plants.

Further chapters delve into molecular phylogeny and phylogeography to trace evolutionary lineages and adaptive divergence among alpine taxa. The role of epigenetic mechanisms, regulation of oxidative and antioxidative enzymes, and cold acclimation processes are explored to reveal the biochemical and regulatory frameworks underlying stress tolerance. In addition, the volume integrates recent developments in remote sensing for vegetation monitoring and innovative tissue culture methodologies to enhance conservation and propagation strategies. Recognizing the complex interdependencies within alpine ecosystems, the book also emphasizes microbial interactions, metabolic pathway elucidation, and the validation of traditional ethnobotanical knowledge. The concluding chapters highlight the socioeconomic and legal dimensions of plant commercialization, intellectual property rights, and sustainable resource management.

This book seeks to address the critical need for scientific understanding of how alpine medicinal plants can be utilized and conserved under rapidly changing climatic regimes. By integrating ecophysiological insights with molecular and omics technologies, it bridges traditional knowledge systems with advanced research approaches. The synthesis presented herein provides a scientific foundation for future studies aimed at improving conservation planning, sustainable exploitation, and biotechnological innovation involving alpine medicinal flora.

We extend our sincere appreciation to the contributing authors and collaborators for their scholarly efforts and commitment. We believe that this edited volume will serve as a valuable reference for researchers, academicians, and practitioners in plant science, ecology, and biotechnology, as well as for policymakers engaged in biodiversity conservation and sustainable management of Himalayan bioresources.

Solan, Himachal Pradesh, India — Pankaj Kumar
Solan, Himachal Pradesh, India — Hemant Sood
Hyderabad, Telangana, India — Rajinder Singh Chauhan

Acknowledgments

I express my sincere gratitude to *Professor Christian Korner*, whose pioneering work *Alpine Plant Life* has been a profound source of scientific inspiration in understanding alpine ecosystems. The opportunity to learn from his enlightening talk during the EMBO Workshop on High Elevation Plant Adaptation in a Changing Climate (HEPACC) held at CSIR-IHBT, Palampur, was truly transformative and pivotal in conceptualizing this edited volume.

I extend heartfelt thanks to my mentors for their invaluable guidance, scientific insight, and constant encouragement. I am especially indebted to *Professor Dinesh Kumar Srivastava*, my PhD mentor, for his sustained academic support, and to my postdoctoral mentors, *Professor (Dr.) Asis Datta, Padma Shri, Padma Bhushan*, and *Shanti Swaroop Bhatnagar Awardee*, Ex-Vice Chancellor, Jawaharlal Nehru University, and Founder Director, National Institute of Plant Genome Research (NIPGR), New Delhi, and *Dr. Sanjay Kumar*, Chairman, ASRB, and former Director, CSIR-IHBT, Palampur, for their distinguished mentorship and motivation. I am also thankful to *Dr. Ashish R. Warghat (Principal Scientist)* for the opportunity to conduct collaborative research in his laboratory at CSIR-IHBT, which enriched my understanding of alpine plant molecular physiology.

I gratefully acknowledge the Director of Research and the Vice-Chancellor, Dr. Y.S. Parmar, University of Horticulture and Forestry, Nauni, Solan, for providing a stimulating and supportive academic environment conducive to the successful completion of this volume. I also extend my appreciation to my colleagues and collaborators in the university for their cooperation and support. I also acknowledge the *Science and Engineering Research Board (SERB)*, Government of India, now reconstituted as the *Anusandhan National Research Foundation (ANRF)*, for supporting my research project as a Young Scientist (NPDF) on high-altitude medicinal plant.

This edited volume represents a collaborative effort between *Prof. Hemant Sood, Prof. Rajinder Singh Chauhan*, and myself (*Dr. Pankaj Kumar*). Our combined expertise in plant biotechnology and molecular biology on alpine ecosystems has resulted in an integrative and comprehensive resource on the climate resilience and molecular adaptation of alpine medicinal plants.

A special note of appreciation goes to *Shagun Sharma (PhD scholar)*, whom I fondly regard as a "*plant scientist gem*," for his dedicated and innovative research on alpine medicinal plants. His scientific rigor, curiosity, and commitment to excellence exemplify the next generation of alpine plant researchers.

Finally, I express my deepest gratitude to my *parents*, *wife* (*Dr Shivanti Negi Sharma*), and my *daughter* (*Jiva Sharma*), for their unwavering love, patience, and moral support. Their understanding and encouragement have been the true foundation of my scientific endeavors and the completion of this book.

Pankaj Kumar

Contents

Editors and Contributors

About the Editors

Pankaj Kumar is presently working as Assistant Professor (Biotechnology), Department of Biotechnology, Dr. Yashwant Singh Parmar University of Horticulture and Forestry, Solan, Himachal Pradesh, India. With more than nine years of specialized expertise in Medicinal plant biotechnology, he has significantly contributed to understanding the high-altitude adaptation of medicinal plants and secondary metabolite production through advanced cell and tissue engineering techniques. His research delves into the molecular mechanisms underlying metabolite production, utilizing cutting-edge comparative transcriptomics and metabolomics approaches. His academic credentials are further bolstered by his post-doctoral research at prestigious institutions such as the National Institute of Plant Genome Research and the Institute of Himalayan Bioresource Technology. He is a prolific author with over 70 research and review papers, 25 book chapters, and 7 notable books, including *Plant Stress Memory, Methods and Protocol* (Springer); *Agricultural Biotechnology- Latest Research and Trends* (Springer); and *Ornamental Horticulture: Latest Cultivation Practices and Breeding* (Springer). His innovative contributions have led to the publishing of five Indian patents, and he has been honored with numerous awards, such as the Young Scientists Fellowship Award-2021 by ICMR and the Scientist's Pool Scheme Award-2020 by CSIR.

Hemant Sood is a Professor at the Department of Biotechnology and Bioinformatics, Jaypee University of Information Technology, Solan, Himachal Pradesh, India. She is an internationally recognized expert in plant biotechnology, with significant contributions to alpine medicinal flora research. She brings over 18 years of academic and research excellence, specializing in plant tissue culture, genetic transformation, and the sustainable utilization of high-altitude medicinal flora. Her achievements include the development of micropropagation technologies for 16 medicinal plant species, many of which are critical for biodiversity conservation. Her global exposure includes receiving an International Scholarship from Israel's Ministry of Foreign Affairs (MASHAV) for training at the Hebrew University of Jerusalem and serving as a session chair at leading international conferences in Italy, Thailand, and the United States. Her accomplishments have earned her numerous accolades, including the Indian Science Congress Young Scientist Award (2011) and recognition as one of 125 women luminaries in STEM by the Confederation of Indian Industry (CII). She is also the holder of four granted patents, including an international patent, showcasing her innovative contributions to plant-based biotechnology. Furthermore, she has a strong publishing record, contributing book chapters as well as authoring numerous high-impact research articles in the journals of national and international repute.

Rajinder Singh Chauhan, a highly accomplished academic and global scientific expert, currently holds the position of Dean of Life Sciences at Mahindra University, Hyderabad, and has over 30 years of academic and research experience in plant biotechnology, medicinal plant ecophysiology, and metabolism. His extensive work on alpine medicinal flora, including leading multi-institutional, government-funded projects on endangered herbs of the Himalayas, underscores his authority in this domain. He has authored over 200 peer-reviewed publications and holds 6 granted patents, including innovative processes related to *Picrorhiza kurroa* and *Swertia chirayita*, two prominent medicinal herbs of alpine regions. His leadership roles in prestigious institutions and collaborations with global researchers from the United States, South Africa, and Europe further demonstrate his international impact and global diversity standards.

Contributors

Rashmi Kishor Agashe Center for Life Sciences, Mahindra University, Hyderabad, Telangana, India

Indesh Attri Sardar Beant State University, Gurdaspur, Punjab, India

Anwesha Bandyopadhyay School of Biotechnology, Faculty of Applied Sciences & Biotechnology, Shoolini University, Bajhol, Himachal Pradesh, India

Ashun Chaudhary Department of Plant Sciences, School of Life Sciences, Central University of Himachal Pradesh, Shahpur Campus, Kangra, Himachal Pradesh, India

Rajinder Singh Chauhan Center for Life Sciences, Mahindra University, Hyderabad, Telangana, India

Deepika Department of Biotechnology, College of Horticulture, Dr. Yashwant Singh Parmar University of Horticulture and Forestry, Solan, Himachal Pradesh, India

Arti Devi Department of Bio-Sciences, Himachal Pradesh University, Shimla, Himachal Pradesh, India

Bandana Dhiman RHRSS & Krishi Vigyan Kendra Tabo, Lahaul & Spiti, Himachal Pradesh, India

Bhawna Dipta ICAR-Central Potato Research Institute, Shimla, Himachal Pradesh, India

Deepti Gupta Department of Bio-Sciences, Himachal Pradesh University, Shimla, Himachal Pradesh, India

Rolika Gupta Division of Microbiology, Career Point University, Hamirpur, Himachal Pradesh, India

Nishant Jain Department of Electronics and Communication Engineering, Jaypee University of Information Technology, Solan, Himachal Pradesh, India

Jerin James Department of Biotechnology, School of Applied and Life Sciences, Uttaranchal University, Dehradun, Uttarakhand, India

Bhuvnesh Kapoor Department of Biotechnology, University Institute of Biotechnology (UIBT), Chandigarh University, Gharuan, Mohali, Punjab, India

Rutvik Sameer Kulkarni Center for Life Sciences, Mahindra University, Hyderabad, Telangana, India

Bhawna Kumari Center for Life Sciences, Mahindra University, Hyderabad, Telangana, India

Neha Kumari School of Biological and Environmental Sciences, Shoolini University, Solan, Himachal Pradesh, India

Ruchika Kumari Department of Plant Sciences, School of Life Sciences, Central University of Himachal Pradesh, Shahpur Campus, Kangra, Himachal Pradesh, India

Pankaj Kumar Department of Biotechnology, College of Horticulture, Dr. Yashwant Singh Parmar University of Horticulture and Forestry, Solan, Himachal Pradesh, India

Suresh Kumar Department of Bio-Sciences, Himachal Pradesh University, Shimla, Himachal Pradesh, India

Varun Kumar Center for Life Sciences, Mahindra University, Hyderabad, Telangana, India

Roop Lal Department of Botany, Government College Chowari, Chamba, Himachal Pradesh, India

Siddhant Diliprao Mahabale Center for Life Sciences, Mahindra University, Hyderabad, Telangana, India

Nikhil Malhotra BRIC-National Institute of Plant Genome Research, Aruna Asaf Ali Marg, New Delhi, India

E. Manikandan Department of Biotechnology, College of Horticulture, Dr. Yashwant Singh Parmar University of Horticulture and Forestry, Solan, Himachal Pradesh, India

Poonam Meena Department of Biotechnology, School of Applied and Life Sciences, Uttaranchal University, Dehradun, Uttarakhand, India

Sambuddha Mukherjee Department of Economics and Sociology, Punjab Agricultural University (PAU), Ludhiana, Punjab, India

Ashiya Naaz Department of Biotechnology, School of Applied and Life Sciences, Uttaranchal University, Dehradun, Uttarakhand, India

Manisha Nanda Department of Microbiology, Graphic Era deemed to be University, Dehradun, Uttarakhand, India

Shivanti Negi Department of Biotechnology, College of Horticulture, Dr. Yashwant Singh Parmar University of Horticulture and Forestry, Solan, Himachal Pradesh, India

Tarun Pal School of Bioengineering and Food Technology, Faculty of Applied Sciences & Biotechnology, Shoolini University, Bajhol, Himachal Pradesh, India

Shiv Rattan Department of Ornamental Plants and Agricultural Biotechnology, Agricultural Research Organization (ARO) – Volcani Institute, Rishon LeZion, Israel

Ankush Saini School of Biological and Environmental Sciences, Shoolini University, Solan, Himachal Pradesh, India

ICAR-Central Potato Research Institute, Shimla, Himachal Pradesh, India

Muskaan Sharma School of Biotechnology, Shoolini University, Solan, Himachal Pradesh, India

Neha Sharma ICAR-Central Potato Research Institute, Shimla, Himachal Pradesh, India

Nidhi Sharma MS Swaminathan School of Agriculture Sciences, Shoolini University, Solan, Himachal Pradesh, India

Nishesh Sharma Department of Biotechnology, School of Applied and Life Sciences, Uttaranchal University, Dehradun, Uttarakhand, India

Rohit Sharma Department of Forest Products, College of Forestry, Dr. Yashwant Singh Parmar University of Horticulture and Forestry, Solan, Himachal Pradesh, India

Shagun Sharma Department of Biotechnology, College of Horticulture, Dr. Yashwant Singh Parmar University of Horticulture and Forestry, Solan, Himachal Pradesh, India

Shubham Sharma Institute of Nano Science and Technology (INST), Sector-81, Knowledge City, Sahibzada Ajit Singh Nagar, Mohali, Punjab, India

Vishal Sharma Department of Forest Products, College of Forestry, Dr. Yashwant Singh Parmar University of Horticulture and Forestry, Solan, Himachal Pradesh, India

Yash Pal Sharma Department of Forest Products, College of Forestry, Dr. Yashwant Singh Parmar University of Horticulture and Forestry, Solan, Himachal Pradesh, India

Hemant Sood Department of Biotechnology & Bioinformatics, Jaypee University of Information Technology, Solan, Himachal Pradesh, India

Naresh Thakur College of Horticulture, Dr. Yashwant Singh Parmar University of Horticulture and Forestry, Neri, Hamirpur, India

Palak Thakur Department of Plant Sciences, School of Life Sciences, Central University of Himachal Pradesh, Shahpur Campus, Kangra, Himachal Pradesh, India

Saurabh Thakur Department of Forest Products, College of Forestry, Dr. Yashwant Singh Parmar University of Horticulture and Forestry, Solan, Himachal Pradesh, India

Shriyanshu Thakur Division of Microbiology, Career Point University, Hamirpur, Himachal Pradesh, India

Ira Vashisht School of Biotechnology, Jawaharlal Nehru University, New Delhi, India

Ashish R. Warghat Biotechnology Division, CSIR-Institute of Himalayan Bioresource Technology, Palampur, Himachal Pradesh, India

Academy of Scientific and Innovative Research, Ghaziabad, India

Rasna Zinta ICAR-Central Potato Research Institute, Shimla, Himachal Pradesh, India

Introduction to Alpine Medicinal Flora and Climate Resilience

1

Vishal Sharma, Ruchika Kumari, Saurabh Thakur, Rohit Sharma, Ashun Chaudhary, Yash Pal Sharma, and Pankaj Kumar

Abstract

The Alpine biome is home to a diverse array of medicinal herbs with noteworthy biological, ethnobotanical, and medicinal importance. The uniqueness of alpine biome medicinal herbs is attributed to their higher altitude, shorter growth periods, and low temperatures. With the intensification of the changing climatic conditions, the alpine zone ecosystem is experiencing increasing annual temperatures, fluctuating rainfall dynamics, melting glaciers and elevated occurrence of severe weather events. These climatic changes endanger the existence, distribution and metabolic makeup of alpine flora in the Himalayan region. This chapter highlights the climate-induced stress impact on the growth dynamics, habitat and plant diversity destruction resulting in elevation shifts for the distribution of important alpine flora, also providing an outline of floral biodiversity in the alpine zone of the Himalayas, emphasizing their survival characteristics, niche functions and involvement in indigenous health systems, specifically in alpine regions. Major focus is on the biological potential of these plants, where majority of them act as ecological markers and are important in village-based livelihoods via indigenous medicinal utilisation and trade. They are vital element of alpine

Authors Vishal Sharma and Ruchika Kumari have equally contributed to this chapter.

V. Sharma · S. Thakur · R. Sharma (✉) · Y. P. Sharma
Department of Forest Products, College of Forestry, Dr. Yashwant Singh Parmar University of Horticulture and Forestry, Solan, Himachal Pradesh, India

R. Kumari · A. Chaudhary
Department of Plant Sciences, School of Life Sciences, Central University of Himachal Pradesh, Kangra, Himachal Pradesh, India

P. Kumar
Department of Biotechnology, College of Horticulture, Dr. Yashwant Singh Parmar University of Horticulture and Forestry, Solan, Himachal Pradesh, India

P. Kumar et al. (eds.), *Climate Resilience and Molecular Adaptation in Alpine Medicinal Plants*, https://doi.org/10.1007/978-981-95-6923-6_1

habitats, and their unique characteristics acts as a foundation for understanding climate impact-related processes. Rising temperatures allow lower-altitude species to infiltrate alpine ecosystems, increase the competition and potentially replacing native flora and fauna. Conservation measures are critical for conserving alpine biodiversity despite these challenges.

Keywords

Alpine biome · Climate resilience · Medicinal Flora · Phytoconstituents · Western Himalayas · Habitat shift

1 Introduction

Medicinal plants play an essential part in various healthcare systems around the world due to their convenience of availability and low cost. Besides that, raw materials derived from medicinal plants account for 70–90% of demand in most developing countries, and this demand is expanding daily (Robinson & Zhang, 2011). Several authors highlighted the essential role of medicinal plants in current medical and pharmaceutical systems, which led to fulfilling the needs of people around the world for the treatment of several chronic diseases (Heinrich et al., 2018). High altitude regions in several parts of the world lack access to modern treatments, so they rely only on available important native medicinal plants. It is considered that plant material or plants taken from higher altitudes are more biologically active than those collected from lower altitudes (Turner et al., 2011). The plants growing in alpine and subalpine regions and in extreme environment adapts to a particular morphological and physiological traits like short plants found in cold-stressed areas. Climate change has the potential to impact both the quantitative and qualitative characteristics of plants, such as phytochemical composition, biomass yield etc. Phytochemical composition has a wide range of direct and indirect effects on human health (Hou et al., 2024). The unpredictable pattern of weather, receding ice masses, increasing sea levels are the important climate change factors, both locally and globally (Kumar et al., 2021b) (Fig. 1.1).

Himalayas are younger and still developing mountain range and renowned as biological diversity hotspot and home for diverse flora and fauna (Saxena & Rao, 2024). The monsoonal and non–monsoonal systems are wide weather systems of Himalayan region. Whereas, the greater Himalayas befalls in the Indian monsoon system categorized by glacier landscapes, higher elevation cold desert in extreme north and equidistant to the Himalayan belt signifies the non-monsoonal climate (Sekar et al., 2024). The soil type is sandy or sandy-loam which is characterized by low content of nitrogen, organic matter with 7–11 pH range (Körner, 2007). Alpine zone exhibits a broad range of ecological characteristics with specific habitat *viz.*, turf forming, low-statured and bushy (Pandey et al., 2018). Flora in alpine biome starts between altitude range of 3200–3500 m amsl forming alpine and temperate flora controlled by species *viz.*, *Pinus wallichiana*, *Cedrus deodara*, *Abies pindrow*,

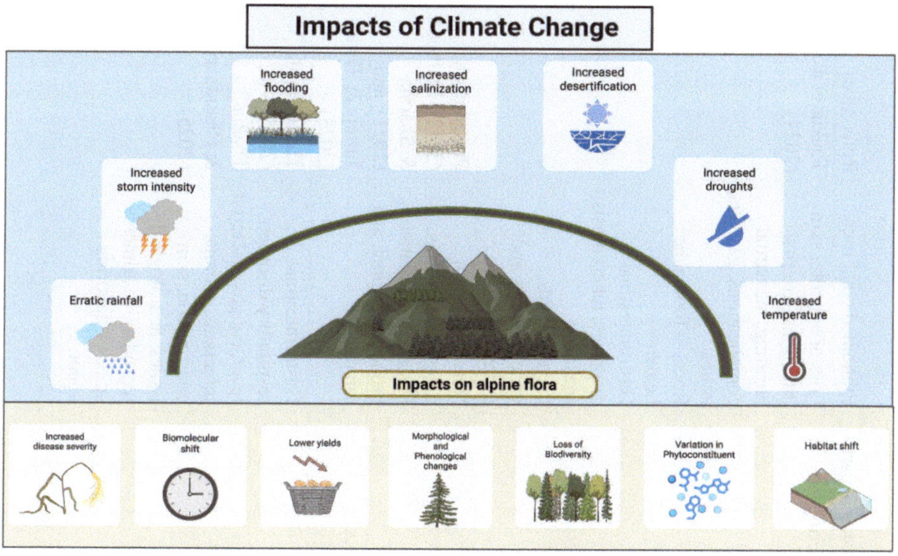

Fig. 1.1 Showing the impact of climate change on high-altitude alpine flora. (Created in BioRender. Science, P. (2025) https://BioRender.com/wk6nquz)

Saussurea involucrata, Rhodiola rosea, Arnebia euchroma, Selinum spp., *Angelica* spp., *Aconitum* spp., *Taxus* spp., *Ephedra* spp., *Dactylorhiza* spp., *Fritillaria* spp., *Polygonatum* spp., *Nardostachys* spp., *Picrorhiza* spp. etc. (Thakur et al., 2024; Kumar et al., 2025; Sharma et al., 2025). At higher altitude, these tree species have ability to bear ecological stress, become dwarfed and create the treeline beyond which they are not able to flourish. Above the treeline, the ecosystem led by bushy habitat, well-developed deep root systems and woody branching such as *Lonicera* spp., *Juniperus* spp. and *Caragana* spp., shadowed by alpine pastures in high altitude which includes herbaceous species (Sekar et al., 2024).

The higher altitudes, plant species are medicinally vital for humans and are being damaged by climate change, as well as due to human-caused environmental destruction and overexploitation. It has become necessary to accumulate well-researched data and observations to sustain the impact of various factors, such as rising CO_2, temperature changes, and rainfall patterns on high-altitude plants. It is important to take necessary measures for determining the true impact of climate change on medicinal plants and other high-value useful flora at higher elevations, and to mitigate these effects on alpine flora.

2 Himalayan Alpine Medicinal Flora

Following are some important high-altitude plants: Their Introduction, Phytochemistry & Ethnomedicinal uses (Table 1.1).

Table 1.1 Important alpine medicinal plants of North Western Himlayan Region

Sr. no.	Botanical name	Common name	Altitudinal range	Economical part	Ethnomedicinal uses	References
1.	*Achillea millefolium* L.	Yarrow or Milfoil	1050–3600 m amsl	Flowering top	Loss of appetite, skin infections, external bleeding, menstrual complications, gastritis issues. Roots—Toothache Leaves—Gastrointestinal disorders, tuberculosis, snakebite, haemorrhoids, varicose veins and dysmenorrhea	Thakur et al. (2024)
2.	*Aconitum heterophyllum* Wall.	Atis & Wolf's bane	2000–5000 m amsl	Root	Nervous system, digestive disorders, hysteria, rheumatism, bronchial asthma, and tumours.	Prasad et al. (2012), Malhotra & Sharma (2021), and Talreja & Tiwari (2023)
3.	*Angelica glauca* Edgew.	Chora, Gandrayan	3000–3700 m amsl	Root	Gastrointestinal, respiratory disorders, fever, coughs, bacterial infections, asthma, constipation, skin issues, neurological, reproductive ailments, spice or flavoring agent.	Thakur et al. (2025) and Thakur & Chaudhary (2025)

4.	*Arnebia euchroma* (Royle) Johnston	Ratanjot	3000–4800 m amsl	Root	Root powder—Coughs, lung ailments, menorrhagia, skin diseases cold, blood vomiting, backache Mixing with apricot/ mustard oil used as hair oils to cure dandruff and baldness.	Kumar et al. (2021a, b)
5.	*Dactylorhiza hatagirea* (D. Don) Soo	Salam panja	2172–4928 m amsl	Tuber	Dysentery, seminal weakness, diarrhoea, leucorrhoea, colic pain, emaciation, urinary tract infections, diabetes, cough, fever, and neurasthenia	Ved & Tandon (2009) and Thakur et al. (2021)
6.	*Fritillaria cirrhosa* D. Don	Kakoli	3000–4200 m amsl	Bulb	Respiratory disorders such as asthma, extreme phlegm, and prolonged cough.	Mathela et al. (2021) and Kumar et al. (2025)
7.	*Nardostachys jatamansi* (D. Don) DC.	Jatamansi	3000–5000 m amsl.	Root and rhizome	Gastrointestinal, neurological, circulatory, respiratory, urinary, reproductive, and skin issues. Used in Tibetan system of medicine. Rhizome—Food poisoning, gastritis, cough and cold, chronic fever and wound healing.	Tanaka & Komatsu (2008) and Mabberley & Noltie (2014)

(continued)

Table 1.1 (continued)

Sr. no.	Botanical name	Common name	Altitudinal range	Economical part	Ethnomedicinal uses	References
8.	*Podophyllum hexandrum* Royle	May Apple	2500–4000 m amsl	Root and rhizome	Powdered roots and leaves—Skin diseases, brain cancer, genital warts, Hodgkin's disease, non-Hodgkin lymphoma, monocytoid leukemia. Fruit—Gastric problems.	Anand et al. (2022)
9.	*Rheum australe* D. Don.	Himalayan Rhubarb	1600–4200 m amsl	Root and rhizome	Hepatic disorders, gastrointestinal problems, blood purifier.	Rokaya et al. (2012)
10.	*Saussurea costus* (Falch.) Lipsch.	Kuth	3300–4000 m amsl	Root	Dysentery, ulcer, and abdominal pain. In Kullu (Himachal Pradesh), root decoction is used for coughs and colds.	Butola & Samant (2010) and Kumari et al. (2024)
11.	*Saussurea obvallata* (DC.) Edgew.	Brahma Kamal	3000–4800 m amsl	Rhizome, whole plant	Hepatic stimulant, digestive properties, urogenital disorders, wound healing. Utilized in the Tibetan medicine system "Amchi" to treat limb paralysis and ischemic brain injury.	Ghosh (2017) and Semwal et al. (2020)

12.	*Selinum* L.	Bhutakeshi/Bhootkeshi	2000–4000 m amsl	Rhizome, fruit	Sedative, aphrodisiac. Skin conditions and muscular oedema, hysteria, epilepsy, mental disorders, cough, asthma, evil spirit, allergies, burns, lactations, insect repellent.	Thakur et al. (2024) and Kumari et al. (2025)
13.	*Taxus contorta* Griff.	West Himalayan Yew	1800–3300 m amsl	Bark and leaves	The drug 'Zaṛnab' from leaves—Cardiac diseases in the Unani system of medicine. Tuberculosis, diarrhoea, inflammation, and fever.	Sajwan et al. (2015) and Majeed et al. (2021)
14.	*Trillium govanianum* Wall. Ex D. Don	Nagchhatri	2400–3500 m amsl	Rhizome	Rhizomes—Inflammation, backache, dysentery, wounds, reproductive disorders, and skin boils. Tribes in Jammu & Kashmir documented that roots beverage consumption cures headache.	Polunin & Stainton (1984) and Kundra et al. (2020)

2.1 *Achillea millefolium* L.

A. millefolium (Common name—'yarrow' & 'milfoil'), member of the family Asteraceae is an alpine medicinal plant, which has its distribution throughout Europe, North Africa and Asia (Vitalini et al., 2011). *A. millefolium* extract possesses potential bioactivities such as antifungal, antioxidant, anticancer, antibacterial, anti-inflammatory and anti-arthritic, antispasmodic, hypotensive, cardiovascular and bronchodilatory effects (Khan & Gilani, 2011). Furthermore, essential oil of this plant is valuable and renowned for its anti-inflammatory and antiseptic properties (Nadim et al., 2011). Phytochemical investigation of this plant revealed the presence of flavonoids *viz.,* Achillinin (Li et al., 2011), Millifolide A, B & C, lutelin, apeginenin, rutin and campherol (Bimbiraite et al., 2008). Chemistry of its essential oil is intricate because of various chemical constituents (Gudaitytė & Venskutonis, 2007). Isolated primary essential oil constituents are β-pinene, sabinene, linalool, chamazulene, 1,8-cineole, α-thujone, ocimene, β-eudesmol, β-thujone, caryophyllene oxide, ascaridole, α-bisabolol and camphor (Mockute & Judzentiene, 2003).

2.2 *Aconitum heterophyllum* Wall

A. heterophyllum (Common name—Atis & wolf's bane) is an important medicinal herb distributed along alpine and subalpine regions (Chandra et al., 2022) and found at elevations of 2000–5000 m amsl in the Western and Eastern Himalayas covering India, Nepal, Bhutan, South Tibet, China, Pakistan and Japan (Malhotra & Sharma, 2021). In India, it is mainly distributed along Jammu & Kashmir, Himachal Pradesh and Uttarakhand (Wani et al., 2022). At present, the consumption of *A. heterophyllum* is estimated to be greater than 200 metric tons per annum and valued at 4000 rupees/kg (Gupta & Souravi, 2020). Due to its high medicinal potential and high market value, it has been exploited unsystematically and is characterised as an endangered species in the IUCN Red List (Ved et al., 2015). Various bioactive compounds obtained from the roots of *Aconitum heterophyllum* mainly alkaloids like aconitine, heteratisine, atisine, heterophylline, mesaconitine, hypaconitine, benzoylaconine, benzoylmesaconine, and benzoylhypaconine (Talreja & Tiwari, 2023). Due to the presence of numerous bioactive compounds, its roots possess various therapeutic benefits (Prasad et al., 2012).

2.3 *Angelica glauca* Edgew

A. glauca belongs to the family Apiaceae is an important aromatic plant commonly known as Chora. It is endemic to the Indian Himalayan Region and distributed throughout the temperate to alpine areas of Jammu & Kashmir, Himachal Pradesh and Uttarakhand (Thakur & Chaudhary, 2024). The suitable natural habitat of *A. glauca* occurs between 3000–3700 m amsl in shady, moist north-facing slopes found in the shady canopy of *Quercus semecarpifolia*, *Rhododendron campanulatum* and *Betula utilis*. It grows and flourishes in humus-rich soil near water springs and forest edges. The whole plant of *A. glauca* possesses medicinal properties. However, its root is primarily responsible for bioactive potential. Roots being the main economic part of the plant, it is susceptible to destructive harvesting and thus it is characterised as an endangered plant species in the IUCN Red List (Thakur & Chaudhary, 2025).

2.4 *Arnebia euchroma* (Royle) Johnston

A. euchroma (Royle) Johnston commonly known as "Ratanjot" is an endangered medicinal herb belongs to family Boraginaceae (Devi et al., 2023). This species is locally known as Demok in Spiti & Zanskar, and Khuammed in Bhoti. It has its habitat on rocky, arid and sandy edaphic conditions at the elevation range between 3000 and 4800 m amsl across temperate to subalpine zones of the Western Himalayas in India. Worldwide, it is distributed in high-altitude zone of North Africa, Afghanistan, and Central Asia (Kumar et al., 2021a). In India, it is distributed in cold alpine region of Himachal Pradesh (Lahaul & Spiti), Uttarakhand, Leh & Ladakh, and Jammu & Kashmir (Ambrish & Srivastava, 2014). *A. euchroma* has purple to pale pink flowers, narrow upper leaves, and large, toothed base leaves. The calyx is very open and enfolds the base. The flowering period is July to September (Malik et al., 2011). *A. euchroma* extract and its phytoconstituents possesses potential bioactivities such as anti-cancer (Xiong et al., 2009), anti-inflammatory (Ou et al., 2017), anti-obesity (Siavash et al., 2016), and anti-diabetic (Noorafshan et al., 2017). The majorly reported phytoconstituents in *A. euchroma* are shikonin, isobutyryi-shikonin, coumarins, acetyl-shikonin, isovaleryl-shikonin, deoxy-shikonin, β,β-di-methylacryl-shikonin, β-hydroxy-isovaleryl-shikonin, arnebin-7, arnebinone and stigmasterol which are primarily renowned for their immunomodulatory, antimicrobial and anti-cancerous effects (Kumar et al., 2021a).

2.5 *Dactylorhiza hatagirea* (D. Don) Soo

D. hatagirea is a critically endangered medicinal orchid (Dhiman et al., 2019) belongs to the family Orchidaceae. Its flowering takes place between July to August, and fruiting commences from August to September (Rawal et al., 2018). It is endemic to the Western Himalayas and distributed along an altitude between 2172 and 4928 m amsl in the subalpine and alpine biome of China, Afghanistan, Bhutan, India, Nepal, Pakistan, and Tibet (Thakur et al., 2021). Tubers of this plant are sold by the names 'Salam panja' and 'Hath panja' locally known under the vernacular names such as 'Hathajari' and 'Salam panja' in Jammu & Kashmir, 'Angulagpa' and 'Wanglak' in Ladakh, 'Hathpanja' in Uttarakhand, and 'Panchaule' in Nepal (Ved & Tandon, 2009). Various phytoconstituents such as dactyloses A, B, dactylorhin A, B, E, resveratrol, loroglossin and millitarine have been reported in the tuber of *D. hatagirea* (Kizu et al., 1999).

2.6 *Fritillaria cirrhosa* D. Don

F. cirrhosa (Common name—Kakoli) is a medicinal herb of Astavarga group native to Asia belonging to the family Liliaceae. Globally, it has wide distribution in India, Nepal, Myanmar, China, Bhutan and Pakistan. While, it has distribution zone in Uttarakhand, Himachal Pradesh and Jammu & Kashmir in India (Chauhan et al., 2011; Sharma et al., 2023). It flourishes primarily on grassy and rocky habitat in subalpine and alpine zone, frequently has association with shrubs such as *Rosa* spp., *Salix* spp. and *Lonicera* spp. at elevation range of 3000–4200 m amsl (Mathela et al., 2021). The phytoconstituents of *F. cirrhosa* bulbs have important pharmacological characteristics, the most distinguished of which are anti-tumorous and anti-inflammatory activities (Wang et al., 2014).

2.7 *Nardostachys jatamansi* (D. Don) DC

N. jatamansi (Common name—'Jatamansi' or 'Spikenard') is a perennial medicinal plant, member of the family Caprifoliaceae and originated in subalpine and alpine Himalayan region. As per the IUCN Red List of Threatened Species, it is categorized as Critically Endangered (Mabberley & Noltie, 2014). Owing to the illegal and massive harvesting, *N. jatamansi* falls under Appendix II of the Convention on International Trade in Endangered Species of Wild Fauna and Flora (CITES). Furthermore, this plant is categorized as Critically Endangered for the North-Western Himalayas by the Conservation Assessment and Management Plan (CAMP) (Molur & Walker, 1998). It has wide distribution in Alpine Himalayan Region and has its origin in Uttarakhand, Sikkim, Himachal Pradesh and Arunachal Pradesh in India. It is distributed in Nepal and Bhutan at an elevation range between 3000–5000 m amsl. Essential oils obtained from root and rhizome of this plant consists of various phytoconstituents such as sesquiterpenoids and derivatives of coumarin, which have excellent pharmacological importance and medicinal uses (Rekha et al., 2013). Globally, this plant is marketed and sold under different local names *viz.*, Gansong in China, Indische Valerian in Netherlands, Kanshoko in Japan, Valeriane Indienne in France and Indische Baldrian in Germany (Tanaka & Komatsu, 2008). In Himachal Pradesh (Parvati valley), the root powder is mixed with tobacco and utilized for the treatment of neurological and cardiac disorders (Sharma et al., 2004). The rhizome and root of this plant contain different phytoconstituents *viz.*, pyranocoumarins (jatamansin), furanocoumarins including oroselol (Shanbhag et al., 1964), xanthogalin, and jatamansinol (Shanbhag et al., 1965), valeranol, and patchouli alcohol (Rucker et al., 1976), and sesquiterpenoids. Nardostachysin is also reported from its rhizomes (Chatterjee et al., 2000).

2.8 *Podophyllum hexandrum* Royle

P. hexandrum is an important high-altitude medicinal plant belonging to the family Berberidaceae, mostly found in the Himalayan region of India and China upto 2500–4000 m amsl (Anand et al., 2022). In India, it is mainly distributed in the higher regions of Himachal Pradesh, Uttarakhand and Jammu & Kashmir. The roots and rhizomes of *P. hexandrum* are the source of resin known as podophyllin, which includes podophyllotoxone, podophyllotoxin, and epipodophyllotoxin (Broomhead & Dewick, 1990). Anti-cancer drugs etoposide and teniposide are semi-synthetic derivatives of natural podophyllotoxin (Zhang et al., 2018). It is harvested unscientifically by uprooting in large quantities from the wild. It is placed in the endangered plant category in the IUCN Red List (Chauhan, 2024). Due to various important bioactive constituents, this plant is considered as a divine drug in traditional Indian medicine and is used for its anti-tumour, antibacterial, antioxidant, and anti-cancerous properties (Sofi et al., 2025).

2.9 *Rheum australe* D. Don

R. australe (Common name—Himalayan Rhubarb) belongs to family Polygonaceae is an endangered alpine medicinal plant endemic to North-Western Himalayan region (Pandith et al., 2018). It is distributed in India, Bhutan, Nepal, China, Pakistan and Myanmar. It has its distribution in India from Jammu & Kashmir to Sikkim at altitude range of 1600–4200 m amsl (Press et al., 2000). This species is propagated by seeds. Flowering occurs from June to August, while fruiting occurs from August to September (Bhattarai & Ghimire, 2006). Economical parts of this plant are roots and rhizomes, harvested in October to November. Tribal communities in Jammu & Kashmir cultivating this plant from around 5000 years owing to its pharmacological importance and utilized in curing cancer (Kumar et al., 2015). The phytoconstituents reported in this plant majorly 1,8-dihydroxyanthraquinones *viz.*,emodin, rhein, physcion, aloe emodin and chrysophanol (Arvindekar et al., 2015), maesopsin and carpusin (Krenn et al., 2003), Revandchinone 1–4 (Babu et al., 2003), cascaroside C & D, cassialoin (Krenn et al., 2004). Stilbenes are the principal components of *Rheum* spp. (Ye et al., 2007). Principal phenolic constituents in roots *viz.*, quercetin, eugenol, rutin, mesopsin, gallic acid, rhapontigenin, epicatechin and desoxyrhapontigenin are accountable for antioxidant activity (Singh et al., 2013).

2.10 *Saussurea costus* (Falch.) Lipsch

S. costus (Common name—"Kuth") belongs to family Asteraceae is endangered medicinal herb and has its origin in Indian Himalayas (Kumari et al., 2024). It has its distribution range across subalpine range of Himalayas and found in Lahaul & Spiti of Himachal Pradesh, Uttarkashi and Barsu in Uttarakhand and Kashmir in Jammu & Kashmir of India. Worldwide, it has its distribution and occurrence in

China's Xinzhu, Ludian, Yulong and Yunnan region and also in Pakistan at an altitude between 3300–4000 m amsl (Butola & Samant, 2010). *S. costus* has a conservation status of endangered in Red Data Book of Indian Plants and is a critically endangered in International Union for Conservation of Nature (IUCN) owing to its scarce distribution range and over-exploitation (Walter & Gillett, 1998). *S. costus* is enlisted among 24 important medicinal plants under the objective of research and development support on *In*-situ and *Ex*-situ conservation methodologies with strategic aim of medicinal plant industry by the National Medicinal Plants Board (NMPB) and Planning Commission of India (Kala et al., 2006). Root of this herb is primary economic part and renowned for its different medicinal properties *viz.*, hepatoprotective, antibacterial and antifungal, anti-inflammatory, and anticonvulsant. Its essential oil colour is yellow to brown and employed broadly in production of perfumes, hair oil, incense and insect repellents because of its excellent blending ability with sandalwood, patchouli, rose and vetiver (Kumari et al., 2024). Phytochemical studies of its roots unveiled the presence of bioactive constituents such as monoterpenes, triterpenes, sesquiterpenoids, glycosides, lignans, steroids and flavonoids. The main bioactive compounds are dehydrocostus lactone, costunolide and saussucotusosides A & B (Liu et al., 2012; Hanh et al., 2021).

2.11 *Saussurea obvallata* (DC.) Edgew

S. obvallata (Common name 'Brahma Kamal'), a member of Asteraceae family is a sacred herb that is endemic to Indian Himalayas. It has a wide distribution in India, Nepal, China, Pakistan, Myanmar, Bhutan, and Tibet at elevations of 3000–4800 m amsl (Kirtikar & Basu, 1984). In India, it is distributed in Jammu & Kashmir, Himachal Pradesh, Uttarakhand and Sikkim. It thrives predominantly in alpine pastures, rocky surface and close to rivers (Ghosh, 2017). The principle phytoconstituents reported in this plant are palmitic acid, linoleic acid, stearic acid, dehydrocostus lactone, pregnanolone, litsomentol, bupleuronol and deconexent (Semwal & Painuli, 2019). This ethnic plant is employed as a tribute to deities in Uttarakhand at Badrinath, Tungnath and Kedarnath. On the occasion of 'Nanda Asthami' festival which is celebrated from September to October, *S. obvallata* ('Brahma Kamal') is tributed to deities and shared among everyone (Samant & Pant, 2003).

2.12 *Selinum* spp.

Genus *Selinum* is an important medicinal and aromatic herb belongs to the Apiaceae family and commonly called Bhutakeshi/Bhootkeshi. This plant grows at an altitude between 2000 and 4000 m amsl and there are two main species found in Himachal Pradesh, *viz., Selinum vaginatum* C. B. Clarke and *Ligusticopsis wallichiana* (DC.) Pimenov & Kljuykov (Syn. *Selinum wallichianum* (DC.) Raizada & H. O. Saxena and *Selinum tenuifolium* Wall. ex DC.). This plant is found in alpine zones of South Africa, China, Nepal, Bhutan, and West Pakistan and in India it is found in Himachal

Pradesh, Uttarakhand, Sikkim, and Kashmir. Essential oils derived from the roots of this plant possess antioxidant, antimicrobial, diuretic, anti-inflammatory, antidiabetic, and insecticidal activity. Selinidin (coumarin) which is derived from the roots has diuretic property. The important phytoconstituents present in *S. vaginatum* are bornyl acetate, camphene, valerenic acid, oroselol, selinitin, elemol, angelicin, and limonene whereas the important phytochemicals present in *L. wallichiana* are quercetin, 4′-O-methyl, rutin, scopoletin, falcarindiol, Bhutkesoside A, pumilaside A, eugenol, umbelliferone, cnidioside, and ferulic acid etc. Due to similarities in the root morphology, sometimes the plant is adulterated with the roots of *N. jatamansi* due to high demand, less supply, and restricted cultivation of rhizomes of *N. jatamansi* (Thakur et al., 2024; Kumari et al., 2025).

2.13 *Taxus contorta* Griff

T. contorta (Common name—West Himalayan Yew) belongs to the family Taxaceae is an endangered plant as per IUCN and enlisted in CITES Appendix—II (Thomas, 2011). It is endemic to the Himalayas and has distribution in Afghanistan, Pakistan, India, China, Tibet and Nepal at altitude of 1800–3300 m amsl (Majeed et al., 2021). In India, it has its distribution in Jammu & Kashmir, Uttarakhand and Himachal Pradesh. Taxine is one of major phytoconstituent present in leaves and bark of this plant, which is a precursor of drug Paclitaxel (Traded in the product name Taxol). It is utilized in curing various types of cancer (Lanker et al., 2010). This plant possesses various medicinal properties such as anti-microbial and immunomodulatory (Gul et al., 2020).

2.14 *Trillium govanianum* Wall. Ex D. Don

T. govanianum (Common name—'Nagchhatri'), alpine herb is a member of the family Trilliaceae which has its origin in Himalayas (Samant et al., 1998). It has its distribution in subalpine region of Bhutan, India, Nepal, China, Pakistan (Chauhan et al., 2020). *T. govanianum* is widely distributed in India across moist and shady habitat of Kashmir, Himachal Pradesh, Uttarakhand and Sikkim at altitude of 2400–3500 m amsl (Polunin & Stainton, 1984). Investigations on pharmacological aspects of this plant unveiled to have various biological activities such as antioxidant, anticancer, antifungal, anti-inflammatory and anti-diabetic (Kumar & Kumari, 2023). The phytoconstituents *viz.*, glycosides, terpenoids, saponins, flavonoids and phenolics holds primary biological activities (Sharma & Parashar, 2016). The phytochemical investigations on *T. govanianum* rhizome revealed the presence of steroidal type saponins such as govanoside A (Ismail et al., 2015), pennogenin, 20-hydroxyecdysone and 5,20-dihydroxyecdysone (Rahman et al., 2017). Previously conducted investigations revealed the presence of Borassoside E (Singh et al., 2020), which is a precursor of corticosteroids utilized in the formation of sex hormones and steroids and hydrolysis yields diosgenin (Chaudhary et al., 2015).

3 Impact of Climate Change on Alpine Medicinal Plants

3.1 Impact of Climate Change on Habitat and Distribution in Alpine Medicinal Plants

Impact of increased CO_2 levels enhanced the rate of photosynthesis, biomass accretion etc. in various species. The study conducted by Chandra et al. (2022) on impact of increased CO_2 levels on *A. heterophyllum* and *A. balfourii* resulted in decreased secondary metabolites (Alkaloids), antibacterial activity, plant nitrogen and starch content. Whereas, enhancement in proline content, antioxidant activity and lipid peroxidation was noticed under elevated CO_2 level. Effect of climate change on present, forthcoming habitat and distribution prospective of *T. contorta* and *T. wallichiana* in Indian Himalayas employing MaxEnt model in different climate change circumstances (SSP126, SSP345, SSP370 and SSP585) was analysed. Overall habitat of these important plants is estimated to decrease considerably by 2050 in lowest and highest discharge trajectories (SSP126 & SSP585), while modest circumstances (SSP370) revealed moderately constant or increasing favourability. These findings highlight the climate sensitivity of both species and underscore the necessity for conservation and habitat management (Rajlaxmi et al., 2025). The effect of climate change on *Hippophae salicifolia* habitat predicted that the plant's habitat will decline to 87% in the Himalayan region by 2050 because of variability in climatic conditions, with upward transfer of altitude from 2813 to 4485 m by 2070 (RCP8.5) (Dhyani et al., 2018). The upcoming shift in habitat of important medicinal plants in the Kashmir region of Himalayas under varying climatic patterns using MaxEnt model which resulted in escalating habitat of *Fritillaria cirrhosa* and *Meconopsis aculeata* because of increased temperature, whereas due to increasing temperature and low rainfall, *A. heterophyllum* and *R. webbianum* are suspected at risk of habitat damage. Plant specific responses underline the susceptibility of constricted-range species to climate change and emphasize the requirement of conservation strategies (Dad & Rashid, 2022). Khan et al. (2025) investigated the present and upcoming distribution of alpine medicinal plant *Dolomiaea costus* employing MaxEnt model which unveiled that the plant inhabits high elevation in North-western Himalayas. Habitat of this species is predicted to reduce in adequate climate change (SSP370) by the year 2100, whereas minor extension might take place in extreme warming (SSP585), however high-grade habitats are predicted to reduce. Rather et al. (2021) studied habitat patterns and shifts in distribution of alpine medicinal plant *Ephedra gerardiana* due to upcoming variations in climate because of increasing temperatures, resulted that the species inhabits approximately 4.8% of the study zone, but its appropriate habitat is estimated to decline under forthcoming climate predictions. Seasonal temperature was observed as a main reason impacting distribution of this species. Arrangement of various population, morphology and phytochemistry of alpine medicinal plant *Trillium govanianum* was investigated by Chauhan et al. (2021) in Indian Himalayan Region which revealed constant morphological characteristics (leaf area and shoot length) but dissimilarity in root characteristics and phytoconstituents (antioxidants, flavonoids, tannins and phenols) content.

Paudel et al. (2025) studied the effect of climatic variation on the distribution of *Rheum australe* in Nepal employing MaxEnt model which unveiled noteworthy habitat shrinkage along upcoming climatic variation, with major habitat loss estimated under SSP585 by the year 2100. Assessment of the present and upcoming possible habitat distribution of *Picrorhiza kurroa* was investigated in Himalayan region employing the MaxEnt model which resulted that the rainfall (33.8%) in dry month (Bio 14), isothermally (Bio 3; 20.2%), average temperature of hottest section (Bio 10; 12.7%), and yearly temperature series (Bio 7; 12.2%) were the key bioclimatic factors affecting habitat of the species. Generally, under consequences of climate change, there is a decline in the habitat of *Picrorhiza kurroa* (Rawat et al., 2022). Distribution prediction studies in *Fritillaria cirrhosa* and *Lilium nepalense* of Himalayan region was conducted by Rana et al. (2017) employing MaxEnt model under four circumstances of climate variation. Results revealed that presently, both the plants have appropriate habitat condition in the middle and east hill zone. Habitats of these plants are estimated to move north by upcoming climate forecasts. The key aspects affecting distribution were rainfall and temperature. For sustained survival of this species, this investigation highlights incorporating climate forecasts into cultivation and conservation policies. The detailed evaluations of the community and distribution mechanism of *Arnebia euchroma* in the Ladakh region of Himalayas were observed under impact of climate change. *A. euchroma* revealed the maximum Importance Value Index, highlighting its predominance in ecosystem. Due to human interferences and ecological severity, maximum populations had low development indices. Thus, investigation underscores this species as a vital ecological indicator and emphasizes its part in local livelihoods, encouraging for immediate conservation owing to habitat deprivation and climate change (Sofi et al., 2022).

3.2 Impact of Climate Change on Phytoconstituents in Alpine Medicinal Plants

Biosynthesis of phytoconstituents in alpine medicinal plants *viz., Primula macrophylla, Poa alpina, Artemisia rupestris* and *Potentilla hololeuca* was affected by climate warming employing open top chambers which resulted in increased total phenolic and flavonoid content. Major phytoconstituents such as quercetin, myricetin, kaempferol, rutin and gallic acid were enhanced in climate warming. Minor phytoconstituents such as ferulic and syringic acid were maximum in *Potentilla hololeuca* and *Poa alpina* (Karimi et al., 2025). Impact of elevation on essential oil constituents of *Artemisia brevifolia* unveiled the increased synthesis of essential oil constituents, majorly low volatile constituents. Essential oil constituents enhanced from 2 phytoconstituent/sample (2897 m amsl) to 4.05 phytoconstituent/sample (4446 m amsl). Production of essential oil constituents in *A. brevifolia* increased with altitude for resilience to climate change (Nataraj et al., 2022). Two alpine plants, *Plantago major* and *Pedicularis punctata* were investigated at varying elevations in the Himalayan region which revealed that with increasing elevation and declined temperature resulted in increment of protein, proline, ABA and sugar content in both species (Rahman et al., 2020). Jan et al. (2016) examined the impact of elevation on synthesis of phytoconstituents in *Rumex dentatus* and *Lavandula officinalis* in the Indian Himalayan Region employing LC-MS method which resulted in enhanced content of anthraquinone glucosides in *R. dentatus* and monoterpenoids (linalool) in *L. officinalis* under the impact of increasing elevation. Jan et al. (2018) studied the effect of edaphic and environmental factors on the phytoconstituent in field-grown and wild plants of *Origanum vulgare* in the North-Western Himalayan region which revealed that plants collected from wild had considerably higher phytochemical content *viz.*, *β*-carotene, ascorbic acid and catechin in comparison to field-grown plants which had maximum arbutin content. Current findings suggested the significance of edaphic and microclimatic factors in determination of Oregano's nutritive value. Investigation on phytoconstituent differences in high-ploidy groups of *Hyoscyamus pusillus* was conducted by Tantray et al. (2021) in which wild populations collected from different elevation in alpine region of Ladakh were analysed utilizing HPTLC method. The findings unveiled that phytoconstituents such as atropine, chlorogenic acid and caffeic acid were observed to be maximum along with antioxidant activity, phenolic and flavonoid content in the plants collected from higher elevation (3300 m) in comparison to low elevation population (2700 m). Kumari et al. (2020) examined adaptation of *Picrorhiza kurroa* in different altitudes of Himalayan region by tissue specific biochemical adaptaions. Metabolic characterization of populations from three elevations (3400, 3800 and 4100 m amsl) was done which characterized 220 phytoconstituents and unveiled noteworthy elevation and tissue specific differences in amino acids, iridoid glycosides (picrosides I–IV), sugars and phenolics. These variations suggested that remodelling of metabolic pathways helps in adaptation of this important medicinal species to ecological adversity in alpine zone. This study underscores the biochemical plasticity that shows high elevation resilience of *P. kurroa*.

4 Conclusion

Medicinal flora in the alpine region flourished under harsh climate conditions and established traditional medicinal systems that played an important role in environmental stability, health, and regional subsistence. However, the current climate change scenario is drastically modifying the niche, endangering the viability of therapeutic flora and the rigidity of the related human-nature relationships. The study emphasises the importance of implementing climate adaptive management plans, adapting studies on biochemical reactions to changing climates, and community-led cultivation practices. The conservation of genetically diverse flora in the alpine region is critical not only for ecosystem stability, but also for the preservation of traditional knowledge and future therapeutic value.

Acknowledgements We are thankful to Hon'ble Vice Chancellor of Dr. YS Parmar University of Horticulture and Forestry for providing support during preparation of this manuscript.

Author's Contributions VS wrote the manuscript and prepared the tables. RK corrected the manuscript, prepared figures and wrote some portions. VS and ST conducted literature mining. RS and PK conceived the idea of manuscript. RS, PK, AC and YPS proof read and corrected the manuscript, figures and tables.

Declaration of Competing Interest The authors declare that they have no known competing financial interests or personal relationships that could have appeared to influence the work reported in this paper.

References

Ambrish, K., & Srivastava, S. K. (2014). Taxonomic studies on the genus "*Arnebia*" Forssk. (Boraginaceae) in India. *Taiwania, 59*(4), 315–325.

Anand, U., Biswas, P., Kumar, V., Ray, D., Ray, P., Loake, V. I., Kandimalla, R., Chaudhary, A., Singh, B., Routhu, N. K., & Chen, Z. S. (2022). *Podophyllum hexandrum* and its active constituents: Novel radioprotectants. *Biomedicine & Pharmacotherapy, 146*, 10–16.

Arvindekar, A., More, T., Payghan, P. V., Laddha, K., Ghoshal, N., & Arvindekar, A. (2015). Evaluation of anti-diabetic and alpha glucosidase inhibitory action of anthraquinones from *Rheum emodi. Food & Function, 6*(8), 2693–2700.

Babu, K. S., Srinivas, P. V., Praveen, B., Kishore, K. H., Murty, U. S., & Rao, J. M. (2003). Antimicrobial constituents from the rhizomes of *Rheum emodi. Phytochemistry, 62*(2), 203–207.

Bhattarai, K., & Ghimire, M. (2006). Cultivation and sustainable harvesting of commercially important medicinal and aromatic plants of Nepal. In *Heritage Research and Development Forum*, Nepal (pp. 369–372).

Bimbiraite, K., Ragažinskienė, O., Maruška, A., & Kornyšova, O. (2008). Comparison of the chemical composition of four yarrow (*Achillea millefolium* L.) morphotypes. *Biologija, 54*(3), 208–212.

Broomhead, A. J., & Dewick, P. M. (1990). Tumour-inhibitory aryltetralin lignans in *Podophyllum versipelle, Diphylleia cymosa* and *Diphylleia grayi. Phytochemistry, 29*(12), 3831–3837.

Butola, J. S., & Samant, S. S. (2010). *Saussurea* species in Indian Himalayan region: Diversity, distribution and indigenous uses. *International Journal of Plant Biology, 1*(1), 43–51.

Chandra, S., Chandola, V., Sultan, Z., Singh, C. P., Purohit, V. K., Nautiyal, B. P., & Nautiyal, M. C. (2022). Climate change adversely affects the medicinal value of *Aconitum* species in Alpine region of Indian Himalaya. *Industrial Crops and Products, 186*, 115277.

Chatterjee, A., Basak, B., Saha, M., Dutta, U., Mukhopadhyay, C., Banerji, J., Konda, Y., & Harigaya, Y. (2000). Structure and stereochemistry of Nardostachysin, a new terpenoid ester constituent of the rhizomes of *Nardostachys jatamansi. Journal of Natural Products, 63*(11), 1531–1533.

Chaudhary, S., Chikara, S. K., Sharma, M. C., Chaudhary, A., Alam Syed, B., Chaudhary, P. S., & Iriti, M. (2015). Elicitation of diosgenin production in *Trigonella foenum-graecum* (fenugreek) seedlings by methyl jasmonate. *International Journal of Molecular Sciences, 16*(12), 29889–29899.

Chauhan, H. K. (2024). *Podophyllum hexandrum.* The IUCN Red List of threatened species 2024: e.T61985348A61985350. https://doi.org/10.2305/IUCN.UK.2024-1.RLTS.T61985348A61985350.en. Accessed on 20 May 2025.

Chauhan, R. S., Nautiyal, M. C., da Silva, J. A. T., Prasad, P., & Purohit, H. (2011). Habitat preference, ecological parameters and conservation of *Fritillaria roylei* Hook., an endangered medicinal herb of the Astavarga Group. *Bioremediation, Biodiversity and Bioavailability, 5*(1), 73–76.

Chauhan, H. K., Bhatt, I. D., & Bisht, A. K. (2020). Biology, uses and conservation of *Trillium govanianum*. In *Socio-economic and eco-biological dimensions in resource use and conservation: Strategies for sustainability* (pp. 235–247). Springer.

Chauhan, H. K., Gallacher, D., Bisht, A. K., Bhatt, I. D., Bhatt, A., Dhyani, P., & Kewlani, P. (2021). Variations in phytochemistry, morphology, and population structure in *Trillium govanianum* (Melanthiaceae). *Botany, 99*(10), 651–664.

Dad, J. M., & Rashid, I. (2022). Differential responses of Kashmir Himalayan threatened medicinal plants to anticipated climate change. *Environmental Conservation, 49*(1), 33–41.

Devi, S., Sharma, P., Sharma, R., & Thakur, M. (2023). Current status and medicinal prominence of *Arnebia euchroma* (Ratanjot): A critically endangered plant of trans-Himalayan region. *Recent Patents on Biotechnology, 17*(1), 92–102.

Dhiman, N., Sharma, N. K., Thapa, P., Sharma, I., Kumar Swarnkar, M., Chawla, A., & Bhattacharya, A. (2019). De novo transcriptome provides insights into the growth behaviour and resveratrol and trans-stilbenes biosynthesis in *Dactylorhiza hatagirea* – an endangered alpine terrestrial orchid of western Himalaya. *Scientific Reports, 9*(1), 1–13.

Dhyani, S., Kadaverugu, R., Dhyani, D., Verma, P., & Pujari, P. (2018). Predicting impacts of climate variability on habitats of *Hippophae salicifolia* (D. Don) (Seabuckthorn) in Central Himalayas: Future challenges. *Ecological Informatics, 48*, 135–146.

Ghosh, D. (2017). Brahma Kamal: The Himalayan beauty. *Resonance, 22*(4), 377–387.

Gudaitytė, O., & Venskutonis, P. R. (2007). Chemotypes of *Achillea millefolium* transferred from 14 different locations in Lithuania to the controlled environment. *Biochemical Systematics and Ecology, 35*(9), 582–592.

Gul, N., Baig, S., Ahmed, R., Shahzadi, I., Zaman, I., Shah, M. M., & Baig, A. (2020). Conservation of an endangered medicinal tree species *Taxus wallichiana* through callus induction and shoot regeneration. *Plant Tissue Culture & Biotechnology, 30*(1), 161–166.

Gupta, A. K., & Souravi, K. (2020). Access and benefit sharing and threatened medicinal plants. In *Conservation and utilization of threatened medicinal plants* (pp. 513–529). Springer.

Hanh, T. T. H., Cham, P. T., My, N. T. T., Cuong, N. T., Dang, N. H., Quang, T. H., & Minh, C. V. (2021). Sesquiterpenoids from *Saussurea costus*. *Natural Product Research, 35*(9), 1399–1405.

Heinrich, M., Lardos, A., Leonti, M., Weckerle, C., Willcox, M., Applequist, W., Ladio, A., Long, C. L., Mukherjee, P., & Stafford, G. (2018). Best practice in research: Consensus statement on ethnopharmacological field studies–ConSEFS. *Journal of Ethnopharmacology, 211*, 329–339.

Hou, M., Leng, C., Zhu, J., Yang, M., Yin, Y., Xing, Y., & Chen, J. (2024). Alpine and subalpine plant microbiome mediated plants adapt to the cold environment: A systematic review. *Environmental Microbiomes, 19*(1), 1–18.

Ismail, M., Shah, M. R., Adhikari, A., Anis, I., Ahmad, M. S., & Khurram, M. (2015). Govanoside A, a new steroidal saponin from rhizomes of *Trillium govanianum*. *Steroids, 104*, 270–275.

Jan, S., Kamili, A. N., Parray, J. A., & Bedi, Y. S. (2016). Differential response of terpenes and anthraquinones derivatives in *Rumex dentatus* and *Lavandula officinalis* to harsh winters across north-western Himalaya. *Natural Product Research, 30*(5), 608–612.

Jan, S., Mir, J. I., Singh, D. B., Faktoo, S. Z., Sharma, A., Alyemeni, M. N., & Ahmad, P. (2018). Effect of environmental variables on phytonutrients of *Origanum vulgare* L. in the sub-humid region of the northwestern Himalayas. *Environmental Monitoring and Assessment, 190*(10), 1–15.

Kala, C. P., Dhyani, P. P., & Sajwan, B. S. (2006). Developing the medicinal plants sector in northern India: Challenges and opportunities. *Journal of Ethnobiology and Ethnomedicine, 2*(1), 1–15.

Karimi, S., Nawaz, M. A., Naseem, S., & Ali, Z. (2025). Influence of open-top chambers induced climate warming on secondary metabolic profile of culturally and medicinally important plants of Himalaya, Karakoram and Hindukush. *PLoS One, 20*(5), 1–15.

Khan, A. U., & Gilani, A. H. (2011). Blood pressure lowering, cardiovascular inhibitory and bronchodilatory actions of *Achillea millefolium*. *Phytotherapy Research, 25*(4), 577–583.

Khan, R. W. A., Shaheen, H., Qayyum, S., Awan, S. N., Shah, T. A., Alsahli, A. A., Younous, Y. A., & El-Sheikh, M. A. (2025). Predicting habitat refugia of the medicinal plant *Dolomiaea costus* under two climate change scenarios. *Discover Life, 55*(1), 1–15.

Kirtikar, K. R., & Basu, B. D. (1984). *Indian medicinal plants*. Bishen Singh Mahendra Pal Singh.

Kizu, H., Kaneko, E. I., & Tomimori, T. (1999). Studies on nepalese crude drugs. XXVI. Chemical constituents of Panch Aunle, the roots of *Dactylorhiza hatagirea* D. Don. *Chemical and Pharmaceutical Bulletin, 47*(11), 1618–1625.

Körner, C. (2007). The use of 'altitude' in ecological research. *Trends in Ecology & Evolution, 22*(11), 569–574.

Krenn, L., Presser, A., Pradhan, R., Bahr, B., Paper, D. H., Mayer, K. K., & Kopp, B. (2003). Sulfemodin 8-O-*β*-d-glucoside, a new sulfated anthraquinone glycoside, and antioxidant phenolic compounds from *Rheum emodi*. *Journal of Natural Products, 66*(8), 1107–1109.

Krenn, L., Pradhan, R., Presser, A., Reznicek, G., & Kopp, B. (2004). Anthrone C-glucosides from *Rheum emodi*. *Chemical and Pharmaceutical Bulletin, 52*(4), 391–393.

Kumar, D., & Kumari, V. (2023). Metabolite profiling, antidiabetic, and antioxidant potential of different tissues of *Trillium govanianum* Wall. ex D. Don. *South African Journal of Botany, 153*, 102–108.

Kumar, D. R. N., George, V. C., Suresh, P. K., & Kumar, R. A. (2015). Cancer-specific chemoprevention and anti-metastatic potentials of *Rheum emodi* rhizome ethyl acetate extracts and identification of active principles through HPLC and GC-MS analysis. *Pakistan Journal of Pharmaceutical Sciences, 28*(1), 83–93.

Kumar, A., Shashni, S., Kumar, P., Pant, D., Singh, A., & Verma, R. K. (2021a). Phytochemical constituents, distributions and traditional usages of *Arnebia euchroma*: A review. *Journal of Ethnopharmacology, 271*, 1–16.

Kumar, V., Ranjan, D., & Verma, K. (2021b). Global climate change: The loop between cause and impact. In *Global climate change* (pp. 187–211). Elsevier.

Kumar, P., Kumar, V., Sharma, S., Sharma, R., & Warghat, A. R. (2025). *Fritillaria* steroidal alkaloids and their multi-target therapeutic mechanisms: Insights from network pharmacology. *Naunyn-Schmiedeberg's Archives of Pharmacology, 398*(3), 2209–2228.

Kumari, M., Joshi, R., & Kumar, R. (2020). Metabolic signatures provide novel insights to *Picrorhiza kurroa* adaptation along the altitude in Himalayan region. *Metabolomics, 16*(7), 1–17.

Kumari, R., Negi, M., Thakur, P., Mahajan, H., Raina, K., Sharma, R., Anand, V., Ming, L. C., Goh, K. W., Calina, D., Sharifi-Rad, J., & Chaudhary, A. (2024). *Saussurea costus* (Falc.) Lipsch.: A comprehensive review of its pharmacology, phytochemicals, ethnobotanical uses, and therapeutic potential. *Naunyn-Schmiedeberg's Archives of Pharmacology, 397*(3), 1505–1524.

Kumari, R., Thakur, P., Sharma, V., & Chaudhary, A. (2025). Exploring two medicinally important *Selinum* species: Bridging traditional knowledge with modern pharmacological insights. *Drug Metabolism and Personalized Therapy, 40*, 199–215.

Kundra, R., Samant, S. S., & Sharma, R. K. (2020). Assessment of antioxidant potential of *Trillium govanianum* Wall. ex D. Don, a critically endangered medicinal plant of Northwestern Indian Himalaya. *Proceedings of the National Academy of Sciences, India Section B: Biological Sciences, 90*(1), 95–101.

Lanker, U., Malik, A. R., Gupta, N. K., & Butola, J. S. (2010). Natural regeneration status of the endangered medicinal plant, *Taxus baccata* Hook. F. syn. *T. wallichiana*, in northwest Himalaya. *International Journal of Biodiversity Science, Ecosystem Services & Management, 6*(1–2), 20–27.

Li, Y., Zhang, M. L., Cong, B., Wang, S. M., Dong, M., Sauriol, F., & Kiyota, H. (2011). Achillinin A, a cytotoxic guaianolide from the flower of Yarrow, *Achillea millefolium*. *Bioscience, Biotechnology, and Biochemistry, 75*(8), 1554–1556.

Liu, Z. L., He, Q., Chu, S. S., Wang, C. F., Du, S. S., & Deng, Z. W. (2012). Essential oil composition and larvicidal activity of *Saussurea lappa* roots against the mosquito *Aedes albopictus* (Diptera: Culicidae). *Parasitology Research, 110*(6), 2125–2130.

Mabberley, D. J., & Noltie, H. J. (2014). A note on *Valeriana jatamansi* Jones (Caprifoliaceae sl). *Blumea-Biodiversity, Evolution and Biogeography of Plants, 59*(1), 37–41.

Majeed, A., Singh, A., & Bhardwaj, P. (2021). Transcribing molecular and climatic data into conservation management for the Himalayan endangered species, *Taxus contorta* (Griff.). *Conservation Genetics, 22*(1), 53–66.

Malhotra, N., & Sharma, S. (2021). *Aconitum heterophyllum*. In *Himalayan medicinal plants* (pp. 5–25). Academic Press.

Malik, S., Bhushan, S., Sharma, M., & Ahuja, P. S. (2011). Physico-chemical factors influencing the shikonin derivatives production in cell suspension cultures of *Arnebia euchroma* (Royle) Johnston, a medicinally important plant species. *Cell Biology International, 35*(2), 153–158.

Mathela, M., Kumar, A., Sharma, M., & Goraya, G. S. (2021). Hue and cry for *Fritillaria cirrhosa* D. Don, a threatened medicinal plant in the Western Himalaya. *Discover Sustainability, 2*(1), 1–7.

Mockute, D., & Judzentiene, A. (2003). Variability of the essential oils composition of *Achillea millefolium* ssp. *millefolium* growing wild in Lithuania. *Biochemical Systematics and Ecology, 31*(9), 1033–1045.

Molur, S., & Walker, S. (1998). *Report of the workshop "Conservation assessment and management plan for selected medicinal plant species of northern, north eastern and central India" (BCPP-endangered species project)* (p. 62). Zoo Outreach Organisation, Conservation Breeding Specialist Group.

Nadim, M. M., Malik, A. A., Ahmad, J., & Bakshi, S. K. (2011). The essential oil composition of *Achillea millefolium* L. cultivated under tropical condition in India. *World Journal of Agricultural Sciences, 7*(5), 561–565.

Nataraj, N., Hussain, M., Ibrahim, M., Hausmann, A. E., Rao, S., Kaur, S., Khazir, J., Mir, B. A., & Olsson, S. B. (2022). Effect of altitude on volatile organic and phenolic compounds of *Artemisia brevifolia* Wall ex Dc. from the Western Himalayas. *Frontiers in Ecology and Evolution, 10*, 1–10.

Noorafshan, A., Ebrahimi, S., Esmaeilzadeh, E., Arabzadeh, H., Bahmani-Jahromi, M., & Ashkani-Esfahani, S. J. A. E. (2017). Effects of *Arnebia euchroma* extract on streptozotocin induced diabetes in rats: A stereological study. *Acta Endocrinologica, 13*(3), 272–277.

Ou, Y. Y., Jiang, Y., Li, H., Zhang, Y. Y., Lu, Y., & Chen, D. F. (2017). Polysaccharides from *Arnebia euchroma* ameliorated endotoxic fever and acute lung injury in rats through inhibiting complement system. *Inflammation, 40*(1), 275–284.

Pandey, A., Rai, S., & Kumar, D. (2018). Changes in vegetation attributes along an elevation gradient towards timberline in Khangchendzonga National Park, Sikkim. *Tropical Ecology, 59*(2), 259–271.

Pandith, S. A., Dar, R. A., Lattoo, S. K., Shah, M. A., & Reshi, Z. A. (2018). *Rheum australe*, an endangered high-value medicinal herb of North Western Himalayas: A review of its botany, ethnomedical uses, phytochemistry and pharmacology. *Phytochemistry Reviews, 17*(3), 573–609.

Paudel, B. R., Subedi, C. K., Ghimire, S. K., Pyakurel, D., Rajbhandari, M., & Chaudhary, R. P. (2025). Impacts of climate change on the distribution pattern of Himalayan Rhubarb (*Rheum australe* D. Don) in Nepal Himalaya. *PLoS One, 20*(5), 1–27.

Polunin, O., & Stainton, A. (1984). *Flowers of the Himalaya* (p. 580). Oxford University Press.

Prasad, S. K., Kumar, R., Patel, D. K., Sahu, A. N., & Hemalatha, S. (2012). Physicochemical standardization and evaluation of *in-vitro* antioxidant activity of *Aconitum heterophyllum* Wall. *Asian Pacific Journal of Tropical Biomedicine, 2*(2), S526–S531.

Press, J. R., Shrestha, K. K., & Sutton, D. A. (2000). *Annotated checklist of the flowering plants of Nepal* (p. 430). Natural History Museum Publications.

Rahman, S. U., Adhikari, A., Ismail, M., Shah, M. R., Khurram, M., Anis, I., & Ali, F. (2017). A new trihydroxylated fatty acid and phytoecdysteroids from rhizomes of *Trillium govanianum*. *Antiseptic, 3*(5), 323–327.

Rahman, I. U., Afzal, A., Iqbal, Z., Hart, R., Abd-Allah, E. F., Alqarawi, A. A., Alsubeie, M. S., Calixto, E. S., Ijaz, F., Ali, N., Kausar, R., Shah, M., & Bussmann, R. W. (2020). Response of plant physiological attributes to altitudinal gradient: Plant adaptation to temperature variation in the Himalayan region. *Science of the Total Environment, 706*, 135714.

Rajlaxmi, A., Chawla, A., & Kumar, M. (2025). Predicting the current and future potential habitat of *Taxus* species over Indian Himalayan Region using MaxEnt model. *Tropical Ecology, 66*(1), 14–34.

Rana, S. K., Rana, H. K., Ghimire, S. K., Shrestha, K. K., & Ranjitkar, S. (2017). Predicting the impact of climate change on the distribution of two threatened Himalayan medicinal plants of Liliaceae in Nepal. *Journal of Mountain Science, 14*(3), 558–570.

Rather, Z. A., Ahmad, R., Dar, A. R., Dar, T. U. H., & Khuroo, A. A. (2021). Predicting shifts in distribution range and niche breadth of plant species in contrasting arid environments under climate change. *Environmental Monitoring and Assessment, 193*(7), 1–17.

Rawal, R. S., Rawal, R., Rawat, B., Negi, V. S., & Pathak, R. (2018). Plant species diversity and rarity patterns along altitude range covering treeline ecotone in Uttarakhand: Conservation implications. *Tropical Ecology, 59*(2), 225–239.

Rawat, N., Purohit, S., Painuly, V., Negi, G. S., & Bisht, M. P. S. (2022). Habitat distribution modeling of endangered medicinal plant *Picrorhiza kurroa* (Royle ex Benth) under climate change scenarios in Uttarakhand Himalaya, India. *Ecological Informatics, 68*, 101550.

Rekha, K., Rao, R. R., Pandey, R., Prasad, K. R., Babu, K. S., Vangala, J. R., & Rao, J. M. (2013). Two new sesquiterpenoids from the rhizomes of *Nardostachys jatamansi. Journal of Asian Natural Products Research, 15*(2), 111–116.

Robinson, M. M., & Zhang, X. (2011). The world medicines situation. In *Traditional medicines: Global situation, issues and challenges*. World Health Organization.

Rokaya, M. B., Münzbergová, Z., Timsina, B., & Bhattarai, K. R. (2012). *Rheum australe* D. Don: A review of its botany, ethnobotany, phytochemistry and pharmacology. *Journal of Ethnopharmacology, 141*(3), 761–774.

Rucker, G., Tautges, J., & Maheswari, M. L. (1976). Norseychelanone, α- and β-patchoulenes and patchouli alcohol from *Nardostachys jatamansi. Phytochemistry, 15*(1), 224.

Sajwan, S., Sajwan, K., Negi, R. K., Hasmi, S. A. A., & Sehar, N. (2015). Pharmacognostical and chromatographic studies on the drug "Zarnab" a cardiac remedy. *International Journal of Engineering and Management Research, 5*(4), 112–114.

Samant, S. S., & Pant, S. (2003). Diversity, distribution pattern and traditional knowledge of sacred plants in Indian Himalayan region. *Indian Journal of Forestry, 26*(3), 222–234.

Samant, S. S., Dhar, U., & Palni, L. M. S. (1998). *Medicinal plants of Indian Himalaya: Diversity distribution potential values*. Gyanodaya Prakashan.

Saxena, K. G., & Rao, K. S. (2024). The himalayan biodiversity hotspot: Geomorphology, biogeography, and climate. In *Biodiversity hotspot of the Himalaya* (pp. 1–9). Apple Academic Press.

Sekar, K. C., Thapliyal, N., Pandey, A., Joshi, B., Mukherjee, S., Bhojak, P., & Bahukhandi, A. (2024). Plant species diversity and density patterns along altitude gradient covering high-altitude alpine regions of west Himalaya, India. *Geology, Ecology, and Landscapes, 8*(4), 559–573.

Semwal, P., & Painuli, S. (2019). Antioxidant, antimicrobial, and GC-MS profiling of *Saussurea obvallata* (Brahma Kamal) from Uttarakhand Himalaya. *Clinical Phytoscience, 5*(1), 1–11.

Semwal, P., Painuli, S., Jugran, A., Palni, L. M. S., Verma, S., Thapliyal, A., & Tewari, D. (2020). Genetic diversity of scanty available Himalayan *Saussurea obvallata* (DC.) Edgew. *Iranian Journal of Science and Technology, Transactions A: Science, 44*(3), 587–594.

Shanbhag, S. N., Mesta, C. K., Maheshwari, M. L., Paknikar, S. K., & Bhattacharyya, S. C. (1964). Terpenoids—LII: Jatamansin, a new terpenic coumarin from *Nardostachys jatamansi. Tetrahedron, 20*(11), 2605–2615.

Shanbhag, S. N., Mesta, C. K., Maheshwari, M. L., & Bhattacharyya, S. C. (1965). Terpenoids—LXXV: Constituents of *Nardostachys jatamansi* and synthesis of (±) dihydrosamidin and visnadin from jatamansin. *Tetrahedron, 21*(12), 3591–3597.

Sharma, A., & Parashar, B. (2016). A review on *Trillium govanianum. World Journal of Pharmaceutical Research, 6*(2), 500–511.

Sharma, P. K., Chauhan, N. S., & Lal, B. (2004). Observations on the traditional phytotherapy among the inhabitants of Parvati valley in western Himalaya, India. *Journal of Ethnopharmacology, 92*(2–3), 167–176.

Sharma, S., Kumar, P., Sharma, R., & Warghat, A. R. (2023). *In vitro* propagation and omics breakthroughs for understanding specialized metabolite production in high-value Himalayan *Fritillaria* species. *Industrial Crops and Products, 205*, 117541.

Sharma, R., Sharma, R., Thakur, V., Randhawa, A., Kumar, R., Dutt, B., & Kumar, S. (2025). Insights into *Pinus* species: Phytochemistry, pharmacology and industrial potential of Indian *Pinus* species. *Phytochemistry Reviews, 24*, 1–31.

Siavash, M., Naseri, M., & Rahimi, M. (2016). *Arnebia euchroma* ointment can reduce abdominal fat thickness and abdominal circumference of overweight women: A randomized controlled study. *Journal of Research in Medical Sciences, 21*(1), 1–4.

Singh, P. P., Ambika, & Chauhan, S. M. S. (2013). Activity-guided isolation of antioxidants from the roots of *Rheum emodi*. *Natural Product Research, 27*(10), 946–949.

Singh, P. P., Bora, P. S., Suresh, P. S., Bhatt, V., & Sharma, U. (2020). Qualitative and quantitative determination of steroidal saponins in *Trillium govanianum* by UHPLC-QTOF-MS/MS and UHPLC-ELSD. *Phytochemical Analysis, 31*(6), 861–873.

Sofi, I. I., Zargar, S. A., Ganie, A. H., & Shah, M. A. (2022). Distribution dynamics of *Arnebia euchroma* (Royle) IM Johnst. and associated plant communities in Trans-Himalayan Ladakh region in relation to local livelihoods under climate change. *Trees, Forests and People, 7*, 1–14.

Sofi, M. A., Sofi, M. A., Nanda, A., Nayak, B. K., Othman, Z., & Sadikan, M. Z. (2025). Exploring the therapeutic potential of *Podophyllum hexandrum* root extract: Chemical composition, antimicrobial efficacy, and antioxidant and anticancer activities. *Scientifica, 2025*(1), 1–13.

Talreja, S., & Tiwari, S. (2023). A comprehensive review of *Aconitum heterophyllum*. *Journal of Ayurveda and Integrated Medical Sciences, 8*(10), 195–201.

Tanaka, K., & Komatsu, K. (2008). Comparative study on volatile components of *Nardostachys* rhizome. *Journal of Natural Medicines, 62*(1), 112–116.

Tantray, Y. R., Wani, M. S., Pradhan, S. K., Jan, I., Singhal, V. K., & Gupta, R. C. (2021). Variation in phytochemical content in polyploid populations of *Hyoscyamus pusillus* thriving at different altitudes in cold deserts of Ladakh, India. *Analytical Chemistry Letters, 11*(2), 255–270.

Thakur, P., & Chaudhary, A. (2024). An overview of *Angelica* species market potential in herbal and medicinal industry. *Journal of Ethnopharmacology and Toxicology, 2*(1), 19–26.

Thakur, P., & Chaudhary, A. (2025). An underexplored treasure of the Himalayas: Ethnobotany, phytochemical, and pharmacological relevance of *Angelica glauca* Edgew. *Biochemical Systematics and Ecology, 123*, 105100.

Thakur, D., Rathore, N., Sharma, M. K., Parkash, O., & Chawla, A. (2021). Identification of ecological factors affecting the occurrence and abundance of *Dactylorhiza hatagirea* (D. Don) Soo in the Himalaya. *Journal of Applied Research on Medicinal and Aromatic Plants, 20*, 1–10.

Thakur, A., Kumari, R., Kumar, A., & Chaudhary, A. (2024). Quantitative ethnobotanical study of medicinal plants used by native people of selected areas of Chhota Bhangal, Himachal Pradesh. *Ethnobotany Research and Applications, 28*, 1–34.

Thakur, P., Kumari, R., Kumar, A., Bhatia, A., Sharma, U., & Chaudhary, A. (2025). Uncovering the mechanisms of *Angelica glauca* Edgew. In *Breast cancer: A combined in vitro and in silico approach* (p. e202402554). Chemistry & Biodiversity. https://doi.org/10.1002/cbdv.202402554

Thomas, P. (2011). *Taxus contorta*. The IUCN Red List of threatened species 2011: e.T39147A10170545. https://doi.org/10.2305/IUCN.UK.20112.RLTS.T39147A10170545.en

Turner, N. J., Deur, D., & Mellott, C. R. (2011). "Up on the mountain": Ethnobotanical importance of montane sites in Pacific coastal North America. *Journal of Ethnobiology, 31*(1), 4–43.

Ved, D. K., & Tandon, V. (2009). *CAMP report for high altitude medicinal plants of Jammu-Kashmir and Himachal Pradesh*. FRLHT.

Ved, D., Saha, D., Ravikumar, K., & Haridasan, K. (2015). *Aconitum heterophyllum*. The IUCN Red List of threatened species 2015: e.T50126560A79579556. https://doi.org/10.2305/IUCN.UK.20153.RLTS.T50126560A79579556.en. Accessed on 15 May 2025.

Vitalini, S., Beretta, G., Iriti, M., Orsenigo, S., Basilico, N., Dall'Acqua, S., & Fico, G. (2011). Phenolic compounds from *Achillea millefolium* L. and their bioactivity. *Acta Biochimica Polonica, 58*(2), 203–219.

Walter, K. S., & Gillett, H. J. (1998). *1997 IUCN Redlist of threatened plants compiled by the World Conservation Monitoring Centre*. IUCN – The World Conservation Union.

Wang, D., Wang, S., Du, Q., Wang, N., Liu, S., Wang, X., & Jiang, J. (2014). Optimization of extraction and enrichment of steroidal alkaloids from bulbs of cultivated *Fritillaria cirrhosa*. *BioMed Research International, 2014*(1), 1–15.

Wani, T. A., Kaloo, Z. A., & Dangroo, N. A. (2022). *Aconitum heterophyllum* Wall. ex Royle: A critically endangered medicinal herb with rich potential for use in medicine. *Journal of Integrative Medicine, 20*(2), 104–113.

Xiong, W., Luo, G., Zhou, L., Zeng, Y., & Yang, W. (2009). *In vitro* and *in vivo* antitumor effects of acetylshikonin isolated from *Arnebia euchroma* (Royle) Johnst (Ruanzicao) cell suspension cultures. *Chinese Medicine, 4*(1), 1–7.

Ye, M., Han, J., Chen, H., Zheng, J., & Guo, D. (2007). Analysis of phenolic compounds in rhubarbs using liquid chromatography coupled with electrospray ionization mass spectrometry. *Journal of the American Society for Mass Spectrometry, 18*(1), 82–91.

Zhang, X., Rakesh, K. P., Shantharam, C. S., Manukumar, H. M., Asiri, A. M., Marwani, H. M., & Qin, H. L. (2018). Podophyllotoxin derivatives as an excellent anticancer aspirant for future chemotherapy: A key current imminent needs. *Bioorganic & Medicinal Chemistry, 26*(2), 340–355.

Resource Use Patterns and Ecosystem Services in Alpine Regions

2

Bandana Dhiman, Indesh Attri, and Sambuddha Mukherjee

Abstract

The resource use pattern of Alpine-Himalayan region as well as its vulnerability can be determined by geological, ecological, climatic, edaphic characteristics, as well as biotic influenced factors—mainly human and their associated interaction temporally. Natural resources from the Himalayan Mountain region are both extensively tangible and numerous in intangible forms. Considering the fact that Alpin-forest ecosystem already has limited resources at fragile capacity of the Himalayas where tribal sociology shows paradigm of co-adaptive. Despite these natural constraints, indigenous communities have evolved context-specific, time-tested land management systems that maintain a delicate balance between ecological limitations and human sustenance reflecting a deep-rooted adaptation to their harsh but resilient alpine environment. However, the changing condition led pressure on alpine highland ecosystems, which are especially sensitive to climate change. The glaciated upper Alpine region has already witnessed the vulnerability consequences due to climate change. Both precipitation and temperature variations affect the socioeconomic structure of the area, affecting tourism, agriculture, cattle, forestry, and human health in the Himalayan regions including its resource use pattern. These changes directly affect the ecosystem services provided by Alpine regions, such as environment amelioration, natural resources, fresh water sources etc. Deviation from the traditional livelihood system and

B. Dhiman (✉)
RHRSS & Krishi Vigyan Kendra Tabo, Lahaul & Spiti, Himachal Pradesh, India

I. Attri
Sardar Beant State University, Gurdaspur, Punjab, India

S. Mukherjee
Department of Economics and Sociology, Punjab Agricultural University (PAU), Ludhiana, Punjab, India

P. Kumar et al. (eds.), *Climate Resilience and Molecular Adaptation in Alpine Medicinal Plants*, https://doi.org/10.1007/978-981-95-6923-6_2

cropping pattern is increasingly observed as per the availability of various ecosystem services and changing weather conditions in that particular region. This chapter presents a framework and the analysis on changing land use pattern in upper Alpine-Himalayan region as a result of socioeconomic human activities and climate change. Conservation is needed to sustain the inherited Traditional Resource use Pattern of upper Himalayan range. Taking the note of SDG, the global agenda for sustainable development has brought mountains into extreme focus and in priority for conservation.

Keywords

Natural resources · Resource use pattern · Alpine-Himalayan region · Ecosystem services · Climate change

1 Introduction

From arctic to tropical latitudes, the alpine zones with higher mountain ranges are throughout the globe. Alpine ecosystem is present in different latitude of globes form 3% of the total geographical area (Körner, 1995) and own 4% floral richness of the total count of higher plant species (Spehn & Körner, 2005) available on planet Earth. It covers maximum area in Asia continent (73%), South America (15%), North America (9%), Europe (2%) and less than 1% at Oceania and African continent. Globally, alpine biomass is estimated as 3.56 Mkm^2 (Testolin et al., 2020). The human settlement patterns in the Alpine region are profoundly influenced by its topographical extremities, climatic severity and limitedness of natural resources. Despite these constraints, alpine landscapes provide a wide range of ecosystem goods and services (ES) that sustain both local and downstream populations. The modern development, often driven by cost-benefit consideration, usually fails to understand these ecosystem services. Mankind depends upon variety of ES provided by forest ecosystem, which are generated as by-product of the interaction of biotic (products from plants and animals) and abiotic (environmental linked).

In mountain ecosystem human settlements are exploiting vegetation through several activities primarily farming, herding, fuel wood collection and trade in medicinal as well as timber plants. The alpine region of western Himalaya is important for economic growth and wellbeing of human well-being. The alpine meadows of western Himalaya mostly associated with the pastoral practices, like livestock grazing by agro-pastoral communities other than that extraction of wild medicinal plants for folk medicinal and commercial purposes. An alpine ecosystem is unique due to presence of rare under-explored medicinal plants have high degree of endemism and well adapted to cold, nutrient-poor environments. They contribute significantly to global biogeochemical cycles by facilitating carbon sequestration and regulating local and regional climate. Ecosystem services such as water regulation, carbon sequestration, and serve as indicators of climate change impacts (Pauli et al., 2012) in the high land such as alpine area.

2 Alpine Himalayas Ecosystem Services and Their Status

The Himalayan range is mainly composed of volcanic formation with the addition of sedimentary rocks, fragile clayey hillock which can be found as patches (Bhatnagar et al., 2016), These ancient Lurian rocks include schists, slate, shale, Balaini conglomerates, and sandstone (Singh et al., 1995). The alpine valleys are dry and freezing. Various peaks in alpine mountainous region are covered by permanent glaciers. Precipitation in the greater Himalayan region primarily falls as snow, while the middle Himalayas receive a mix of snow and rain, and the outer Himalayas experience light to moderate rainfall (Jain et al., 2009). This precipitation is mainly influenced by western disturbances from October to May and by the southwest monsoon from July to September. However, the scenario has been changed due to shift in climatic conditions. The demand for and use of ecosystem services (ES) are rapidly increasing in the current era of development. While responses such as plantation efforts and ex-situ conservation are being implemented, their effectiveness and actual progress require careful assessment and quantification. Intangible ecosystem services, which provide countless benefits freely to society, remain less recognized and are often undervalued. These hidden contributions are foundational to development, construction, and manufacturing. Overall, the integrity of mountain landscapes is increasingly compromised by various disturbances, including changes in glacier and water status, shifting precipitation patterns, and soil degradation. The ecosystem services at alpine region can be broadly categorized on the basis of three sub-ecosystems, such as Forest ecosystem, Agro ecosystem, and Freshwater ecosystem (Fig. 2.1).

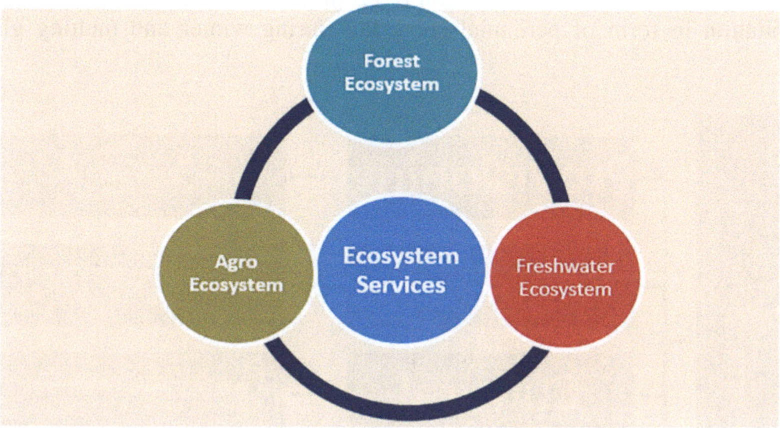

Fig. 2.1 Ecosystem services in forest system, agro-ecosystem and freshwater ecosystem

3 Ecosystem Services (ES) from Alpine Ecosystems

The Millennium Ecosystem Assessment has identified four major categories of ES, provisioning, regulating, cultural and supporting services. These all have different varieties of provisioning Services which are interconnected have been summarized below (Fig. 2.2).

3.1 Provisioning Services

It includes services, such as food, water, climate regulation, and biodiversity. The Millennium Ecosystem Assessment (MA) was conducted from 2001 to 2005 to evaluate how ecosystem changes impact human and shows that human activities have significantly degraded many ecosystems.

3.2 Water

Water is perhaps the most crucial ES provided by mountains, especially freshwater supply to plains, which justifies their label as the 'water towers' of the world (Viviroli et al., 2007). Alpine region is also associated with seasonal water regulation as wetlands and permafrost act as natural reservoirs, releasing water gradually during summer season (melting of glaciers). Data suggests that Himalayan glaciers receding with the rate of 18–20 m per year. The warming of Himalayan region has become a grave issue to provide water to 1.3 billion people and its detrimental effects on downstream habitats are more prominent. The main source of water is precipitation in form of perennial snow fall during winter and melting glaciers

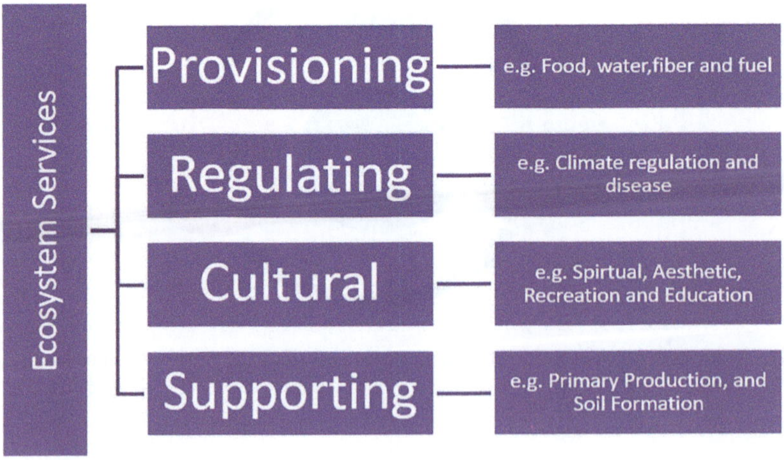

Fig. 2.2 Ecosystem services

during summers. This vital provisioning service is under threat due to retreating glaciers. Snow/glacier is a dominant land cover type in the region, investigate the variability in weather conditions focusing on rainfall and temperature -monthly, seasonal, and annual basis from 1981 to 2021 in the cold desert region of Himalaya (Kumar et al., 2025), having fragile ecosystem with dynamic geomorphological structural activity in response to or affected by proneness to natural disaster. Because it already has few natural resources, this glaciated area is particularly vulnerable to climate change. With an average increase in atmospheric temperature, snowy precipitation turning into erratic drizzling rainfall has also been witnessed in areas of altitude above 3500 amsl. As a clear sign of global warming consequences, snowy crystals fall-down toward the ground and transform into drizzling rain. Thus, lower areas of the regions are now experiencing rains as the warming effect prevails and breeching the effect developed as rain shadow area which is responsible for 'dry conditions in the cold desert'. As it is located on the leeward side of the Himalayas, the region becomes rain-shadow and turns out to be inaccessible for annual south-east monsoon winds which create dry conditions and altitudinal position leads to coldness in the region. These harsh winter conditions force locals who are aged, pregnant ladies, sensitive children as well as sick patients to migrate in the lower adjoining districts to Lahaul and Spiti. Increase in rain intensity and frequency affect topography due to landslide of naked desert mountains. Also cloud blocking creates cloud burst situation at higher altitude.

3.3 Non-Timber Forest Products-Fuel Wood/Fodder/Medicinal and Aromatic Plants (MAPs)

NTFP of alpine and subalpine zone of western Himalaya is an important component of local people and tribal communities. Medicinal and aromatic plants constitute the most important provisioning services offered by the alpine zones of the western Himalaya. The alpine regions of the Western Himalaya possess a remarkable diversity of medicinal plants, with studies reporting 77–125 species employed in the treatment of a broad spectrum of ailments, particularly dermatological, gastrointestinal, and respiratory conditions (Malik et al., 2015; Singh et al., 2017). Local communities possess rich ethnomedicinal knowledge that enables the sustainable use of these alpine resources. However, this traditional knowledge base is increasingly threatened by generational gaps, socio-economic changes, and migration, which risk disrupting the transmission of ethnobotanical practices essential for both cultural heritage and ecosystem-based healthcare. MAPs help maintain pollinator diversity and overall ecosystem functioning, while their sustainable use and conservation are crucial for preserving biodiversity hotspots threatened by climate and land-use changes. Key species in moist sub-alpine zone include *Angelica cyclocarpa, Aconitum heterophyllum, Nardostachys jatamansi, Picrorhiza kurroa, Saussurea costus, Trillium govanianum* etc.

Pastoralism was the predominant occupation of people in the alpine region of western Himalayas. Pastoralists mostly depend upon natural resources specifically for fuel wood and fodder. This region has a significant population of livestock that includes sheep, goats, cattle, yak and yak hybrids etc. Though livestock grazing has positive effect on the species diversity and enhances the productivity yet rational grazing practices are required to preserve these pastures (Ingty, 2021). A significant amount of energy is needed for cooking and heating during the harsh winter months. Consequently, residents of the region utilize various plant species as fuelwood (Badoni & Bhatt, 1989). In the western Himalayan alpine and subalpine regions, several plants species are used as fuelwood, especially in areas where access to other energy sources is limited. Fuelwood species commonly used in the Western Himalayan alpine region *Betula utilis*, *Juniperus* spp., *Salix* spp., *Hippophae rhamnoides*, and *Ephedra gerardiana*. In areas where fuelwood is scarce, alternatives such as crop residues and dung cakes are commonly used as fuel sources. The majority of this kind of plant can be found in regions with prominent bushes/Shrubs are highly valuable medicinally such as species of *Ribes, Myricaria, Ephedra*, and Himalayan juniper (Shukpa) have good calorific value.

3.4 Regulating Services

Regulating services—such as watershed services, climate stabilization, prevention of soil erosion and air purification are crucial in the alpine landscape of western Himalaya. As these rangelands and pastures in Himachal Pradesh, Uttarakhand acts as buffer against landslides and avalanches by protecting the soil on the other hand vegetation helps in microclimate regulation. The Himalayan lakes and rivers paly vital role in the hydrological cycle like ground water recharge and constant supple in perennial streams. Supporting the traditional irrigation systems like *khuls* in Himachal Pradesh. Though these services are important for civilization yet limited reports are there for instance 5.4% respondents in Spiti valley recognized it. As per data 69% of total ecosystem services are from regulating service alone in Spiti (Murali et al., 2017).

3.5 Cultural Services

Cultural services in these high mountain regions are deeply imbedded, offering non-material benefits for human kind like spiritual identity, heritage and recreation. One can find century old traditions still preserved in the form of monasteries (Tabo, Kee, Dhankar). Local folk lore running through generation celebrated as festivals bringing nature, tradition and communities together. The traditional practices of conservation and protection of small forest patches by dedicating them to the local deities or ancestral spirits, by various indigenous communities of the region is forest patches and sacred grove (Fig. 2.3) (Tables 2.1 and 2.2).

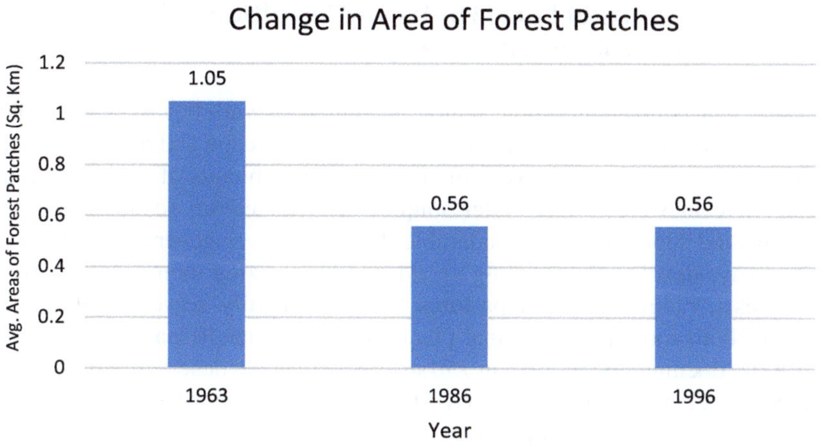

Fig. 2.3 Graphical presentation of change in area of forest patches. (Rao and Pant, 2001)

Table 2.1 Change in Forest patches, area, average parmeter along with shape complexity index (Rao and Pant, 2001)

Year	Forest patches	Area (km. sq.) Avg.	Min.	Max.	Avg. perimeter (km)	Shape complexity index[a] (km/sq. km)
1963	35	1.05	0.009	32.45	5.88	5.6
1986	60	0.56	0.008	25.04	4.78	8.5
1996	56	0.56	0.001	24.35	5.19	9.3

[a]Shape complexity index (SCI) is a parameter used to quantify a polygon's shape; in Forestry and Environmental Science it refers the ratio of average perimeter of a forest patch to its average area

Table 2.2 Change in Natural Forest and Agricultural land area in 3 time periods: 1963, 1986, 1996 (Rao and Pant, 2001)

Year	Land use	Area (Km. sq.)	Area %-age	Changes (Km. sq.)	Changes %-age	Deforestation rate (%-age/year)
1963	Natural forest	36.63	46.2			
	Agriculture land[a]	42.7	53.8			
1986	Natural forest	31.85	40.2	−4.78	−13.1	0.57
	Agriculture land[a]	47.48	59.9			
1996	Natural forest	31.49	39.7	−0.13[b], −5.14[c]	−1.1[b], -14[c]	0.113[b] 0.425[c]
	Agriculture Land[a]	47.84	60.3			

[a]Agriculture land area includes farming, pasture, shrubs, gardens, settlement and riverine terraces
[b]Changes and annual deforestation between 1986 and 1996. [c]Changes and annual deforestation between 1963 and 1996. Values within the parentheses pertains to changes and deforestation between 1963 and 1996

4 Importance of Local Herbs and Traditional Crops in Traditional Healthcare System

The valleys in Himalayan alpine region encompasses varieties of wild medicinal and aromatic plants (MAPs) that adorn its fragile landscape and are further tagged as a unique ecosystem in the Western Himalayas. Inhabitants of the alpine regions typically live in valleys near water sources, as villages are scattered across the mountains and separated by considerable distances. Over generations, local communities have developed lifesaving knowledge by utilizing their natural surroundings. Forests, rangelands, and pastures have historically served as the primary sources of both food and medicinal plants with ethno-medicinal values. Chaursia et al. (2007) documented more than 750 species in the cold desert Himalayan region, while Singh et al. (2009) identified 76 plant families comprising 208 genera. Approximately 400 medicinal and aromatic plants have been found in the alpine regions of Uttarakhand, and 30–35 species with therapeutic potential are widely traded commercially (Chandra et al., 2021). These products have commercial value in the international market and thus under constant pressure and overexploitation. Some of the most valuable NTFPs are medicinal plants e.g. *Picrorhiza kurroa, Aconitum heterophyllum.* Other than that, edible nutritional fungi *Morchella esculenta* (Morel or Gucchi mushroom) has significant international demands due to its medicinal properties.

5 Supporting Services

The Himalayan Forest not only support native residents but countless community residing in the Indo-Gangetic plains through water cycle regulation (Bruijnzeel & Bremmer, 1989). The water as the provision services from Himalayas is essential for Highland-Lowland sustainability in Asia. The highland of the Himalayas is storehouse of water, many perennial rivers such as Indus and Ganges, western Himalayan region, which impact a large population in India. These rivers are fed by glaciers and melting of snow in summer. Additionally, these landscapes play a role in soil stabilization, nutrient cycling, and serve as indicators of climate change. The conservation of alpine ecosystems is therefore imperative for sustaining ecological integrity and securing the provisioning, regulating, and cultural services they offer.

In India's Western Himalayas, changes in altitude are intense and tend to produce a very specific pattern of vegetation types that include subtropical forests, alluvial grasslands, conifer mountain forests, and alpine meadows (Singh & Chatterjee, 2020).

6 Categories of Forests in Alpine Region

Forest Types in India, in accordance to the Champion & Seth Classification of Forest Types (1968), the Alpine ecology under sub group 14/C1a and 14/C1b persists above altitude of 3300 amsl which is influenced by cold deserts, glaciers, and perpetual snow prevent growing of trees in the extreme harsh parts of the Alpine Himalayas. The term alpine refers only to zone found above the tree line with unclear boundary with sub alpine zone. Alpine meadows above tree lines are refereed as Alpine pastures (15/C3) and lower Alpine scrub includes Dry alpine scrub (16/C1), deciduous subalpine scrub (14/1S2), and Dwarf Rhododendron scrub (15/C2/E1) forests in Himachal Pradesh in the Trans-Himalayan region, the area covered by alpine meadows is approximately three times larger than that of alpine scrub forests. Subalpine Forests with mixed vegetation occupy at elevations between 3500 m and up to zone of the alpine scrub forest where can be witnessed. The alpine region in Himachal Pradesh spans approximately 10,052 km^2, accounting for about 17% of the state's total geographical area. At lower elevations, the vegetation mainly consists of dry scrub woodlands, whereas at higher elevations, moist alpine scrubs with meadows and pastures are prevalent. These upper alpine regions are typically treeless, although sparse stands of juniper and birch may occur along riverbanks or in moist, rocky areas. Due to low rainfall and arid conditions, the dry alpine scrub found at elevations up to 6000 m is dominated by pastures and meadows. This zone lacks true hydrophytes, epiphytes, strong hardwood trees, and lower herbaceous plants, especially in areas such as Leh, Kashmir, Lahaul and Spiti, and upper Kinnaur (H.P.). Despite these harsh conditions, the region supports a rich diversity of plant species, particularly medicinal herbs that flourish in narrow ecological zones near melting glaciers.

7 Floristic Status

The studies floristic status of Himalayan alpine is always a curious aspect among ecologists. Numerous researchers have studied the vegetation and published updated accounts over time. For instance, Stewart (Stewart, 1916–1917) recorded about 66 flora families comprising 831 plant species at the western Tibetian plateau and Ladakh area. Kachroo et al. (1977), with his fellow researchers, estimated 611 plant species (540 dicotyledons, 65 monocotyledons, and 2 gymnosperms) from the Ladakh region. Where H C Chaudhary mentioned 477 species in 218 genera, belongs to 50 families (23 are monogerenic & 11 are monospecific), and Asteraceae is largest family with recorded 72 species, then Fabaceae with 43 species and third most is Poaceae having 34 plants species. Subsequently, Murti (2001) estimated that there were 1268 taxa in the floristic record of the western Himalaya's cold desert, including—Pteridophytes plants of 22 species, 1 subspecies, (16 genera, and 10 families), Gymnosperms (15 species, 1 variety, 6 genera and 3 families) and Angiosperms (1158 species, 29 subspecies, 41 varieties, one forma distributed under 384 genera and 72 families). As per to a survey conducted by Dutt (2018), documented 489 plant species comprising 218 Genera and 50 families comprises the flora of the Himalayan valley. Among the monocot flora, Poaceae and Cyperaceae are dominant, followed by Alliaceae and Liliaceae. The dicot flora is most abundant in the Asteraceae family throughout the Himalayan range (Hooker 1886) followed by Lamiaceae, then Ranunculaceae) along with Brassicaceae, Fabaceae, Graminae, etc. are also prevalent in western Himalaya (Kala, 2006). Under phyto-sociological analysis, the change in floristic diversity with shifting patterns of vegetation along the altitudinal gradient can be accessed by the qualitative and quantitative methodology. Phyto-diversity based on species rich ness in alpine Himalayan region is comparatively assess more in term of Shannon–Wiener index (H) *i.e.,* 6.55 in lower Alpine scrub than Alpine meadows having value of 5.41 (Chandrashekhar et al., 2003). The ecological gradients significantly influence the structuring and shaping of plant species richness along altitudinal gradients.

The Tibetan medicine system, Sowa-Rigpa (science of healing), employs about 162 plant species. Some important herbs include *Aconitum* spp., *Artemisia* spp., *Arnebea* spp., *Angelica glauca, Bergenia* spp., *Capparis spinosa, Carum carvi, Dactylorhiza hatagirea, Ephedra* spp., *Eremurus himalaicus, Geranium* spp., *Gentiana* spp., *Hyoscyamus niger, Juniperus indica, Thymus* spp., *Rheum* spp., *Rhodiola tibetica, Saussurea thomsonii, Senecio chrysanthemoides, etc.* Folk healers, known as Amchi or Larji, traditionally trek into the mountains to collect these medicinal herbs as raw materials for herbal medicine formulation (Dhiman et al., 2025). However, there is a noticeable decline in this traditional profession. Interaction between folk healers and the surrounding forests has decreased, partly because fewer young people in their families are willing to undertake the laborious and risky task of herb collection. Additionally, land use is shifting toward cash and horticultural crops, with about 60% of traditional farms abandoned for business-oriented farming. Local livestock are increasingly neglected, pasture routines are often discontinued, and the younger generation is gravitating toward homestays and hotel construction, embracing modern livelihoods. Given these changes, it is crucial to document knowledge of ethno-medicinal values local herbs and traditional agro-diversity of Himalayan region aware community about their nutritional security as well as value addition.

8 High-Altitude Pastures

Alpine pastures are highly prevalent in this region, locally known as 'bugyals' in Uttarakhand and 'margs' in Kashmir and 'kanda/dongri/dhum' in Kinnaur as well as Lahual & Spiti district of Himachal Pradesh. These may be marked as the connecting links between alpine ecology and traditional ranching. Proper management of these landscapes is immensely necessary to maintain the resilience of mountain ecosystem as well as inhabitant pastoral community and their ecological niche. High altitude pasture support various significant grasses, herbs and cold tolerant alpine flora to grow: adapted for short growing seasons and harsh variation of the climate. Although these ecosystems are seasonally rejuvenated by snowmelt and monsoon rains, which contribute to enhance the lushness of the meadows, which acts as a crucial summer grazing grounds for ranching and pastoral communities (Ram, 2005).

Subsequently, transhumant pastoralism is an intrinsic biogeographical characteristic noticed in the alpine Himalayan ecology to regenerate pasture when population migrate at lower elevation during winter. It also helps to maintain greater plant diversity and species richness (Ingty, 2021) and improve carbon sequestration (Mancilla-Leytón et al., 2022). Transhumance is rightly recognised by UNESCO as the 'Intangible Cultural Heritage of Humanity' highlighting its stewardship in delicate ecological balancing and preserving indigenous and traditional knowledge (UNESCO Report, 2019). Transhumant Pastoralism is not a naturalised way of balancing the fragile Himalayan ecology but also provides financially viable livelihood strategies and income generation patterns. Seasonal cattle migration accounts for 71–84% of household income from livestock products in Bhutan, livestock products like cheese and butter were not only produced for initial consumption but are also tradeable (Moktan et al., 2008). Integrated crop-livestock cropping system has taken immensely important role to strengthen the economic resilience as well as ecological sustainability in Lang-tang valley at the Himalayan region of Nepal (McVeigh, 2004).

9 Sacred Groves and Protected Forest Patches: Biocultural Landscapes

The ecologically fragile and climatically unfavourable nature of Alpine Himalayan region hosts a promising tapestry of spiritual cum cultural relationship between forest dwellers and the forested and semi-forested regions. One of the most enduring manifestations of this relationship are sacred groves and community-protected forest patches: serving both spiritual and ecological functions representing a harmonious convergence of belief systems, conservation ethics and traditional stewardship.

The ecological values of these patches include:

- Conserve relic species of medicinal herbs, wild edibles and alpine flora (Singh et al., 2011).
- Provide microhabitats for birds, pollinators and small mammals.
- Act as seed banks and gene pools, aiding ecological resilience.
- Protect critical water sources and minimises soil erosion.

An important observation from Social Forestry and Community Science standpoint is this forest patches are typically maintained through local customs and informal traditions, not by formal laws or under the purview of traditional administration system (Negi, 2010). Elders, religious custodians, local forest councils (*e.g., Van Panchayats* in Uttarakhand) etc. are entrusted with the management and enforcements of the rules, along with intergenerational transfer of senses of ecological protection (Nagahama, 2016).

10 Changing Traditional Resource Use Pattern and Its Implications

The general land use category includes human production activities mainly agriculture, forestry, animal husbandry, and urbanisation as in latest scenario, while land cover refers to features like roads, soil, ice, water bodies, farmland, forestland, and grassland (Zhao et al., 2020). The socioeconomic structure of the area is strongly impacted by changing climate parameters mainly temperature and precipitation patterns which, in turn, influence industries like forestry, agriculture, animal husbandry, tourism, and human health. The negative implications on water availability, soil fertility, and disease dynamics of insect and fungus, invasion of foreign weeds which pose a major danger to the fragile balance of the Alpine ecosystem as well as challenge on traditional farming practice (Kumar & Kumar, 2025). As per information compiled in 'Statistical Hand Book' (2004–2005), traditional agriculture and animal husbandry remains the mainstay of the economy in the alpine area of the Western Himalayas centuries ago. For livelihood, majority of their population depends upon low-input traditional farming system and conservatively pattern by growing crops such as Droh, Gandam (*Triticum aestivum*), Neh-Dolo (Nanga Jau), Jau-Dolo *(Hordeum himalayense*), Black pea, mountain potatoes in active growing season and were dried as processed as source of stored food-stock for winters. The local leafy vegetables include Rheum, Amaranths, Chenopodium, Hey-pommo, Allium scape (laabat) wild mint, new tender twigs of capers, Mau saag, Ta-sae etc. However, the abandonment of traditional farming practices has become increasingly common. Cultivation of traditional crop is declining, while agricultural activities are shifting toward cash crops such as green pea and a variety of exotic vegetables like cauliflower, tomato, pea, cabbage, squash, beetroot, and broccoli. These commercial crops have attracted tribal farmers due to their business potential. This transition has led to the deterioration of the natural genetic diversity of traditional crops, including Kala-matar (black pea, *Pisum* spp.), Pahari potato, Nanga-jau, and other local Alpine millets.

Yields and net cropping areas of traditional landraces are generally lower compared to introduce high-yielding varieties, whether in cereals, pulses, vegetables, or horticultural crops. Substantial research evidence points to the extinction and degradation of traditional germplasm within agro-ecosystems (Nautiyal et al., 2000). Traditional crops such as *Amaranthus* spp. (Bathu/yer), *Phaseolus vulgaris* (kidney beans), *P. lunatus* (chhimi), *Eleusine coracana* (finger millet), *Panicum miliaceum* (hog-millet), *Solanum tuberosum* (potato), *Triticum aestivum* (wheat), *Hordeum vulgare* (jau), and *H. himalayense* (naked barley) in the Uttarakhand Himalayan region are all in decline status.

11 Adaptation Tactics for Sustaining Use Pattern in Alpine Ecosystem

Community-based initiatives can help mitigate adverse effects by encouraging the cultivation of traditional crops (Nautiyal & Kaechele, 2007), adopting resilient crop varieties in response to climate change, and implementing effective water management techniques (Kumar & Kumar, 2025), thereby promoting sustainable land use development. General trends among indigenous communities also reflect a degradation of the cultural landscape, with a shift towards modern construction. Therefore, conducting transparent Environmental Impact Assessments (EIA) before undertaking projects in sensitive zones can help reduce negative impacts.

12 Eco-species Status and Upward Movement to Check the Adaptation Miseries

Alpine Forest above 3300 amsl with dry condition sparsely covered flora having xerophytes adaptation. Prominent adapted flora with dwarf growth, thick and prickled leaves, spiny habit, cushiony stems etc. which can be seen in species of *Astragalus, Arnebia, Artemisia, Caragana, Capparis Cousinia, Clematis, Chesneya, Lactuca, Herbacious, Hyoscyamus Oxytropis, Cotoneaster, Cicer, Lindelophia, Nepeta longibracteata, Heracleum, Lepidium, Taraxacum, Crepis flexuosa, Saussurea present* in scrub forest of Cold desert forest ecosystem. The floristic status of Himalayan cold desert species has been affected by numerous factors such as habitat degradation, bioprospecting, blind extraction, unscientific harvesting, population growth, climate change, etc. (Zou et al., 2023). Under the influence of climate change, remaining hardy species have been shifting further toward high altitudes for adaptation as well as survival. This has influenced the species in terms of habitat, dominance as well as frequency.

There is a gradual, often unseen, conflict between lower and upper alpine species, resulting in a subtle yet continuous reorganization of the ecosystem. Hence infinite sources are tumbling down and it is needed to manage the finite sources efficiently to keep in regulation under the norm of CBD *i.e.*, equitable/sustainable sharing and development while striving to restore ecological health. Previous research highlights the effectiveness of sustainable ecosystem and resource management strategies and emphasizes their positive impacts when applied collectively.

Mountains, due to their varied topography and pronounced climatic gradients, serve as highly suitable sites for studying climate change, with early signs often appearing there first (Beniston, 2003). It is pertinent to evaluate the changes in the climatic condition, prominently temperature and precipitation trend in that specific region. How the changes affecting the socioeconomic conditions and how developmental activities are changing the microclimate for that region. The loss in ecosystem both visible and intangible is irreversible and create emergency to preserve the ecosystem entities.

13 Socioecological Transitions in Changing Land Use Regime: Linking Socioecology with Land Economy

As the Alpine Himalayan region experiences intensifying exposures to socioeconomic modernization, demographic transitions and climatic uncertainties, the human–environment relationships are undergoing profound and dynamic transformation. These shifts are not merely external disruptions but are actively reconstituting the core structures, foundational norms and ecological feedback systems that have historically governed the land use and resource stewardship in high-altitude landscape.

Traditional socioecological systems once marked by cyclical land use, collective water governance and sacred ecological stewardship are now being reconstituted by emerging economic aspirations, evolving labour dynamics and shifting relationships with the land. This is reflected in several ongoing transitions:

- *From collective to individualized resource use*: Pasturelands, once governed by community norms and shared access, are increasingly enclosed or left untended due to outmigration and breakdown of customary institutions.
- *From subsistence to commodified landscapes*: Terraced fields, alpine meadows and even sacred sites are reoriented toward oriented activities such as tourism, herb extraction, commercial monoculture etc.
- *From TEK-centred practices to expert-driven interventions*: Traditional knowledge systems are being sidelined in favor of top-down, externally designed agricultural and conservation models, often misaligned with local realities.
- *From socio-spiritual land values to utilitarian land logic*: The deep spiritual and cultural meanings, once attached to sacred groves, forest springs or seasonal migrations, are eventually eroded by a land valuation system purely by economic returns and land productivity.

These transitions are not intrinsically detrimental but demand nuanced, participatory governance frameworks that safeguard ecological resilience and cultural continuity. Ensuring that development trajectories remain compatible with the region's unique socioecological identity is critical for fostering sustainable mountain futures.

14 Resource Use Patterns Evaluation

Under land use pattern analysis, land use and cover are important components to study. Their variations can create impact on degree of supply ecosystem services and affects the landscape function (Balthazar et al., 2015). The optimization of land use and land cover requires consideration of inherent laws, spatiotemporal variations in biophysical conditions, and the structure and function of ecosystems (Baral et al., 2016). Numerical modelling and statistical methods including general descriptive statistics, time-series analysis, regression analysis, and cost-benefit ratio analysis can be used to evaluate resource use patterns for effective management and sustainable utilization. Various study paradigms have addressed these approaches (Gardner et al., 1987). Other important analytical tools include carbon footprint assessment, carbon sequestration measurement, Geographic Information Systems (GIS), spatial analysis, and landscape fragmentation models. A novel framework for studying large-scale ecosystem spatial patterns has been developed by integrating satellite remote sensing (RS), GIS, and landscape ecology theory (O'neill et al., 1999).

Using Information Resource management for alpine areas of Himalayas, the annual data is compiled and assessed by concerned departments having the Environmental Information System (ENVIS) Centre such as at Himalayan ecology of the Institute of G.B. Pant Institute of Himalayan Environment and Development (GBPIHED).That acquired data along with historical or previous databases on natural resource, human resource, infrastructure and socio-economic aspects, etc., is integrated them with maps based on the Survey of India. On analysis it was witnessed that the unstable socio-economic system is impacted by unmanaged natural resources where degradation of natural resources starts. Hence there is spark in issue for conservation and policy working and need of community engagement, various awareness program by UNDP, IUCN, etc. conducted for people become more aware of how important the natural resources found in mountainous areas are to both lowland and upland populations.

15 Integration with Water Management

The alpine Himalayan region necessitates intricate and adaptive water management strategies due to steep slopes, limited glacial sources and seasonality of precipitation. Local communities have traditionally developed ingenious methods to integrate water management with land use plannings, to ensure the sustainability and survival at the mountain regions and for the inhabiting hilly communities in these ecologically fragile and high-altitude environments.

Traditional Water Management Systems

While in the utmost level of fluctuating water availability, communities have devised sophisticated, community-driven techniques such as:

- Kulhs
  - Gravity-fed Irrigation Channels
  - traditional irrigation canals, prominent in Himachal Pradesh and Uttarakhand
  - divert glacial meltwater or spring water through sloping channels to terraces of agricultural fields.
  - Often maintained collectively by village communities
  - kulhs exemplify decentralized, low-impact water governance.
- Zings and Guls
  - Storage Ponds and Streams
  - Noticed in high-altitude cold desert regions, *e.g.*, Ladakh, northern Sikkim

  - small storage tanks or ponds called *zings*: they collect the snowmelt or rainwater for later
  - often connected to *guls:* artificial rills that distribute water across agricultural land.
- Jab Systems
  - Stone Water Troughs
  - In pastoral zones and grazing corridors, natural springs and seasonal streams are channelised into stone-lined water troughs
  - locally called *jabs* in some areas
  - ensuring both livestock access and groundwater recharge.

16 Agroforestry—Alpine Himalaya

There is no doubt, that agroforestry with inheriting traditional methods acts as sustainable practice to make the agriculture and farming resilient. When there was surge in demand for food, wood, fuel and fodder requirement from every corner of the globe, 'Agroforestry' has always responded as multipurpose beholder to absorb their insecurity. The term "agroforestry" refers to a system and technique of land use wherein woody perennials are purposefully utilised on the same land management unit as agricultural crops and/or animals in some kind of spatial arrangement or temporal sequence. There are interactions between the many components of an agroforestry system on both an ecological and economic level. (Lundgren & Raintree, 1982).

Water and soil fertility management can conserve and sustained the characteristic of parent soil in fragile ecosystem. This can be enhanced by adopting traditional practice-based agroforestry using local species which could thrive in limited water resources. Moreover, the coexistence of interacting flora and fauna promotes biodiversity, supporting predator-prey relationships and contributing organic matter typically scarce in dry areas. Leaves of tree and fodder species in Agroforestry can be a valuable source of organic matter, enhancing nutrient cycling in the soil, source of food to livestock. The alpine region, naturally devoid of dense tree canopies except for forest plantations and scattered native tree species, benefits greatly from agroforestry, which creates a conducive habitat for both native and planted trees. Trees in the alpine zone must adapt to limiting factors such as moisture scarcity, high-intensity winds, intense sunlight, nutrient competition, extreme winters, fluctuating diurnal temperatures, and anthropogenic disturbances. Five agroforestry systems (AFS) were identified as Agri-horticulture, Agri-silviculture, Agri-silvi-pastoral, Pastoral-silvi-culture, Pastoral-horticulture (Anita Kumari et al., 2008) and practiced in Lahual and Spiti region, where Salix species is only tree component in three agroforestry system rest is dominated by Pea, potato, and apple and followed by playhouses vegetables, grasses/fodders etc. Native natural species of *Salix* i.e., *S. alba, S. excelsa, S. angustifolia, and S. fragilis* in cold deserts are a particular type of ecosystem where water is scarce. These species with good growth and vegetative potential survive in the harsh circumstances of a cold desert climate. In the area of northwest Himalayan region, most prevalent trees are common willow species are *Salix fragilis* Linn and *Salix alba* Linn are widely cultivated as local agroforestry system (Rawat et al., 2006) and Pahari peepal (*Populus ciliata*). But local community prefer *Salix* (Changma), is the mostly planted agroforestry tree at their farm boundaries. *Salix fragilis,* a willow species widely cultivated in the indigenous agroforestry system as well as at forestry settlement in Lahaul Valley (Rawat et al. 2006). Plantation of *Rubina* spp. (Black locust) has profusely growing in Spiti valley were planted by forest department having aromatic potential in flowers and patches of *Betula* spp. are visible at side edges of the alpine villages.

Apart from trees various shrubs like *Rosa webbiana* (Sia), *Hippophae rhamnoides* (Chharma), *Juniperus* spp. (Shukpa) and native grasses are found naturally in the field farm. These native species are crucial components of cold desert agroforestry systems along with assisting in reduction of rate of desertification. These species are socially and environmentally suitable for forestry operations due to their ability to flourish in shoot-cutting plantations in the harsh, xeric climate of cold deserts. *Salix* spp. has been cultivated over the past century in various northern-western part of Himalayan region, particularly for the production of biomass and bioremediation (Karp et al., 2011). Farmers increasingly choose these multipurpose trees along agricultural boundaries, as they provide significant food and fuel resources and offer vegetation cover essential to the dry environment. Regarding alpine grasses, about 141 taxa of Poaceace family recorded in Desert, Himachal Pradesh, India having 27 species, '*Poa*' is the most dominant genera of alpine grasses in the region which is followed by *Festuca* spp., *Stipa* spp., *Bromus* spp., *Elymus* spp., *Calamagrostis* spp., *Agrostis* spp., *etc.* Among fodder species *Poa koelzii* and *P. lahaulensis* are under grazing pressure and declared as endemic (Negi et al., 2019) (Fig. 2.4).

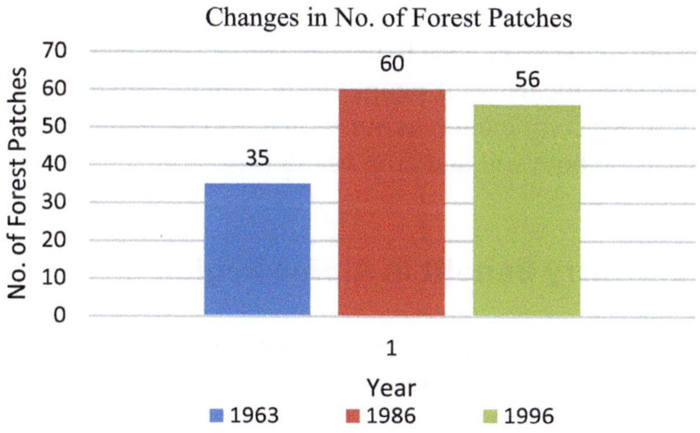

Fig. 2.4 Graphical presentation of change in no. of forest patches. (Rao and Pant, 2001)

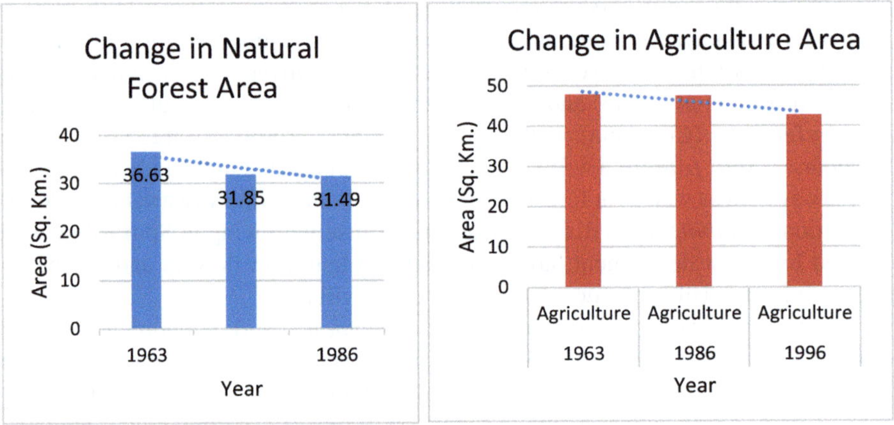

Fig. 2.5 Graphical presentation of change in natural forest and agricultural land. (Rao and Pant, 2001)

17 Awareness on Soil Fertility

The future of human existence and livelihood is threatened by global climate change, particularly for those who currently reside delicate biospheres. In most of the agroforestry system main agricultural crop is Matar or Pea (both local and green pea). Nowadays farmers are also growing various pulses and millets on recommendation of state government schemes. Their continuous cultivation on same unit of land has increased the nitrogen content in soil which is really required as alkaline soil of alpine regions devoid of Nitrogen. Local observers have noticed that apart from Nitrogen, other nutrients like Potassium, Phosphorus, Sodium, Calcium etc. increase but lead to deficiency of other micro-nutrients due to repetition of same cultivar. Hence, there in agroforestry system crop rotation is very important to maintain the soil health of the farm. Due to unawareness of soil health and recommendation on soil nutrient, locals aged people were using chemical fertilizers such as Urea, CAN, 12:32:16 combination as nutrition to soil without considering the negative consequences on both soil and crop health (Fig. 2.5).

18 Agroforestry Benefit in Alpine Region

Various research efforts are focused on developing sustainable agroforestry technologies for sensitive regions such as the cold desert Himalayan region as well as arid, semi-arid zones and coastal areas. These ecosystems need to be sustained for persuading productivity through adopting natural resource management like Agroforestry (Fig. 2.6).

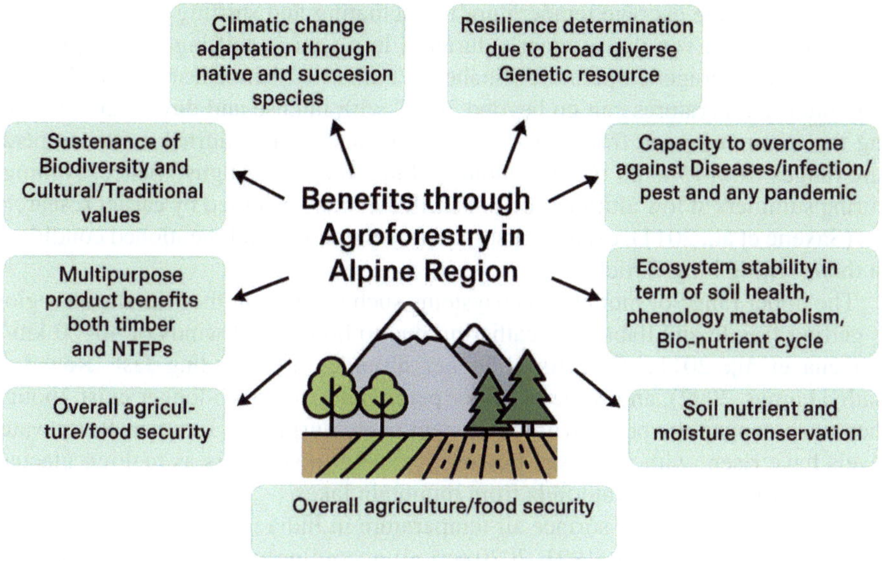

Fig. 2.6 Benefits through agroforestry in Alpine Region

19 Hydro-Ecological Integration with Land Use and Settlement

Water systems of the Himalayan Alpine regions are not isolated infrastructural elements but intricately woven into the broader fabric of socioecological landscape of the area. Traditional irrigation channels are carefully integrated with the terraces to maximize water flow and minimize water losses (Ray et al., 2014). The sacred groves serve as protective buffers for spring catchments. This integrated approach reflects a deep ecological consciousness, where water is both a physical necessity and a sociocultural integration of mountain livelihood (Jana et al., 2017).

Factually, water management in the alpine Himalayan region is deeply embedded with land, culture and ecology. Protecting, revitalizing and enhancing these systems is essential not only for the sustainability of mountain communities but also for broader, long-term hydrological resilience of the Himalayan watershed.

20 Climatic Status

A latitudinal gradient of mean annual temperature affects alpine ecosystems, which are climatically distinguished by seasonality and continentally across the globe. Taking example of Alpine Himalayan region which is at east of the cold mountain desert of the Tibetan Plateau. During winters, cold desert region exhibits sub-zero temperature during day time which could drop up to −35 °C (Mishra, 2001) at midnights in the extreme winter months (December to February). But at lower altitudes of alpine area, the average temperature is fluctuating and shifting toward the positive warm side. Even the summer's duration in a year is shifting and progressing with raise in average temperature (Shaheen et al., 2013) since last decade. Diurnal maximum temperatures can go beyond 30 °C with intense and direct sunlight during the summer season from mid-June to August. Average diurnal-mean temperature ranges from 15 to 20 °C for 3 months (June, July, and August) hence, evenings during summers at low altitudes are not chilled as were reported by earlier researchers (Saxene et al., 2011), excursionists and travellers who had mentioned conditions in their writing blogs/articles.

The upper range of mountain ecosystems such as those in the Himalayan region is getting fragile and liable to weathering due to high-speed winds of 40–60 km/h (Saxena et al., 2011). Glaciers at higher altitudes are receding fast (Rawat & Sathyakumar, 2002), and some of minor perennial glaciers no longer exist, though they once capped the peaks with permanent snow during the summer. River water levels have risen, with increased flowing intensity in summers as melting glaciers form water nallahs and waterfalls from mountain lakes.

The annual mean land surface air temperature in India in 2024 was 0.65° higher than the long-term average (1991–2020 period), according to the India Meteorological Department (IMD). The average temperature recorded as 25.75 °C. The obtained SD of 0.75 °C indicates that temperature had varied by 0.75 °C from the annual mean temperature. The temperature trend in higher altitude Himalayan region using Mann–Kendall and Sen's slope trend tests from January 1981 to December 2021. (Kumar et al., 2025).

21 Socioeconomic Drivers of Change

The socioecological equilibrium is increasingly disrupted by various changes due to some external and internal drivers:

- Youth migration to urban areas reduces traditional labour pools and also erodes the local knowledge systems (Basnet et al., 2025)
- The commodification of high-altitude resources, particularly medicinal herbs [*e.g., Yarsagumba* (*Ophiocordyceps sinensis*)] has led to overharvesting and overexploitation of the environment.
- Expansion of tourism and infrastructure development alters land use priorities, shifts labour patterns, increases ecological footprints thus imputing the ecosystemic fabric of that region (Vijay et al., 2016) (Fig. 2.7).
- Capital-intensive agricultural models are displacing traditional, low-impact practices in favour of cash crops and intensified livestock rearing, thereby fragmenting the region's ecological mosaic (Behera et al., 2015) (Fig. 2.8).
- These transformations are reconfiguring long-standing human–nature relationships, often undermining the finely tuned socioecological systems cultivated over generations.

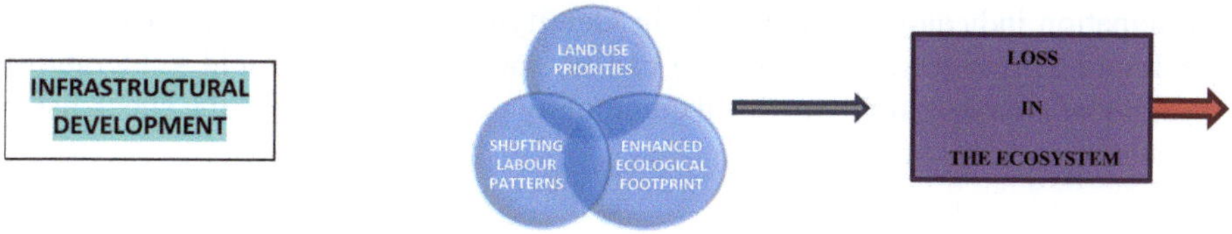

Fig. 2.7 Infrastructural development

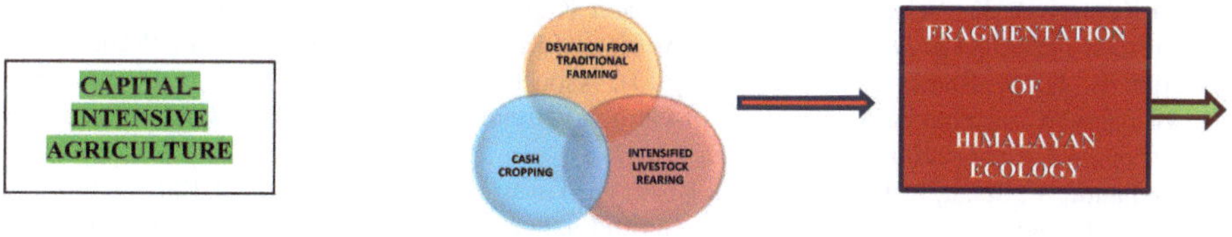

Fig. 2.8 Capital intensive agriculture

Emerging Adaptive Practices

Despite these challenges, communities across the Himalayan alpine belt are actively innovating and adapting: These adaptive strategies reflect a socioecological resilience rooted in flexibility, collaboration, and cultural continuity.

Prior Assessments for Conservation Strategies

The fundamental knowledge of species composition and diversity along the mountain elevation gradient was first made possible by Alexander von Humboldt's, on that base the 'altitudinal gradient of vegetation' mentioned by Fischer et al. (2011). Following the similar approach, Bahukhandi et al. (2024) in their study titled 'Floristic diversity and species composition along altitudinal gradient in the alpine ecosystem of the cold desert region in Western Himalaya, India' outlined a multistage methodology. The preliminary stage is to 'explore the ecological condition of the region focusing on floristic diversity and species composition, next to explore the trend of species richness along the elevation gradient, and finally to assess the threatened and at-risk native MAPs of alpine region. Rathore and Shashni (2023) elaborated and correlated the traditional knowledge with the livelihood of Himalayan communities to adapt and sustain.

Rapid vulnerable assessment (RVA) is urgently needed in this delicate ecosystem to assess the plant species to climate risk. Raising awareness among tribal communities through RVA tools can help to implement the foremost solution to preserve wild species and make them more resilience w.r.t. prevailing environment. By educating forest residing populations about the Wildlife Protection Act and their inclusion- community in-situ conservation and balanced usage, developing herbal garden by herbalist.

Vulnerability assessment helps to prioritising selection of prominent indicators along with identify area that are most likely to be at verge of risk and create and distribute adaptive solutions in accordance with those findings. With community participation indicator variables help in identifying climate-sensitive sites. Hence, the Rapid Threat Assessment (RTA) can be enclosed using a rapid vulnerability assessment method employing local community participation to get the check list of threatened species.

References

Anita Kumari, A. K., Sehgal, R. N., & Shailender Kumar, S. K. (2008). *Traditional agroforestry systems practiced in Lahaul (Lahaul & Spiti) and Kinnaur districts of Himachal Pradesh.*

Badoni, A. K., & Bhatt, B. P. (1989, August). Aspects and prospects of energy Plantation in Himalayan wastelands. A case study of Garhwal Himalaya. In *National workshop of economics of energy plantation on wastelands* (pp. 7–8).

Bahukhandi, A., Sekar, K. C., Negi, V. S., Bisht, K., Tiwari, D. C., Mehta, P., et al. (2024). Floristic diversity and species composition along altitudinal gradient in the alpine ecosystem of the cold desert region in Western Himalaya, India. *Frontiers in Plant Science, 15*, 1469579. https://doi.org/10.3389/fpls.2024.1469579

Balthazar, V., Vanacker, V., Molina, A., & Lambin, E. F. (2015). Impacts of forest cover change on ecosystem services in high Andean mountains. *Ecological Indicators, 48*, 63–75. https://doi.org/10.1016/j.ecolind.2014.07.043

Baral, H., Guariguata, M. R., & Keenan, R. J. (2016). A proposed framework for assessing ecosystem goods and services from planted forests. *Ecosystem Services, 22*, 260–268. https://doi.org/10.1016/j.ecoser.2016.10.002

Basnet, P., & Khatiwada, S.P. (2025). Exploring the socio-cultural effects of youth migration in rural areas. *The Third Pole: Journal of Geography, 25*, 51–63. https://doi.org/10.3126/ttp.v25i1.87775

Behera, M. K., Pradhan, T. R., & Sahoo, J. (2015). Role of sacred groves in the conservation and management of medicinal plants. *Journal of Medicinal Plants Research, 9*(29), 792–798. WGBIS/CES, Indian Institute of Science. https://doi.org/10.5897/JMPR2013.5781

Beniston, M. (2003). Climatic change in mountain regions: a review of possible impacts. *Climatic Change, 59*(1), 5–31. https://doi.org/10.1023/A:1024458411589

Bhatnagar, Y. V., Mathur, V. B., Sathyakumar, S., Ghoshal, A., Sharma, R. K., Bijoor, A., Raghunath, R., Timbadia, R., & Lal, P. (2016). South Asia: India. In *Snow leopards* (pp. 457–469). Academic. https://doi.org/10.1016/B978-0-12-802213-9.00036-5

Bruijnzeel, L. A., & Bremmer, C. N. (1989). *Highland-lowland interactions in the Ganges Brahmaputra river basin: a review of published literature* (ICIMOD Occasional Paper no. 11, 136 pp). ICIMOD.

Chandra, N., Singh, G., Lingwal, S., Rai, I. D., & Tewari, L. M. (2021). Alpine medicinal and aromatic plants in the Western Himalaya, India: An ecological review. *Indian Journal of Ecology, 48*(2), 319–331.

Chandrashekhar, M. B., Singh, S., & Roy, P. S. (2003). Geospatial modelling techniques for rapid assessment of phytodiversity at landscape level in western Himalayas, Himachal Pradesh. *Current Science, 84*, 663–670.

Chaursia, O.P., Ahmed & Ballabh. (2007). *Ethnobotany and plants of trans-Himalaya, Field research laboratory (DRDO) Leh Ladakh*

Dhiman, B., Spehia, R. S., Attri, I., Dev, I., & Chandel, R. S. (2025). An analytical ethnobotanical survey-study and its contribution to 'Amchi System' of Spiti Valley, India: Amchi System'-A traditional healthcare system of Spiti Valley. *Indian Journal of Traditional Knowledge (IJTK), 24*(12), 1164-1174. https://doi.org/10.56042/ijtk.v24i12.21131

Dutt, B. (2018). Diversity of medicinal herbs in cold desert region of Himachal Pradesh. In *Perspectives on biodiversity of India* (Vol. 20).

Fischer, A., Blaschke, M., & Bässler, C. (2011). Altitudinal gradients in biodiversity research: the state of the art and future perspectives under climate change aspects. *Waldökologie, Landschaftsforschung und Naturschutz, 11*, 35–47

Gardner, R. H., Milne, B. T., Turnei, M. G., & O'Neill, R. V. (1987). Neutral models for the analysis of broad-scale landscape pattern. *Landscape Ecology, 1*(1), 19–28.

Hooker, J. D. (1886). *Flora of British India* (Vol. 5, p. 489). L. Reeve and Co.

Ingty, T. (2021). Pastoralism in the highest peaks: Role of the traditional grazing systems in maintaining biodiversity and ecosystem function in the alpine Himalaya. *PloS One, 16*(1), e0245221. https://doi.org/10.1371/journal.pone.0245221

Jain, S. K., Goswami, A., & Saraf, A. K. (2009). Role of elevation and aspect in snow distribution in Western Himalaya. *Water Resources Management, 23*(1), 71–83.

Jana, P., Dasgupta, S., & Todaria, N. P. (2017). Impact and ecosystem service of forest and sacred grove as saviour of water quantity and quality in Garhwal Himalaya, India. *Environmental Monitoring and Assessment, 189*(9), 477. https://doi.org/10.1007/s10661-017-6173-2

Kachroo, P., Sapru, B. L., & Dhar, U. (1977). *Flora of Ladakh* (Vol. 15, pp. 4253–4270). Bishen Singh Mahendra Pal Singh.

Kala, C. P. (2006). Medicinal plants of the high-altitude cold desert in India: diversity, distribution and traditional uses. *The International Journal of Biodiversity Science and Management, 2*(1), 43–56.

Karp, A., Hanley, S. J., Trybush, S. O., Macalpine, W., Pei, M., & Shield, I. (2011). Genetic improvement of willow for bioenergy and biofuels free access. *Journal of Integrative Plant Biology, 53*(2), 151–165. https://doi.org/10.1111/j.1744-7909.2010.01015.x

Körner, C. H. (1995). Alpine plant diversity: a global survey and functional interpretations. In *Arctic and alpine biodiversity: patterns, causes and ecosystem consequences* (pp. 45–62). Springer Berlin Heidelberg.

Kumar, A., & Kumar, S. (2025). Impact of climate change on Himalayan ecosystem and traditional crops. In *Climate change impact on Himalayan biodiversity* (pp. 393–413).

Kumar, P., Deka, D., Husain, M. A., Kumar, M., Choudhari, P., Singh, S., & Banerjee, A. (2025). Trend analysis of precipitation and temperature in Lahaul-Spiti district, Himachal Pradesh, India. *Frontiers in Climate, 6*, 1348132. https://doi.org/10.3389/fclim.2024.1348132

Lundgren, B. O., & Raintree, J. B. (1982). Sustained agroforestry. In B. Nestel (Ed.), *Agricultural Research for Development: Potentials and Challenges in Asia* (pp. 37–49). ISNAR.

Malik, Z. A., Bhat, J. A., Ballabha, R., Bussmann, R. W., & Bhatt, A. B. (2015). Ethnomedicinal plants traditionally used in health care practices by inhabitants of Western Himalaya. *Journal of Ethnopharmacology, 172*, 133–144. https://doi.org/10.1016/j.jep.2015.06.002

Mancilla-Leytón, J. M., Gribis, D., Pozo-Campos, C., Morales-Jerrett, E., Mena, Y., Cambrollé, J., & Vicente, Á. M. (2022). Ecosystem services provided by pastoral husbandry: A bibliometric analysis. *Land, 11*(11), 2083. https://doi.org/10.3390/land11112083

McVeigh, C. E. (2004). *Can pastoralists protect our parks? Linking biodiversity conservation and sustainable use in Nepal's Langtang Valley*. Cornell University.

Mishra, C. (2001). *High altitude survival: conflicts between pastoralism and wildlife in the Trans-Himalaya*. Wageningen University and Research.

Moktan, M. R., Norbu, L., Nirola, H., Dukpa, K., Rai, T. B., & Dorji, R. (2008). Ecological and social aspects of Transhumant Herding in Bhutan. *Mountain Research and Development, 28*(1), 41–48. https://doi.org/10.1659/mrd.0802

Murali, R., Redpath, S., & Mishra, C. (2017). The value of ecosystem services in the high altitude Spiti Valley, Indian Trans-Himalaya. *Ecosystem Services, 28*, 115–123.

Murti, S. K. (2001). *Flora of cold deserts of western Himalaya* (Vol. 2). Botanical survey of India.

Nagahama, K., Satya, L., & Saito K. (2016). The Van Panchayat movement and struggle for achieving sustainable management of the forest: A case study of Uttarakhand in North India. *SDRP Journal of Earth Sciences & Environmental Studies. 1*(3), 61–70. https://doi.org/10.15436/JESES.1.3.1

Nautiyal, S., & Kaechele, H. (2007). Conservation of crop diversity for sustainable landscape development in the mountains of the Indian Himalayan region. *Management of Environmental Quality: An International Journal, 18*(5), 514–530. https://doi.org/10.1108/14777830710778283

Nautiyal, S., Rao, K. S., Maikhuri, R. K., & Negi, K. S. (2000). Apne hi ghar main kho gai mukhmar Mukhmar. *Envis Bulletin, Himalayan Ecology and Development, 8*(2), 83–84.

Negi, C. S. (2010). Traditional culture and biodiversity conservation: Examples from Uttarakhand, Central Himalaya. *Mountain Research and Development, 30*(3), 259–265. https://doi.org/10.1659/MRD-JOURNAL-D-09-00040.1

Negi, R., Verma, P. K., Baig, S., Chandra, A., Naithani, H. B., Verma, R., & Kumar, A. (2019). Checklist of family Poaceae in Lahaul and Spiti district (Cold Desert), Himachal Pradesh, India. *Plant Science Today, 6*(2), 270–274. https://doi.org/10.14719/pst.2019.6.2.500

O'neill, R. V., Riitters, K. H., Wickham, J. D., & Jones, K. B. (1999). Landscape pattern metrics and regional assessment. *Ecosystem Health, 5*(4), 225–233.

Pauli, H., Gottfried, M., Dullinger, S., Abdaladze, O., Akhalkatsi, M., Alonso, J. L. B., Coldea, G., Dick, J., Erschbamer, B., Calzado, R. F., & Ghosn, D. (2012). Recent plant diversity changes on Europe's Mountain summits. *Science, 336*(6079), 353–355. https://doi.org/10.1126/science.1219033

Ram, J. (2005). Biodiversity and conservation of high-altitude meadows of Uttaranchal Himalaya.

Rao, K. S., & Pant, R. (2001). Land use dynamics and landscape change pattern in a typical micro watershed in the mid elevation zone of central Himalaya, India. *Agriculture, Ecosystems & Environment, 86*(2), 113–124. https://doi.org/10.1016/S0167-8809(00)00274-7

Rathore, S., & Shashni, S. (2023). Indigenous knowledge system and livelihood option of natives of Lahaul and Spiti district, Himachal Pradesh. *In Climate change adaptation, risk management and sustainable practices in the Himalaya* (pp. 213–235). Cham: Springer International Publishing

Rawat, G. S., & Sathyakumar, S. (2002). Conservation issues in the Himalayan Regions of India (IN). In S. Sathyakumar & Y. V. Bhatnagar (Eds.), *ENVIS Bulletin: Wildlife and protected areas* (Vol. 1, No. 1) (pp. 50–56).

Rawat, Y. S., Oinam, S. S., Vishvakarma, S. C., Kuniyal, C. P., & Kuniyal, J. C. (2006). Willow (Salix fragilis Linn.): a multipurpose tree species under pest attack in the cold desert of Lahaul valley, northwestern Himalaya, India. *Ambio: A Journal of the Human Environment, 35*(1), 43–48. https://doi.org/10.1579/0044-7447-35.1.43

Ray, R., Chandran, M.D.S., Ramachandra, T.V. (2014). Biodiversity and ecological assessments of Indian sacred groves. *Journal of Forestry Research. 25*(1), 21–28. https://doi.org/10.1007/s11676-014-0429-2

Saxena, K. G., Liang, L., & Xue, X. (2011). *Global change, biodiversity, and livelihoods in cold desert region of Asia*. Bishen Singh Mahendra Pal Singh.

Shaheen, F. A., Wani, M. H., Wani, S. A., & Norphel, C. (2013). Climate change impact in cold arid desert of North–Western Himalaya: Community based adaptations and mitigations. In *Knowledge systems of societies for adaptation and mitigation of impacts of climate change* (pp. 239–256). Springer Berlin Heidelberg.

Singh, A., Manohar, L., & Samant, S. S. (2009). Diversity, indigenous uses and conservation of medicinal plants in Lahaul Valley, Proposed Cold Desert Biosphere Reserve, India. *International Journal of Biodiversity Science and Management, 5*(3), 132–154. https://doi.org/10.1080/17451590903230249

Singh, A., Nautiyal, M. C., Kunwar, R. M., & Bussmann, R. W. (2017). Ethnomedicinal plants used by local inhabitants of Jakholi block, Rudraprayag district, western Himalaya, India. *Journal of Ethnobiology and Ethnomedicine, 13*(1), 49.

Singh, H., Agnihotri, P., Pande, P. C., & Husain, T. (2011). Biodiversity conservation through a traditional beliefs system in Indian Himalaya: A case study from Nakuleshwar sacred grove. *The Environmentalist, 31*(3), 246–253. https://doi.org/10.1007/s10669-011-9329-6

Singh, R. P., Bahar, N., & Negi, D. V. (1995). Impact of grazing on soil erosion in forest ecosystems. *Indian For., 121*(8), 717–719.

Singh, S., & Chatterjee, S. (2020). Provisioning ecosystem services of Rhododendron-rich forests in the Western Himalayas. *Environmental Sciences Proceedings, 3*(1), 70. https://doi.org/10.3390/IECF2020-08070

Spehn, E. M., & Körner, C. (2005). A global assessment of mountain biodiversity and its function. In *Global change and mountain regions: An overview of current knowledge* (pp. 393–400). Springer Netherlands.

Stewart, R R. (1916–1917). *The flora of Ladakh, Western Tibet*. Ph.D. Thesis, Columbia University.

Testolin, R., Attorre, F., & Jiménez-Alfaro, B. (2020). Global distribution and bioclimatic characterization of alpine biomes. *Ecography, 43*(6), 779–788. https://doi.org/10.1111/ecog.05012

UNESCO. (2019). Transhumance, the seasonal droving of livestock along migratory routes in the Mediterranean and in the Alps (Representative List of the Intangible Cultural Heritage of Humanity). *UNESCO Intangible Cultural Heritage*. https://ich.unesco.org/en/RL/01470

Vijay, R., Kushwaha, V. K., Chaudhury, A. S., Naik, K., Gupta, I., Kumar, R., & Wate, S. R. (2016). Assessment of tourism impact on land use/land cover and natural slope in Manali, India: A geospatial analysis. *Environmental Earth Sciences, 75*(1), 20. https://doi.org/10.1007/s12665-015-4858-9

Viviroli, D., Dürr, H. H., Messerli, B., Meybeck, M., & Weingartner, R. (2007). Mountains of the world, water towers for humanity: Typology, mapping, and global significance. *Water Resources Research, 43*(7). https://doi.org/10.1029/2006WR005653

Zhao, Q., Wen, Z., Chen, S., Ding, S., & Zhang, M. (2020). Quantifying land use/land cover and landscape pattern changes and impacts on ecosystem services. *International Journal of Environmental Research and Public Health, 17*(1), 126. https://doi.org/10.3390/ijerph17010126

Zou, H., Chen, B., Zhang, B., Zhou, X., Zhang, X., Zhang, X., & Wang, J. (2023). Conservation planning for the endemic and endangered medicinal plants under the climate change and human disturbance: a case study of Gentiana manshurica in China. *Frontiers in Plant Science, 14*, 1184556

3 Functional Diversity and Ecophysiological Traits Along Elevational Gradients

Shiv Rattan

Abstract

This chapter synthesized existing literature on functional diversity and ecophysiological traits of high-altitude plants along elevational gradients in the Western Himalayas and similar mountainous regions. It discussed how environmental stressors, including temperature, precipitation, UV radiation, and nutrient availability, influenced plant adaptations across varying altitudes. Key topics included patterns of species richness and distribution, morpho-anatomical variations such as leaf thickness, stomatal density, and root architecture, and eco-physiological responses like photosynthetic efficiency, nutrient resorption, and biochemical changes involving antioxidants and osmoprotectants. The compilation also covered molecular mechanisms, including gene expression, hormonal signaling, and transcriptomic regulation under multifactorial stresses. Emphasis was placed on the role of phenotypic plasticity and genetic diversity in enabling plant survival and acclimatization to harsh alpine conditions. Overall, the chapter highlighted implications for ecosystem resilience, biodiversity conservation, and vegetation responses to ongoing climate change in these dynamic environments.

The original version of the chapter has been revised. A correction to this chapter can be found at https://doi.org/10.1007/978-981-95-6923-6_18

S. Rattan (✉)
Department of Ornamental Plants and Agricultural Biotechnology, Agricultural Research Organization (ARO) – Volcani Institute, Rishon LeZion, Israel

P. Kumar et al. (eds.), *Climate Resilience and Molecular Adaptation in Alpine Medicinal Plants*, https://doi.org/10.1007/978-981-95-6923-6_3

Keywords

Eco-physiological traits · Diversity · Metabolism · Adaptation · Alpine · Medicinal plants

1 Introduction

Global warming is adversely affecting species dispersals and is thought to be a primary factor for extinction of species caused by variations in phenology, species range shifts, altered biological interaction, and growth pattern changes (Franklin et al., 2016; Thakur & Chawla, 2019). An increase in species richness has been documented in high-altitude mountain environments due to plant upward mobility (Steinbauer et al., 2018). Therefore, the upward migration of vigorous generalist species may cause local extinction of the species already found at high altitudes, resulting in competitive exclusion (Alexander et al., 2015). Over the years, a wealth of information has been gathered on the physiological, molecular, and biochemical responses to biotic and abiotic stress in control environments. Nevertheless, only a limited research is concentrated to develop climate-resilient crops in natural environments (Nuccio et al., 2018; Kumari et al., 2025). This might be attributed to a knowledge gap among unpredictable natural conditions and controlled experiments. Plants in natural environments are frequently exposed to varying seasonal fluctuations such as light intensity, air temperature, soil composition, nutrient availability, and relative humidity making phenotyping more realistic. High-altitude alpine regions exhibit a broad range of harsh environmental conditions, and substantial environmental fluctuations transpire over small spatial distances in mountain ecosystems (Kumari et al., 2025). High-altitude species are frequently subjected to single stresses or combinations of stresses, which would negatively affect the general growth, development, distribution, along with interactions of plants with further organ systems, such as the soil microbiome (Kumari & Kumar, 2024; Kumari et al., 2025). Extreme unregulated habitats in open conditions provide natural laboratories for studies on the adaptive evolution of an organism, even if it is challenging to predict the molecular players of plant adaptation in a natural uncontrolled environment. In alpine environments, where mild to severe stresses work concurrently, it is crucial to comprehend the elements that impact plant survival and successful adaptation. Multifactorial stresacts combinations are more harmful to total plant growth than individual stress (Zandalinas et al., 2021). Because of different stress receptors, downstream signalling pathways, hormone-regulatory network interactions, and other factors, plants under combined stress may adopt a new strategy that is different from those under individual stress. The stress response mechanism is substantially determined by plant genotype, species, developmental stage, and the intensity & extent of stress stimuli (Shaar-Moshe et al., 2017).

Depending on the degree and length of stress stimuli, plants may prioritise one method over the other, attain a collective retort, or develop an entirely new/novel approach (Rivero et al., 2022). For instance, when drought and heat stresses are combined, plant prioritise drought-resulted stomatal regulation above heat-resulted stomatal regulation (Zandalinas et al., 2020). Additionally, the kind and degree of expression of the enzymes that detoxify reactive oxygen species (ROS) during combined drought and heat stress differed from those under specific stress (Pandey et al., 2015). Likewise, proline was the main osmoprotectant in specific stress, but glycine, betaine, and trehalose are the main osmoprotectants in heat and salinity stress (Pandey et al., 2015). Furthermore, compared to individual or specific stress, the relative levels and functions of many hormones, such as jasmonic acid, salicylic acid, abscisic acid (ABA), and ethylene are entirely different in stress combinations (Gupta et al., 2017). In order to adapt to stress and grow, plants may therefore arrange their metabolism to strike a stability among energy levels and resource distribution.

Mountainous ecosystems have diverse zones of vegetation along an altitudinal gradient, which contribute to their unique biodiversity. The distribution and organisation of plant communities are influenced by their notable microclimate diversity, which includes little height dissimilarity (Antonelli et al., 2018; Fatima et al., 2022). As temperatures rise, it is anticipated that the distribution of plant species will move to higher elevations (Freeman et al., 2018). Altitudinal gradients in high altitude were investigated to comprehend the effects of climate change on distribution of plants and soil qualities because significant environmental changes can transpire over short distances (Mangral et al., 2023). Elevation gradients variation has a substantial impact on both abiotic and biotic components, which are intimately tied to constant human interferences. The main factors of vegetation change along altitude are wind, temperature, precipitation, ultraviolet radiations (UV), and nutrition availability. Soil physicochemical qualities and vegetation change greatly according to varied climatic circumstances, which are associated with lower levels of organic carbon, nitrification, and pH (Kewlani et al., 2021; Mangral et al., 2023). One important element that decreases with elevation is temperature; altitudinal gradients work as thermoclines, causing long-term environmental conditions. Because they can adapt to changing environments, plants with higher phenotypic flexibility have been discovered more commonly at higher elevations. Phenotypic flexibility compensates for the short growing season by enabling fast reproduction in many alpine and subalpine species. Plants modify critical physiological, reproductive, and physical characteristics in response to changes in the climate (Wani et al., 2022). Because of their adaptability, plants can live across altitudinal gradients in a range of ecosystems. Investigating these adaptive features and connecting them to environmental factors will help us better understand the limitations that altitudinal gradients place on plant dispersal (Fatima et al., 2022; Mangral et al., 2023).

Understanding how plants respond to environmental stress, particularly in alpine environments where circumstances are harsh and fluctuating, depends heavily on functional diversity and ecophysiological features. Functional diversity refers to the range and value of traits that influence plant performance and ecosystem functioning. This comprises characteristics that assist plants adapt to cold temperatures, brief growing seasons, and scarce resources in alpine habitats, such as leaf morphology, root architecture, and reproductive methods (Stanisci et al., 2020). While sub-alpine communities tend towards conservative strategies, favouring long-term survival under arid conditions, alpine communities frequently display acquisitive strategies-traits that maximise rapid carbon acquisition during brief favourable periods (Stanisci et al., 2020). The morphological, biochemical, and molecular mechanisms that allow plants to withstand abiotic conditions including drought, cold, and intense UV light are referred to as ecophysiological features. These include adaptations that have frequently developed in high-altitude and water-limited habitats, such as crassulacean acid metabolism (CAM) photosynthesis, succulence, and specialised leaf architecture. Natural selection and genetic restrictions form these characteristics, which enable plants to maximise resource utilisation and preserve fitness in the face of stress (Kanta et al., 2024). The plant community's response to climate change is determined by both ecophysiological features and functional diversity. Species composition, ecosystem resilience, and nutrient cycling can all be affected by changes in characteristics such as specific leaf area, photosynthetic rate, and nutrient utilisation efficiency. Predicting vegetation responses and managing ecosystems in a warming environment require an understanding of these dynamics (Yu et al., 2022). So, in this book chapter, we have discussed various aspects of functional diversity and ecophysiological traits of alpine or high-altitude medicinal plants and their significance along elevational gradients.

2 Functional Diversity/ Ecophysiological Traits and Species Richness

Biological invasion is determined not only by the environmental conditions of a given place, but also by the vital properties of the intruder species. Valuation of plant interpretation across multiple altitudinal gradients and habitats provides perceptions into the qualities that contribute to incursion accomplishment, along with habitat compatibility, susceptibility, and processes used by the plants. Known as a "high-altitude cold desert," the Trans-Himalayan zone is located in the rain shadow of India's main Himalayan region. These locations are characterised by sparse yet distinct vegetation composition (Bahukhandi et al., 2024). About 25% of the world's forests and 50% of the world's biodiversity hotspots are found in mountains, which are extraordinarily rich and significant as centres of biological diversity (Bahukhandi et al., 2024). Roughly 10,000 species of vascular plants, or 4% of all higher plant species worldwide, are thought to be present in the alpine life zone (Bahukhandi et al., 2024). The Himalayas constitute the Himalayan Biodiversity Hotspot and are the youngest, most dynamic, and most diverse mountain ecosystems on Earth

(Rawal et al., 2013). The mountain region is the repository of the world's biodiversity because plant communities and species variety often show distinct reactions to elevation gradients.

Despite their great diversity, the Himalayan alpine ecosystems are delicate and dynamic, offering a complex range of successional stages from pioneer lichen-moss communities to established groups (Bahukhandi et al., 2024). Abiotic and biotic variables influence both plant distribution and diversity along altitudinal gradients (Thakur et al., 2022). It is well known that understanding the floristic composition of plant communities, endemism, and their rarity patterns along the environmental gradient are necessary to appreciate the conservation significance in the high-altitude mountain region (Rawal et al., 2024; Negi et al., 2024). The basic understanding of species composition and diversity along elevation gradients was made possible by Alexander von Humboldt's initial definition of the altitudinal gradient of vegetation (Bahukhandi et al., 2024). For identifying long-term ecological responses to environmental changes, altitudinal gradients provide extremely effective natural experimental conditions. Recent studies have revealed that mid-elevations, which differ among plants and mountain ranges in various climatic circumstances, are often where plant variety peaks. Diversity was previously believed to decline linearly with elevation (Rawal et al., 2024; Negi et al., 2024). Based to numerous studies conducted worldwide including in the Himalayan region, ecological gradients significantly influence the structure and shaping of plant species richness along altitudinal gradients (Rawal et al., 2024). The patterns of plant species richness and diversity in the chilly desert regions of the Indian Trans-Himalaya have not been well studied (Saxena & Rao, 2024).

The flora diversity of the Indian Himalayan Region cold desert region has been the subject of few explorations but less is known about Uttarakhand's Trans Himalayan regions, particularly the Nelang Valley (Negi et al., 2018; Bahukhandi et al., 2024). Due to their remoteness and inaccessibility, community structure, distribution, and species richness of cold desert environments are crucial for comprehending ecological dynamics worldwide (Rawal et al., 2024). Although Uttarakhand government has prohibited tourists from visiting the area for over the years, some parts of the territory are now accessible to tourists. Recent studies have concentrated on ecological conditions, species diversity patterns, and the effects of human activity (Rawal et al., 2024; Bahukhandi et al., 2024). Constant loss of floristic variety and shifting vegetation patterns caused by these stresses Bahukhandi et al. (2024) studied this area and hypothesises that exposing the Nelang valley to travellers may change the vegetation composition and upsurge anthropogenic burden, negatively effecting on the region's biodiversity. They studied the vegetation composition in the alpine region of the Nelang valley and found 68 taxa of higher plants, including herbs, shrubs, climbers, and trees. Herbaceous plants contributed the most species diversity; the main genera were Artemisia, Astragalus, and Juniperus, followed by Lonicera, Aster, Poa, Oxytropis, and Salix. The valley has an uneven dispersal of plant species richness over the altitudinal gradient, with the highest taxa occurring among 3500–3600 m in elevation. In the Nelang valley, overall phyto diversity individuals decreased significantly as altitude increased (3100–4300 m). In addition, the investigation identified 33 key medicinal floras employed in the Himalayan region's traditional herbal medical system. *Juniperus semiglobosa* dominates the valleys treeline ecotone, with the highest important value index, followed by *Pinus wallichiana* and *Cedrus deodara.*

Altitudinal variations in soil properties have a considerable effect on plant community composition. Soil properties can be affected by vegetation changes along an elevational gradient, and vice versa (Ahmad et al., 2018; Mehta & Chawla, 2024). Examining how soil conditions and plants interact along an altitudinal gradient can help us better understand plant communities in alpine environments (Mangral et al., 2023). Slopes and other geographic features of soil are significantly influenced by alpine elevations (Bardelli et al., 2017; Mangral et al., 2023). Variations in nutritional status in various places are caused by differences in air temperature with elevation. Other factors, such as vegetation, slope exposure, altitude, and soil temperature can also affect soil quality (Bardelli et al., 2017). Many researches have shown that climate differences along an elevational gradient have a varied effect on soil properties (Zhang et al., 2021; Mangral et al., 2023). The amounts of total organic carbon and nitrogen rose with elevation and were positively correlated with precipitation and the annual mean temperature (Zhang et al., 2021). Electrical conductivity, soil temperature, C:N ratio, and pH all showed an increasing tendency towards increasing altitudinal gradient (Hamid et al., 2020).

As elevation rises, temperature decreases and precipitation increases (Hassan & Nile, 2021). Simon et al. (2018) found a linear association among soil organic carbon content and pH with height. Climate change alters biomass and plant composition resulting in variations in soil properties and taxa (Hamid et al., 2020). Rhodiola roots contain the active substance salidroside (Rattan et al., 2020; Rattan et al., 2022). The salidroside concentration of *R. sachalinensis* varies substantially depending on its natural habitat. The influence of soil nutritional variables on salidroside accumulation in *R. sachalinensis* roots were studied in Chang Bai Mountain, China. The findings demonstrated that the most crucial elements for guaranteeing salidroside productivity were high levels of exchangeable and total nitrogen in the soil, low pH, and abundant organic matter (Yan et al., 2004).

2.1 Morpho Anatomical Traits

Numerous biological levels, including morphological structures, physiological functions, and cellular metabolism, can exhibit intraspecific trait variation (Sharma et al., 2023). Functional trade-offs between resource acquisition and environmental stress tolerance are reflected at the morphological level by variations in leaf size, stomatal density, and specific leaf area. For instance, in high-elevation populations, the growth of smaller leaves with a reduced specific leaf area may minimise water loss and maintain structural integrity when exposed to cold stress or desiccation (Zhou et al., 2023). At the physiological level, lower photosynthetic rates enable plants to conserve water and assimilate carbon in drought conditions. Modifications in secondary metabolite profiles at the cellular level provide improved chemical defence against oxidative stress and herbivory (Ma et al., 2015). In order to maximise adaptive performance, these features are frequently phenotypically integrated into coordinated syndrome. For instance, in order to adapt to the high-elevation conditions, *Potentilla saundersiana* has developed a number of characteristics, such as root architecture, leaf morphology, and antioxidant protein expression (Ma et al., 2015). There aren't enough thorough investigations on multidimensional trait syndromes and how they contribute to ecological resilience.

The Caprifoliaceae family includes the highly endangered medicinal herb *N. jatamansi*, which is found mostly between 3000 and 5000 meters above sea level (Agrawal et al., 2023). Despite being listed as endangered, this plant is found throughout a vast geographic area and is prized for its rhizomes and roots, which are abundant in phenolic compounds, terpenes, and essential oils. Because of its therapeutic potential, it has been used extensively in traditional medicine (Dhiman & Bhattacharya, 2020). Field studies have shown significant morphological diversity between populations, with a particularly noticeable divergence in leaf sheath and shape. *N. jatamansi* can be divided into two morphological types based on these characteristics: type I, which has fibrous sheaths and spatulate leaves, and type II, which has lanceolate leaves and lamellar sheaths. Their ecological relevance and patterns of regional distribution are still unknown. Furthermore, there aren't many research examining the eco-physiological processes that underlie these variations, which are crucial for directing conservation and sustainable use.

A study was conducted by Khatri et al. (2023) in order to comprehend the morphological and functional characteristic differences in *Ageratina adenophora* along the altitudinal gradient in the Kumaun region, India. The findings showed that this weed can exploit accessible invasion windows because of the greater plasticity in the plant features. The accomplishment of *A. adenophora* invasion in a variety of habitats was explained by morphological characteristics like above and belowground biomass, number of leaves, plant height, branches, root length, and capitula; and functional characteristics like stem weight ratio, root: shoot ratio, root weight ratio, leaf weight ratio, inflorescence weight ratio, and seed output. The adaptability of plant characteristics across a range of environments represents the tactics that let *A. adenophora* establish itself as a successful invasive species. The majority of plants use this strategy to harvest sunlight, and more biomass allocation to shoots promotes a rapid growth in plant height. In comparison to forest and cultivated land, *A. adenophora* performed better in streamline, wasteland, wall, and roadside populations, indicating vulnerability and potential for further spread in these habitats. Improved plant performance was observed at mid-altitudinal range (1000–2000 m) across all altitudes.

Plasticity for fitness-related features that allow an invasive species to adapt to its environment determines its capacity for colonisation. The adaptability of plant characteristics across many environments suggested tactics that let *A. adenophora* establish itself as a effective invasive species. Extra biomass was devoted to *A. adenophora* aboveground portions, which facilitated its hasty growth in height. This encouraged the plant to accumulate extra biomass. Furthermore, this method enabled *A. adenophora* to dislocate surrounding native vegetation, as evidenced by the small density of herbs, shrubs, and seedlings in invaded areas (Khatri et al., 2023). Since several gradients coexist on a very small spatial scale, high-altitude environments, like the Himalaya, are the best natural laboratory for researching plant ecological responses (Mangral et al., 2023). Study on the impact of soil physicochemical and ecophysiological qualities on the *Rhododendron anthopogon* along an altitude gradient was conducted by Mangral et al. (2023).

Leaf and soil samples were collected from five distinct locations along an altitudinal gradient (3220–3908 m) for ecophysiological and soil physicochemical investigation, respectively. The results show a considerable change in soil physicochemical and ecophysiological characteristics across the altitudinal gradient. A substantial association was found between the various examined factors and altitude. The PCA of the examined soil physicochemical and ecophysiological constraints discovered that all of the locations were divided into four clusters based on height, indicating a stronger effect of altitude. The study found that soil permeation capacity improved with altitude, which benefited *R. anthopogon* growth and development. They also detected considerable differences in eco-physiological constraints, suggesting its crucial role in *R. anthopogon* adaptation to the severe alpine areas. By providing an adaptive benefit and rapidly extending its range, this work contributes to our understanding of *R. anthopogon* adequate eco-physiological plasticity and soil physicochemical, which would be advantageous for its persistence in imminent environments.

The conclusion that stress on plants increases with altitude is supported by a thorough examination of the morphophysiological and anatomical characteristics, photosynthetic pigments, and metabolomic spectrum of *R. linearifolia* flowers and shoots at different elevations (Terletskaya et al., 2024). However, establishing a strict linear link between altitude and changes in these features is made more difficult by the variety of abiotic elements seen in mountainous areas. Furthermore, the findings imply that elevated stress levels at higher elevations promote the production of secondary antioxidant metabolites, which may be important when choosing plants for medicinal uses. Some examples of morpho anatomical traits variation with altitude gradient is listed in Table 3.1.

2.2 Eco-physiological Traits

Phenotypic plasticity, broad ecological amplitude, and adequate genetic variety of survival-critical features are likely responsible for a species ability to survive in changing environmental conditions (Ding et al., 2019). Understanding the characteristics of species that define their distribution range is of primary interest to many eco-physiologists. Based on the variations in environmental conditions across spatial dimensions, environmental gradients are categorised as either elevational or latitudinal gradients. Understanding plants adaptation over elevation gradient is more significant than understanding plant adaptation along latitudinal gradients since the former have more homogenous macroclimatic conditions. Elevational gradients provide significant changes in climatic circumstances, giving them a natural

Table 3.1 Morpho-anatomical trait variation with altitude gradient

Trait	Plant species	Low altitude	High altitude	Ecological implication	Reference
Leaf thickness	Aveneae grass *Epilobium amurense*, *Pedicularis densispica*, *Potentilla fulgens*	Thinner	Thicker	Enhances protection and photosynthetic efficiency	Ahmad et al., 2016; Liu et al., 2020; Pescador et al., 2015
Sclerification	Aveneae grass	Less intense	More intense	Provides mechanical strength against wind and frost	Ahmad et al., 2016
Trichome density	Aveneae grass	Lower	Higher	Minimizes water loss and UV damage	Ahmad et al., 2016
Stomatal density	*Epilobium amurense*, *Pedicularis densispica*, *Potentilla fulgens*	Lower	Higher	Improves gas exchange under low CO_2 pressure	Liu et al., 2020
Chlorophyll content	*Rhodiola linearifolia* *Rhododendron anthopogon*	Lower	Higher	Enhances its photosynthetic pigment production as a stress adaptation to high-altitude conditions.	Terletskaya et al., 2024; Rathore et al., 2018
Mesophyll tissue thickness	*Epilobium amurense*, *Pedicularis densispica*, *Potentilla fulgens*	Thinner	Thicker	Enhances internal heat retention and photosynthesis	Liu et al., 2020

experimental set-up for studying plants behaviour in various climatic environments. Along elevation gradient, significant changes in climate have been noted over short distances. Although water accessibility and more temperature at lower elevation have a varied effect on species occurrence, this causes the temperature to drop at higher elevations (Ding et al., 2019).

Mehta and Chawla (2024) determined that *Rumex nepalensis* and *Taraxacum officinale* with widespread elevation distribution, several features, primarily photosynthetic trait, assisted by some additional traits, follow a generalised pattern regardless of climatic conditions or species. According to trends seen in biochemical features, plants at higher elevations make greater efforts to survive in challenging environmental situations. However, various plant species and populations in diverse transects preserve a stability of eco-physiological features aimed at survival and growth in a variety of environmental situations. Their research sheds light on pattern of adaptation that species choose in elevational gradients throughout several transects. Although widely dispersed species have gotten less attention due to their excellent adaptability, research on such species is required to fill the gap in understanding the effect of climatic alteration and phylogeographic processes on the dispersal of vast elevational species.

Nutrient resorption efficiency plays a crucial role in understanding nutrient cycling and maintaining ecosystem balance, especially under changing environmental conditions. It involves the retrieval of nutrients from senesced leaves and their subsequent allocation and utilization within the plant. Zhu et al. (2024) highlight that phosphorus (P) and potassium (K) resorption efficiencies are notably higher than nitrogen (N) resorption efficiency in nutrient-deficient tropical forest soils, suggesting that soil nutrient availability significantly influences plant nutrient allocation strategies. Climate factors also affect resorption patterns; for instance, in Chinese fir plantations, P resorption efficiency consistently surpasses that of N across varying climatic conditions. Despite existing research on nutrient status in mountain ecosystems, studies specifically addressing nutrient resorption efficiencies under diverse climate scenarios remain limited. This gap underscores the need for further investigation into how mountain plants adapt their nutrient strategies in response to environmental variability. Some examples of eco-physiological trait variation with altitude gradient is listed in Table 3.2.

2.3 Molecular Traits

Transcriptomic studies have recently been carried out to comprehend the molecular insights of high-altitude adaptation in various plant species, which postulate that higher elevations cause major stress for plant than lower elevation, in addition to the abundance of evidence on the plasticity in eco-physiological and morpho-anatomical traits along the higher elevation gradient (Zhao et al., 2024; Luo et al., 2024; Kumari et al., 2025). Mostly, these molecular investigations have mostly focused on key bioactive accretion and associated gene expression pattern by simply comparing high- and low-elevation species. Nevertheless, investigations with representatives scattered with a continuous gradient/range that include all aspects of plant organisation are rare. Kumari et al. (2025) investigated *Picrorhiza kurroa* to better understand plant molecular pathways under intricate, variable, and multifactorial stress conditions along high-altitude gradient. *P. kurroa* is a high-altitude endangered medicinal plant with outstanding pharmaceutical values (Partap et al., 2022).

This plant species represents an emerging model species for plant adaptation studies since it has effectively adapted to the alpine environment of the Himalayas (3000–5000 m) while frequently experiencing a variety of stress combinations in its natural habitat (Kumari et al., 2021; Kumari & Kumar, 2024). Kumari et al. (2025) examined the transcriptome regulation of *P. kurroa* under complex high-altitude stress conditions (3400, 3800, and 4100 m). The findings indicate that, apart from active calcium signalling with rising high elevation, salicylic acid and auxin signalling also guide the transcriptional regulation at 3800 m, whereas abscisic acid production and signalling favourably regulate transcriptional regulation at 4100 m. The distinct molecular response of *P. kurroa* to the diverse complicated environment of

Table 3.2 Eco-physiological trait variation with altitude gradient

Trait	Plant species	Low altitude	High altitude	Ecological implication	Reference
Leaf size	*Epilobium amurense Pedicularis densispica Potentilla fulgens*	Higher	Lower	Decreased with altitude, indicating adaptation to environment	Liu et al., 2020
Leaf C:N ratio	*Rhododendron anthopogon*	Higher (more carbon, less nitrogen)	Lower (more nitrogen relative to carbon)	Reflects nutrient availability and photosynthetic efficiency	Rathore et al., 2018
Leaf N:P ratio	*Rhododendron anthopogon*	Higher	Lower	Suggests phosphorus limitation at higher elevations	Rathore et al., 2018
Plant height	*Epilobium amurense Pedicularis densispica Potentilla fulgens*	Taller	Shorter	Reduces exposure to wind and cold stress	Liu et al., 2020
Photosynthetic rate	*Artemisia tridentata*	Lower	Higher	Increases with altitude, demonstrating a greater photosynthetic capacity in alpine plants	Reed and Loik (2016)
Stomatal conductance	*Artemisia tridentata*	Lower	Higher	Enhances gas exchange under low CO_2 pressure	Reed & Loik, 2016
Soil moisture	Rhododendron anthopogon	Low	High	Increased soil moisture was found at higher elevation	Rathore et al., 2018

higher elevation is mediated by the convergence and cross-talk of many signalling pathways. Stress-responsive unigenes, key bioactive biosynthesis, and developmental and cellular processes like trichomes, modification of cell wall, and reproductive development-associated unigenes were all identified in the downstream transcriptional profile. These changes appeared to be the consequence of evolutionary adaptations. The study advances our knowledge of the complex and effective molecular mechanisms underlying medicinal plants' responses to multifactorial alpine stress environments. This knowledge can then be applied to enhance plant tolerance in a constantly shifting field environment.

Table 3.3 Molecular trait variation along altitude gradient

Trait	Plant species	Low altitude	High altitude	Ecological implication	Reference
Antioxidant enzyme activity	*Lonicera caerulea* *Coleus forskohlii*	Moderate	Elevated (SOD, CAT, POD, antioxidant activity)	Protects against oxidative stress from UV and cold	Ge et al., 2024; Rana et al., 2020
Secondary metabolite accumulation	*Coleus forskohlii* *Cotoneaster orbicularis*	Lower	Higher (Total phenolics, Total flavonoids)	Enhances stress tolerance and medicinal potency	Rana et al., 2020; Ibrahim et al., 2022
Hormonal crosstalk (abscisic acid, salicylic acid, auxin)	*Picrorhiza kurroa*	SA, auxin has transcriptional control at mid altitude 3800 m	ABA has transcriptional control at high altitude 4100 m	Modulates growth and stress signaling	(Kumari et al., 2025)
Soluble proteins	*Picea likiangensis* *Nepeta septemcrenata*	Moderate	Elevated Found maximum at 1800-2000 m	Supports stress adaptation and metabolic flexibility	Jiang et al., 2021 Sharaf et al., 2013

Durufle et al. (2019) investigated natural disparity in *Arabidopsis thaliana* across an altitudinal gradient in the Pyrenees, analyzing 341 individuals from 30 populations collected between 200 and 1800 meters. Using ribosomal RNA sequences and class III peroxidase as genetic markers, they found significant divergence between these Pyrenean populations and other regional accessions from the 1001 genomes project, such as those from northern Spain and southern France. When grown under controlled sub-optimal temperatures (22 °C vs. 15 °C), individuals exhibited distinct morphological changes, highlighting intraspecific phenotypic plasticity. This integrative approach—combining genetic, phenotypic, and environmental data—offers valuable insights into how geographically proximate populations adapt to temperature variation. Some examples of Molecular traits variation with altitude gradient is listed in Table 3.3.

3 Conclusion

Plants have acquired a variety of resistance features as a result of population or species reorganisation caused by climatic change along the altitude gradient. Plant species with greater genotypic or phenotypic diversity may be better able to withstand these changes. Studying these features and relating them to environmental gradients may provide valuable information about plants' ability to adapt to changing conditions or explain how they are distributed within a specific geographic area. Harsh environmental conditions and less resource availability at higher elevations encourage stress-tolerant species to develop xeric traits such as reduced leaf area and heavily pubescent leaf surface, more prostrate habit, increased lignin deposition from water conservation, or perhaps succulence. Furthermore, in order to improve our understanding of how species adapt to various climatic circumstances, it is essential to comprehend the underlying molecular pathways involved in adaption of such widely distributed species. Thus, by taking into account such widely occurring species and examining them at various times of the year, further research can be conducted in this area and may yield a better knowledge of the adaption mechanisms.

References

Agrawal, A., Devi, M., Dhiman, K., & Dkhar, J. (2023). A note on the identity of the spikenard (*Nardostachys jatamansi*, Caprifoliaceae) based on DNA sequence data. *Phytotaxa, 578*(2), 213–218.

Ahmad, K. S., Hameed, M., Fatima, S., Ashraf, M., Ahmad, F., Naseer, M., & Akhtar, N. (2016). Morpho-anatomical and physiological adaptations to high altitude in some Aveneae grasses from Neelum Valley, Western Himalayan Kashmir. *Acta Physiologiae Plantarum, 38*(4), 93.

Ahmad, K. S., Hameed, M., Hamid, A., Nawaz, F., Kiani, B. H., Ahmad, M. S. A., Deng, J., Ahmad, F., Hussain, I., & Fatima, S. (2018). Beating cold by being tough: Impact of elevation on leaf characteristics in *Phleum himalaicum Mez.* endemic to Himalaya. *Acta Physiologiae Plantarum, 40*(3), 56.

Alexander, J. M., Diez, J. M., & Levine, J. M. (2015). Novel competitors shape species' responses to climate change. *Nature, 525*(7570), 515–518.

Antonelli, A., Kissling, W. D., Flantua, S. G., Bermúdez, M. A., Mulch, A., Muellner-Riehl, A. N., Kreft, H., Linder, H. P., Badgley, C., Fjeldsa, J., & Fritz, S. A. (2018). Geological and climatic influences on mountain biodiversity. *Nature Geoscience, 11*(10), 718–725.

Bahukhandi, A., Sekar, K. C., Negi, V. S., Bisht, K., Tiwari, D. C., Mehta, P., Upadhyay, S., Siddiqui, S., & Ayari-Akkari, A. (2024). Floristic diversity and species composition along altitudinal gradient in the alpine ecosystem of the cold desert region in Western Himalaya, India. *Frontiers in Plant Science, 15*, 1469579.

Bardelli, T., Gómez-Brandón, M., Ascher-Jenull, J., Fornasier, F., Arfaioli, P., Francioli, D., Egli, M., Sartori, G., Insam, H., & Pietramellara, G. (2017). Effects of slope exposure on soil physico-chemical and microbiological properties along an altitudinal climosequence in the Italian Alps. *Science of the Total Environment, 575*, 1041–1055.

Dhiman, N., & Bhattacharya, A. (2020). *Nardostachys jatamansi* (D. Don) DC.-Challenges and opportunities of harnessing the untapped medicinal plant from the Himalayas. *Journal of Ethnopharmacology, 246*, 112211.

Ding, Y., Zang, R., Lu, X., Huang, J., & Xu, Y. (2019). The effect of environmental filtering on variation in functional diversity along a tropical elevational gradient. *Journal of Vegetation Science, 30*(5), 973–983.

Durufle, H., Ranocha, P., Mbadinga Mbadinga, D. L., Déjean, S., Bonhomme, M., San Clemente, H., Viudes, S., Eljebbawi, A., Delorme-Hinoux, V., Sáez-Vásquez, J., & Reichheld, J. P. (2019). Phenotypic trait variation as a response to altitude-related constraints in Arabidopsis populations. *Frontiers in Plant Science, 10*, 430.

Fatima, S., Hameed, M., Ahmad, F., Ahmad, M. S. A., Khalil, S., Munir, M., Ashraf, M., Ahmad, I., Ahmad, K. S., & Kaleem, M. (2022). Structural and functional responses in widespread distribution of some dominant grasses along climatic elevation gradients. *Flora, 289*, 152034.

Franklin, J., Serra-Diaz, J. M., Syphard, A. D., & Regan, H. M. (2016). Global change and terrestrial plant community dynamics. *Proceedings of the National Academy of Sciences of the United States of America, 113*(14), 3725–3734.

Freeman, B. G., Scholer, M. N., Ruiz-Gutierrez, V., & Fitzpatrick, J. W. (2018). Climate change causes upslope shifts and mountaintop extirpations in a tropical bird community. *Proceedings of the National Academy of Sciences of the United States of America, 115*(47), 11982–11987.

Ge, L., Yan, J., Qi, Q., Zhang, Q., Jiang, W., Xue, J., Ju, P., Lin, M., & Lü, C. (2024). Antioxidant enzyme activity in the leaves of *Lonicera caerulea* along an altitude gradient on the north slope of Changbai Mountain. *Polish Journal of Environmental Studies, 33*(3), 2613–2621.

Gupta, A., Hisano, H., Hojo, Y., Matsuura, T., Ikeda, Y., Mori, I. C., & Senthil-Kumar, M. (2017). Global profiling of phytohormone dynamics during combined drought and pathogen stress in *Arabidopsis thaliana* reveals ABA and JA as major regulators. *Scientific Reports, 7*(1), 4017.

Hamid, M., Khuroo, A. A., Malik, A. H., Ahmad, R., & Singh, C. P. (2020). Elevation and aspect determine the differences in soil properties and plant species diversity on Himalayan mountain summits. *Ecological Research, 36*(2), 340–352.

Hassan, W. H., & Nile, B. K. (2021). Climate change and predicting future temperature in Iraq using CanESM2 and HadCM3 modeling. *Modeling Earth Systems and Environment, 7*(2), 737–748.

Ibrahim, I. A., Jabbour, A. A., Abdulmajeed, A. M., Elhady, M. E., Almaroai, Y. A., & Hashim, A. M. (2022). Adaptive responses of four medicinal plants to high altitude oxidative stresses through the regulation of antioxidants and secondary metabolites. *Agronomy, 12*(12), 3032.

Jiang, Y., Cheng, X. M., Zi, H., & Huang, X. (2021). Physiological and ecological responses of an alpine plant *Picea likiangensis* at different altitudinal gradients. *Pakistan Journal of Botany, 53*(6).

Kanta, C., Kumar, A., Chauhan, A., Singh, H., & Sharma, I. P. (2024). The interplay between plant functional traits and climate change. In *Plant functional traits for improving productivity* (pp. 41–58). Springer.

Kewlani, P., Negi, V. S., Bhatt, I. D., Rawal, R. S., & Nandi, S. K. (2021). Soil nutrients concentration along altitudinal gradients in Indian Western Himalaya. *Scandinavian Journal of Forest Research, 36*(2–3), 98–104.

Khatri, K., Negi, B., Bargali, K., & Bargali, S. S. (2023). Phenotypic variation in morphology and associated functional traits in *Ageratina adenophora* along an altitudinal gradient in Kumaun Himalaya, India. *Biologia, 78*(5), 1333–1347.

Kumari, M., & Kumar, R. (2024). Functional trait correlation network and proteomic analysis reveal multifactorial adaptation mechanisms to a climatic gradient associated with high altitude in the Himalayan region. *Plant, Cell & Environment, 47*(5), 1556–1574.

Kumari, M., Kumar, P., Saini, V., Joshi, R., Shankar, R., & Kumar, R. (2025). Transcriptional landscape illustrates the diversified adaptation of medicinal plants to multifactorial stress combinations linked with high altitude. *Planta, 261*(5), 1–25.

Kumari, M., Pradhan, U. K., Joshi, R., Punia, A., Shankar, R., & Kumar, R. (2021). In-depth assembly of organ and development dissected *Picrorhiza kurroa* proteome map using mass spectrometry. *BMC Plant Biology, 21*(1), 604.

Liu, W., Zheng, L., & Qi, D. (2020). Variation in leaf traits at different altitudes reflects the adaptive strategy of plants to environmental changes. *Ecology and Evolution, 10*(15), 8166–8175.

Luo, L. Y., Jin, G. H., Ma, P. F., & Li, D. Z. (2024). Continuous genetic adaptation to high elevations of alpine bamboos in the Hengduan Mountains, Southwest China. *Journal of Systematics and Evolution, 62*(4), 702–714.

Ma, L., Sun, X., Kong, X., Galvan, J. V., Li, X., Yang, S., Yang, Y., Yang, Y., & Hu, X. (2015). Physiological, biochemical and proteomics analysis reveals the adaptation strategies of the alpine plant *Potentilla saundersiana* at altitude gradient of the Northwestern Tibetan Plateau. *Journal of Proteomics, 112*, 63–82.

Mangral, Z. A., Islam, S. U., Tariq, L., Kaur, S., Ahmad, R., Malik, A. H., Goel, S., Baishya, R., Barik, S. K., & Dar, T. U. H. (2023). Altitudinal gradient drives significant changes in soil physico-chemical and eco-physiological properties of *Rhododendron anthopogon*: A case study from himalaya. *Frontiers in Forests and Global Change, 6*.

Mehta, N., & Chawla, A. (2024). Eco-physiological trait variation in widely occurring species of Western Himalaya along elevational gradients reveals their high adaptive potential in stressful conditions. *Photosynthesis Research, 159*(1), 29–59.

Negi, V. S., Pandey, A., Singh, A., Bahukhandi, A., Pharswan, D. S., Gaira, K. S., Wani, Z. A., Bhat, J. A., Siddiqui, S., & Yassin, H. M. (2024). Elevation gradients alter vegetation attributes in mountain ecosystems of eastern Himalaya, India. *Frontiers in Forests and Global Change, 7*, 1381488.

Nuccio, M. L., Paul, M., Bate, N. J., Cohn, J., & Cutler, S. R. (2018). Where are the drought tolerant crops? An assessment of more than two decades of plant biotechnology effort in crop improvement. *Plant Science, 273*, 110–119.

Pandey, P., Ramegowda, V., & Senthil-Kumar, M. (2015). Shared and unique responses of plants to multiple individual stresses and stress combinations: Physiological and molecular mechanisms. *Frontiers in Plant Science, 6*, 723.

Partap, M., Rattan, S., Kumar, D., & Warghat, A. R. (2022). Bioprocess development for enhancing cell biomass, differential picrosides accumulation, and gene expression profiling at shake flask levels in suspension cultures of *Picrorhiza kurroa*. *Industrial Crops and Products, 187*, 115311.

Pescador, D. S., de Bello, F., Valladares, F., & Escudero, A. (2015). Plant trait variation along an altitudinal gradient in mediterranean high mountain grasslands: Controlling the species turnover effect. *PLoS One, 10*(3), 0118876.

Rana, P. S., Saklani, P., & Chandel, C. (2020). Research article influence of altitude on secondary metabolites and antioxidant activity of *Coleus forskohlii* root extracts. *Research Journal of Medicinal Plants, 14*, 43–52.

Rathore, N., Thakur, D., & Chawla, A. (2018). Seasonal variations coupled with elevation gradient drives significant changes in eco-physiological and biogeochemical traits of a high altitude evergreen broadleaf shrub, *Rhododendron anthopogon*. *Plant Physiology and Biochemistry, 132*, 708–719.

Rattan, S., Kumar, P., Kaur, E., Sood, A., Acharya, V., & Warghat, A. R. (2022). Comparative transcriptome and tissue-specific expression analysis of genes reveal tissue-cultured plants as an alternative source for phenylethanoids and phenylpropanoids in *Rhodiola imbricata* (Edgew.). *Gene, 836*, 146672.

Rattan, S., Sood, A., Kumar, P., Kumar, A., Kumar, D., & Warghat, A. R. (2020). Phenylethanoids, phenylpropanoids, and phenolic acids quantification vis-à-vis gene expression profiling in leaf and root derived callus lines of *Rhodiola imbricata* (Edgew.). *Industrial Crops and Products, 154*, 112708.

Rawal, R., Negi, V. S., Bhatt, I. D., & Tiwari, L. M. (2024). Rarities pattern of vascular plants in the high-altitude forests of Indian western Himalaya: Conservation implications. *Journal for Nature Conservation, 79*, 126588.

Rawal, R. S., Bhatt, I. D., Sekar, K. C., & Nandi, S. K. (2013). *The himalayan biodiversity: Richness, representativeness, uniqueness and life-support values* (p. 84). GB Pant Institute of Himalayan Environment and Development (GBPIHED).

Reed, C. C., & Loik, M. E. (2016). Water relations and photosynthesis along an elevation gradient for *Artemisia tridentata* during an historic drought. *Oecologia, 181*(1), 65–76.

Rivero, R. M., Mittler, R., Blumwald, E., & Zandalinas, S. I. (2022). Developing climate-resilient crops: Improving plant tolerance to stress combination. *The Plant Journal, 109*(2), 373–389.

Saxena, K. G., & Rao, K. S. (2024). The himalayan biodiversity hotspot: Geomorphology, biogeography, and climate. In *Biodiversity hotspot of the himalaya* (pp. 1–9). Apple Academic Press.

Shaar-Moshe, L., Blumwald, E., & Peleg, Z. (2017). Unique physiological and transcriptional shifts under combinations of salinity, drought, and heat. *Plant Physiology, 174*(1), 421–434.

Sharaf, A. A., Khafagi, O. M. A., Hatab, E. B. E., & Moursy, M. M. (2013). Effect of altitudinal gradients on the content of carbohydrate, protein, proline and total phenols of some desert plants in Saint Katherine Mountain, South Sinai, Egypt. *Middle-East Journal of Scientific Research, 14*(1), 122–129.

Sharma, P., Wajid, M. A., Fayaz, M., Bhat, S., Nautiyal, A. K., Jeet, S., Yadav, A. K., Singh, D., Shankar, R., Gairola, S., & Misra, P. (2023). Morphological, phytochemical, and transcriptome analyses provide insights into the biosynthesis of monoterpenes in *Monarda citriodora. Planta, 258*(3), 49.

Simon, A., Dhendup, K., Rai, P. B., & Gratzer, G. (2018). Soil carbon stocks along elevational gradients in Eastern Himalayan mountain forests. *Geoderma Regional, 12*, 28–38.

Stanisci, A., Bricca, A., Calabrese, V., Cutini, M., Pauli, H., Steinbauer, K., & Carranza, M. L. (2020). Functional composition and diversity of leaf traits in subalpine versus alpine vegetation in the Apennines. *AoB Plants, 12*(2), 004.

Steinbauer, M. J., Grytnes, J. A., Jurasinski, G., Kulonen, A., Lenoir, J., Pauli, H., Rixen, C., Winkler, M., Bardy-Durchhalter, M., Barni, E., & Bjorkman, A. D. (2018). Accelerated increase in plant species richness on mountain summits is linked to warming. *Nature, 556*(7700), 231–234.

Terletskaya, N. V., Erbay, M., Mamirova, A., Ashimuly, K., Korbozova, N. K., Zorbekova, A. N., Kudrina, N. O., & Hoffmann, M. H. (2024). Altitude-dependent morphophysiological, anatomical, and metabolomic adaptations in *Rhodiola linearifolia Boriss. Plants, 13*(19), 2698.

Thakur, D., & Chawla, A. (2019). Functional diversity along elevational gradients in the high altitude vegetation of the western Himalaya. *Biodiversity and Conservation, 28*(8), 1977–1996.

Thakur, S., Negi, V. S., Dhyani, R., Bhatt, I. D., & Yadava, A. K. (2022). Influence of environmental factors on tree species diversity and composition in the Indian western Himalaya. *Forest Ecology and Management, 503*, 119746.

Wani, I. A., Khan, S., Verma, S., Al-Misned, F. A., Shafik, H. M., & El-Serehy, H. A. (2022). Predicting habitat suitability and niche dynamics of *Dactylorhiza hatagirea* and *Rheum webbianum* in the Himalaya under projected climate change. *Scientific Reports, 12*(1), 13205.

Yan, X., Wu, S., Wang, Y., Shang, X., & Dai, S. (2004). Soil nutrient factors related to salidroside production of *Rhodiola sachalinensis* distributed in Chang Bai Mountain. *Environmental and Experimental Botany, 52*(3), 267–276.

Yu, H., Chen, Y., Zhou, G., & Xu, Z. (2022). Coordination of leaf functional traits under climatic warming in an arid ecosystem. *BMC Plant Biology, 22*(1), 439.

Zandalinas, S. I., Fritschi, F. B., & Mittler, R. (2020). Signal transduction networks during stress combination. *Journal of Experimental Botany, 71*(5), 1734–1741.

Zandalinas, S. I., Sengupta, S., Fritschi, F. B., Azad, R. K., Nechushtai, R., & Mittler, R. (2021). The impact of multifactorial stress combination on plant growth and survival. *New Phytologist, 230*(3), 1034–1048.

Zhang, Y., Ai, J., Sun, Q., Li, Z., Hou, L., Song, L., Tang, G., Li, L., & Shao, G. (2021). Soil organic carbon and total nitrogen stocks as affected by vegetation types and altitude across the mountainous regions in the Yunnan Province, South-Western China. *Catena, 196*, 104872.

Zhao, Q., Zhang, J., Li, Y., Yang, Z., Wang, Q., & Jia, Q. (2024). Integrated metabolomic and transcriptomic analysis of Nitraria berries indicate the role of flavonoids in adaptation to high altitude. *Metabolites, 14*(11), 591.

Zhou, F., Matthew, C., Yang, P., Huang, Y., Nie, B., & Nan, Z. (2023). Leaf morphology, functional trait and altitude response in perennial vetch (*Vicia unijuga* A. Braun), alfalfa (*Medicago sativa* L.) and sainfoin (*Onobrychis viciifolia* Scop.). *Planta, 257*(4), 75.

Zhu, D., Zhou, Y., Peng, S., Hua, W., Luo, B., & Hui, D. (2024). Impacts of altitude on plant green leaf, fresh litter, and soil stoichiometry in subtropical forests. *Frontiers in Forests and Global Change, 7*, 1331623.

High-Altitude Resilience: Environmental Perturbations and the Medicinal Potential of Alpine Flora

4

Pankaj Kumar, Bhuvnesh Kapoor, Shivanti Negi, Deepika, E. Manikandan, Shagun Sharma, and Naresh Thakur

Abstract

Alpine ecosystems, occupying only a small fraction of Earth's land surface and represent some of the most ecologically significant yet fragile landscapes. These regions are characterized by short growing seasons, steep climatic gradients, intense UV radiation, nutrient-poor soils, and frequent abiotic and biotic stresses. Alpine plants have evolved a diverse range of morphological, physiological, and molecular strategies to cope with these challenges. Adaptations include cushion morphology, dense trichomes, antioxidant defenses, hormonal cross-talk, and epigenetic stress memory, all of which ensure survival under extreme environments. Importantly, many alpine species synthesize unique secondary metabolites such as flavonoids, terpenoids, alkaloids, and iridoids, that not only protect against stress but also possess immense medicinal value. Alpine floral species such as *Rhodiola, Saussurea*, *Nardostachys*, and *Picrorhiza* highlight the intersection of ecological resilience and pharmacological potential. However, climate change, overharvesting, habitat loss, and biotic invasions threaten the persistence of these plants. Integrating multi-omics approaches, molecular breeding, conservation programs, and traditional knowledge is crucial for sustaining alpine biodiversity and ensuring the availability of valuable phytochemicals. Alpine

P. Kumar (✉) · S. Negi · Deepika · E. Manikandan · S. Sharma
Department of Biotechnology, College of Horticulture, Dr. Yashwant Singh Parmar University of Horticulture and Forestry, Solan, Himachal Pradesh, India

B. Kapoor
Department of Biotechnology, University Institute of Biotechnology (UIBT), Chandigarh University, Gharuan, Mohali, Punjab, India

N. Thakur
College of Horticulture, Dr. Yashwant Singh Parmar University of Horticulture and Forestry, Neri, Hamirpur, India

P. Kumar et al. (eds.), *Climate Resilience and Molecular Adaptation in Alpine Medicinal Plants*, https://doi.org/10.1007/978-981-95-6923-6_4

ecosystems, as natural laboratories of stress biology, offer both insights into plant resilience and opportunities for sustainable healthcare innovations. Protecting these systems is thus both a scientific necessity and a cultural and moral imperative.

Keywords

Alpine ecosystems · Abiotic stress · Medicinal plants · Secondary metabolites · Conservation

1 Introduction

Alpine ecosystems, positioned above the upper tree line, represent some of the most striking yet fragile landscapes on Earth. Although they occupy only about 3% of the global terrestrial surface, their ecological and cultural significance is disproportionately large (Körner, 2021). These high-altitude systems function as vital freshwater reservoirs, regulate hydrological cycles, and sustain biodiversity hotspots rich in endemic and medicinally important species (Wilhelm et al., 2013). They also hold profound cultural value, providing resources for indigenous communities who rely on alpine plants for food, traditional medicine, and spiritual practices (Kala, 2005).

The uniqueness of alpine ecosystems is shaped by a suite of extreme environmental constraints. Short and unpredictable growing seasons, intense solar radiation, large diurnal temperature fluctuations, and nutrient-poor soils create some of the most demanding habitats for plant survival. Species diversity is typically lower compared to lowland systems, yet alpine plants have evolved remarkable physiological and biochemical strategies to withstand these stresses, including photoprotective pigmentation, antioxidant defense systems, and structural adaptations such as cushion growth forms (Körner, 2003; Cavieres et al., 2016).

In recent decades, global climate change has emerged as the most pressing threat to alpine ecosystems. High-altitude regions are warming at nearly twice the global average, a phenomenon described as elevation-dependent warming (Pepin et al., 2015). This has triggered a cascade of ecological consequences, including upslope shifts of vegetation belts, retreat of glaciers, altered snowmelt regimes, and disruption of water availability in downstream areas (Pauli et al., 2012). These physical changes also interact with biotic pressures: the expansion of herbivores and pathogens into alpine zones, once restricted by cold winters, is reshaping community dynamics and threatening endemic flora (Hodkinson, 2005).

The repercussions extend far beyond ecological processes. Many alpine plants produce high-value phytochemicals that are increasingly exploited by global markets. Species such as *Rhodiola rosea* and *Saussurea obvallata* are renowned for their antioxidant and adaptogenic properties, linking ecological adaptation to pharmaceutical potential (Panossian et al., 2010; Liu et al., 2016). Overharvesting, combined with climate stress, places these species under severe conservation pressure. At the same time, local pastoral and farming communities face reduced grazing capacity, increased vulnerability to natural hazards, and declining access to traditional resources. Thus, alpine ecosystems stand at the intersection of ecological resilience, economic value, and cultural survival.

Understanding how alpine plants perceive and respond to environmental perturbations is therefore critical. Recent research highlights the importance of integrated stress responses involving ROS signaling, hormonal cross-talk, and epigenetic memory that together enhance tolerance to extreme conditions (Avramova, 2015; Rawal et al., 2020). Equally important is the study of medicinal alpine plants, which serve as model systems to link stress adaptation with secondary metabolite biosynthesis. By connecting ecological processes with molecular insights, these case studies reveal how adaptation to stress not only ensures survival but also drives the production of compounds with global health significance.

In this chapter, we examine the abiotic and biotic stressors shaping alpine ecosystems, the sophisticated sensing and signaling mechanisms employed by plants, and case studies of medicinal species that exemplify resilience under extreme conditions. By synthesizing ecological, physiological, and molecular perspectives, we aim to highlight the adaptive capacity of alpine plants and the urgent need for their conservation in the face of accelerating climate change.

2 Abiotic Stressors in Alpine Ecosystems

2.1 Climatic Extremes and Microclimatic Variability

Alpine regions are shaped by some of the most severe climatic fluctuations found in terrestrial ecosystems. Daily and seasonal temperature oscillations can be extreme, with mid-day soil surface temperatures reaching above 40 °C, followed by nocturnal frosts even in midsummer (Körner, 2003). These conditions stem from reduced atmospheric density, thinner air, and high solar irradiance at altitude. Snow cover plays a dual role: it insulates vegetation against sub-zero conditions in winter, yet delays the onset of the growing season by limiting light and access to thawed soils (Inouye, 2000). Early snowmelt may extend growth opportunities, but it simultaneously exposes young shoots to late frost events, a frequent cause of mortality in alpine seedlings.

Microclimate is an essential determinant of species persistence. Small-scale temperature and moisture gradients within alpine habitats influence germination, phenology, and reproductive success (Körner, 2021). For example, cushion plants such as *Silene acaulis* act as "nurse plants" by creating thermally stable microsites that reduce frost incidence and conserve soil moisture (Cavieres et al., 2016). Thermal imaging studies further illustrate how alpine flowers can warm to 10 °C above ambient air temperature, enhancing pollinator activity and seed set (Larcher & Wagner, 2004). Geographic comparisons reveal regional heterogeneity in climatic stressors. In Mediterranean mountains, summer drought and heat dominate, whereas in the Himalayas and Andes, cold and frost remain the principal constraints. This variability underscores the adaptive plasticity of alpine species, with strategies fine-tuned to local stress regimes.

2.2 Photosynthetic Challenges and ROS Dynamics

Photosynthesis in alpine plants operates at the edge of physiological limits. Cold temperatures slow down enzymatic processes in the Calvin cycle, while heat destabilizes the D1 protein of photosystem II (PSII), leading to impaired electron transport (Huner et al., 1998). These stresses result in the overproduction of reactive oxygen species (ROS), including singlet oxygen, superoxide radicals, and hydrogen peroxide. While ROS at basal levels serve as signaling molecules, their uncontrolled accumulation damages membranes, proteins, and nucleic acids (Foyer & Noctor, 2005).

To counteract these challenges, alpine plants rely on robust antioxidant defenses. Enzymatic systems such as superoxide dismutase (SOD), catalase (CAT), and ascorbate peroxidase (APX) are central to detoxification, while non-enzymatic antioxidants—flavonoids, carotenoids, and phenolic compounds—provide additional buffering (Streb et al., 2003). Notably, antioxidant capacity tends to increase with elevation, suggesting adaptive regulation of ROS-scavenging pathways in response to heightened stress exposure (Sharma et al., 2025).

The xanthophyll cycle is particularly critical. Through reversible conversion of violaxanthin to zeaxanthin, plants dissipate excess absorbed light energy as heat, thus preventing photoinhibition (Demmig-Adams & Adams, 1996). *Ranunculus glacialis,* which flowers at altitudes above 4000 m in the Alps, demonstrates exceptional tolerance by maintaining PSII efficiency under sub-zero conditions through stabilization of thylakoid membranes and high antioxidant capacity (Lütz et al., 2012). These examples highlight the fine-tuned coordination between photoprotection and ROS detoxification in alpine photosynthesis.

2.3 Drought and Water Limitation

Although alpine landscapes are often associated with snow and ice, water limitation is a recurring and significant stressor. Snowmelt provides an initial water pulse in spring, but late summer frequently brings prolonged droughts due to high evapotranspiration rates and declining precipitation (Körner, 2003). Alpine forests in South Tyrol, Italy, illustrate the complexity of drought responses: while certain stands showed biomass decline, others experienced enhanced productivity under warming, depending on elevation, soil depth, and exposure (Rammig et al., 2010).

Physiologically, alpine plants respond to drought via multiple strategies. Stomatal closure reduces transpiration, while osmotic adjustment through proline, sugars, and glycine betaine maintains cellular turgor (Zhang et al., 2006). ABA plays a pivotal role in orchestrating these responses, regulating guard cell function, and inducing drought-responsive gene expression. Structural traits, such as reduced leaf area, dense trichomes, and compact cushion forms, further limit water loss (Zahedi et al., 2025). Comparisons across mountain ranges demonstrate evolutionary divergence in drought resilience. For example, Andean cushion plants exhibit higher intrinsic water-use efficiency compared to Himalayan species, reflecting adaptation to chronically arid conditions (Soukup et al., 2021). Such cross-regional differences underscore how water availability has shaped alpine plant evolution.

2.4 UV Radiation and High-Light Stress

High-altitude environments receive 10–30% more ultraviolet-B (UV-B) radiation than lowlands due to reduced atmospheric shielding (Blumthaler et al., 1997). Elevated UV exposure damages DNA, proteins, and photosynthetic pigments, but alpine plants have evolved protective strategies. Many species synthesize UV-absorbing flavonoids and anthocyanins that accumulate in epidermal tissues, shielding underlying cells (Liu et al., 2016). Leaf pubescence and cuticular waxes also reduce UV penetration.

Experimental studies show that alpine plants grown under UV-exclusion treatments exhibit reduced flavonoid accumulation and increased susceptibility to oxidative stress (Rozema et al., 1997). This suggests that UV-induced secondary metabolism is not only protective but also integral to ecological interactions, as many flavonoids confer anti-herbivore and anti-pathogen benefits.

2.5 Frost and Snow Dynamics

Frost events pose recurrent risks, especially when snow cover is absent. Ice crystal formation damages cell membranes and disrupts vascular tissues. Alpine plants mitigate these effects through the synthesis of antifreeze proteins, cryoprotective sugars, and dehydrins that stabilize cellular structures during freezing (Wisniewski et al., 2003). The timing and depth of snowpack are equally critical: thick snow cover insulates vegetation, whereas shallow or intermittent snow exposes plants to lethal frost cycles (Inouye, 2000).

Changes in snowfall patterns under climate change are altering these dynamics. Reduced snowpack in the Himalayas, for instance, threatens overwintering success of cushion plants and early-flowering herb. The interplay of frost resistance and snow cover thus represents a central adaptive axis for alpine vegetation.

2.6 Synergistic Stress Interactions

In nature, stresses rarely occur in isolation. Alpine plants often experience simultaneous combinations of cold, high light, drought, and nutrient scarcity. Such stress combinations elicit non-additive responses, with unique signaling pathways activated under multi-stress conditions (Rasmussen et al., 2013). For example, drought combined with high light amplifies ROS production, requiring synergistic upregulation of antioxidant defenses. Similarly, cold combined with UV-B stress enhances flavonoid accumulation more than either stress alone (Zhou et al., 2017).

These findings highlight the importance of studying alpine stress responses under realistic, multifactorial scenarios rather than isolated treatments. Understanding these interactions is essential for predicting how alpine ecosystems will respond to climate change. Biological invasions further compound the challenge. Non-native species introduced through grazing or tourism disrupt trophic interactions and compete with endemic flora (Pauchard et al., 2009). The combined impact of nutrient scarcity, herbivory, and pathogens presents a multifaceted stress scenario for alpine vegetation.

3 Nutrient and Biotic Challenges

3.1 Soil Constraints and Nutrient Limitations

Alpine soils represent some of the harshest substrates for plant growth. They are generally shallow, stony, and often underlain by permafrost or bedrock, which restricts root penetration and nutrient acquisition. Soil formation is slow due to low temperatures, weak weathering, and frequent cryoturbation—physical mixing of soil layers caused by freeze-thaw cycles (Körner, 2021). As a result, nutrient availability is highly constrained. Nitrogen and phosphorus, the two most limiting macronutrients, are present largely in recalcitrant forms. Nitrogen often exists in complex organic compounds that require microbial mineralization, a process slowed by low temperatures (Chapin et al., 1995). Phosphorus tends to be bound to mineral complexes or occluded in frozen soils, further limiting plant uptake. Consequently, primary productivity in alpine ecosystems is far lower compared to temperate forests or grasslands (Körner, 2003).

Symbiotic associations play a crucial role in mitigating these nutrient limitations. Mycorrhizal fungi enhance nutrient absorption by expanding root surface area and mobilizing phosphorus through enzymatic secretion (Read & Haselwandter, 1981). Nitrogen-fixing legumes such as Trifolium alpinum and Astragalus spp. locally enrich soils, fostering community-level diversity (Carlsson & Huss-Dell, 2003). However, the efficiency of such interactions declines under extreme cold and drought, reducing their buffering potential (Rime et al., 2015). Soil microbial communities, though less diverse than those in lowlands, are key drivers of nutrient turnover. They regulate organic matter decomposition, nitrogen cycling, and phosphorus mobilization. Warming accelerates decomposition rates, potentially increasing nutrient availability but also destabilizing soil carbon pools (Schmidt et al., 2011). This creates a trade-off: nutrient release may initially stimulate growth, but long-term carbon loss undermines ecosystem stability.

Recent studies in the Tibetan Plateau have shown that microbial activity and soil respiration increase under warming, yet the nitrogen made available is not always efficiently utilized by plants due to mismatched phenology and root dynamics (Xu et al., 2014). Thus, nutrient limitation in alpine systems cannot be explained solely by absolute availability but must also consider synchrony between soil processes and plant demand.

3.2 Herbivory and Pathogen Dynamics

In addition to abiotic limitations, alpine plants face increasing biotic pressures. Historically, extreme cold limited the survival of herbivores and pathogens, offering alpine flora relative escape from biological stress. However, climate warming is altering these dynamics. Warmer winters enhance survival rates of insect herbivores such as aphids, grasshoppers, and lepidopteran larvae, facilitating upslope range expansion (Hodkinson, 2005).

In some alpine meadows, herbivory intensity has doubled over the past three decades, with significant reductions in seedling survival and reproductive success (Pellissier et al., 2012). Herbivory alters competitive dynamics, often favouring species with chemical or structural defenses over less-defended taxa. For example, *Saussurea* species with dense trichomes or high flavonoid levels are less affected compared to palatable grasses. Ungulates also exert substantial pressure in alpine zones. Domestic grazing by sheep, yak, and goats is widespread in the Himalayas, contributing to overharvesting and soil degradation (Kala, 2005). While moderate grazing can maintain species diversity, overgrazing reduces biomass, disrupts regeneration, and facilitates invasion by disturbance-tolerant species (Pauchard et al., 2009).

3.2.1 Pathogen Dynamics

Fungal pathogens such as rusts (*Puccinia* spp.) and downy mildews are increasingly reported at higher elevations. In addition, the spread of Phytophthora root rot from subalpine to alpine zones represents a novel threat, as many alpine plants lack co-evolved defenses against such pathogens (Brasier, 2008). Bacterial diseases, though less documented, are likely to expand with warming and increased plant tissue damage due to frost-thaw cycles.

One striking example is the spread of *Colletotrichum* species, which cause anthracnose in alpine medicinal herbs, reducing both yield and metabolite content (Sharma et al., 2010). These infections not only compromise ecological resilience but also reduce the medicinal and economic value of alpine plants.

3.2.2 Biological Invasions and Tropical Interactions

Non-native species introduced by tourism, grazing, or climate-driven dispersal are an emerging challenge. Invasive grasses and forbs compete with native flora for space and nutrients, often outcompeting slower-growing alpine specialists (Pauchard et al., 2009). For instance, *Taraxacum officinale* has spread widely into alpine pastures, altering pollination networks and soil nitrogen dynamics (Seipel et al., 2012).

The combined impact of herbivory, pathogens, and invasive species creates cascading effects on trophic networks. Reduced plant fitness alters pollinator availability, seed dispersal, and predator-prey interactions, leading to long-term shifts in community composition (Pellissier et al., 2012). This highlights the importance of viewing biotic stress not in isolation but as an integrated driver of ecosystem change.

3.3 Synergistic Impacts of Nutrient and Biotic Stressors

Alpine ecosystems often face simultaneous nutrient constraints and biotic pressures. Nutrient-poor conditions can exacerbate vulnerability to herbivory, as plants with low nitrogen status allocate fewer resources to defense (Bryant et al., 1983). Conversely, herbivory and pathogen attack can further reduce nutrient availability by damaging roots and leaves, thereby diminishing photosynthetic and absorptive capacity.

A case study from the Swiss Alps demonstrated that fertilization increased plant growth but simultaneously heightened susceptibility to fungal pathogens, suggesting a trade-off between growth and defense under enriched nutrient regimes (Phoenix et al., 2006). This interplay complicates predictions of ecosystem responses to climate change, as warming simultaneously alters nutrient cycling and biotic pressure.

3.4 Implications for Alpine Medicinal Plants

The dual challenge of nutrient scarcity and biotic stress has profound implications for alpine medicinal species. Many of these plants, such as *Nardostachys jatamansi, Picrorhiza kurroa,* and *Saussurea obvallata*, are slow-growing and restricted to nutrient-poor soils (Kala, 2005). Biotic stressors not only reduce their population viability but also affect the biosynthesis of secondary metabolites critical for their medicinal properties.

For instance, herbivory-induced reduction in photosynthesis directly influences carbon allocation to secondary metabolism, potentially lowering metabolite yield, and therapeutic value (Sharma et al., 2017). Overharvesting, combined with grazing and pathogen infection, has pushed several species toward endangered status, raising urgent conservation concerns.

4 Stress Sensing and Signal Transduction

Alpine plants persist in an environment characterized by extreme fluctuations in temperature, radiation, water availability, and biotic pressure. Their ability to detect stress rapidly and translate it into adaptive responses is mediated by intricate sensing and signaling networks. These mechanisms act at multiple levels; membranes, second messengers, protein cascades, transcription factors, and epigenetic regulators—ensuring timely responses that enhance survival. Understanding these processes provides insight into how plants thrive at high altitudes and informs strategies for improving stress resilience in crops.

4.1 Thermosensing and Membrane Dynamics

Cellular membranes are among the first components to perceive temperature stress. High temperatures increase lipid bilayer fluidity, while low temperatures decrease it, disrupting protein function and membrane-associated processes (Nishida & Murata, 1996). Plants adjust membrane composition through the activity of fatty acid desaturases, which increase unsaturated lipid content, thereby maintaining fluidity under cold stress (Barrero-Sicilia et al., 2017).

In alpine species, membrane remodeling is particularly crucial. *Potentilla saundersiana* from the Tibetan Plateau shows elevated expression of omega-3 fatty acid desaturase genes at high altitude, ensuring stable chloroplast membranes during cold snaps (Ma et al., 2015). Conversely, under heat stress, increased lipid fluidity activates phospholipases that release signaling lipids such as inositol trisphosphate (IP3), which regulate calcium release from intracellular stores (Testerink & Munnik, 2011). Thus, the membrane functions both as a sensor and a signaling hub, linking external cues to intracellular responses.

4.2 Calcium and ROS Signaling Networks

Calcium ions (Ca^{2+}) act as universal second messengers in stress signaling. Stress perception triggers influx of Ca^{2+} into the cytosol through cyclic nucleotide-gated channels (CNGCs), mechanosensitive channels (MCAs), and glutamate receptor-like channels (GLRs) (Wdowiak et al., 2024). The resulting "calcium signatures"—defined by amplitude, frequency, and duration—convey stress-specific information to downstream effectors.

Calcium interacts closely with reactive oxygen species (ROS) in alpine stress responses. Stress-induced ROS, generated mainly by NADPH oxidases (respiratory burst oxidase homologs, RBOHs), activate additional Ca^{2+} channels, creating self-propagating ROS-Ca^{2+} waves across tissues (Gilroy et al., 2016). This integration ensures rapid systemic signaling during sudden environmental changes. Downstream, calcium-binding proteins such as calmodulins (CMLs), calcium-dependent protein kinases (CDPKs), and calcineurin B-like proteins (CBLs) decode Ca^{2+} signals, phosphorylating targets involved in antioxidant defense and transcriptional regulation.

4.3 MAPK Cascades and Transcriptional Reprogramming

Mitogen-activated protein kinase (MAPK) cascades are among the most conserved and versatile signaling modules in plant stress responses. They operate through a three-step phosphorylation relay where MAPKKKs activate MAPKKs, which in turn phosphorylate MAPKs; the activated MAPKs then target transcription factors, enzymes, or regulatory proteins to reprogram gene expression. This mechanism allows alpine plants, constantly exposed to sudden and overlapping stresses, to transform external signals into rapid adaptive responses. In high-altitude environments, stresses rarely occur in isolation: cold snaps often coincide with intense solar radiation and water limitation, meaning plants must process multiple inputs simultaneously. MAPKs act as convergence hubs that integrate these inputs into a unified cellular response.

Evidence from model plants such as *Arabidopsis thaliana* shows that MAPK3 and MAPK6 are rapidly activated by low temperature, oxidative stress, and drought, and regulate cold-responsive (COR) and stress-inducible genes (Zhu, 2016). Comparable dynamics have been reported in alpine taxa. Transcriptome profiling of *Saussurea involucrata* under cold and hypobaric stress revealed induction of multiple MAPKKK and MAPK transcripts, particularly those associated with phenylpropanoid and flavonoid biosynthesis pathways (Quan et al., 2025). This demonstrates that MAPKs not only mediate stress tolerance but also influence metabolite accumulation, directly linking signaling cascades with pharmacologically valuable phytochemistry.

MAPKs also regulate reactive oxygen species (ROS) homeostasis. Stress-induced ROS bursts activate MAPK pathways, which then upregulate antioxidant enzyme genes such as SOD, CAT, and APX, ensuring ROS act as messengers without causing oxidative injury. Crosstalk between MAPKs and hormone pathways further enhances stress adaptation. For example, MAPK6 can phosphorylate ABA-responsive transcription factors, promoting drought tolerance, while interactions with jasmonic acid (JA) signaling link cold responses with biotic defense. In alpine contexts where ROS surges are frequent due to high irradiance and frost-thaw cycles, this feedback between ROS, MAPKs, and hormones provides a crucial buffering mechanism.

Alpine plants appear to have optimized MAPK networks for heightened sensitivity and faster activation. Comparative analyses suggest that MAPK gene families are expanded or differentially regulated in high-altitude species, an adaptation that allows them to "pre-arm" stress responses in anticipation of extreme fluctuations. This rewired signaling not only confers resilience but also contributes to their distinct metabolic signatures, underscoring the central role of MAPKs in alpine stress biology.

4.4 Hormonal Regulation in Stress Adaptation

Phytohormones orchestrate an extensive communication network that enables plants to perceive stress, prioritize responses, and balance growth with survival. In alpine plants, where multiple stresses such as drought, frost, UV radiation, and herbivory often overlap within a single day, hormonal regulation functions as a flexible decision-making system. Unlike linear pathways, hormonal responses are characterized by intricate cross-talk that integrates abiotic and biotic cues.

Abscisic acid (ABA) is a cornerstone of alpine stress signaling. It accumulates rapidly during drought and cold, regulating stomatal closure, promoting osmolyte synthesis, and inducing dehydrins and late embryogenesis abundant proteins that stabilize membranes (Singh & Roychoudhury, 2023). Elevated basal ABA levels have been observed in several high-altitude herbs, suggesting an evolutionary priming strategy that allows quicker activation of ABA-dependent defenses under sudden stress. For example, in *Bistorta affinis*, ABA accumulation during freezing events correlates with enhanced membrane stability and antioxidant activity (Rahman et al., 2019).

Jasmonic acid (JA), though traditionally linked with herbivory and pathogen defense, has gained prominence in abiotic stress tolerance. JA enhances expression of CBF transcription factors and induces flavonoid biosynthesis, both of which are essential for cold and UV stress resistance. In *Saussurea* spp., JA induction under chilling conditions promotes pigment accumulation, shielding tissues from photodamage while generating bioactive flavonoids of medicinal relevance (Zhao et al., 2024).

Salicylic acid (SA), known for activating systemic acquired resistance, also modulates cold and oxidative stress responses. In alpine ecosystems where pathogen prevalence is rising due to warming, SA likely serves as a bridge between biotic and abiotic stress signaling, ensuring broad-spectrum defense. Ethylene plays multifaceted roles, particularly under hypoxic stress induced by snow cover or waterlogging. By promoting aerenchyma formation and regulating root elongation, ethylene aids survival in saturated soils, a condition common after snowmelt.

Auxins and cytokinins, typically growth-promoting, are downregulated during stress, reflecting a shift from growth toward defense. Reduced auxin levels limit elongation growth, while cytokinin suppression reinforces ABA-driven survival responses. This hormonal reallocation is crucial for alpine plants, which must conserve resources under brief and unpredictable growing seasons.

The true resilience of alpine species lies in hormonal cross-talk. ABA and JA act synergistically to enhance freezing tolerance, while ABA and ethylene interact antagonistically to balance drought responses. Crosstalk between SA and ROS signaling ensures coordinated activation of antioxidant defenses. These networks function less like isolated pathways and more like interconnected circuits, capable of fine-tuning responses depending on stress intensity and duration. By maintaining this hormonal plasticity, alpine plants not only survive in harsh climates but also optimize secondary metabolism, producing phytochemicals that reinforce both ecological defense and pharmacological value.

4.5 Stress Memory and Epigenetic Modulation

Stress memory is increasingly recognized as a cornerstone of alpine plant survival, enabling species to anticipate recurrent episodes of freezing, drought, or high UV radiation and mount faster, more efficient responses upon re-exposure. At the molecular level, this memory is largely underpinned by epigenetic mechanisms such as DNA methylation, histone modifications, nucleosome remodeling, and small RNA-mediated regulation, all of which alter chromatin accessibility and stabilize stress-responsive transcriptional programs. In alpine regions, where plants endure repeated freeze–thaw cycles and daily oscillations in temperature and irradiance, epigenetically reinforced priming provides a survival advantage. For example, field studies have shown that phenolic profiles in alpine herbs shift dynamically with temperature fluctuations and altitude, reflecting not only immediate physiological adjustments but also persistent biochemical "signatures" of past stress (Albert et al., 2009; Spitaler et al., 2008; Zidorn, 2010). Experimental manipulations in mountain environments demonstrate that intermittent cold and UV exposure results in stronger photoprotective responses than continuous stress, consistent with the concept of acquired stress memory (Streb et al., 2003; Bascuñán-Godoy et al., 2010).

At the genetic level, stress memory often involves reversible methylation of promoter regions in stress-related genes such as CBFs, allowing plants to "remember" prior activation and re-induce expression more rapidly. Histone acetylation and methylation further reinforce accessibility of stress-inducible loci, while small RNAs modulate post-transcriptional control of genes involved in ROS detoxification, hormone metabolism, and secondary metabolism. Notably, alpine populations exhibit significant altitudinal variation in secondary metabolites such as flavonoids and phenolics, which are not only protective against UV and oxidative stress but also carry pharmacological importance (Bernal et al., 2013; Nenadis et al., 2015). Such metabolite patterns suggest an epigenetically regulated memory effect that ensures faster or stronger accumulation of protective compounds upon repeated stress events.

Transgenerational effects provide another dimension of memory. Alpine species with short growing seasons often rely on rapid reproductive cycles, and evidence from long-term ecological studies indicates that maternal stress can imprint offspring with heightened tolerance to cold or drought (Galen & Stanton, 1993; Giménez-Benavides et al., 2011). By transmitting epigenetic marks through gametes, alpine plants effectively accelerate adaptation without requiring immediate genetic changes. This "soft inheritance" may explain how populations persist in habitats undergoing rapid climatic shifts (Walther et al., 2005; Rubidge et al., 2012). The ecological significance is coupled with biochemical outcomes: plants previously exposed to stress not only survive better but also accumulate higher levels of secondary metabolites such as anthocyanins, terpenoids, and phenolic antioxidants (Peñuelas & Munné-Bosch, 2005; Albert et al., 2009). These compounds contribute to both ecological resilience and medicinal value, reinforcing the dual role of epigenetic memory as a survival mechanism and as a driver of pharmacologically relevant diversity.

4.6 Integration of Signaling Networks

While MAPK cascades, calcium–ROS signaling, hormones, and transcription factors each play vital roles, alpine plant resilience ultimately depends on the seamless integration of these pathways into holistic signaling networks capable of responding to rapid and overlapping environmental stresses. Integration ensures that plants translate diverse external stimuli—such as frost at night, intense UV at midday, and drought in the growing season—into coordinated physiological and metabolic outputs. For instance, cold perception at membranes triggers Ca^{2+} influx, ROS bursts, and lipid remodeling, which then activate MAPK cascades. These cascades phosphorylate transcription factors such as DREBs and NACs, which in turn regulate antioxidant enzymes, osmolyte biosynthesis, and flavonoid production. Concurrently, ABA and JA pathways are activated, orchestrating stomatal control and metabolic reallocation (Körner, 1999, 2003).

Case studies in alpine species provide empirical evidence for this integration. In *Saussurea involucrata,* transcriptome and metabolome analyses under hypobaric and cold stress revealed synchronous activation of MAPKs, ROS-scavenging enzymes, ABA-responsive genes, and flavonoid biosynthetic pathways (Buchner et al., 2015; Dietrich & Körner, 2014). Similarly, *Rhodiola crenulata* populations at higher altitudes show coordinated regulation of salidroside biosynthesis, ROS detoxification, and JA signaling, illustrating how stress integration directly translates into accumulation of medicinally important metabolites (Albert et al., 2009; Zidorn, 2010). These findings underscore the concept that alpine plants treat stress not as isolated events but as complex combinations requiring networked responses.

The ecological consequences of integrated signaling are evident in population-level persistence under shifting climatic extremes. For instance, phenological shifts in alpine herbs reflect not just temperature responses but an integrated adjustment of hormonal, transcriptional, and metabolic pathways (Stinson, 2004; Aragón et al., 2008). Similarly, photoprotective strategies in high-Andean *Colobanthus quitensis* involve the combined action of ROS signaling, flavonoid induction, and leaf micromorphological changes, all coordinated across networks (Bascuñán-Godoy et al., 2010). Network integration also enhances ecological plasticity: plants can modulate flowering time, resource allocation, and chemical defense based on a "multiplexed" interpretation of overlapping signals (Kelly, 1994; Webb & Kelly, 1993).

From a mechanistic standpoint, these integrated responses resemble systems-level "signal multiplexing," where specific combinations of Ca^{2+} spikes, ROS pulses, and hormone ratios encode information about stress type and intensity. Decoding these signatures allows plants to deploy appropriate transcriptional programs. Such complexity ensures that alpine plants are not just tolerant but adaptive, adjusting strategies dynamically depending on the stress cocktail present in any given season (Scherrer & Körner, 2009; Theurillat & Guisan, 2001). Importantly, this integration is also linked to phytochemical diversity: many secondary metabolites in alpine species, such as flavonoids, phenolics, and terpenoids, are induced through multiple pathways simultaneously, explaining their richness in compounds with ecological and medicinal significance (Spitaler et al., 2008; Bernal et al., 2013).

Looking forward, systems biology and multi-omics approaches are essential to fully unravel these networks. Alpine plants provide unique models for studying signaling integration because their habitats impose stress combinations rarely seen in lowlands. Deciphering these networks not only enhances ecological understanding but also offers tools for crop improvement under climate change. By mimicking alpine integration strategies, it may be possible to engineer crops that combine rapid perception, flexible signaling, and enhanced metabolite production. Thus, integrated signaling networks stand as the ultimate expression of alpine resilience, uniting environmental sensing, molecular regulation, and metabolic specialization into survival strategies that double as pharmacological innovation.

5 Case Studies in Alpine Medicinal Plants

Alpine ecosystems harbour numerous medicinal species that exemplify the dual role of adaptation: ecological survival under extreme stress and biosynthesis of bioactive compounds with therapeutic value. The following case studies highlight how physiological, biochemical, and molecular mechanisms of stress tolerance intersect with phytochemistry, linking alpine ecology to pharmacological importance.

5.1 *Rhodiola* spp.: Stress-Induced Adaptogens

The genus *Rhodiola* (Crassulaceae), particularly *Rhodiola rosea* and *R. crenulata*, has been revered as a classical adaptogen in traditional medicine. Its roots contain salidroside, rosavin, and tyrosol, compounds linked to anti-fatigue, antioxidant, and neuroprotective properties (Panossian et al., 2010).

Adaptations:
Rhodiola species inhabit elevations between 3000 and 5000 m across the Himalayas, Altai, and Tibetan Plateau. Morphologically, they form compact rosettes that minimize heat loss, and their thick succulent leaves reduce water loss. Physiologically, they accumulate osmoprotectants such as proline and soluble sugars during chilling stress, stabilizing membranes and proteins.

Molecular Insights:
Transcriptomic analyses reveal strong induction of antioxidant genes such as *RcGPX5* (glutathione peroxidase), cold-responsive transcription factors, and enzymes of the phenylpropanoid pathway. Overexpression of *RcGPX5* in heterologous systems enhances cold tolerance and increases secondary metabolite production, linking ROS detoxification to bioactive compound biosynthesis (Zhang et al., 2019).

Phytochemistry and Stress Response:
Cold stress enhances salidroside and rosavin accumulation, suggesting metabolite production is ecologically coupled with stress adaptation. Populations from higher altitudes display greater antioxidant capacity than lowland populations, reflecting environmental selection.

Conservation:
Overharvesting for pharmaceutical demand threatens wild populations. Cultivation and biotechnological approaches, such as hairy root cultures and metabolic engineering, are being explored to ensure sustainable supply.

5.2 *Saussurea* spp.: The Photoprotective Specialists

The genus *Saussurea* (Asteraceae) comprises over 300 species, many adapted to alpine and subnival zones. *Saussurea obvallata* (Brahma Kamal), sacred in Himalayan traditions, and *Saussurea involucrata*, used in Chinese medicine, exemplify photoprotective and cold-adaptive strategies.

Adaptations:
These species often develop dense woolly trichomes and translucent bracts that act as optical filters, reducing UV penetration and retaining heat (Körner, 2021). Leaves accumulate anthocyanins and flavonoids, functioning as sunscreens.

Molecular Insights:
Transcriptomic analyses in *S. involucrata* under hypobaric stress identified various differentially expressed genes, particularly in flavonoid biosynthesis pathways (Quan et al., 2025). *MYB-bHLH-WD40* complexes regulate anthocyanin accumulation, while cold and UV stress enhance transcription of chalcone synthase (*CHS*) and flavonol synthase (*FLS*) genes (Li, 2014).

Phytochemistry:
Saussurea species produce acacetin, apigenin, and rutin derivatives, compounds with anti-inflammatory, hepatoprotective, and antioxidant effects (Elshaer et al., 2024). Flavonoid accumulation is strongly stress-induced, highlighting the ecological-pharmacological nexus.

Conservation:
Sacred *S. obvallata* is endangered due to habitat loss and overcollection for rituals. *Ex situ* conservation, tissue culture propagation, and community-based management are crucial to prevent extinction (Kala, 2005).

5.3 *Nardostachys jatamansi*: Neuroprotective Rhizomatous Herb

Nardostachys jatamansi (Valerianaceae), endemic to Himalayan alpine meadows (3000–5000 m), produces jatamansone, patchouli alcohol, and phenolic antioxidants used for neuroprotection and hepatoprotection (Pathak & Godela, 2024).

Adaptations:
The plant grows as a rhizomatous perennial, enabling belowground storage of nutrients and water. Rhizomes possess corky tissues reducing water loss, while dense trichomes on aerial parts reflect solar radiation.

Molecular Insights:
Transcriptomic studies revealed upregulation of terpenoid and sesquiterpenoids pathway genes is tissue specific manner in roots and rhizome. Enhanced expression of 28 terpenoid synthesis genes along with 43 CYPs found involved in increased metabolite accumulation (Feng et al., 2022).

Phytochemistry:
Secondary metabolites accumulate preferentially in rhizomes, with levels peaking during cold and drought stress. This stress-metabolite coupling underscores the dual ecological and pharmacological significance of Nardostachys.

Conservation:
The species is critically endangered due to overharvesting for its rhizomes. Cultivation in alpine fields, micropropagation, and metabolic engineering are being investigated as conservation strategies (Chauhan et al., 2021).

5.4 *Picrorhiza kurroa*: Cold-Responsive Iridoid Producer

Picrorhiza kurroa (Plantaginaceae), native to 3000–5000 m in the Himalayas, produces iridoid glycosides such as picrosides I and II, which have hepatoprotective, immunomodulatory, and antioxidant properties (Sharma, 2021).

Adaptations:
The species develops extensive lateral roots to optimize water uptake from shallow soils. Biomass allocation favors rhizomes during stress, ensuring survival under short growing seasons.

Molecular Insights:
Cold stress enhances expression of key biosynthetic genes including *PAL* (phenylalanine ammonia-lyase), *DXS* (1-deoxy-D-xylulose-5-phosphate synthase), and *HMGR* (3-hydroxy-3-methylglutaryl CoA reductase), increasing picroside accumulation (Barsain & Yadav, 2019).

Phytochemistry:
Iridoid glycosides exhibit strong ecological functions, including defense against herbivores and oxidative stress, while also being highly valued medicinally.

Conservation:
Overharvesting and habitat degradation have endangered *P. kurroa*. Cultivation, cryopreservation, and *in vitro* root cultures offer promising avenues for conservation and metabolite production (Das et al., 2024).

5.5 Other Alpine Medicinal Plants

Several additional alpine species illustrate diverse adaptive and pharmacological strategies:

Aconitum heterophyllum: Produces diterpenoid alkaloids under cold stress; alkaloid biosynthesis is linked to CBF transcription factor activity (Rai et al., 2025).
Berberis aristata: Synthesizes berberine, an antimicrobial alkaloid, which accumulates more at high altitudes due to UV and cold stress (Malhotra et al., 2021).
Podophyllum hexandrum: Increases podophyllotoxin levels under chilling stress, reflecting activation of phenylpropanoid metabolism (Sun & Li, 2016).
Inula racemosa: Produces sesquiterpene lactones such as alantolactone under drought stress, used in respiratory and digestive ailments (Mudassir et al., 2022).
Hippophae rhamnoides (sea buckthorn): Berries accumulate high levels of vitamin C, carotenoids, and flavonoids, conferring both antioxidant protection and nutritional benefits (Du et al., 2022).

These examples emphasize how abiotic stress drives phytochemical diversity in alpine flora, generating metabolites that serve ecological defense and human health.

5.6 Integrative Lessons from Alpine Medicinal Plants

Collectively, alpine medicinal plants reveal several unifying themes:

Stress-driven secondary metabolism: Phytochemical accumulation is often enhanced by cold, UV, drought, or nutrient stress, linking ecological survival to pharmacological potency.
Multi-level adaptation: Morphological traits (cushion forms, trichomes), physiological responses (osmolytes, antioxidants), and molecular regulation (TFs, MAPKs, hormones) converge to ensure resilience.
Conservation urgency: Overharvesting and climate change threaten many alpine medicinal plants, underscoring the need for integrative conservation strategies.
Biotechnological opportunities: Omics tools, metabolic engineering, and tissue culture provide promising avenues for sustainable metabolite production.

6 Future Perspectives

Alpine ecosystems, though covering only a small fraction of Earth's land surface, carry immense ecological, cultural, and medicinal importance. They act as critical water reservoirs, regulate climatic and hydrological processes, and serve as refuges of biodiversity. Many of their plants are also vital sources of traditional remedies and modern pharmaceuticals.

Yet these ecosystems are fragile. Climatic extremes, nutrient scarcity, herbivory, pathogens, and human pressures are converging to challenge their resilience. Alpine plants endure such stresses through structural innovations like cushion morphology and dense trichomes, and through finely tuned physiological and molecular mechanisms. These include antioxidant defenses, hormonal cross-talk, transcriptional reprogramming, and epigenetic stress memory. The dual significance of alpine plants is striking. On one hand, their adaptive strategies sustain ecosystem functioning and biodiversity. On the other, their specialized metabolites—salidrosides from *Rhodiola*, flavonoids from *Saussurea*, iridoids from *Picrorhiza*, offer therapeutic benefits for human health. This duality positions alpine ecosystems as both ecological refuges and evolutionary pharmacies.

6.1 Key Themes Emerging from Alpine Research

First, stress-driven phytochemistry is central. Abiotic factors such as cold, UV radiation, and drought consistently trigger enhanced accumulation of flavonoids, terpenoids, and alkaloids. These compounds serve dual roles, protecting plants from oxidative damage and providing pharmacological potency. Second, multi-level adaptation defines alpine resilience. Survival strategies operate at morphological, physiological, biochemical, and molecular levels, ensuring plants can cope with complex and overlapping stresses. Third, plasticity and stress memory allow alpine plants to prepare for future exposure. Epigenetic modifications and primed gene expression equip them to withstand repeated or prolonged stress. In some cases, these benefits extend across generations, supporting evolutionary persistence. Finally, fragility and conservation urgency are clear. Many alpine plants are slow-growing, habitat-restricted, and heavily harvested for medicinal use. Without immediate conservation efforts, these species face high extinction risk.

6.2 Research Frontiers

While much has been learned, important knowledge gaps remain. One priority is multi-omics integration. Genomic, transcriptomic, proteomic, metabolomic, and epigenomic datasets need to be combined to fully unravel the regulatory networks behind stress resilience and metabolite biosynthesis. Such systems-level studies are especially valuable for connecting stress sensing with secondary metabolism. Another area is climate modelling. Mechanistic models that link plant physiology with regional climate projections will help predict future shifts in species distribution and community composition. These insights can guide conservation planning and resource management. Molecular breeding and biotechnology also hold promise. By identifying stress-related transcription factors, enzymes, and regulatory RNAs, researchers can target metabolic pathways for crop improvement and metabolite enhancement. Modern tools—synthetic biology, hairy root cultures, and CRISPR-based genome editing—offer practical solutions for sustainable production. Equally important is epigenetic inheritance. Further study is needed to determine how stress memory is passed across generations in alpine plants, and whether similar mechanisms could be applied to enhance resilience in crops.

Finally, the integration of traditional knowledge with modern science must not be overlooked. Alpine communities have relied on these plants for centuries. Documenting and validating ethnobotanical wisdom will protect cultural heritage while expanding scientific understanding.

6.3 Conservation and Policy Perspectives

In situ conservation is essential, ensuring that alpine plants persist in their natural habitats. Biosphere reserves, protected areas, and community-managed landscapes can safeguard ecological processes that cannot be replicated elsewhere. *Ex situ* conservation provides backup support. Tissue culture, seed banking, and cryopreservation help preserve genetic diversity, particularly for endangered species such as *Saussurea obvallata* and *Picrorhiza kurroa.*

Equally critical are sustainable harvesting and cultivation programs. Encouraging community-based cultivation reduces pressure on wild populations while supporting local livelihoods. This is particularly important in the Himalayas, where medicinal plants are integral to cultural and economic life. At the policy level, stronger regulatory frameworks and benefit-sharing mechanisms are needed. Because alpine ranges often cross-national borders, international collaboration will be vital for protecting transboundary biodiversity.

6.4 Broader Implications

The lessons from alpine ecosystems extend well beyond high mountains. As global climate change intensifies stress in all ecosystems, alpine strategies antioxidant defenses, hormonal integration, epigenetic memory offer valuable models for enhancing resilience in crops and lowland plants. Studying alpine taxa thus contributes not only to biodiversity conservation but also to global food and health security. These plants remind us that stress is not merely a constraint but also a creative force. It drives the evolution of novel metabolic pathways and the production of compounds with life-saving potential for humans.

7 Conclusion

Alpine ecosystems function as natural laboratories of stress biology. Their plants embody resilience shaped by millennia of selective pressure, producing metabolites that sustain both ecological integrity and human well-being. But they are also fragile systems, acutely sensitive to climate change and unsustainable exploitation. Protecting alpine ecosystems requires integrative approaches that combine molecular biology, ecology, ethnobotany, biotechnology, and policy. Such efforts are not simply about preserving plants for their own sake. They are about safeguarding cultural traditions, sustaining biodiversity, ensuring ecosystem services, and unlocking new therapeutic possibilities for humanity.

Ultimately, protecting these ecosystems is both a scientific responsibility and a moral imperative. The survival of alpine plants secures not only biodiversity and ecological resilience but also the promise of resilience strategies and medicines for an uncertain future.

References

Albert, A., Sareedenchai, V., Heller, W., Seidlitz, H., & Zidorn, C. (2009). Temperature is the key to altitudinal variation of phenolics in Arnica montana L. cv. ARBO. *Oecologia, 160*, 1–8.

Aragón, C., Escudero, A., & Valladares, F. (2008). Stress-induced dynamic adjustments of reproduction differentially affect fitness components of a semi-arid plant. *J Eco, 96*, 222–222.

Avramova, Z. (2015). Transcriptional 'memory' of a stress: Transient chromatin and memory (epigenetic) marks at stress-response genes. *The Plant Journal, 83*(1), 149–159. https://doi.org/10.1111/tpj.12832

Barrero-Sicilia, C., Silvestre, S., Haslam, R. P., & Michaelson, L. V. (2017). Lipid remodelling: Unravelling the response to cold stress in Arabidopsis and its extremophile relative *Eutrema salsugineum*. *Plant Science, 263*, 194–200.

Barsain, B. L., & Yadav, S. K. (2019). Picrorhiza kurrooa Royle ex Benth., an endangered Himalayan elixir-medicinal importance and exploration of biotechnological approaches in picroside production. *Current Traditional Medicine, 5*(4), 298–320.

Bascuñán-Godoy, L., García-Plazaola, J., Bravo, L., & Corcuera, L. (2010). Leaf functional and micromorphological photo protective attributes in two ecotypes of Colobanthus quitensis from the Andes and Maritime Antarctic. *Polar Biology, 33*, 885–896.

Bernal, M., Llorens, L., Julkunen-Tiitto, R., Badosa, J., & Verdaguer, D. (2013). Altitudinal and seasonal changes of phenolic compounds in Buxus sempervirens leaves and cuticles. *Plant Physiology and Biochemistry, 70*, 471–482.

Blumthaler, M., Ambach, W., & Ellinger, R. (1997). Increase in solar UV radiation with altitude. *Journal of Photochemistry and Photobiology, B: Biology, 39*(2–3), 130–134. https://doi.org/10.1016/S1011-1344(96)07145-X

Brasier, C. M. (2008). The biosecurity threat to the UK and global environment from international trade in plants. *Plant Pathology, 57*(5), 792–808. https://doi.org/10.1111/j.1365-3059.2008.01887.x

Bryant, J. P., Chapin, F. S., & Klein, D. R. (1983). Carbon/nutrient balance of boreal plants in relation to vertebrate herbivory. *Oikos, 40*(3), 357–368. https://doi.org/10.2307/3544304

Buchner, O., Stoll, M., Karadar, M., Kranner, I., & Neuner, G. (2015). Application of heat stress in situ demonstrates a protective role of irradiation on photosynthetic performance in alpine plants. *Plant, Cell & Environment, 38*, 812–826.

Carlsson, G., & Huss-Dell, K. (2003). Nitrogen fixation in natural and managed ecosystems. *Plant and Soil, 249*(1), 1–14. https://doi.org/10.1023/A:1022830128575

Cavieres, L. A., Hernández-Fuentes, C., Sierra-Almeida, A., & Kikvidze, Z. (2016). Facilitation among plants as an insurance policy for diversity in Alpine communities. *Functional Ecology, 30*(1), 52–59. https://doi.org/10.1111/1365-2435.12545

Chapin, F. S., III, Shaver, G. R., Giblin, A. E., Nadelhoffer, K. J., & Laundre, J. A. (1995). Responses of arctic tundra to experimental and observed changes in climate. *Ecology, 76*(3), 694–711. https://doi.org/10.2307/1939337

Chauhan, H. K., Oli, S., Bisht, A. K., Meredith, C., & Leaman, D. (2021). Review of the biology, uses and conservation of the critically endangered endemic Himalayan species Nardostachys jatamansi (Caprifoliaceae). *Biodiversity and Conservation, 30*(12), 3315–3333.

Das, S., Shukla, S., Debnath, S., & Kalyani, M. I. (2024). Synergistic role of plant tissue culture and plant growth-promoting micro-organisms as conservation strategies for Himalayan medicinal plants amidst climate change. In *Threatened medicinal plants in the Indian Himalayan Region: Sustainability challenges and conservation strategies* (pp. 133–165). Springer.

Demmig-Adams, B., & Adams, W. W., III. (1996). The xanthophyll cycle. *Australian Journal of Plant Physiology, 23*(6), 631–640. https://doi.org/10.1071/PP9960631

Dietrich, L., & Körner, C. (2014). Thermal imaging reveals massive heat accumulation in flowers across a broad spectrum of alpine taxa. *Alpine Botany, 124*, 27–35.

Du, W., Ding, J., Lu, S., Wen, X., Hu, J., & Ruan, C. (2022). Identification of the key flavonoid and lipid synthesis proteins in the pulp of two sea buckthorn cultivars at different developmental stages. *BMC Plant Biology, 22*(1), 299.

Elshaer, S. E., Hamad, G. M., Sobhy, S. E., Darwish, A. M. G., Baghdadi, H. H., H. Abo Nahas, H., et al. (2024). Supplementation of Saussurea costus root alleviates sodium nitrite-induced hepatorenal toxicity by modulating metabolic profile, inflammation, and apoptosis. *Frontiers in Pharmacology, 15*, 1378249.

Feng, M., Chen, C., Qu-Bie, J., Qu-Bie, A., Bao, X., Cui, Q., et al. (2022). Metabolome and transcriptome associated analysis of sesquiterpenoid metabolism in *Nardostachys jatamansi. Frontiers in Plant Science, 13*, 1041321.

Foyer, C. H., & Noctor, G. (2005). Redox homeostasis and antioxidant signaling: A metabolic interface between stress perception and physiological responses. *The Plant Cell, 17*(7), 1866–1875. https://doi.org/10.1105/tpc.105.033589

Galen, C., & Stanton, M. L. (1993). Short-term responses of alpine buttercups to experimental manipulations of growing-season length. *Ecology, 74*, 1052–1058.

Gilroy, S., Białasek, M., Suzuki, N., Górecka, M., Devireddy, A. R., Karpiński, S., & Mittler, R. (2016). ROS, calcium, and electric signals: key mediators of rapid systemic signaling in plants. *Plant Physiology,* 171(3), 1606–1615.

Giménez-Benavides, L., García-Camacho, R., Iriondo, J. M., & Escudero, A. (2011). Selection on flowering time in Mediterranean high-mountain plants under global warming. *Evolutionary Ecology, 25*, 777–794.

Hodkinson, I. D. (2005). Terrestrial insects along elevation gradients: Species and community responses to altitude. *Biological Reviews, 80*(3), 489–513. https://doi.org/10.1017/S1464793105006767

Huner, N. P. A., Öquist, G., Sarhan, F., & Ivanov, A. G. (1998). Photosynthesis in the cold: An adaptation of the photosynthetic apparatus to low temperatures. *Photosynthesis Research, 56*(3), 365–375. https://doi.org/10.1023/A:1006129200496

Inouye, D. W. (2000). The ecological significance of snow cover in alpine plant communities. *Arctic, Antarctic, and Alpine Research, 32*(2), 128–136. https://doi.org/10.1080/15230430.2000.12003349

Kala, C. P. (2005). Indigenous uses, population density, and conservation of threatened medicinal plants in protected areas of the Indian Himalayas. *Conservation Biology, 19*(2), 368–378. https://doi.org/10.1111/j.1523-1739.2005.00602.x

Kelly, D. (1994). The evolutionary ecology of mast seeding. *Trends in Ecology and Evolution, 9*, 465–470.

Körner, C. (1999). *Alpine plant life: Functional plant ecology of High Mountain ecosystems*. Springer.
Körner, C. (2003). *Alpine plant life: Functional plant ecology of high mountain ecosystems* (2nd ed.). Springer. https://doi.org/10.1007/978-3-642-18970-8
Körner, C. (2021). *Alpine plant life: Functional plant ecology of high mountain ecosystems* (3rd ed.). Springer. https://doi.org/10.1007/978-3-030-59538-8
Larcher, W., & Wagner, J. (2004). High temperature stress and plant survival in the Central Alps. *Phyton, 44*(2), 237–252.
Li, S. (2014). Transcriptional control of flavonoid biosynthesis: Fine-tuning of the MYB-bHLH-WD40 (MBW) complex. *Plant Signaling & Behavior, 9*(1), e27522.
Liu, J., Wang, Y., Hu, Z., Zhang, L., & Wang, J. (2016). Phytochemical composition and antioxidant activities of Saussurea obvallata in the Indian Himalaya. *Journal of Ethnopharmacology, 191*, 274–282. https://doi.org/10.1016/j.jep.2016.07.030
Lütz, C., Engel, S., & Roser, C. (2012). Functional and structural adaptations of photosynthesis in Ranunculus glacialis under simulated alpine conditions. *Environmental and Experimental Botany, 82*, 95–104. https://doi.org/10.1016/j.envexpbot.2012.05.008
Ma, L., Sun, X., Kong, X., Galvan, J. V., Li, X., Yang, S., et al. (2015). Physiological, biochemical and proteomics analysis reveals the adaptation strategies of the alpine plant Potentilla saundersiana at altitude gradient of the Northwestern Tibetan Plateau. *Journal of Proteomics, 112*, 63–82.
Malhotra, B., Kulkarni, G. T., Dhiman, N., Joshi, D. D., Chander, S., Kharkwal, A., et al. (2021). Recent advances on Berberis aristata emphasizing berberine alkaloid including phytochemistry, pharmacology and drug delivery system. *Journal of Herbal Medicine, 27*, 100433.
Mudassir Jeelani, S., Shahnawaz, M., Prakash Gupta, A., & Lattoo, S. K. (2022). Phytochemical diversity in relation to cytogenetic variability in Inula racemosa Hook. f., an endangered medicinal plant of Himalayas. *Chemistry & Biodiversity, 19*(12), e202200486.
Nenadis, N., Llorens, L., Koufogianni, A., Díaz, L., Font, J., Gonzalez, J., & Verdaguer, D. (2015). Interactive effects of UV radiation and reduced precipitation on the seasonal leaf phenolic content/composition and the antioxidant activity of naturally growing Arbutus unedo plants. *Journal of Photochemistry and Photobiology, B: Biology, 153*, 435–444.
Nishida, I., & Murata, N. (1996). Chilling sensitivity in plants and cyanobacteria: The crucial contribution of membrane lipids. *Annual Review of Plant Physiology and Plant Molecular Biology, 47*, 541–568. https://doi.org/10.1146/annurev.arplant.47.1.541
Panossian, A., Wikman, G., & Wagner, H. (2010). Effects of adaptogens on the central nervous system and the molecular mechanisms associated with their stress-protective activity. *Pharmaceuticals, 3*(1), 188–224. https://doi.org/10.3390/ph3010188
Pathak, S., & Godela, R. (2024). Nardostachys jatamansi: Phytochemistry, ethnomedicinal uses, and pharmacological activities: A comprehensive review. *Fitoterapia, 172*, 105764.
Pauchard, A., Kueffer, C., Dietz, H., Daehler, C. C., Alexander, J., Edwards, P. J., Guisan, A., Haider, S., et al. (2009). Ain't no mountain high enough: Plant invasions reaching new elevations. *Frontiers in Ecology and the Environment, 7*(9), 479–486. https://doi.org/10.1890/080072
Pauli, H., Gottfried, M., Dullinger, et al. (2012). Recent plant diversity changes on Europe's mountain summits. *Science, 336*(6079), 353–355. https://doi.org/10.1126/science.1219033
Pellissier, L., Roger, A., Bilat, J., & Conti, E. (2012). Long-term impacts of increased herbivory on plant–pollinator interactions and community composition in alpine ecosystems. *Ecology Letters, 15*(6), 615–622. https://doi.org/10.1111/j.1461-0248.2012.01766.x
Peñuelas, J., & Munné-Bosch, S. (2005). Isoprenoids: An evolutionary pool for photoprotection. *Trends in Plant Science, 10*, 166–169.
Pepin, N., Bradley, R. S., Diaz, H. F., et al. (2015). Elevation-dependent warming in mountain regions of the world. *Nature Climate Change, 5*(5), 424–430. https://doi.org/10.1038/nclimate2563
Phoenix, G. K., Lee, J. A., & Power, S. A. (2006). Effects of nutrient availability on the susceptibility of upland plants to herbivores and pathogens. *Plant Ecology, 186*(1), 35–46. https://doi.org/10.1007/s11258-006-9167-6

Quan, X., Fan, F., Cao, H., Tang, N., Xu, C., & Wang, C. (2025). Transcriptome and metabolome analysis of low-pressure regulation in *Saussurea involucrata* leaves. *Genes, 16*(3), 328.

Rahman, I. U., Hart, R., Afzal, A., Iqbal, Z., Alqarawi, A. A., Abd-Allah, E. F., et al. (2019). Ecophysiological plasticity and cold stress adaptation in Himalayan alpine herbs: Bistorta affinis and Sibbaldia procumbens. *Plants, 8*(10), 378.

Rai, M., Rai, A., Mori, T., Nakabayashi, R., Nakamura, M., Kojoma, M., et al. (2025). Multi-omics analysis reveals tissue-specific biosynthesis and accumulation of diterpene alkaloids in Aconitum japonicum. *Journal of Natural Medicines, 79*(3), 499–516.

Rammig, G., Buchmann, N., et al. (2010). Climate change and drought impacts on growth and biomass of Alpine forests in South Tyrol, Italy. *Global Change Biology, 16*(2), 619–631. https://doi.org/10.1111/j.1365-2486.2009.02142.x

Rasmussen, S., Barah, P., Suarez-Rodriguez, M. C., Bressendorff, S., Friis, P., Costantino, P., Bones, A. M., & Mundy, J. (2013). Transcriptome responses to combinations of stresses in Arabidopsis thaliana. *Plant Physiology, 161*(4), 1783–1794. https://doi.org/10.1104/pp.112.214324

Rawal, R., Rawat, R., & Panwar, R. (2020). Integrated stress responses in alpine plants: Roles of reactive oxygen species signaling, hormonal cross-talk, and epigenetic memory. *Journal of Experimental Botany, 71*(14), 4153–4170. https://doi.org/10.1093/jxb/eraa174

Read, D. J., & Haselwandter, K. (1981). Mycorrhizal symbiosis and ecological adaptation in alpine and arctic environments. In *Arctic and Alpine ecosystems* (pp. 297–320). Springer.

Rime, T., Hartmann, M., Brunner, I., Widmer, F., & Zeyer, J. (2015). Decay dynamics of fungal necromass partitioned with molecular and isotopic tracers. *Soil Biology and Biochemistry, 83*, 192–203. https://doi.org/10.1016/j.soilbio.2015.01.025

Rozema, J., van de Staaij, J., Björn, L. O., & Caldwell, M. M. (1997). UV-B as an environmental factor in plant life: Stress and regulation. *Trends in Ecology & Evolution, 12*(1), 22–28. https://doi.org/10.1016/S0169-5347(96)10040-4

Rubidge, E. M., Patton, J. L., Lim, M., Burton, A. C., Brashares, J. S., & Moritz, C. (2012). Climate-induced range contraction drives genetic erosion in an alpine mammal. *Nature Climate Change, 2*, 285–288.

Scherrer, D., & Körner, C. (2009). Infra-red thermometry of alpine landscapes challenges climatic warming projections. *Global Change Biology, 16*, 2602–2613.

Schmidt, M. W. I., Torn, M. S., Abiven, S., Dittmar, T., Guggenberger, G., Janssens, I. A., et al. (2011). Persistence of soil organic matter as an ecosystem property. *Nature, 478*(7367), 49–56. https://doi.org/10.1038/nature10386

Seipel, C., Winkler, A., Lins, I., Yanar, S., Stunner, H., & Zidorn, C. (2012). The ecological impact of Taraxacum officinale invasion in Alpine pastures: Effects on pollination networks and soil nitrogen dynamics. *Ecological Monographs, 82*(3), 347–364. https://doi.org/10.1890/11-1602.1

Sharma, A. D., Sharma, S., & Shikha. (2010). Incidence and severity of anthracnose (Colletotrichum sp.) in alpine medicinal plants of western Himalaya. *Journal of Mycology and Plant Pathology, 40*(1), 91–94.

Sharma, E., Anand, G., & Kapoor, R. (2017). Terpenoids in plant and arbuscular mycorrhiza-reinforced defence against herbivorous insects. *Annals of Botany, 119*(5), 791–801.

Sharma, N. (2021). Picrorhiza kurroa. In *Himalayan medicinal plants* (pp. 67–83). Academic.

Sharma, S., Negi, S., Kumar, P., & Irfan, M. (2025). Nitric oxide dynamics in high-altitude medicinal plants: Role in stress adaptation, signaling, and phytohormonal interactions. *Physiologia Plantarum, 177*(3), e70342.

Singh, A., & Roychoudhury, A. (2023). Abscisic acid in plants under abiotic stress: Crosstalk with major phytohormones. *Plant Cell Reports, 42*(6), 961–974.

Soukup, A., Pecková, E., Ježková, B., & Sklenář, P. (2021). Structural adaptations in plants from the humid equatorial Andes indicate a trade-off between hydraulic transport efficiency and safety. *American Journal of Botany, 108*(11), 2127–2142.

Spitaler, R., Winkler, A., Lins, I., Yanar, S., Stunner, H., & Zidorn, C. (2008). Altitudinal variation of phenolic contents in flowering heads of Arnica montana cv. ARBO: A 3-year comparison. *Journal of Chemical Ecology, 34*, 369–375.

Stinson, K. A. (2004). Natural selection favors rapid reproductive phenology in Potentilla pulcherrima (Rosaceae) at opposite ends of a subalpine snowmelt gradient. *American Journal of Botany, 91*, 531–539.

Streb, P., Aubert, S., Gout, E., & Bligny, R. (2003). Reversibility of cold- and light-stress tolerance and accompanying changes of metabolite and antioxidant levels in the two high mountain plant species Soldanella alpina and Ranunculus glacialis. *Journal of Experimental Botany, 54*(381), 405–418. https://doi.org/10.1093/jxb/erg048

Sun, P., & Li, M. F. (2016). Chilling temperature stimulates growth, gene over-expression and podophyllotoxin biosynthesis in Podophyllum hexandrum Royle. *Plant Physiology and Biochemistry, 107*, 197–203.

Testerink, C., & Munnik, T. (2011). Molecular, cellular, and physiological responses to phosphatidic acid formation in plants. *Journal of Experimental Botany, 62*(7), 2349–2361. https://doi.org/10.1093/jxb/erq414

Theurillat, J.-P., & Guisan, A. (2001). Potential impact of climate change on vegetation in the European Alps: A review. *Climatic Change, 50*, 77–109.

Walther, G. R., Beißner, S., & Burga, C. A. (2005). Trends in the upward shift of alpine plants. *Journal of Vegetation Science, 16*(5), 541–548.

Wdowiak, A., Podgórska, A., & Szal, B. (2024). Calcium in plants: An important element of cell physiology and structure, signaling, and stress responses. *Acta Physiologiae Plantarum, 46*(12), 108.

Webb, C. J., & Kelly, D. (1993). The reproductive biology of the New Zealand flora. *Trends in Ecology and Evolution, 8*, 442–447.

Wilhelm, L., Singer, G. A., Faschling, C., Battin, T. J., & Besemer, K. (2013). Microbial biodiversity in glacier-fed streams. *The ISME Journal, 7*, 1651–1660.

Wisniewski, M., Fuller, M., & Palta, J. P. (2003). Ice recrystallization inhibition and the role of antifreeze proteins in cold-hardy plants. *Physiologia Plantarum, 119*(1), 63–73. https://doi.org/10.1034/j.1399-3054.2003.1190110

Xu, W., Baoyin, T., Ayodele, O. E., Mi, Z., Zhang, Y., Gong, J., & Han, Y. (2014). Response of soil microbial activity and nutrient cycling to experimental warming in an alpine meadow on the Tibetan Plateau. *Applied Soil Ecology, 83*, 89–97. https://doi.org/10.1016/j.apsoil.2014.04.006

Zahedi, S. M., Karimi, M., Venditti, A., Zahra, N., Siddique, K. H., & Farooq, M. (2025). Plant adaptation to drought stress: The role of anatomical and morphological characteristics in maintaining the water status. *Journal of Soil Science and Plant Nutrition, 25*(1), 409–427.

Zhang, L., Wu, M., Teng, Y., Jia, S., Yu, D., Wei, T., Chen, C., & Song, W. (2019). Overexpression of the glutathione peroxidase 5 (*RcGPX5*) gene from *Rhodiola crenulata* increases drought tolerance in *Salvia miltiorrhiza*. *Frontiers in Plant Science, 9*, 1950.

Zhang, J., Huang, B., & Yu, X. (2006). Osmotic adjustment and antioxidative responses to drought in the grass species Festuca arundinacea. *Plant Physiology and Biochemistry, 44*(7–9), 602–610. https://doi.org/10.1016/j.plaphy.2006.04.004

Zhao, X., Wang, Z., Li, R., Liu, Y., Chen, Y., & Lu, C. (2024). Insights into the molecular mechanism underlying cold-induced flavonoids biosynthesis in callus of a Tibetan medicinal plant Saussurea laniceps. *Plant Cell, Tissue and Organ Culture, 158*(3), 46.

Zhou, Z., Xu, S., Liu, J., Bian, M., Sun, X., & Holbrook, N. M. (2017). Cold and UV-B light interactions induce synergistic flavonoid accumulation in Arabidopsis thaliana. *New Phytologist, 214*(3), 1120–1132. https://doi.org/10.1111/nph.14305

Zhu, J. K. (2016). Abiotic stress signaling and responses in plants. *Cell, 167*(2), 313–324.

Zidorn, C. (2010). Altitudinal variation of secondary metabolites in flowering heads of the Asteraceae: Trends and causes. *Phytochemistry Reviews, 9*, 197–203.

Advances in Remote Sensing for Monitoring Alpine Vegetation

5

Nishant Jain and Hemant Sood

Abstract

Advances in remote sensing have revolutionized the monitoring of alpine vegetation, providing crucial insights into these fragile mountain ecosystems and their medicinal plant species. This chapter examines the evolution of remote sensing technologies ranging from satellite, UAV, airborne, and ground-based platforms to advanced sensor types including optical, multispectral, hyperspectral, radar, and thermal systems for comprehensive alpine vegetation analysis. Emphasis is placed on the integration of diverse datasets, such as high-resolution imagery, LiDAR, gridded climate data, and derived vegetation indices, which enable detailed mapping, species discrimination, habitat assessment, and phenological tracking at multiple spatial and temporal scales. The chapter further explores the implementation of cutting-edge AI-driven image processing, feature extraction, classification, and data fusion methodologies, highlighting developments in machine learning, deep learning, and fusion algorithms that enhance the accuracy and ecological interpretation of remote sensing outputs for alpine environments. Key applications for effective ecosystem management and conservation included are biodiversity assessment, monitoring vegetation dynamics in response to climate change, high-precision vegetation cover estimation, productivity modeling, and habitat classification. The chapter concludes by addressing prevailing challenges such as data acquisition constraints, technical and cost barriers, and the need for open-access, standardized approaches, outlining innova-

N. Jain (✉)
Department of Electronics and Communication Engineering, Jaypee University of Information Technology, Solan, Himachal Pradesh, India
e-mail: nishant.jain@juitsolan.in

H. Sood
Department of Biotechnology & Bioinformatics, Jaypee University of Information Technology, Solan, Himachal Pradesh, India

P. Kumar et al. (eds.), *Climate Resilience and Molecular Adaptation in Alpine Medicinal Plants*, https://doi.org/10.1007/978-981-95-6923-6_5

tive future directions to ensure remote sensing fulfills its transformative potential in safeguarding alpine vegetation.

Keywords

Alpine vegetation · Remote sensing · Hyperspectral imagery · Machine learning · UAV monitoring · Vegetation indices · Biodiversity assessment · Climate change impacts

1 Introduction

Alpine vegetation at high altitude represents a very important component of mountain ecosystems. These plants are characterized by harsh climatic conditions, steep terrain, and ecological fragility. Their contribution to biodiversity, watershed protection, carbon storage, and soil stabilization in alpine environments is very significant. Among alpine flora, medicinal plants have very high ecological and economic importance. These medicinal plants contain unique bioactive compounds and hence are used extensively in traditional medicine and modern pharmaceuticals. They are a big support to health care systems worldwide.

Medicinal plants such as Neopicrorhiza scrophulariiflora and others endemic found in the alpine and subalpine zones produce secondary metabolites. These secondary metabolites help these plants to withstand extreme conditions and also form the biochemical basis for their therapeutic properties. More than 1700 medicinal plant species are found in the high-altitude Himalayan region. These also significantly help local livelihoods through harvesting, trade, and by manufacturing medicinal products. Using these plants in domestic use as well as in export markets helps in supporting rural economies. The continuous increase in the demand of natural and herbal remedies increases the demand for alpine medicinal plants globally.

Due to global high demand of alpine plants, especially medicinal species, the monitoring of these plants are highly required to ensure sustainable use and conservation. However, the rugged and often inaccessible terrain, harsh weather conditions, persistent snow and cloud cover makes field-based surveys a very challenging task. These extreme conditions create difficulties in observing temporal and spatial opportunities in the alpine vegetation area. Change in the climatic conditions in the high altitudes causes shifts in phenology, in distribution of species within the area, and in the quality of habitat. These changes further add complexity in monitoring of alpine plants and require frequent and detailed observation. Spectral similarity between species, small plant stature, and patchy distribution complicate identification of plant species and their mapping with high accuracy.

Alpine vegetation, especially medicinal plants face threats from these challenges of overharvesting, habitat degradation, and climate stress, and hence effective conservation and sustainable management of alpine vegetation becomes difficult. More efficient monitoring tools are required which can assess population status and detect habitat loss. Monitoring tools should be capable of monitoring the effect of changes in the environment and the activities of the humans in the alpine vegetation (Awasthi & Varade, 2021; Xie et al., 2008; Chen et al., 2015; Li et al., 2018).

To address these challenges of monitoring alpine vegetation, remote sensing technology has emerged with tools that help to monitor alpine vegetation without a field-based survey. Sensors and cameras mounted on satellites, UAVs (drones), aircraft, and ground-based systems help in capturing the growth of the alpine vegetation and in capturing terrain and weather information remotely without going physically for the ground surveys. These sensors provide high-resolution data of the inaccessible alpine landscapes continuously and repeatedly.

Advancement in imaging technologies like optical, multispectral, hyperspectral, radar, and thermal sensors has further improved the capabilities of the remote sensing platforms. Digital images acquired with these imaging technologies enables mapping alpine vegetation, discriminating different species, assessing the health of the plant, and in tracking phenology of the plants. Digital image processing techniques, and machine learning models implemented on these images help in classifying plant species, monitoring their growth and in ecological interpretation. These techniques have the capability to integrate information from multiple sensors and to provide assessment of the entire alpine landscape with high accuracy and details. Advancements in data acquisition sensors and software capabilities have made real time monitoring possible.

Together, data acquisition sensors, remote sensing platforms, image processing techniques and machine learning models, represents a transformative approach to monitor and preserve the alpine vegetation ecosystems. These advancements ensure the preservation of ecological functions and alpine vegetation by enabling effective monitoring, management and conservation efforts.

2 Remote Sensing Technologies for Alpine Vegetation Monitoring

Remote sensing is a technique that involves collecting data and information about an object or area from a distance without going physically to the field. Unlike traditional ground-based methods that require direct physical contact for the measurements, remote sensing uses sensors that detect and record electromagnetic radiation reflected or emitted by the target. Electromagnetic radiations like visible light, infrared, microwave, and other wavelengths are sensed by sensors to acquire the information of the inaccessible regions. The sensors are mounted on the aircraft, satellites, drones, or other airborne platforms. These platforms like satellites and drones are used to carry the sensors to the field. Remote sensing provides comprehensive, efficient, and cost effective solutions to observe and analyze the Earth's surface and atmosphere without the need for physical presence. It is widely used in various fields such as environmental monitoring, agriculture, urban planning, disaster management, and climate studies (Awasthi & Varade, 2021).

The data captured using remote sensing can provide insights across multiple dimensions. These multiple dimensions enable comprehensive observation and analysis of the alpine vegetation. The key dimensions of data collected through remote sensing are:

(a) *Spatial Scope:* It defines the geographic area covered by remote sensing technologies. It is used to monitor very small areas, like a small patch of vegetation, to a very vast region like a complete region of alpine vegetation. Images captured to scan the alpine region spatially may vary from high-resolution to a low resolution. High resolution satellite images can capture detailed features at the meter level, whereas low-resolution sensors may provide details at a few Km. Low resolution images are suitable for large-scale studies, such as climate change or deforestation assessments.
(b) *Temporal Scope:* It defines the change in the geographic area with respect to time. These allow remote sensing technologies to capture time series data in the form of videos with very less frame rates. Changes and trends over time can be analysed using time series data. It helps in analyzing dynamic environmental processes, climate patterns and temporal variations in the alpine vegetation.
(c) *Spectral Scope:* It defines the acquisition of data using electromagnetic radiation of different wavelengths. Data acquired with infrared radiations is useful for analysing the health of alpine vegetation. Similarly microwave radiations can penetrate through the clouds and hence provide information about soil moisture or ice. Thus data captured using broad spectral ranges contains information of features and phenomena that are not visible to the naked eye. This allows for detailed analysis of alpine vegetation by capturing information beyond visible light. It helps in identifying plant health, differentiating different species, and assessing environmental conditions.

Together, data capturing through these multiple dimensions made remote sensing a powerful tool for resource management, and protection of alpine vegetation. Remote sensing has the ability to access remote regions for monitoring the alpine zone.

2.1 Types of Sensors

Sensors serve as the fundamental devices in remote sensing that detect and record electromagnetic signals reflected or emitted from Earth's surface. The choice of sensor technology depends on the specific information required and the environmental conditions of the study area. Details of some of the main types of sensors used for remote sensing are given below:

(a) *Optical Sensors or LiDAR:* Optical sensors operate by capturing reflected sunlight in the visible and near-infrared (NIR) spectra, functioning similarly to conventional cameras but often with enhanced capabilities. LiDAR (Light Detection and Ranging), although technically a laser-based active sensor, falls into this category due to its use of light to measure distances. These sensors are extensively used for:

(i) Mapping land cover types.
(ii) Monitoring vegetation greenness.
(iii) Assessing phenological changes (such as leaf unfolding and senescence).

Their high spatial resolution and spectral sensitivity make them invaluable for detailed topographic and vegetation studies (Coops et al., 2013).

(b) *Radar Sensors (Synthetic Aperture Radar (SAR)):* Radar sensors, particularly SAR systems, operate using microwave wavelengths that can penetrate cloud cover and operate effectively regardless of lighting conditions (day or night) (Vaglio Laurin et al., 2013). This capability is essential in regions with frequent cloud cover or snow, such as alpine zones. SAR provides:

(i) Structural information on vegetation, including biomass and canopy structure.
(ii) Surface roughness data.
(iii) Digital elevation models (DEMs) and terrain analysis.
(iv) These features make radar sensors ideal for environmental monitoring in challenging weather conditions.

(c) *Multispectral Sensors:* Multispectral sensors capture data across several broad wavelength bands typically ranging from visible to shortwave infrared. They enable detailed analysis of:

(i) Vegetation health and vigor.
(ii) Land use and land cover classification.
(iii) Spectral indices like NDVI (Normalized Difference Vegetation Index), which indicate plant condition.

These sensors are commonly used for ecological and agricultural applications, providing a balance between spectral detail and data volume (Lin et al., 2022).

(d) *Hyperspectral Sensors:* Hyperspectral sensors collect data across hundreds of narrow, contiguous spectral bands, offering highly detailed spectral information. This allows for:

(i) Precise identification of plant species.
(ii) Detection of plant stress or disease.
(iii) Subtle environmental variations that multispectral sensors might miss.

Hyperspectral data facilitate advanced environmental analysis, enabling scientists to distinguish minute differences in surface materials and biological conditions (Wu et al., 2008).

(e) *Thermal Sensors:* Thermal sensors measure emitted infrared radiation from Earth's surface, providing surface temperature data. They are used to:

(i) Assess plant water stress and soil moisture.
(ii) Monitor microclimate variations.

(iii) Detects heat fluxes associated with biological or geological processes.

Thermal imagery is particularly useful in understanding heat dynamics in complex terrains like alpine environments, where temperature variations can influence ecological processes.

Selecting the appropriate sensor type depends on the specific research objectives, the environmental conditions of the study area, and the nature of the information required. Combining multiple sensor types often provides a more comprehensive understanding of complex ecological and geological systems.

2.2 Platforms for Remote Sensing

Physical platforms that carry sensors into the remote environment, that is to be monitored, are known as remote sensing platforms. Sensors are mounted on these remote sensing platforms and hence these platforms act as a vehicle to sensors. Sensors are able to sense the remote environment only with the support of these platforms. There exist different types of remote sensing platforms that carry the sensors to remote locations. Selection of remote sensing platform is crucial and depends on the following factors:

- The coverage area.
- Spatial resolution required.
- Temporal resolution (frequency of data collection) required.
- Sensor type and capabilities (such as multispectral or hyperspectral imaging).
- Accessibility and operational constraints of the platform.
- Specific project objectives like monitoring vegetation health, urban development, or disaster management.
- Budget involved.

Details of the main types of remote sensing platforms are given below:

(a) *Satellites:* Satellites are platforms that orbit the Earth. Since they revolve around the earth at various altitudes and revisit areas multiple times, they are capable of covering an extensive area for the acquisition of data repeatedly. This makes them ideal for long-term, and large-scale monitoring of remote locations.

(b) *UAVs (Unmanned Aerial Vehicles or Drones):* UAVs are low-altitude remote sensing platforms that can be operated either manually or automatically. They can provide high resolution data for the specific area. Sensors mounted on UAVs are capable of capturing detailed imagery of small or complex alpine vegetation. They can be deployed in challenging terrains for targeted ecological mapping, habitat assessment, and monitoring of localized environmental changes.

(c) *Aircraft:* Fixed wing planes and helicopters are also used as remote sensing platforms. They cover larger areas than drones and have lower resolution than UAVs. They have heavy payload capacity and can carry advanced sensor systems.
(d) *Ground-Based Systems:* Stationary sensors, towers, vehicles, or portable instruments installed in the remote location are known as ground based systems. They provide precise, high-resolution data on local features such as soil moisture, temperature, or vegetation health. These measurements complement the observations of other remote sensing platforms. They also help in validating remote sensing data from satellites, aircraft, or drones.

The Table 5.1 provides a comprehensive comparison of different remote sensing platforms listed above. Integrating the information provided by multiple platforms

Table 5.1 Comparison of remote sensing platforms

Factors	Satellites	UAVs	Aircraft	Ground-based systems
The coverage area	Extensive, covering entire mountain ranges or continents	Small to medium areas (up to several km^2)	Medium to large regions, depending on flight range	Local, specific sites or points
Spatial resolution	Low (10 m to several km)	Very high (~1 cm to 1 m)	High (~10 cm to 1 m)	Very high (sub-millimeter to cm)
Temporal resolution	Varies from daily (Sentinel-2) to weekly or bi-weekly (Landsat)	Minutes to hours (depends on UAV flight schedule)	Several times daily, weather permitting	Continuous or periodic, depending on system
Sensor type and capabilities	Multispectral, hyperspectral, thermal, radar sensors.	RGB, multispectral, thermal, LiDAR, hyperspectral	Multispectral, thermal, LiDAR, hyperspectral	Spectrometers, sensors for specific measurements
Accessibility and operational constraints	Affected by orbital paths, cloud covers.	Limited by battery life, weather, regulations	Weather-dependent, requires aircraft operation permits	Usually fixed, limited by site access
Specific project objectives	Monitoring glacier retreat, vegetation phenology, and land-use changes in remote or inaccessible alpine regions.	Precision agriculture, detailed mapping, research	Aerial surveys, environmental monitoring	In-situ measurements, calibration, sampling
Budget involved	Moderate to high (data costs, processing)	Low to moderate (equipment and flight costs)	Moderate to high (aircraft operation, maintenance)	Low to moderate (equipment and personnel costs)

provides a comprehensive understanding of alpine ecosystems from broader range to detailed surveys.

2.3 Type of Dataset

Data obtained from different sensors that sense the environment with different prospects are required for proper monitoring of alpine vegetation. For accurate analysis of alpine vegetation, these diverse datasets are required to capture the complex spatial, temporal, and structural characteristics of alpine ecosystems. Inclusion of high-resolution and very-high-resolution data in the dataset is crucial for detailed mapping of vegetation types, structural attributes, and species distribution for the monitoring of heterogeneous alpine environments. For example, airborne and ground-based data, such as LiDAR and photogrammetry, provides precise structural information and ground truth validation. Environmental datasets like climate variables from ERA5 (European Centre for Medium-Range Weather Forecasts Reanalysis fifth Generation) provides details of vegetation dynamics within the influence of temperature, precipitation, and radiation patterns. Parameters like NDVI (Normalized Difference Vegetation Index) and MSAVI2 (Modified Soil-Adjusted Vegetation Index 2), obtained from the data collected by remote sensing platforms, provide details of plant health, chlorophyll content, and phenological stages of the plants. Collectively, these datasets form an integrated framework essential for understanding, monitoring, and managing alpine vegetation under changing environmental conditions. Combining several data sources (satellite, airborne, ground-based, climate) yields the most robust monitoring and modelling outcomes for alpine vegetation. Table 5.2 summarises the remote sensing datasets utilized for monitoring alpine vegetation.

The success of global vegetation monitoring relies heavily on international collaboration and open access to data. Embracing open data ecosystems, more high-resolution satellite platforms are expected to adopt policies that promote data sharing, similar to ESA's Sentinel and NASA's Landsat programs, but with an extension toward sub-meter resolution where feasible. This approach will help reduce dependence on costly commercial imagery, making ecological monitoring more accessible and sustainable.

Furthermore, the development of cross-border alpine research networks will facilitate large-scale, continent-spanning monitoring efforts. These initiatives will involve pooling datasets from multiple countries and standardizing methodologies, enabling comprehensive studies of mountain ecosystems across regions such as the Himalayas, the Alps, and the Tien Shan. Coordinated global observatories dedicated to alpine vegetation will enhance our understanding of these sensitive environments.

Table 5.2 Summary of remote sensing datasets utilized for monitoring alpine vegetation

Platforms	Sensors	Spatial resolution	Frequency	DataTypes/Dataset
Satellite imagery	MODIS (moderate resolution imaging spectroradiometer)	250 m to 1 km	Daily to 16-day composites	Multispectral, thermal, reflectance sensors
	Sentinel-2 (Ali et al., 2019)	10–20 m across multiple spectral bands.	Every 5 days	Multispectral, RGB, NDVI sensors
	Landsat series	30 m	16 days (Landsat 8/9), with revisit variations	Multispectral, panchromatic, thermal sensors
Very-high-resolution imagery	Commercial satellite: WorldView Eg. WV-2 satellite	0.31 m (WorldView-3/4) 0.41 m (GeoEye-1) WV-2: 2 m	Varies from daily to weekly depending on task. WV2: 1–3 days	Panchromatic, multispectral, hyperspectral sensors.
	Drone-based imagery	Typically 1–10 centimeters	As frequently as needed, often real-time	RGB, multispectral, hyperspectral, thermal sensors
Airborne and ground-based imagery	Airborne laser scanning (LiDAR)	Typically 10 cm to 1 m resolution	As needed, often seasonal.	Laser scanner sensors capturing 3D point clouds
	Ground-based photogrammetry	From a few millimeters to a few centimeters.	As needed, usually project-based	High-resolution cameras, digital photogrammetric systems
Gridded climate and environmental datasets	ERA5	Spatial resolution varies; approximately 31 km	Hourly intervals across variables	Climate variables (temperature, precipitation, humidity, radiation)

3 AI Tools for the Monitoring of Alpine Vegetation

For monitoring alpine vegetation, diverse dataset collected using different sensors mounted on the different remote sensing platforms are analysed sequentially using multiple Artificial Intelligence (AI) tools. These tools range from enhancement of data or images within the dataset for better visual perception, to extracting relevant features. Extracted features are used to train different machine learning models with the dataset, enabling the model to map different types of vegetation and predict the health of the alpine ecosystem.

AI tools used for analysing remote sensing data for monitoring alpine vegetation can be categorised into: image processing methods, feature extraction techniques, feature selection and dimensionality reduction methods, classification algorithms, and data fusion techniques. Brief details of each of these categories are provided in the following subsections:

3.1 Image Processing Methods

- *Radiometric Correction:* Due to sensor noise like calibration error, or due to the presence of atmospheric particles, gases and aerosols that absorb scatter radiations, data or images captured from remote sensing techniques contains noise or distortions in the image. In few cases, images captured may contain corrupted or missing scan lines in the image data. Due to these distortions and bad scan lines, acquired images will not represent the exact information of the alpine vegetation accurately. Degraded images can be enhanced using image processing methods. Raw digital data or images can also be converted to standardized reflectance values, which enables easy comparison across different images (Wu et al., 2020).
- Background subtraction methods like Dark Object Subtraction (DOS) removes atmospheric scattering effects by subtracting the minimum pixel values. Algorithms like FLAASH (Fast Line-of-sight Atmospheric Analysis of Spectral Hypercubes) and 6S (Second Simulation of the Satellite Signal in the Solar Spectrum) also helps in removing distortions due to scattering of radiation by the atmospheric particles, gases or aerosols. To remove bad scan lines, interpolation techniques can be used that consider neighboring pixels to replace pixels that belong to bad lines. This image processing ensures that subsequent analyses are based on physically meaningful data rather than sensor or atmospheric artifacts.
- *Geometric Correction and Orthorectification:* Due to the motion of the sensors and the angles at which they are mounted on the remote sensing platforms, and due to the high altitude of the satellites, spatial alignment of the images got distorted which distorts the the geometry of the alpine vegetation in the captured images. For the accurate analysis of alpine vegetation, geometry correction of the images are required. Image processing methods such as polynomial transformations, affine transformations, rubber sheeting, and resectioning using Ground Control Points (GCPs) are commonly used to remove these distortions. Digital Elevation Models (DEMs) are also used to ensure that each pixel's location in the image corresponds precisely to geographic coordinates. Image processing methods known as image registration methods are used to align multiple images taken at different times or using different sensors to a common coordinate system. Registering images to a common coordinate enables temporal analysis of data and also helps in fusion of information obtained from different sources. Image registration methods are done on the basis of matching similar features, using cross-correlation techniques, or with algorithms that use the concept of mutual information. Collectively, these methods, important for achieving spatial consistency, for detecting changes, and for integration with Geographic Information

Systems (GIS), ensures the positional accuracy necessary for various remote sensing applications.

- *Masking/Filtering:* Images captured using different sources may suffer from poor contrast due to which it may become difficult to differentiate between different species of the alpine vegetation easily. Images may also consist of irrelevant non-vegetation areas covering clouds, water bodies, and snow. Due to the presence of some noise during acquisition, a noisy pixel may be mistaken with a pixel of vegetation and may misdirect the complete analysis of alpine vegetation. To enhance the quality of the image and to focus on the most important areas of the image consisting of vegetation pixels, filtering and masking techniques of image processing are used. With masking and filtering, pixels belonging to water bodies or snow which are not required to differentiate different species are defocused by reducing the contrast. Image processing algorithms extract spectral signatures and contextual information from the image to generate masks that are capable of removing or ignoring irrelevant features from the image during analysis. Techniques like Gaussian filters, median filters, and other smoothing methods, are used to reduce the level of noise in the image. By combining masking to eliminate interfering features and filtering to reduce noise, the focus on the vegetation cover can be increased more effectively. This also helps in calculating different vegetation indices and features like the Normalized Difference Vegetation Index (NDVI) more precisely.
- *Temporal Smoothing (for time series):* For the determination of the growth of alpine plants with respect to different seasons, for analysing the trends in vegetation indices, and for the detection of phenological metrics, time series dataset of high quality is required. Similar to spatial data, time series data also degrade with atmospheric conditions, sensor inconsistencies, and other anomalies. To enhance the quality of the time series dataset, noise and fluctuations in the dataset are reduced using smoothing filters. Common smoothing methods include moving average filters, Gaussian smoothing, and median filtering. These methods help to highlight trends, seasonal patterns, and phenological events such as leaf-out, or flowering, making it easier to analyze vegetation health, productivity, and responses to environmental changes (Kuchler et al., 2020).
- *Normalization:* Normalization, also known as feature scaling, is done to ensure uniformity among data features. It prevents any single feature with larger numeric values from disproportionately influencing the analysis. It is very important for analysing and comparing multiple images of the same area captured from different sources or/and captured at different times. Normalisation adjusts the range and distribution of pixel values or feature measurements to standardize data. Different commonly used normalisation techniques are min-max and z-score normalisation. Min-max scaling rescales data to a specific range (usually 0 to 1), whereas z-score normalization adjusts data zero mean and unit standard deviation. Overall, normalization ensures that features are on a similar scale, facilitating more accurate, stable, and meaningful analysis (Wang et al., 2025).

3.2 Feature Extraction

Feature extraction is a method that quantifies the textures and patterns in the image. It analyses the images to detect structural changes within satellite or aerial images and represents smoothness, roughness, or regularity present in the image quantitatively with a texture parameter. Feature extraction methods examine the spatial arrangement, distribution, orientation and frequency of pixel intensities in the image. They not only detect spatial textures but also detect temporal changes in the time series data. Commonly used feature extraction techniques are first order statistics like mean, standard deviation, and skewness, second order statistics like Gray Level Co-occurrence Matrices (GLCM), Local Binary Patterns (LBP), wavelet transforms, and Fourier analysis. These extracted features are used to differentiate between different vegetation types, land cover classes, or structural features. These parameters also help to detect changes in the health of vegetation, and structural modifications over time. Overall, texture and pattern analysis enhances the ability to interpret complex spatial information, facilitating accurate classification, change detection, and monitoring of landscape dynamics (Wang et al., 2025).

Texture features detected in alpine vegetation are also termed as vegetation indices. Common indices obtained from multispectral or hyperspectral images are NDVI (Normalized Difference Vegetation Index), EVI (Enhanced Vegetation Index), and NDWI (Normalized Difference Water Index). Distinct reflectance properties of vegetation and water in different spectral bands are quantified with these indices, and hence estimates vegetation health, biomass, chlorophyll content, and water content quantitatively. For example, vegetation strongly reflects near infrared and absorbs red light. NDVI utilizes the difference between near-infrared and red light to gauge plant vigor and density. Similarly, NDWI quantifies the reflectance in the near-infrared and shortwave infrared bands to monitor water content in vegetation. EVI corrects atmospheric and soil influences to enhance sensitivity in dense vegetation areas.

3.3 Feature Selection and Dimensionality Reduction

Feature selection are the methods that remove irrelevant and redundant data or features from the dataset. It also helps in reducing the size of the dataset for better computational complexity, especially the multispectral and hyperspectral images that contain hundreds of spectral bands leading to high data volume and computational complexity. Commonly used feature selection and reduction techniques are Principal Component Analysis (PCA) and Nonnegative Matrix Factorization (NMF). PCA identifies the maximum variance within the data in the form of principal components and projects the data onto these components. The components with less or very less variance are removed and hence removes the redundancy and noise for better classification and analysis. Similarly, NMF decomposes the data into nonnegative factors that correspond to physical components or endmembers in the hyperspectral data. Thus this method enhances interpretability with spectral features relevant to specific land cover types. Integrating these feature selection techniques into classification pipelines improves processing speed, and often enhances classification accuracy.

3.4 Machine Learning for Classification and Regression

Classification and regression is one of the most important steps that provides final decisions regarding different types of alpine vegetation, maps the complete alpine region, analyse the health of the plants and predict the future condition of the alpine region. Classification is the machine learning (ML) model that makes a decision regarding the category of the input data. For example, alpine plants are classified as per their names, plant health is categorised as normal or abnormal, and the complete alpine region is mapped into different categories. Regression is a machine learning model that predicts the future trends as per the input data. These help in assessing the pattern change in biodiversity in the alpine region, aid in ecological monitoring, and improve land management practices by providing detailed, accurate, and scalable vegetation maps (Du et al., 2025; Constantin et al., 2022; Meng et al., 2020; Cheshkova, 2017).

To achieve high efficiency in terms of accuracy and precision, classification and regression models are provided with large and complex datasets, such as hyperspectral and multispectral images. They are also provided with the features extracted from the dataset and with vegetation indices obtained in the previous steps.

Commonly used ML models are Random Forests, Support Vector Machines (SVM), and Gradient Boosting. They effectively handle high-dimensional data and complex patterns. A subcategory of ML models are deep learning (DL) models. DL models like convolutional neural networks (CNNs) enhance classification accuracy by automatically learning hierarchical features from raw spectral data, capturing complex relationships and spatial context.

Recent advances in machine learning methodologies introduce ensemble learning and transfer learning approaches. In ensemble learning, learning of different machine learning models are combined together to create an ensembled machine learning models having characteristics of all the models combined together. In transfer learning, learning of the machine learning models trained with a dataset other than that of alpine vegetation may be transferred to a dataset of alpine vegetation for making classification or regression decisions. Details of commonly used machine learning models are given below:

- *Support Vector Machines (SVM):* SVM is a machine learning model consisting of kernel functions and regularization parameters. Fine-tuning of the kernel functions and regularization parameters gives SVM capability to learn complex, and high-dimensional data to create optimal decision boundaries. Hyperspectral data of the alpine vegetation often consists of overlapping spectral signatures among different plant species. SVMs can adapt to the spectral variability caused by snow cover, soil moisture, and seasonal changes. It classifies these overlapped and complex signatures very effectively. SVMs have the potential to distinguish these subtle differences, and hence classify species or vegetation types very accurately and precisely. SVM are also robust to limited training data. This makes SVMs suitable for remote alpine areas, where collecting extensive ground truth data can be challenging.

- *Random Forest (RF):* Random Forests (RFs) are algorithms that consist of multiple decision trees to improve classification stability and accuracy. RFs handle high-dimensional data efficiently and are less sensitive to overfitting compared to single decision trees. They effectively manage the spectral variability of alpine vegetation, such as differences in vegetation phenology, soil background, and snow cover. They require relatively few parameters to tune with the dataset. They can provide the details of important features for distinguishing alpine species like mosses, lichens, and flowering plants. RF classifiers are widely used in remote sensing applications for alpine ecological monitoring due to their robustness and ease of implementation (Yin et al., 2020).
- *Convolutional Neural Networks (CNN):* CNN models consist of multiple convolutional layers that extracts features from the input images followed by activation functions to introduce non-linearity. Pooling layer after the activation function reduces the spatial dimensions and helps with translation invariance. The final layer of CNN model is a fully connected layer that makes a decision to classify the image. CNNs have demonstrated remarkable proficiency in analyzing hyperspectral images of alpine vegetation. Alpine environments often contain dense, heterogeneous vegetation patches such as grasses, shrubs, and specialized alpine species. CNNs capture the intricate spectral and spatial heterogeneity of alpine ecosystems. CNNs has the capability to automatically learn features from spectral and spatial data, and hence is able to discriminate between similar species or vegetation types precisely. CNNs trained on Indian Pines or specific alpine regions have shown high accuracy in identifying species such as sedges, alpine grasses, and cushion plants. Fine-tuning of hyperparameters like filter sizes, network depth, and dropout rates ensures the models are robust for the variability and complexity of alpine terrains (Guerri et al., 2024; Xu & Guo, 2015).
- *Transfer Learning Models:* Due to the difficulty in acquiring data from remote locations of alpine vegetation, it becomes difficult to obtain robust machine learning models with less datasets. Transfer learning offers a valuable solution for this problem of having less datasets. It uses pre-trained models developed on similar or dissimilar, large and well-annotated datasets. These models are fine-tuned with alpine vegetation data, so that these models adapt to the specific spectral and spatial characteristics of alpine ecosystems present in the dataset. For example, a CNN model trained on extensive forest or grassland datasets can be transferred to classify alpine vegetation. This approach significantly reduces the need for exhaustive ground surveys. It also expedites ecological mapping and enhances the accuracy of vegetation classification. Transfer learning is very useful for rare or endangered species like Gentiana kurroo, where obtaining large datasets is difficult.

- *Ensemble Learning:* Alpine vegetations are characterized by complex spectral signatures and high heterogeneity. Different machine learning models capture different characteristics of this heterogeneous vegetation. Integration of different models like CNNs, SVM and RF into one model known as ensemble model, can better handle the variability and spectral differences among alpine species. For example, an ensemble model can accurately distinguish between closely related alpine species or different growth forms. These ensemble models are more resilient to noise and spectral overlaps.
- *Automated Pixel-level Classification:* In alpine ecosystems, where vegetation is often patchy and interspersed with snow, rocks, and soil, pixel-level classification is especially valuable. Machine learning models can analyze each pixel's spectral feature vector independently, enabling highly detailed maps that capture the fine-scale heterogeneity of mountain vegetation. This approach facilitates species-level discrimination within dense alpine meadows, cushion plant communities, or sparse shrublands. Such detailed classification is essential for monitoring ecological changes, assessing habitat health, and supporting conservation efforts in fragile mountain environments. It allows researchers and land managers to detect subtle shifts in vegetation distributions over time, which can be indicative of climate change impacts or human disturbance.
- *Explainable and Trustworthy AI:* As conservation decisions increasingly depend on AI-generated insights, the importance of explainable AI becomes paramount, especially in sensitive and complex alpine ecosystems. In this context, models will not only provide classifications of vegetation types such as mosses, lichens, or alpine shrubs but will also elucidate which spectral bands, structural features, or temporal patterns contributed most to each decision. For example, an explainable AI model might reveal that a particular spectral signature associated with snow cover or specific phenological timing influenced the identification of a certain alpine plant species. This transparency enables ecologists and conservationists to validate the AI's reasoning process, ensuring that classifications are based on ecologically meaningful features rather than spurious correlations. Consequently, explainable AI helps avoid "black box" decision-making, fostering greater trust and facilitating more informed management strategies in fragile alpine environments where accurate identification and monitoring are critical for conservation efforts.
- *Predictive Ecosystem Modelling:* Next-generation AI will integrate time-series remote sensing data with climate projections to simulate future changes in alpine vegetation distribution and health under various warming and land-use scenarios. By analyzing patterns over multiple seasons and years, these models can predict how temperature increases, snow cover variation, and human activities may impact species like cushion plants, lichens, and alpine shrubs. Such insights will be invaluable for climate adaptation planning, allowing ecologists and policymakers to identify potential refugia (areas that may remain suitable for vulnerable species despite climate shifts). These advanced models will enable proactive conservation strategies, helping safeguard alpine biodiversity in the face of rapid environmental change.

3.5 Data Fusion

Fusion techniques in remote sensing represent a transformative advance for alpine vegetation monitoring and ecological applications. By combining data from multiple sensors like multispectral, hyperspectral, and radar, fusion leverages the strengths of each to overcome their individual limitations, resulting in enhanced accuracy, spatial detail, and spectral information for improved vegetation analysis (Vivone, 2023; Kahraman & Bacher, 2021; Fu et al., 2022).

Multispectral sensors offer relatively high spatial resolution but with fewer spectral bands, suitable for broad land cover discrimination. Hyperspectral sensors provide hundreds of narrow and contiguous spectral bands, ideal for detailed species identification and physiological status assessment but often at coarser spatial resolution. Radar sensors operate in the microwave domain, enabling penetration of clouds and vegetation canopy, providing structural and moisture information independent of lighting conditions. Fusion integrates these complementary data types to produce a fused dataset combining high spatial detail (from multispectral), rich spectral resolution (from hyperspectral), and structural insights (from radar).

Fusion enhances spatial resolution of hyperspectral data, improves classification accuracy for vegetation types and condition, and integrates structural and moisture-related information from radar, facilitating thorough vegetation and habitat characterization. It also mitigates limitations of cloud cover in optical data through radar synergy. State-of-the-art fusion methods include:

- *Component Substitution (CS):* Techniques replace certain multispectral or hyperspectral components with high-resolution spatial components to enhance spatial detail while preserving spectral fidelity.
- *Multiresolution Analysis (MRA):* Wavelet or Laplacian pyramid transforms decompose images into spatial frequency components before fusing spatial details from one image with spectral details from another.
- *Spectral Unmixing-Based Fusion:* Applies linear or nonlinear models to unmix the hyperspectral data into endmembers and abundance matrices and then sharpens abundances with higher-resolution data for accurate reconstruction.
- *Bayesian and Matrix Factorization Techniques:* Use probabilistic models or factorization algorithms (e.g., nonnegative matrix factorization) to optimally combine data while accounting for sensor-specific properties like spectral response and point spread functions.
- *Machine Learning and Deep Learning:* Emerging approaches use convolutional neural networks and other learning models to learn complex nonlinear mappings for fusion tasks, improving fusion accuracy and generalizability across diverse landscapes.

Fusion of multispectral, hyperspectral, and radar data represents a cutting-edge approach in remote sensing, maximizing information extraction from each sensor type. Advanced fusion algorithms, including component substitution, multiresolution analysis, spectral unmixing, and machine learning, enable refined alpine vegetation monitoring, critical for mapping sensitive medicinal plant habitats and comprehensive ecological observations. This integrated approach overcomes individual sensor limitations, delivering improved spatial-spectral resolution and resilience against environmental challenges like cloud cover. Hence, fusion techniques are indispensable tools for high-fidelity, scalable alpine ecosystem management and conservation under environmental change.

3.6 Accuracy Assessment and Validation of AI System

This process involves validating remote sensing results by comparing them with field-based observations or expert-labeled samples. In the context of alpine vegetation, ground truth data might include detailed surveys of plant species, cover types, and structural features collected through fieldwork. By aligning these on-the-ground observations with the classification maps generated from remote sensing data, ecologists can assess the accuracy of the models. This validation step ensures that the vegetation maps are reliable, precise, and suitable for informing conservation decisions, ultimately leading to high-quality, trustworthy classification outputs.

Effective monitoring of alpine vegetation through remote sensing relies on robust preprocessing, advanced classification algorithms, feature extraction methods (vegetation indices, texture), and integration of multi-source data. Techniques such as radiometric/geometric correction, image masking, and machine/deep learning significantly improve map quality and ecological interpretation for alpine environments.

4 Applications of Remote Sensing in Alpine Vegetation

Remote sensing has become essential for in-depth, multi-scale vegetation research and ecosystem management in the demanding conditions of alpine environments. Its development, driven by advanced sensors, innovative algorithms, and integrative analytical approaches, is revolutionizing both scientific investigation and conservation efforts. Some key applications of remote sensing in alpine vegetation include:

4.1 Mapping Vegetation Cover and Species Distribution

Remote sensing is essential for mapping vegetation cover and delineating species distribution in alpine regions, where challenging terrain and environmental conditions complicate traditional survey methods. Recent technological advances enable detailed and large-scale mapping by integrating multi-source satellite and aerial imagery.

High-resolution satellite and aerial images facilitate the discrimination of different vegetation types and even individual species by analyzing their unique spectral and structural characteristics. Automated analysis techniques such as linear discriminant analysis, synthetic feature extraction, and graph-cut algorithms enhance efficiency, allowing for precise mapping at spatial resolutions as fine as 30 m. These methods are validated using drone imagery and commercial high-resolution data to ensure reliability.

Spectral signatures derived from optical and multispectral imagery, often combined through image fusion and sophisticated classification algorithms, improve the accuracy of vegetation mapping. This approach supports the identification of features like alpine grasslines and other habitat boundaries, providing valuable insights into vegetation patterns and ecosystem changes in high-altitude environments.

4.2 Monitoring Vegetation Dynamics and Phenology Changes Due to Climate Shifts

Remote sensing is an essential tool for monitoring temporal changes in alpine vegetation brought about by climate change. By utilizing time-series analysis with repeated observations, researchers can effectively track trends in vegetation greenness, growth stages, and seasonal patterns, shedding light on how warming temperatures, altered precipitation, and other climatic factors influence high-altitude ecosystems. Multi-temporal imagery supports the observation of seasonal dynamics such as growth, flowering, senescence, and long-term shifts in vegetation boundaries, serving as indicators of environmental response to climate signals. Phenological monitoring primarily employs vegetation indices like NDVI, derived from red and near-infrared radiances, to detect changes in photosynthetic activity and plant health over time. Additionally, high temporal resolution data reveal spatial variations in the timing of phenological events across the landscape, providing compelling evidence of climate-induced transformations in alpine environments and aiding in the development of adaptive conservation and management strategies (Liu et al., 2024).

4.3 Assessment of Biodiversity and Habitat Heterogeneity Using Advanced Algorithms

Assessment of biodiversity and habitat heterogeneity in alpine environments has been significantly advanced through the application of cutting-edge algorithms and sophisticated data processing techniques in remote sensing. By leveraging advanced machine learning and statistical methods, researchers can integrate multi-source imagery to identify species-level diversity and ecological gradients across complex landscapes. Classification systems are developed by correlating specific vegetation types with discernible spectral, structural, or thermal characteristics captured in remote sensing data, enabling detailed and accurate habitat mapping. The use of hyperspectral data analysis and the fusion of multiple imagery sources further enhances classification accuracy, allowing for the discrimination of diverse and rare alpine habitats that might otherwise be difficult to distinguish. These objective and reproducible remote sensing approaches facilitate the quantification of habitat heterogeneity, monitoring of rare or sensitive species, and assessment of ecological processes over large and often inaccessible areas. Overall, remote sensing technologies have revolutionized alpine vegetation studies by providing efficient, scalable methods for mapping, monitoring, and ecological assessment, which are crucial for informing long-term conservation and management efforts amidst ongoing environmental changes (Ren et al., 2025).

4.4 Productivity and Ecosystem Function Assessment

Advanced remote sensing models integrate multiple satellite sources such as MODIS and Sentinel-2 with ground measurements and digital terrain models to accurately quantify Gross Primary Productivity (GPP) across diverse alpine grassland communities. By combining data from different sensors, these models leverage the strengths of high spatial and temporal resolution to capture complex ecological dynamics in challenging mountain environments.

Hybrid modeling approaches utilizing vegetation indices like MCARI2 and MSAVI2 enable researchers to estimate light use efficiency effectively. These indices help account for variations in plant health, canopy structure, and soil background, providing insights into the spatial variability of productivity under extreme mountain conditions. Such methods are crucial for representing the wide range of alpine communities and abiotic factors such as slope, aspect, and soil moisture that influence ecosystem function at high altitudes (Vicario et al., 2024).

4.5 High-Precision Fractional Vegetation Cover (FVC)

High-precision fractional vegetation cover (FVC) estimation has been enhanced through the integration of optimized algorithms that leverage multi-dimensional features. By combining UAV-acquired data with satellite imagery and topographical information, researchers can generate highly accurate, pixel-level FVC maps. Modern machine learning techniques, such as optimized random forest models and PI-RFE (Feature Selection), utilize extensive datasets that include spectral bands, numerous vegetation indices (up to 22), and topographical factors to achieve remarkable estimation accuracy, with R^2 values reaching as high as 0.92. These advanced methods are particularly valuable for sensitive and vulnerable sites like the Qinghai-Tibetan Plateau and other fragile alpine meadows, as they demonstrate high sensitivity to sample size and efficiency, enabling precise monitoring and assessment of vegetation cover in challenging environments (Liu et al., 2024).

4.6 Advanced Environmental Monitoring

Phenological dynamics and the impacts of climate change are effectively monitored using multi-temporal datasets such as MODIS NDVI series. These datasets enable researchers to track shifts in key phenological events, including the timing of maximum greenness and spring phenology, in response to rising temperatures. Over periods spanning 20 years or more, such analyses have documented phenomena like the upward migration of treelines and alterations in growth dynamics within alpine ecosystems. These insights are crucial for supporting adaptive management strategies in the face of asymmetric warming patterns, providing robust evidence of climate-related ecological changes and informing conservation and mitigation efforts in sensitive high-altitude environments (Zheng et al., 2022).

4.7 Biodiversity, Structural Complexity, and Habitat Classification

Biodiversity, structural complexity, and habitat classification are significantly enhanced through the integration of spectral, structural, and thermal data. By employing machine learning and data fusion techniques that combine multispectral, hyperspectral, thermal, and LiDAR datasets, researchers can objectively quantify habitat heterogeneity and identify biodiversity hotspots. Advanced algorithms trained on spectral and textural features are capable of discriminating rare or unique communities, facilitating repeatable and reliable assessments even in inaccessible and rugged mountain terrains. This integrated approach improves understanding of ecosystem diversity and supports targeted conservation efforts in complex high-altitude landscapes.

4.8 Decision Support and Resource Management Enhancement

Decision support and resource management in alpine landscapes are greatly advanced through the integration of GIS and AI technologies. Utilizing open-access remote sensing data alongside geographic information systems and artificial intelligence enables capabilities such as glacier danger assessment, snowpack evaluation, and fire event detection. These are key applications for effective landscape management. AI-driven cloud computing workflows enhance the speed, efficiency, and reproducibility of ecosystem monitoring tasks, establishing remote sensing as a vital tool for informed conservation decisions in mountainous regions. This technological synergy facilitates proactive management and helps address the complexities of high-altitude environmental challenges (Jombo et al., 2023).

4.9 Applications in Identifying Microsite Preferences and Refugia for Endangered Species

High-resolution imagery plays a crucial role in identifying microsite preferences and refugia for endangered alpine species by revealing fine-scale vegetation patterns, terrain features, and habitat heterogeneity that provide shelter from harsh conditions. This detailed spatial data enable precise mapping of refugia, facilitating conservation prioritization, and allows for the analysis of habitat attributes such as slope, aspect, soil texture, and snow cover that influence species distribution. Additionally, it supports monitoring small, isolated populations over time, detecting subtle shifts or encroachment, and offers insights into ecological processes like vegetation responses to climate variability and disturbances. Together, these capabilities enhance targeted conservation efforts and deepen understanding of the mechanisms driving alpine biodiversity resilience.

4.10 Water, Snow, and Fire Detection

Remote sensing is instrumental in monitoring vital environmental parameters such as water availability, snowpack levels, land surface temperature, and fire events, all of which directly influence alpine vegetation dynamics. By providing real-time and spatially detailed data, it allows researchers and conservationists to track changes in water extraction, assess snow cover variability crucial for hydrological and ecological processes, and detect the occurrence and extent of fires that can dramatically alter habitat conditions. These insights support proactive management strategies aimed at preserving fragile alpine ecosystems amid climate change and increasing anthropogenic pressures.

5 Conclusion, Limitations and Future Directions

For the monitoring and management of alpine vegetation, remote sensing has emerged as a very effective tool. Using different sensors and remote sensing platforms, it has the capability to acquire the complete details of the alpine vegetation without going physically to remote locations. With the advancement of acquisition tools, high-resolution images are acquired using optical sensors, hyperspectral sensors, and multispectral sensors. These images capture the detailed information of the alpine vegetation at different spectral bands. AI tools including image processing and machine learning algorithms show the capability to process the sensor data and images for removing any type of noise or distortions present. These tools are also used to enhance the visual perception of alpine vegetation. Machine learning tools show their capability to map alpine vegetation for the types of vegetation, to detect phenological changes, and to assess biomass and health. Fusion of data obtained from diverse data sources also shows their potential to provide more precise decisions regarding alpine vegetation. Consequently, remote sensing has become central to efforts aimed at conserving biodiversity, supporting sustainable resource use, and adapting to climate change impacts in alpine environments.

References

Ali, A. M., Darvishzadeh, R., Rafiezadeh Shahi, K., & Skidmore, A. (2019). Validating the predictive power of statistical models in retrieving leaf dry matter content of a coastal wetland from a Sentinel-2 image. *Remote Sensing, 11*(16), 1936.

Awasthi, S., & Varade, D. (2021). Recent advances in the remote sensing of alpine snow: A review. *GIScience & Remote Sensing, 58*(6), 852–888. https://doi.org/10.1080/15481603.2021.1946938

Chen, X., An, S., Inouye, D. W., & Schwartz, M. D. (2015). Temperature and snowfall trigger alpine vegetation green-up on the world's roof. *Global Change Biology, 21*(10), 3635–3646.

Cheshkova, A. F. (2017). A review of hyperspectral image analysis techniques for the detection and classification of the early onset of plant disease and stress. *Plant Methods, 13*(1), 23–33. https://doi.org/10.1186/s13007-017-0233-z

Constantin, A., Fauvel, M., & Girard, S. (2022). Joint supervised classification and reconstruction of irregularly sampled satellite image time series. *IEEE Transactions on Geoscience and Remote Sensing, 60*, 1–13.

Coops, N. C., Morsdorf, F., Schaepman, M. E., & Zimmermann, N. E. (2013). Characterization of an alpine tree line using airborne LiDAR data and physiological modeling. *Global Change Biology, 19*(12), 3808–3821.

Du, K., Shao, Y., Yao, N., Yu, H., Ma, S., Mao, X., Wang, L., & Wang, J. (2025). Alpine meadow fractional vegetation cover estimation using UAV-aided Sentinel-2 imagery. *Sensors (Basel), 25*(14), 4506. https://doi.org/10.3390/s25144506

Fu, B., et al. (2022). Classifying vegetation communities karst wetland synergistic use of image fusion and object-based machine learning algorithm with Jilin-1 and UAV multispectral images. *Ecological Indicators, 141*, 109001.

Guerri, M. F., Sharma, S. R., Melgarejo, A. D., & Koutroumpas, A. (2024). Deep learning techniques for hyperspectral image analysis in agriculture: A review. *Patterns, 4*(3), 100812. https://doi.org/10.1016/j.patter.2024.100812

Jombo, S., Abd Elbasit, M. A. M., Gumbo, A. D., & Nethengwe, N. S. (2023). Remote sensing application in mountainous environments: A bibliographic analysis. *International Journal of Environmental Research and Public Health, 20*(4), 3538. https://doi.org/10.3390/ijerph20043538

Kahraman, S., & Bacher, R. (2021). A comprehensive review of hyperspectral data fusion with lidar and SAR data. *Annual Reviews in Control, 51*, 236–253. https://doi.org/10.1016/j.arcontrol.2021.03.003

Kuchler, P. C., Bégué, A., Simões, M., Gaetano, R., Arvor, D., & Ferraz, R. P. D. (2020). Assessing the optimal preprocessing steps of MODIS time series to map cropping systems in Mato Grosso, Brazil. *Computers and Electronics in Agriculture, 169*, 105184.

Li, C., Wulf, H., Schmid, B., He, J.-S., & Schaepman, M. E. (2018). Estimating plant traits of alpine grasslands on the Qinghai-Tibetan Plateau using remote sensing. *IEEE Journal of Selected Topics in Applied Earth Observations and Remote Sensing, 11*(7), 2263–2275. https://doi.org/10.1109/JSTARS.2018.2824901

Lin, X., et al. (2022). Quantification of alpine grassland fractional vegetation cover retrieval uncertainty based on multiscale remote sensing data. *IEEE Geoscience and Remote Sensing Letters, 19*, 1–5. https://doi.org/10.1109/LGRS.2021.3109725

Liu, Y., Wang, W., Mu, X., Zhou, H., & Wang, L. (2024). Spatiotemporal changes and driving factors of alpine land cover in Tianshan world natural heritage sites. *Scientific Reports, 14*(1), 12345. https://doi.org/10.1038/s41598-024-71788-3

Meng, B., et al. (2020). Modeling alpine grassland above-ground biomass based on remote sensing data and machine learning algorithm: A case study in east of the Tibetan Plateau, China. *IEEE Journal of Selected Topics in Applied Earth Observations and Remote Sensing, 13*, 2986–2995. https://doi.org/10.1109/JSTARS.2020.2999348

Ren, Z., Zhang, L., Wang, Q., Hu, W., & Shi, Z. (2025). A novel framework for assessing shrublines and their geophysical constraints in alpine regions through probabilistic vegetation mapping and seed-filling algorithm. *Ecological Indicators, 177*, 113788. https://doi.org/10.1016/j.ecolind.2025.113788

Vaglio Laurin, G., Del Frate, F., Pasolli, L., Notarnicola, C., Guerriero, L., & Valentini, R. (2013). Discrimination of vegetation types in alpine sites with ALOS PALSAR-, RADARSAT-2-, and lidar-derived information. *International Journal of Remote Sensing, 34*(19), 6898–6913.

Vicario, S., Magnani, M., Adamo, M., Vivaldo, G., Richiardi, C., Giamberini, M., & Provenzale, A. (2024). Predicting the productivity of alpine grasslands using remote sensing information. *EGUsphere* [Preprint]. https://doi.org/10.5194/egusphere-2023-2824

Vivone, G. (2023). Multispectral and hyperspectral image fusion in remote sensing: A survey. *Information Fusion, 89*, 405–417. https://doi.org/10.1016/j.inffus.2022.08.032

Wang, D., Yang, F., Jia, S., Wang, Z., Dong, C., Lang, M., et al. (2025). Alpine wetlands information extraction using optimized multifeatures and random forest algorithm. *IEEE Journal of Selected Topics in Applied Earth Observations and Remote Sensing., 18*, 7347–7363.

Wu, C., Niu, Z., Tang, Q., & Huang, W. (2008). Estimating chlorophyll content from hyperspectral vegetation indices: Modeling and validation. *Agricultural and Forest Meteorology, 148*(8–9), 1230–1241.

Wu, J., Chen, B., Reynolds, G., Xie, J., Liang, S., O'Brien, M. J., & Hector, A. (2020). Monitoring tropical forest degradation and restoration with satellite remote sensing: A test using Sabah Biodiversity Experiment. In *Tropical ecosystems in the 21st century* (Vol. 62, p. 117).

Xie, Y., Sha, Z., & Yu, M. (2008). Remote sensing imagery in vegetation mapping: A review. *Journal of Plant Ecology, 1*(1), 9–23. https://doi.org/10.1093/jpe/rtm005

Xu, D., & Guo, X. (2015). Some insights on grassland health assessment based on remote sensing. *Sensors, 15*(2), 3070–3089.

Yin, J., et al. (2020). Estimation of grassland height based on the random forest algorithm and remote sensing in the Tibetan Plateau. *IEEE Journal of Selected Topics in Applied Earth Observations and Remote Sensing, 13*, 178–186. https://doi.org/10.1109/JSTARS.2019.2954696

Zheng, L., Li, D., Xu, J., Xia, Z., Hao, H., & Chen, Z. (2022). A twenty-years remote sensing study reveals changes to alpine pastures under asymmetric climate warming. *ISPRS Journal of Photogrammetry and Remote Sensing, 190*, 69–78. https://doi.org/10.1016/j.isprsjprs.2022.06.001

Next Generation Sequencing (NGS) and Genomics in Alpine Plants

6

Ankush Saini, Neha Sharma, Rasna Zinta, Bhawna Dipta, Nidhi Sharma, Muskaan Sharma, and Neha Kumari

Abstract

Alpine regions, characterised by naturally treeless belts above the montane zone, harbour unique plant communities comprising dwarf shrubs, rosette herbs, cushion species and tussock-forming graminoids. These plants exhibit specialised adaptations, including small stature, near-ground meristem protection and rapid seasonal growth, which allow survival under extreme climatic conditions. They contribute significantly to biodiversity, demonstrate high levels of endemism and provide models for evolutionary research due to frequent hybridisation, polyploidy and niche differentiation. Next-generation sequencing (NGS) technologies, including Illumina, Roche 454, Sequencing by oligonucleotide ligation and detection (SOLiD), PacBio single-molecule real-time (SMRT) and Oxford Nanopore, have revolutionised alpine plant research by enabling high-throughput genome sequencing, transcriptomics, methylation mapping and small RNA analyses. NGS facilitates marker development, population genomics, phylogeography, microbiome profiling, medicinal plant functional genomics and endophytic

A. Saini (✉)
ICAR-Central Potato Research Institute, Shimla, Himachal Pradesh, India

N. Sharma (✉) · R. Zinta · B. Dipta
ICAR-Central Potato Research Institute, Shimla, Himachal Pradesh, India

N. Sharma
MS Swaminathan School of Agriculture Sciences, Shoolini University, Solan, Himachal Pradesh, India

M. Sharma
School of Biotechnology, Shoolini University, Solan, Himachal Pradesh, India

N. Kumari
School of Biological and Environmental Sciences, Shoolini University, Solan, Himachal Pradesh, India

P. Kumar et al. (eds.), *Climate Resilience and Molecular Adaptation in Alpine Medicinal Plants*, https://doi.org/10.1007/978-981-95-6923-6_6

community characterisation, providing insights into adaptation, evolutionary dynamics and ecosystem functioning. Public genomic repositories such as The International Nucleotide Sequence Database Collaboration (INSDC), GenBank, DNA Data Bank of Japan (DDBJ) and ENA ensure permanent accessibility of genome assemblies, organellar sequences, transcriptomes and functional annotations, supporting comparative genomics and conservation strategies. Despite these advances, challenges remain, including short-read limitations, complex polyploid genomes, sequencing errors, data processing demands and climate-driven habitat shifts. Integrating genomic tools with ecological and remote-sensing approaches is essential to understand alpine plant diversity, adaptation and resilience under rapidly changing environments, thereby guiding conservation and management efforts.

Keywords

Illumina · Roche 454 pyrosequencing · SOLiD · PacBio · Oxford Nanopore · Genome assembly · Transcriptome analysis · Bioinformatics

1 Introduction

The alpine region refers to the naturally treeless belt situated above the montane zone, demarcated by the high-elevation climatic treeline. Within this zone, alpine plants are characterized by their inherently small stature, primarily comprising dwarf shrubs, tussock-forming graminoids, rosette herbs and cushion species. Their distinctive architecture reflects evolutionary adaptations that allow them to aerodynamically decouple from harsh atmospheric conditions, safeguard apical meristems near or below the soil surface and ensure rapid seasonal growth and reproduction in a challenging environment (Körner, 2023). Ecologically, alpine plants sustain biodiversity by linking the physical environment with physiological and life history processes, where microclimatic influences shape their persistence, while human activities such as land-use change and global warming increasingly disrupt these delicate systems (Johnstone, 2021). Furthermore, alpine plants exhibit exceptionally high levels of vascular plant endemism, with 569 endemic taxa (including 43 subendemics) forming about 62% of the alpine flora, primarily inhabiting screes, rocks and thorn-cushion grasslands, with their richness peaking near 3000 m above sea level, thereby emphasizing the need for continued botanical exploration and taxonomic revision (Noroozi et al., 2016). In addition to their endemism and ecological uniqueness, alpine plants form historically isolated, disturbance resistant communities that are now threatened by rapid climate warming and intensified human influence, as many native and nonnative species shift to higher elevations and latitudes, creating novel assemblages that challenge biodiversity conservation and require prevention, monitoring and adaptive management (Rew et al., 2020). Connectivity dynamics in alpine ecosystems, shaped by pronounced climatic fluctuations over the past ~2.6 million years, indicate that allopatric speciation and range fragmentation alone cannot fully explain the accelerated build-up of endemism and species richness in mountains (Muellner-Riehl, 2019). Most plant species thus responded to temperature fluctuations by altering their ranges, with cold-adapted taxa and alpine communities descending during glacial phases and ascending during interglacials. Depending on mountain morphology, these climatic shifts either expanded habitat availability or enhanced connectivity for alpine flora, indicating that diversification driven by connectivity may have played a greater role in shaping mountain biodiversity than isolation alone (Muellner-Riehl, 2019). Alpine plants also serve as important models in evolutionary research: phylogenomic evidence indicates multiple colonizations (probably from the late Miocene), frequent homoploid hybridization with subsequent introgression and widespread polyploidy ($\approx$40% of species), thereby illuminating speciation, biogeography and genome evolution (Wagner et al., 2021). Alpine plants highlight key evolutionary processes, as repeated colonisations, overlapping distributions enabling frequent hybridisation, ecological niche differentiation across elevation and habitat gradients and genomic barriers such as ploidy variation collectively influence speciation, while only a few hybrid lineages persist to form populations of lasting evolutionary relevance (Wagner et al., 2021). Alpine plants are invaluable for evolutionary studies as they mirror the global range of plant functional traits, with a predominant emphasis on resource use strategies and exhibit strong functional convergence across regions. Macroclimate exerts only a limited effect, mainly at the community level, while environmental and phylogenetic dissimilarities contribute equally to functional differences, leaving alpine vegetation largely independent of the surrounding vegetation zones (Testolin et al., 2021). Alpine plants are also important in terms of productivity, as their net primary production, averaging about 2.2 g dry matter m^{-2} per day or nearly 185 g m^{-2} over an 84-day growing season, can equal or even exceed that of humid lowland vegetation, underscoring the pivotal role of growth timing and ground cover in shaping alpine ecosystem functioning (Körner, 2023). Studying alpine flora is challenging because available land area decreases with elevation and low temperatures constrain species, while multiple interacting factors such as altered snowpack, moisture regimes, heatwave-induced drought, nitrogen deposition and habitat modification through treeline advance and glacier retreat further affect development, growth and survival. Confined above the climatic treeline, alpine plants adapt to such extremes by maintaining small stature, synchronised seasonal development and regulating near-ground microclimates, with their diversity shaped by a mosaic of adjacent thermal niches that exceed projected warming scenarios, although habitat contraction from treeline expansion is expected to far outweigh the global land gains from glacier retreat (Körner & Hiltbrunner, 2021). Additional challenges in studying alpine plants arise from habitat insularity and environmental heterogeneity, which limit accessibility and obscure comprehensive understanding of species distributions and community dynamics (Hotaling et al., 2017). Furthermore, habitat isolation, small and scattered populations and extreme physiological stresses limit species accessibility and complicate ecological assessments (Rundel & Millar, 2016). The advent of next-generation sequencing enables the detection of cryptic alpine plant lineages by capturing genome-wide variation, revealing genetic and ecological differentiation even in the absence of morphological differences (Mikulášková et al., 2024). Integrating genomic tools can refine knowledge of evolutionary processes and improve conservation strategies in these complex ecosystems (Hotaling et al., 2017). Genomic tools, including population genomics, enable the assessment of genetic variation, local adaptation and climate response in alpine plant species, providing insights for conservation and management (Long et al., 2025). The use of next-generation sequencing and genomic tools enables the detection of cryptic alpine plant lineages, assessment of genetic variation, local adaptation and climate response, thereby refining evolutionary understanding and guiding conservation strategies.

2 NGS Technologies

A suite of innovative sequencing platforms, collectively referred to as "next-generation sequencing" (NGS), has revolutionized life sciences by enabling high-throughput analyses while substantially reducing the time, effort and cost of sequencing. NGS has driven a scientific breakthrough, evidenced by numerous genome publications across diverse organisms, despite inherent challenges such as large genome sizes, high GC content and homopolymer issues. Strategies to overcome these limitations have been developed and sequencing techniques refined, sustaining this ongoing scientific trend. In plants, the most extensive application of NGS is whole genome sequencing (WGS), aimed at uncovering entire plant genome sequences and their genetic composition (Hodzic et al., 2017). The rise of high-throughput next-generation sequencing has shifted polymorphism studies from fragment-based methods to sequence-oriented SNP detection, thereby enhancing marker availability for population genetics and accelerating the construction of dense genetic maps along with the recognition of genes and quantitative trait loci (QTLs). The steady decline in sequencing expenses has also supported genomic consortia in releasing reference genomes for a broad spectrum of plant species (Martina et al., 2023). Genome sequencing technologies have witnessed rapid refinement, marked by reduced costs, greater data output and improved speed. The limitations of first-generation dideoxy sequencing in handling large-scale, high-quality genome projects led to the emergence of second-generation approaches. These next-generation sequencing (NGS) platforms enabled the parallel analysis of millions of nucleic acid sequences, fundamentally transforming sequencing practices. Advances in this domain introduced systems such as Illumina/Solexa, Roche 454 and ABI SOLiD, which bypassed the reliance on electrophoretic visualization with radioactively or fluorescently labelled nucleotides, thereby offering a more efficient and scalable alternative for genome analysis (Hong et al., 2023).

2.1 Illumina Sequencing-By-Synthesis

Illumina, a leading second-generation platform, initiates sequencing with DNA fragmentation, end-repair, adapter ligation, surface immobilisation and in-situ amplification. As a short-read technology, it enables massively parallel sequencing of millions of reads that must be reassembled across long genomic regions, which poses challenges in resolving structural variants and low-complexity sequences. Illumina's sequencing-by-synthesis employs fluorescently labelled reversible-terminator chemistry, where clonal clusters are generated on the flow cell through bridge amplification. Each cycle incorporates a single labelled dNTP, imaged optically, followed by cleavage of the terminator and dye to permit the next addition. Paired-end sequencing, reading both ends of DNA fragments, ensures deep coverage and accurate phasing of distal polymorphisms, with MiSeq commonly using 350–500 bp fragments, though longer fragments (600–900 bp or more) are preferred for extended phasing. This technology offers high accuracy with an inaccuracy level of ~0.1%, predominantly substitution errors. Illumina's portfolio includes benchtop sequencers (iSeq, MiniSeq, MiSeq, NextSeq 550, NextSeq 1000, NextSeq 2000) and diagnostic systems (MiSeqDx, NextSeq 550Dx). While MiSeq runs may extend up to 56 h using a 2 × 300 bp kit, newer models such as MiniSeq, NextSeq 550 and NovaSeq 6000 employ two-channel SBS, minimizing data processing and sequencing time while sustaining quality (Hu et al., 2021). Illumina/Solexa platforms represent the most widely used NGS technology in plant genome sequencing, employed either as a sole platform in 40.42% of published genomes or in combination with other sequencing methods in 48.64% of studies. Their broad applicability across genome sizes, affordability and widespread availability make them the preferred choice for plant genome projects, including the estimated sequencing of *Helichrysium arenarium* in regions with limited resources (Hodzic et al., 2017). The Solexa approach, later taken over by Illumina and appearing after 454 sequencing, employs sequencing-by-synthesis (SBS) founded on reversible terminal termination and DNA clustering. DNA fragments are immobilised on the flow cell via cleaved adapters and complementary oligonucleotides rather than by bead-based PCR and solid-phase PCR generates bridged, duplicated templates that form clusters through bridge amplification. Sequencing uses fluorescent reversible-terminator dNTPs that block the 3′ hydroxyl and prevent extension until the fluorophore is eliminated prior to polymerization. Modified dNTPs and DNA polymerases are then introduced in a cyclic manner and washed across single-stranded, surface-bound clusters while a CCD detects activated fluorophores. Subsequently, enzymatic cleavage of the blocked fluorescent moiety advances the reaction to the next cycle. Illumina commercialised Solexa in 2006 with the Initial Genome Analyzer (GA) and later released Hiseq models including the X-ten system, Hiseq2000, Hiseq2500, Hiseq3000 and Hiseq4000, subsequently adding the compact, lower-throughput Miseq and the high-throughput Nextseq; Miseq offers lower cost, faster turnaround and longer read lengths (Hong et al., 2023). Furthermore, Illumina's Nextera technology provides an substitute library-preparation approach using on bead tagmentation, which combines DNA normalisation, fragmentation and size selection. Subsequently to tagmentation, a limited PCR incorporates sequencing adapters and barcodes for sample indexing. The workflow is rapid and straightforward, yielding sequencing-ready libraries in under 90 min with less than 15 min hands-on time (Hu et al., 2021).

2.2 Roche 454 Pyrosequencing

This technique is a sequencing-by-synthesis method, utilises water-in-oil PCR and pyrosequencing chemistry. PCR is performed in small wells with each cycle adding a single dNTP; incorporation releases pyrophosphate which, via ATP sulfate chemase, yields ATP. Oxidation of the newly generated ATP by luciferase produces light that is recorded by a CCD camera. Exact measurement of homopolymer lengths remains difficult with this technology, causing potential errors. Owing to such accuracy limitations that constrained platform upgrades and application scope, Roche discontinued its 454 business in 2013 (Hong et al., 2023). Roche/454, though offering the longest read lengths and greatest speed among NGS platforms, has not been employed as a standalone tool and is used only in combination with other methods (for example with Sanger sequencing in five projects), accounting for 25.53% of published plant genome studies that employed mixed platforms. This limited use is presumably due to its high per-run cost, estimated at USD 80,000–180,000 for the *Helichrysium arenarium* genome size range, whereas Illumina/Solexa, being more affordable and more widely available, remains the preferred platform (Hodzic et al., 2017). Thus, Illumina sequencing, with its high accuracy, broad applicability and cost-effectiveness, holds strong potential for studying alpine plants; however, challenges associated with short-read assemblies, structural variants and extended run times highlight the need for continued advancements such as longer read lengths, improved chemistries and faster platforms to better resolve complex, resource-limited alpine genomes.

2.3 SOLiD

SOLiD operates on the principle of "sequencing while ligating," utilising water-in-oil PCR and ligase-based chemistry. Library preparation mirrors the Solexa approach by attaching DNA to beads; beads, PCR reagents and oil form isolated microreactors that enhance template amplification. In the subsequent sequencing step, beads are deposited onto a SOLiD slide for detection and decoding depends on fluorescent probes ligating to target sequences to emit signals that reveal base order. The platform attains ~99.99% accuracy, surpassing earlier second-generation methods, through dual-reading and linked chemistry that mitigate PCR-related issues in GC-rich regions, with reports of improved performance on high-GC samples. From 2004 to 2015, DNA sequencing capacity rose far faster than Moore's Law: while transistor density typically doubles every 2 years, sequencing output doubled about every 5 months. The distinct functional and chemical attributes of these technologies thus provide researchers with multiple platforms to suit specific experimental needs (Hong et al., 2023). ABI/SOLiD had the lowest representation, appearing in only two publications (4.25%) and used exclusively in combination with other platforms such as Roche/454 and Illumina/Solexa; it was not employed as a standalone tool. For the estimated *Helichrysium arenarium* genome size, the calculated cost for a single ABI/SOLiD run was USD 1040–2340 and platform combinations were a common practice in published studies (Hodzic et al., 2017). Thus, SOLiD technology, with its exceptionally high accuracy and capacity to handle GC-rich sequences, holds potential for studying alpine plants, though its limited standalone use, high costs and reliance on platform combinations highlight the need for continued advancements in reducing expenses and simplifying workflows to broaden its applicability in such genomic research.

2.4 Pacific Biosciences SMRT

Pacific Biosciences's (PacBio's) SMRT sequencing, a third-generation advance, obviates PCR and enables sequencing of very long nucleic acid molecules, thereby substantially reducing single-base errors and overcoming earlier constraints. SMRT is unbiased, unaffected by palindromic sequences and resistant to human-induced alterations; it accurately detects mutations, lowers false positives, ensures even coverage and permits direct methylation detection. The method employs zero-mode waveguides (ZMWs), nanoscale apertures that allow DNA polymerase to incorporate labelled dNTPs base-by-base in real time, with fluorescence from incorporated nucleotides detected and extinguished by label removal. A salient technical advantage is the capture of polymerase kinetic data during sequencing, which facilitates base-modification detection and supports de novo genome assembly, thus accelerating single-molecule sequencing while providing detailed information on genetic variation and structure (Hong et al., 2023). PacBio SMRT sequencing employs circular SMRTbell templates and zero-mode waveguides (ZMWs), where a single DNA polymerase incorporates fluorescently labelled nucleotides in real time, producing continuous long reads (CLRs). Multiple passes generate subreads that can be collapsed into highly accurate circular consensus sequences (CCS/HiFi), with Sequel II platforms achieving up to 99.8% accuracy and reliable detection of SNVs, indels and structural variants. HiFi reads, though shorter than the longest PacBio reads (>60 kb), remain considerably longer than short-read methods (~13.5 kb). PacBio also enables Iso-Seq for full-length transcript analysis. Data analysis uses tools such as HGAP, Falcon, Canu, Sprai, Celera Assembler, MHAP and the SMRT Analysis suite, with additional pipelines like SQANTI, TAMA and LoReAn for RNA. Advantages include very long reads, real-time sequencing, PCR-free libraries and detection of base modifications (m6A, m4C). Earlier high error rates (~14%) have been mitigated by HiFi sequencing, though high costs and large instrument size remain limitations. Use of unique molecular identifiers (UMIs) has further reduced consensus error rates to as low as 0.0041% (Hu et al., 2021). PacBio SMRT sequencing, with its ability to generate long, high-accuracy reads and detect structural variants and base modifications, holds strong potential for exploring alpine plant genomes; however, challenges such as high cost and large instrument size limit broader application, which may be mitigated in future through advances in efficiency, portability and cost reduction.

2.5 Oxford Nanopore Sequencing

Oxford nanopore sequencing, a concept dating to the 1980s, detects signal changes as biomolecules traverse microscopic pores, thereby enabling single molecule sensing and analysis. ONT's MinION, released in 2014, exemplifies this single molecule nanopore platform. A suite of tools now supports base detection, data processing, read mapping, de novo assembly and variant discovery; examples include Poretools for quality control and format conversion, Genopo for Android-based nanopore analysis, SquiggleKit for signal extraction and visualization and ModPhred for embedding DNA and RNA modification information into FASTQ and BAM files for easier inspection. By simplifying access to genetic and epigenetic variation and their roles in gene expression, nanopore sequencing has diverse applications in plant genomics (Hong et al., 2023). ONT sequences single-stranded DNA/RNA by translocating molecules through protein nanopores such as α-hemolysin, assisted by adapter ligation, a 5′ motor enzyme and applied ion current. Each nucleotide generates a distinct current disruption, enabling read lengths of 10–100 kb, ultra-long reads over 100 kb and stitched reads exceeding 2 Mb. Raw accuracy ranges from 87–98%, with homopolymer resolution around 91–93% for ≥5-base runs; base-calling algorithms largely determine performance. Devices include MinION (portable, 2–20 Gb), GridION (five flow cells, 2–20 Gb each), PromethION (up to 48 flow cells, 50–100 Gb per flow cell) and Flongle (low-throughput, 126 nanopores). Data processing involves MinKNOW and FAST5 output with base-callers such as Albacore, Guppy and Scrappie. Strengths include long-read sequencing, real-time analysis, structural variant resolution, portability and low cost, while limitations involve error rates of 2–15%, systematic biases and dependence on enzyme kinetics and base-calling, though advances in pore design and algorithms are improving accuracy (Hu et al., 2021). Thus, Oxford Nanopore sequencing, with its capacity for ultra-long reads, real-time analysis, portability and affordability, offers great promise for exploring the complex genomes of alpine plants, though challenges such as error rates and systematic biases remain; ongoing refinements in pore design, enzyme kinetics and base-calling algorithms provide effective strategies to overcome these limitations and enhance its future utility.

Collectively, NGS techniques including Illumina, Roche 454, SOLiD, PacBio SMRT and Oxford Nanopore have each contributed distinct strengths to plant genomics, ranging from Illumina with its high accuracy and affordability, Roche 454 with longer reads, SOLiD with exceptional precision in GC rich regions, PacBio with highly accurate long reads and base modification detection, to Oxford Nanopore with real time ultra long and portable sequencing, as illustrated in figure (Fig. 6.1). For alpine plant research, Illumina remains the most widely adopted due to its cost effectiveness and accessibility, while PacBio and Oxford Nanopore provide superior capabilities for resolving complex large or repetitive genomes despite limitations of cost, instrument size and error rates. Future advancements in extending read lengths, reducing costs, improving base calling algorithms and enhancing portability will be important for overcoming these challenges and unlocking deeper insights into the unique adaptations and genomic complexity of alpine plants.

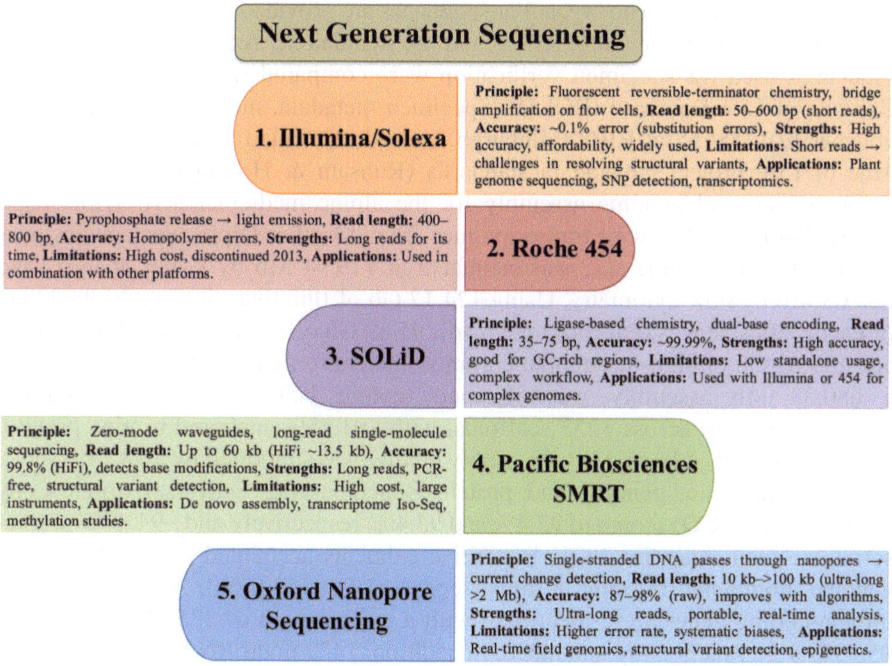

Fig. 6.1 Key NGS technologies with principles, read lengths, accuracy, strengths, limitations and applications in plant genomics

3 Genomic Resources and Databases

INSDC is a long-standing global initiative that ensures unrestricted public access to primary nucleotide-sequence data. Since the early 1980s, it has implemented uniform standards, formats and submission protocols, along with the internationally recognised accession-number system for journals and data contributors. Its three partners, the DDBJ, the European Nucleotide Archive (ENA) and GenBank, capture, exchange on a daily basis and preserve comprehensive records ranging from raw next-generation sequencing reads to assemblies, annotations and associated sample or project information, accommodating vast data volumes (about 2400 trillion bases in 2015). The INSDC policy makes submission a prerequisite for publication, places accountability for accuracy on authors and ensures data permanence and quality through standardised vocabularies such as the Feature Table Definitions and the INSDC country list, structured missing-value reporting and validator-supported submission systems (Cochrane et al., 2016). This open and permanent access to nucleotide sequence archives enables alpine researchers to plan experiments, evaluate published results and utilise the data as an enduring scientific resource. Genomic resources for *Astragalus alpinus* include PacBio HiFi assemblies, Hi-C scaffolded chromosomes, RNA-seq data and the assembled mitochondrial genome (313.83 kb) and plastid genome (123.2 kb), accessible through public sequence repositories. The organellar sequences are provided both as contigs within the multifasta genome submission and as standalone records. DNA barcode sequences used for specimen verification were compared with GenBank and the Barcode of Life Database (BOLD). Specimen metadata, including the herbarium voucher E01358555 at RBG Edinburgh, were documented following the Darwin Tree of Life project's metadata standards (Ruhsam & Hollingsworth, 2025). A chromosome-level genome assembly for the alpine medicinal herb *Triplostegia glandulifera*, together with transcriptomes from 13 other Dipsacales species, was generated. The genome size was estimated at 719.94 Mb by k-mer analysis and 668.05 Mb by flow cytometry. Using 124.17 Gb of Illumina short reads, 98.30 Gb of Pacific Biosciences (PacBio) long reads, 95.89 Gb of 10x Genomics linked reads and 149.61 Gb of high-throughput chromosome conformation capture (Hi-C) data, a 680.38 Mb assembly was obtained (contig N50 = 1.81 Mb; scaffold N50 = 66.68 Mb) across 1233 scaffolds, with ~91.25% anchored to nine pseudomolecules (36.35–110.58 Mb). Assembly quality was supported by a 98.14% short-read mapping rate, genome and protein Benchmarking Universal Single-Copy Orthologs (BUSCO) scores of 97.4% and 95.4%, respectively and ~94.47% of transcriptome reads mapping back. Repetitive elements accounted for 62.88% of the genome, dominated by long terminal repeats (LTRs, 52.09%). A total of 32,123 protein-coding genes were identified, with a mean length of 3351 base pairs and 95.53% were functionally annotated. In addition, 1427 microRNAs (miRNAs), 755 transfer RNAs (tRNAs), 6905 ribosomal RNAs (rRNAs) and 552 small nuclear RNAs (snRNAs) were identified (Zhang et al., 2024). Together, these genomic resources and databases provide a permanent, standardised and accessible foundation for advancing alpine plant research and facilitating future evolutionary, ecological and functional studies.

Moreover, the reference genomes of 2-alpine species, *Rheum alexandrae* (Polygonaceae) and *Saussurea obvallata* (Asteraceae), were assembled (61,463 and 37,938 protein-coding genes, respectively) and integrated with five additional alpine genomes to probe convergence in high-altitude adaptation. Comparative analyses revealed convergent contraction of disease-resistance gene families, signatures of positive selection on respiration and reproduction genes (e.g., MMD1, NBS1, HPR) and molecular convergence in loci for stress tolerance, DNA repair, cell-wall modification and self-incompatibility. Transcriptomic evidence of higher expression of flavonoid biosynthetic and cuticular-wax genes in leafy bracts supports a convergent "greenhouse" morphology, together illuminating shared genomic routes to alpine survival (Zhang et al., 2023). While next-generation sequencing has reduced genome sequencing costs, de novo assembly is limited by short reads, repetitive sequences, polymorphisms and errors. An extended reference-guided de novo approach, which incorporates information from a related genome, consistently outperforms standalone de novo methods, even across species and under low coverage. This pipeline integrates quality trimming, reference mapping, block partitioning, local assemblies, redundancy removal, consensus building and scaffolding. Since single indices like N50 can be misleading, evaluation based on multiple statistics provides a more reliable measure of both continuity and accuracy (Lischer &

Shimizu, 2017). Collectively, these genomic assemblies and methodological advances highlight both the shared evolutionary strategies of alpine plants and the effectiveness of integrated reference-guided approaches in overcoming assembly challenges, thereby providing a robust framework for studying adaptation in extreme environments.

4 Applications of NGS in Alpine Plant Studies

NGS systems rapidly sequence whole genomes in the form of millions of short DNA fragments in an economical, high-throughput manner. In plants, this technique has been extensively utilised for *de novo* genome sequencing, for resequencing to characterise genetic variation, for methylation sequencing to map DNA-methylation patterns and for transcriptome and small-RNA sequencing, thereby enabling a wide range of functional and comparative genomics analyses (Wei et al., 2013). These NGS approaches have been increasingly applied to diverse alpine flora, providing critical insights into their genetic diversity, evolutionary history and adaptive mechanisms in harsh mountain environments.

4.1 Population Genomics and Marker Development

Next-generation sequencing has been employed to develop genetic markers for invasive alpine plants, exemplified by *Heracleum persicum* in Nordic regions. A microsatellite library was generated and 164 primer pairs were designed and validated using a cost-effective PCR strategy. Of the 30 markers screened across geographically diverse *H. persicum*samples, 25 were polymorphic, showing 2–4 alleles per locus observed and expected heterozygosity ranging from 0.0–1.0 and 0.06–0.84, respectively. Cross-species amplification with related *Heracleum* taxa and *Anthriscus sylvestris*was highly efficient (84–100%), with a substantial proportion of markers remaining polymorphic. Genetic ordination revealed clear structure among *Heracleum* taxa, demonstrating the utility of these NGS-derived markers for studying taxonomy, population genetics, phylogeny and informing invasive species management (Rijal et al., 2015). Thus, next-generation sequencing-derived markers provide a robust tool for investigating taxonomy, population structure and phylogenetic relationships in alpine plants, while also aiding the management of invasive species.

4.2 Microbiome Studies and Rhizosphere Analysis

The rhizosphere of *Leontopodium nivale* subsp. *alpinum* (Edelweiss) harbours nutrient-driven microbial diversity, including Actinobacteria known for bioactive metabolite production. From pre-treated rhizosphere soil, 77 isolates were identified by 16S rRNA barcoding, spanning several genera such as Streptomyces and

Micromonospora. Parallel metagenomic 16S rRNA amplicon sequencing (Illumina HiSeq2500, V3–V4 region) across roots, rhizomes, leaves and inflorescences generated 30,000–50,000 reads per tissue. Data processing with the mothur MiSeq SOP confirmed overlap between isolates and metagenomic OTUs, suggesting plant-mediated recruitment of soil bacteria, while also revealing uncultured actinobacterial groups enriched in belowground tissues (Oberhofer et al., 2019). Thus, next-generation sequencing effectively uncovers the taxonomic diversity and tissue-specific distribution of rhizosphere and endophytic microbiota in alpine plants, providing insights into microbial recruitment and hidden actinobacterial lineages.

4.3 Population Structure and Phylogeography

Next-generation sequencing was employed to explore the population structure and genetic diversity of alpine clematis (*Clematis alpina* subsp. *alpina*) across its European range. Using sequence capture of 1429 orthologous loci, including Angiosperms353 and Ranunculaceae-specific targets, 148 samples were sequenced on an Illumina NextSeq 1000 platform. Reads were processed and filtered to generate high-quality consensus sequences, which were used for variant calling, SNP pruning and downstream analyses. Population structure analyses, including FastStructure, PCA and coalescent-based phylogenetic inference, revealed four genetically distinct groups corresponding to subspecies and geographically segregated populations, with Prealps populations forming a unique cluster. Measures of nucleotide diversity, heterozygosity, FST, AMOVA and isolation-by-distance tests highlighted both divergence and admixture among populations. These results demonstrate that NGS-based sequence capture provides a robust approach to assess intraspecific variation, identify unique marginal lineages and inform the conservation and evolutionary study of alpine plant species (Stefani et al., 2025). Next-generation sequencing approaches, such as pooled genotyping-by-sequencing (GBS), provide a rapid and cost-effective means to assess genetic diversity and population structure in alpine flora. Application of GBS across multiple plant species from Australian alpine summits revealed deep evolutionary splits between regions, consistent with patterns observed in other taxa, while also highlighting higher local connectivity in plants. These genome-wide analyses enable characterization of lineage diversification and endemism, offering valuable insights for conservation planning and understanding evolutionary processes at both multispecies and landscape scales (Bell et al., 2018). Next-generation sequencing-based sequence capture offers a powerful tool to resolve population structure, detect unique lineages and evaluate genetic diversity in alpine plants, thereby supporting conservation and evolutionary research.

4.4 Medicinal and Functional Genomics

Next-generation sequencing and other advanced sequencing technologies have transformed the study of medicinal plants by enabling comprehensive identification of genes engaged in the biosynthesis of active compounds. Transcriptome (RNA-seq) analyses have elucidated biosynthetic pathways and the regulatory roles of miRNAs, while in silico studies have revealed the potential inhibitory effects of plant-derived compounds on pathogenic genes. Coupled with analytical techniques such as spectroscopy and chromatography for metabolite validation, these molecular approaches provide powerful tools to explore and characterize the bioactive potential of traditionally important, yet understudied, medicinal plants (Singh et al., 2023). The application of these sequencing and analytical methods in medicinal plants provides a valuable framework for investigating the genetic and metabolic potential of traditionally important, yet unexplored, alpine medicinal species.

4.5 Pollen and Biodiversity Studies

Next-generation sequencing, combined with classical microscopic pollen analysis, has enabled high-resolution identification of airborne pollen in alpine habitats. By constructing reference databases and optimizing DNA extraction and metabarcoding protocols, NGS revealed greater taxonomic detail than microscopy, distinguishing genera and seasonal diversity patterns. These approaches provide valuable tools for biodiversity assessments, floral surveys and monitoring vegetation changes in alpine ecosystems, complementing traditional methods (Leontidou, 2018). Next-generation sequencing enhances pollen-based biodiversity assessments in alpine ecosystems, providing finer taxonomic resolution and complementing traditional microscopic methods for effective vegetation monitoring.

4.6 Genomic Resources and Adaptation Studies

Next-generation sequencing has facilitated the assembly of *Arabis alpina* genomic fragments, covering 50% of its genome and capturing 96% of conserved eukaryotic genes. Comparative analysis revealed high sequence identity with 77% of Arabidopsis thaliana genes across functional categories. Resequencing of 16 samples identified 127,041 single-nucleotide polymorphisms, providing a valuable resource to study the ecological genomics and local adaptation of this alpine plant (Lobréaux et al., 2014). Next-generation sequencing, specifically 454 sequencing of barcoded amplicon mixtures, enabled the analysis of polyploid Australian alpine Poa species. Using 64 barcodes, three chloroplast and two nuclear gene regions were sequenced across 60 individuals of 11 species, capturing all homeologous nuclear alleles with high accuracy. This method revealed extensive allele sharing, detected hypermutation in the trnH-psbA marker and identified spatial genetic structure, demonstrating its utility in phylogenetic and phylogeographic studies of alpine plants (Griffin et al., 2011). Similarly, 454 pyrosequencing enabled the development of 19 polymorphic microsatellite primers for the alpine perennial *Arabis alpina*. Tested on 60 individuals from three Swiss Alpine sub-populations, these primers successfully amplified di- and trinucleotide repeats with two to five alleles per locus. This approach overcame the limitations of previous enrichment-library and cross-amplification methods, providing a reliable tool to advance *A. alpina* as a model for eco-genomic research (Buehler et al., 2011). Next-generation sequencing provides a robust framework for generating genomic resources, identifying genetic variation and developing molecular markers, thereby advancing the study of ecological genomics, local adaptation and phylogenetic relationships in alpine plants.

4.7 Endophytic Communities

Next-generation sequencing of 16S rRNA amplicons in *Leontopodium nivale* subsp. *alpinum* (Edelweiss) revealed a rich diversity of actinobacterial operational taxonomic units across roots, rhizomes, leaves and inflorescences. Comparison with cultured isolates demonstrated near-complete recovery of taxonomic diversity, highlighting the recruitment of soil Actinobacteria by the plant. NGS data also uncovered differences in taxonomic abundance between aboveground and belowground tissues, providing insights into endophytic community composition and potential targets for bioactive compound discovery (Oberhofer et al., 2019). Furthermore, next-generation sequencing enabled the characterization of endophytic fungal communities in alpine conifers, including *Thuja koraiensis*, *Taxus cuspidata* var. *nana, Pinus pumila* and *Abies nephrolepis*. A total of 4618 and 2268 reads were obtained for *A. nephrolepis* and *T. koraiensis*, revealing genus-level dominance by Ophiostomataceae_uc, Nectriaceae_uc and Magnaporthaceae_uc, respectively. The study highlights distinct fungal assemblages among different host species, emphasizing the role of ecological niches in alpine conifer restoration (Eo & Eom, 2022). Next-generation sequencing enables comprehensive profiling of endophytic bacterial and fungal communities in alpine plants, revealing tissue-specific diversity, host-specific assemblages and ecological interactions critical for plant-microbe studies and alpine ecosystem restoration.

Moreover, next-generation sequencing provides a powerful and precise approach to unravel the genetic, microbial and ecological complexity of alpine plants, facilitating insights into adaptation, biodiversity, population structure and endophytic interactions essential for conservation and functional studies, as illustrated in figure (Fig. 6.2).

5 Challenges and Limitations

Despite the significant progress and benefits of NGS in advancing the genomics of alpine flora, its application remains constrained by several challenges and limitations that warrant careful consideration. Although NGS offers high throughput and

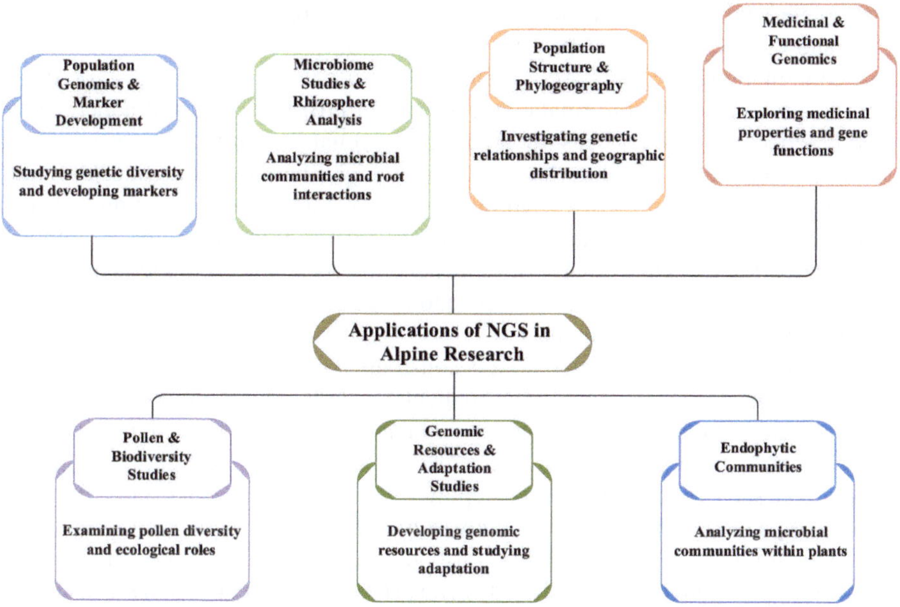

Fig. 6.2 Applications of NGS in alpine plant research, from population genomics to adaptation and endophytic studies

cost and time-efficiency, it poses distinct challenges for plant genomics: it generates large numbers of short reads (typically 25–400 bp) and, in plants with complex, often polyploid genomes, complicates analysis of genomes and traits, creating difficulties for genome resequencing, transcriptome sequencing, SNP marker development and novel microRNA identification (Wei et al., 2013). Even though RNA-Seq and transcriptome sequencing have expanded exponentially across important medicinal-plant genera, a major challenge is that comparatively fewer genomes have been sequenced or are being addressed; this imbalance underscores the need for a comprehensive, updated assessment of sequencing generations, the model and non-model plants sequenced up to 2022 and the attendant status, opportunities and challenges of these efforts (Singh et al., 2025). Sequencing by synthesis (Illumina, Roche/454) and sequencing by ligation (SOLiD) platforms generate vast amounts of plant genomic data, yet pose several challenges. PCR amplification can introduce errors and bias, while instrument scanning and washing cycles make data generation time-consuming. Short read lengths, particularly from Illumina and SOLiD, complicate sequence alignment and genome assembly, and the large volume of data creates demands for storage, transfer and computational analysis. Despite these limitations, high-throughput SGS platforms remain widely applied in plants for genome sequencing, transcriptomics, gene expression profiling and marker discovery (Thudi et al., 2012). Illumina sequencing has enabled rapid and cost-effective generation of large volumes of plant genomic data, yet several challenges remain. Data quality is highly influenced by library preparation, sample handling, instrument settings and sequencing chemistry. Short read lengths, increased error rates and the presence of adapter sequences or chimeras at molecule ends complicate computational analyses. Addressing these issues requires specialist expertise at both laboratory and bioinformatic stages, careful project design and adherence to best-practice principles to ensure accurate and reliable results in plant genomics studies (Kircher et al., 2011). while NGS technologies have revolutionized the study of alpine plant genomics by enabling high-throughput, cost- and time-efficient sequencing, their effective application requires careful consideration of genome complexity, read length limitations, sequencing errors and computational demands, alongside rigorous experimental design and specialist expertise to ensure accurate, reliable and meaningful insights.

Additionally, studying alpine flora is constrained by multiple interrelated drivers of non-native spread along elevation gradients, including biotic interactions, evolutionary change environmental limits on population growth and propagule availability. Although the highest elevations remain relatively free of non-native plants, nearly 200 such species have been reported worldwide, of which only three are specifically cold-adapted and only a few exert notable impacts on native communities. This limited invasion is largely attributed to low propagule pressure and low invasibility compared to lowland ecosystems. However, evidence from some regions shows that alpine systems are not inherently resistant and increasing pressures from pre-adapted introductions, climate change and native range expansions underline the importance of preventative measures to mitigate future impacts (Alexander et al., 2016). Furthermore, rapid climate change is reshaping alpine ecosystems as receding glaciers and snowfields, combined with altered precipitation regimes, drive shifts in hydrology, species distributions and basal resources while threatening habitats and biota. Alpine streams contain substantial genetic and species diversity due to environmental heterogeneity and habitat insularity and these changes are expected to influence biodiversity from genes to communities. Addressing the resulting challenges in studying alpine flora requires greater synthesis of established and emerging approaches, temporal comparison of existing datasets, integration of remote sensing and GIS and application of genomic tools to guide conservation and mitigate rapid environmental change (Hotaling et al., 2017). Likewise, Climate change is projected to alter vegetation and snow distributions in mountains, yet the coupling of hydrological processes and ecological shifts within alpine zones has received little scientific attention. In the Himalayan alpine (4100–6000 m a.s.l.) this gap is urgent because up to 1.6 billion people depend on mountain water; monitoring stations are scarce and specific mechanisms and functional relationships explaining interactions between snow and dwarf plants are missing. Consequently, the direction and strength of ecohydrological relationships remain unknown, calling for purposeful monitoring below glaciers and above the treeline, experiments to query the role of small plants in the alpine hydrological system and community engagement. New snow and vegetation products derived from remote sensing observations have the potential to improve scientific understanding of these interacting effects (Leng et al., 2023). Studying alpine flora under climate warming is challenging because plant responses are multidimensional and vary across space and time.

While higher temperatures can enhance growth and leaf production in cold-adapted species such as Rumex alpinus, warming also induces water limitation, particularly during periods of low summer precipitation. Growth and reproductive performance thus shift from being primarily temperature-limited to water-limited, with effects differing among alpine, subalpine and montane populations, highlighting the complexity of predicting plant responses in changing mountain environments (Dolezal et al., 2020). Thus, studying alpine flora is increasingly challenged by the combined effects of non-native species spread, habitat insularity and rapid climate change, which alter hydrology, species distributions and resource availability, while plant responses remain multidimensional and spatially variable, underscoring the urgent need for integrated monitoring, genomic tools, remote sensing and targeted conservation strategies to understand and mitigate these complex environmental impacts.

6 Conclusion

The combination of multi-omics strategies, including metabolomics, transcriptomics and genomics, together with advancements in machine learning and artificial intelligence is profoundly transforming the study of alpine plant diversity and adaptation. These comprehensive frameworks enable predictive modelling of adaptive traits and evolutionary dynamics, significantly enhancing the resolution of ecological and evolutionary studies in high-altitude ecosystems. Next-generation sequencing (NGS) has become a cornerstone technology, dramatically accelerating the acquisition and analysis of large-scale genomic datasets. By employing cutting-edge sequencing platforms such as Illumina, PacBio and Oxford Nanopore, researchers can now assemble complex genomes, analyse population structure, annotate gene functions and perform comparative genomics with exceptional accuracy and depth. This revolution in data generation has revealed multilayered adaptive traits and cryptic genetic variation among alpine species, providing a strong scientific foundation for conservation planning and adaptive biodiversity management under environmental stress. Global repositories and collaborative initiatives, including INSDC, enhance the accessibility and standardisation of genomic data, supporting reproducibility and long-term research strategies in mountain ecosystems. Nevertheless, significant challenges remain, including short read lengths, sequencing errors, computational requirements and the impacts of rapidly changing alpine climates, necessitating continued methodological refinement and specialised expertise in laboratory and bioinformatic analyses. The convergence of NGS technologies, multi-omics integration and artificial intelligence marks a new era in alpine genomics, offering unparalleled opportunities to safeguard threatened plant species, elucidate complex evolutionary and adaptive pathways and inform sustainable management of vulnerable mountain biodiversity.

References

Alexander, J. M., Lembrechts, J. J., Cavieres, L. A., Daehler, C., Haider, S., Kueffer, C., Liu, G., McDougall, K., Milbau, A., Pauchard, A., Rew, L. J., & Seipel, T. (2016). Plant invasions into mountains and alpine ecosystems: Current status and future challenges. *Alp Botany, 126*, 89–103. https://doi.org/10.1007/s00035-016-0172-8

Bell, N., Griffin, P. C., Hoffmann, A. A., & Miller, A. D. (2018). Spatial patterns of genetic diversity among Australian alpine flora communities revealed by comparative phylogenomics. *Journal of Biogeography, 45*, 177–189. https://doi.org/10.1111/jbi.13120

Buehler, D., Graf, R., Holderegger, R., & Gugerli, F. (2011). Using the 454 pyrosequencing-based technique in the development of nuclear microsatellite loci in the alpine plant Arabis alpina (Brassicaceae). *American Journal of Botany, 98*, e103–e105. https://doi.org/10.3732/ajb.1000488

Cochrane, G., Karsch-Mizrachi, I., Takagi, T., & International Nucleotide Sequence Database Collaboration. (2016). The International Nucleotide Sequence Database Collaboration. *Nucleic Acids Research, 44*, D48–D50. https://doi.org/10.1093/nar/gkv1323

Dolezal, J., Kurnotova, M., Stastna, P., & Klimesova, J. (2020). Alpine plant growth and reproduction dynamics in a warmer world. *New Phytologist, 228*, 1295–1305. https://doi.org/10.1111/nph.16790

Eo, J.-K., & Eom, A.-H. (2022). Community of endophytic fungi from alpine conifers on Mt. Seorak. *Mycobiology, 50*, 317–325. https://doi.org/10.1080/12298093.2022.2135832

Griffin, P. C., Robin, C., & Hoffmann, A. A. (2011). A next-generation sequencing method for overcoming the multiple gene copy problem in polyploid phylogenetics, applied to Poa grasses. *BMC Biology, 9*, 19. https://doi.org/10.1186/1741-7007-9-19

Hodzic, J., Gurbeta, L., Omanovic-Miklicanin, E., & Badnjevic, A. (2017). Overview of next-generation sequencing platforms used in published draft plant genomes in light of genotypization of immortelle plant (Helichrysium Arenarium). *Medical Archives, 71*, 288–292. https://doi.org/10.5455/medarh.2017.71.288-292

Hong, K., Radian, Y., Manda, T., Xu, H., & Luo, Y. (2023). The development of plant genome sequencing technology and its conservation and application in endangered gymnosperms. *Plants, 12*, 4006. https://doi.org/10.3390/plants12234006

Hotaling, S., Finn, D. S., Joseph Giersch, J., Weisrock, D. W., & Jacobsen, D. (2017). Climate change and alpine stream biology: Progress, challenges, and opportunities for the future. *Biological Reviews, 92*, 2024–2045. https://doi.org/10.1111/brv.12319

Hu, T., Chitnis, N., Monos, D., & Dinh, A. (2021). Next-generation sequencing technologies: An overview. *Human Immunology, 82*, 801–811. https://doi.org/10.1016/j.humimm.2021.02.012

Johnstone, J. (2021). Review of alpine plant life: functional plant ecology of high mountain ecosystems. *Mountain Research and Development, 41*, M1–M2.

Kircher, M., Heyn, P., & Kelso, J. (2011). Addressing challenges in the production and analysis of illumina sequencing data. *Bmc Genomics, 12*, 382. https://doi.org/10.1186/1471-2164-12-382

Körner, C. (2023). Concepts in alpine plant ecology. *Plants, 12*, 2666. https://doi.org/10.3390/plants12142666

Körner, C., & Hiltbrunner, E. (2021). Why is the alpine flora comparatively robust against climatic warming? *Diversity, 13*, 383. https://doi.org/10.3390/d13080383

Leng, R., Harrison, S., & Anderson, K. (2023). Himalayan alpine ecohydrology: An urgent scientific concern in a changing climate. *Ambio, 52*, 390–410. https://doi.org/10.1007/s13280-022-01792-2

Leontidou, K. (2018). Airborne pollen patterns along an altitudinal gradient of the Italian Alps: Combination of classical pollen identification methods and next generation sequencing on environmental DNA (Doctoral Thesis). *Aristotle University of Thessaloniki.*

Lischer, H. E. L., & Shimizu, K. K. (2017). Reference-guided de novo assembly approach improves genome reconstruction for related species. *Bmc Bioinformatics, 18*, 474. https://doi.org/10.1186/s12859-017-1911-6

Lobréaux, S., Manel, S., & Melodelima, C. (2014). Development of an Arabis alpina genomic contig sequence data set and application to single nucleotide polymorphisms discovery. *Molecular Ecology Resources, 14*, 411–418. https://doi.org/10.1111/1755-0998.12189

Long, Z., Sang, Y., Feng, J., Zhang, X., Shi, T., Zhang, L., Mao, K., Rieseberg, L. H., Liu, J., & Wang, J. (2025). Evolutionary genomics unravels the responses and adaptation to climate change in a key alpine forest tree species. *Molecular Biology and Evolution, 42*, msaf116. https://doi.org/10.1093/molbev/msaf116

Martina, M., Acquadro, A., Portis, E., Barchi, L., & Lanteri, S. (2023). Diversity analyses in two ornamental and large-genome Ranunculaceae species based on a low-cost Klenow NGS-based protocol. *Frontiers in Plant Science, 14*. https://doi.org/10.3389/fpls.2023.1187205

Mikulášková, E., Peterka, T., Šmerda, J., & Hájek, M. (2024). Next-generation sequencing reveals hidden genomic diversity in glacial relicts: A case study of Meesia triquetra. *Journal of Systematics and Evolution, 62*, 475–488. https://doi.org/10.1111/jse.13005

Muellner-Riehl, A. N. (2019). Mountains as evolutionary arenas: Patterns, emerging approaches, paradigm shifts, and their implications for plant phylogeographic research in the Tibeto-Himalayan region. *Frontiers in Plant Science, 10*. https://doi.org/10.3389/fpls.2019.00195

Noroozi, J., Moser, D., & Essl, F. (2016). Diversity, distribution, ecology and description rates of alpine endemic plant species from Iranian mountains. *Alp Botany, 126*, 1–9. https://doi.org/10.1007/s00035-015-0160-4

Oberhofer, M., Hess, J., Leutgeb, M., Gössnitzer, F., Rattei, T., Wawrosch, C., & Zotchev, S. B. (2019). Exploring actinobacteria associated with rhizosphere and endosphere of the native alpine medicinal plant Leontopodium nivale subspecies alpinum. *Frontiers in Microbiology, 10*. https://doi.org/10.3389/fmicb.2019.02531

Rew, L. J., McDougall, K. L., Alexander, J. M., Daehler, C. C., Essl, F., Haider, S., Kueffer, C., Lenoir, J., Milbau, A., Nuñez, M. A., Pauchard, A., & Rabitsch, W. (2020). Moving up and over: Redistribution of plants in alpine, Arctic, and Antarctic ecosystems under global change. *Arctic, Antarctic, and Alpine Research, 52*, 651–665. https://doi.org/10.1080/15230430.2020.1845919

Rijal, D. P., Falahati-Anbaran, M., Alm, T., & Alsos, I. G. (2015). Microsatellite markers for Heracleum persicum (Apiaceae) and allied taxa: Application of next-generation sequencing to develop genetic resources for invasive species management. *Plant Molecular Biology Reporter, 33*, 1381–1390. https://doi.org/10.1007/s11105-014-0841-y

Ruhsam, M., & Hollingsworth, P. M. (2025). The genome sequence of the alpine milkvetch, Astragalus alpinus L. (Fabales: Fabaceae). *Wellcome Open Research, 10*, 366. https://doi.org/10.12688/wellcomeopenres.24576.1

Rundel, P. W., & Millar, C. I. (2016). Chapter 29. Alpine ecosystems. In E. Zavaleta & H. Mooney (Eds.), *Ecosystems of California* (pp. 613–634). University of California Press.

Singh, D., Mathur, S., Prasad, M., & Ranjan, R. (2025). Next-generation sequencing in medicinal plants: Recent progress, opportunities, and challenges. *Journal of Plant Growth Regulation, 44*, 1448–1464. https://doi.org/10.1007/s00344-024-11238-6

Singh, D., Mittal, N., Verma, S., Singh, A., & Siddiqui, M. H. (2023). Applications of some advanced sequencing, analytical, and computational approaches in medicinal plant research: A review. *Molecular Biology Reports, 51*, 23. https://doi.org/10.1007/s11033-023-09057-1

Stefani, S., Champoud, L., Fazan, L., Ronikier, M., Perret, M., Kozlowski, G., & Christe, C. (2025). Genetic divergence of Clematis alpina in the Swiss Prealps: A tale of the margins. *Alp Botany*. https://doi.org/10.1007/s00035-025-00333-3

Testolin, R., Carmona, C. P., Attorre, F., Borchardt, P., Bruelheide, H., Dolezal, J., Finckh, M., Haider, S., Hemp, A., Jandt, U., Korolyuk, A. Y., Lenoir, J., Makunina, N., Malanson, G. P., Mucina, L., Noroozi, J., Nowak, A., Peet, R. K., Peyre, G., Sabatini, F. M., Šibík, J., Sklenář, P., Vassilev, K., Virtanen, R., Wiser, S. K., Zibzeev, E. G., & Jiménez-Alfaro, B. (2021). Global functional variation in alpine vegetation. *Journal of Vegetation Science, 32*, e13000. https://doi.org/10.1111/jvs.13000

Thudi, M., Li, Y., Jackson, S. A., May, G. D., & Varshney, R. K. (2012). Current state-of-art of sequencing technologies for plant genomics research. *Briefings in Functional Genomics, 11*, 3–11. https://doi.org/10.1093/bfgp/elr045

Wagner, N. D., He, L., & Hörandl, E. (2021). The evolutionary history, diversity, and ecology of willows (Salix L.) in the European Alps. *Diversity, 13*, 146. https://doi.org/10.3390/d13040146

Wei, L., Xiao, M., Hayward, A., & Fu, D. (2013). Applications and challenges of next-generation sequencing in Brassica species. *Planta, 238*, 1005–1024. https://doi.org/10.1007/s00425-013-1961-6

Zhang, J., Dong, K.-L., Ren, M.-Z., Wang, Z.-W., Li, J.-H., Sun, W.-J., Zhao, X., Fu, X.-X., Ye, J.-F., Liu, B., Zhang, D.-M., Wang, M.-Z., Zeng, G., Niu, Y.-T., Lu, L.-M., Su, J.-X., Liu, Z.-J., Soltis, P. S., Soltis, D. E., & Chen, Z.-D. (2024). Coping with alpine habitats: Genomic insights into the adaptation strategies of Triplostegia glandulifera (Caprifoliaceae). *Horticulture Research, 11*, uhae077. https://doi.org/10.1093/hr/uhae077

Zhang, X., Kuang, T., Dong, W., Qian, Z., Zhang, H., Landis, J. B., Feng, T., Li, L., Sun, Y., Huang, J., Deng, T., Wang, H., & Sun, H. (2023). Genomic convergence underlying high-altitude adaptation in alpine plants. *Journal of Integrative Plant Biology, 65*, 1620–1635. https://doi.org/10.1111/jipb.13485

7 Time-Series Transcriptomics in Understanding Cold Acclimation in Alpine Plants

Bhuvnesh Kapoor

Abstract

Cold acclimation is a critical adaptive strategy that enables alpine plants to survive and function under extreme and fluctuating seasonal conditions. Until recently, progress in understanding this process was limited by both technical and biological challenges. With the advent of time-series transcriptomics, however, it has become possible to unravel the temporal dynamics of gene expression underlying cold responses. Comparative studies in model systems and alpine species have highlighted the importance of transcription factors, circadian regulation, alternative splicing, and epitranscriptomic modifications, alongside key pathways such as the CBF cascade, lipid remodeling, and hormone-mediated signaling. Emerging approaches in single-cell and spatial transcriptomics now open access to cell- and tissue-specific temporal responses, offering unprecedented resolution. These insights have practical applications ranging from crop improvement through marker-assisted selection and genome editing to refining ecological models of alpine adaptation. Looking ahead, integrating multi-omics datasets, capturing organ- and tissue-resolved temporal changes, and standardizing experimental frameworks will be crucial for advancing cross-study comparability. This chapter underscores the transformative role of time-series transcriptomics in decoding cold acclimation, outlining its methodological foundations, analytical strategies, and contributions to understanding regulatory networks that sustain alpine plant resilience.

B. Kapoor (✉)
Department of Biotechnology, University Institute of Biotechnology (UIBT), Chandigarh University, Gharuan, Mohali, Punjab, India

P. Kumar et al. (eds.), *Climate Resilience and Molecular Adaptation in Alpine Medicinal Plants*, https://doi.org/10.1007/978-981-95-6923-6_7

Keywords

Cold acclimation · Time-series transcriptomics · Alpine plants · Gene regulatory networks · Stress-responsive pathways

1 Introduction

The ability of alpine and high-altitude plants to withstand fluctuating and prolonged cold stress arises from finely regulated adaptive mechanism collectively called as cold acclimation response. This physiological plasticity is induced by non-lethal exposure to cold conditions. At molecular level, this response involves a cascade of signaling molecules, biochemical, and cellular modifications that enhance subsequent cold tolerance. With the current scenario of climate change intensifications and fluctuations in temperature regimes further induce the changing altitudinal gradient in alpine regions. Hence it became pertinent to understand the spatio-temporal regulation of cold acclimation that could ensure the ecological sustainability and pave the way for improvement of cold resilient crops.

Recent advancements in time-series transcriptomics have allowed a developing understanding of underlying dynamics involved in cold response. Unlike static transcriptome snapshots, time-series RNA-Seq aims for more detailed dynamics of successive transcriptional events such as early activation of sensory molecules, intermediate molecular responses, late acclimation followed by de-acclimation collectively represents cold tolerance response. In concordance, this approach was able to decipher that acclimation and de-acclimation represents distinct transcriptional trajectories, orchestrated by dynamic shifts in gene regulatory networks (GRNs) rather than simple reversals (Rathore et al., 2022).

Broadly, cold tolerance mechanism involves factors such as cold-responsive (COR) genes, biosynthesis, and accumulation of cryoprotective metabolites (such as sugars, flavonoids, and polyols), reinforcement of cellular membranes, and modulation of the cytoskeleton. Among all pathways, the central role is played by factors such as C-repeat binding factors (CBFs), which operate as master regulators of downstream cold-responsive genes (Ding et al., 2019; Tang et al., 2020).

Emerging studies, however, highlight that post-transcriptional regulation, chromatin remodelling, and epigenetic memory also play key roles in long-term acclimation (Xu et al., 2022). Work on alpine species illustrates the temporal precision of cold signaling. For instance, in *Rhododendron anthopogon*, high-resolution time-series transcriptomics identified 9844 differentially expressed genes, revealing a distinct transcriptionally active de-acclimation stage, underscoring that recovery from cold stress is governed by specialized molecular programs (Rathore et al., 2022).

Integrating transcriptomics with metabolome and proteome profiling has further demonstrated that cold acclimation involves a tightly linked regulatory network. In tobacco, coordinated changes in ~7000 transcripts and >30 metabolites highlight the direct alignment of gene expression and metabolic phenotypes (Xu et al., 2020).

In recent cold-acclimation studies emphasize the role of non-coding RNAs, alternative splicing, and RNA modifications in fine-tuning stress signaling (Cai et al., 2025).

Importantly, advances in computational approaches are expanding the potential of time-series transcriptomics. The integration of machine learning, single-cell transcriptomics, and dynamic network inference now enables causal modelling of gene hierarchies and cross-talk among cold-responsive pathways (Jain, 2024). Despite these advances, major challenges remain—particularly in resolving organ-specific responses, multi-omics integration, and translating lab observations into field-relevant outcomes.

This chapter synthesizes transcriptomic research, with particular emphasis on alpine plants as natural models of cold tolerance. By combining insights from high-resolution temporal datasets, integrative omics platforms, and computational modelling, we highlight both conserved and species-specific strategies of cold adaptation (Fig. 7.1). Finally, we discuss methodological innovations, current gaps, and future pathways for leveraging time-series data to engineer climate-resilient crops—a necessity as global climate volatility accelerates.

2 Methodological Approaches in Time-Series Transcriptomics

Over the past two decades, major developments in transcriptomics have provided new tools to study cold acclimation in alpine plants. The choice of transcriptomic platforms significantly influences the resolution, scope, and interpretability of acclimation-associated transcriptional trajectories particularly in alpine ecosystems, where physiological and gene expression responses are governed by gradual environmental cues.

2.1 Early Methods (Microarray, SAGE and MPSS)

The earliest transcriptomic investigations of plant responses to cold relied heavily on microarray technologies, which offered the ability to monitor thousands of transcripts simultaneously. Despite limitations in dynamic range and probe specificity, microarrays provided valuable insights into the temporal activation of cold-responsive genes, particularly in model species such as *Arabidopsis thaliana* (Oono et al., 2006).

In alpine plants, microarrays found limited but notable application due to the scarcity of species-specific arrays and reference genomes. However, they helped establish core signatures of cold-responsive transcription factors (e.g., CBFs/DREBs), promoters, dehydrins, and late embryogenesis abundant (LEA) proteins. For instance, Grene et al. (2012) mapped microarray and metabolic data to reveal the link between ethylene signaling biosynthesis and targeting of cell wall reinforcement molecules during winter hardening in *Picea sitchensis.*

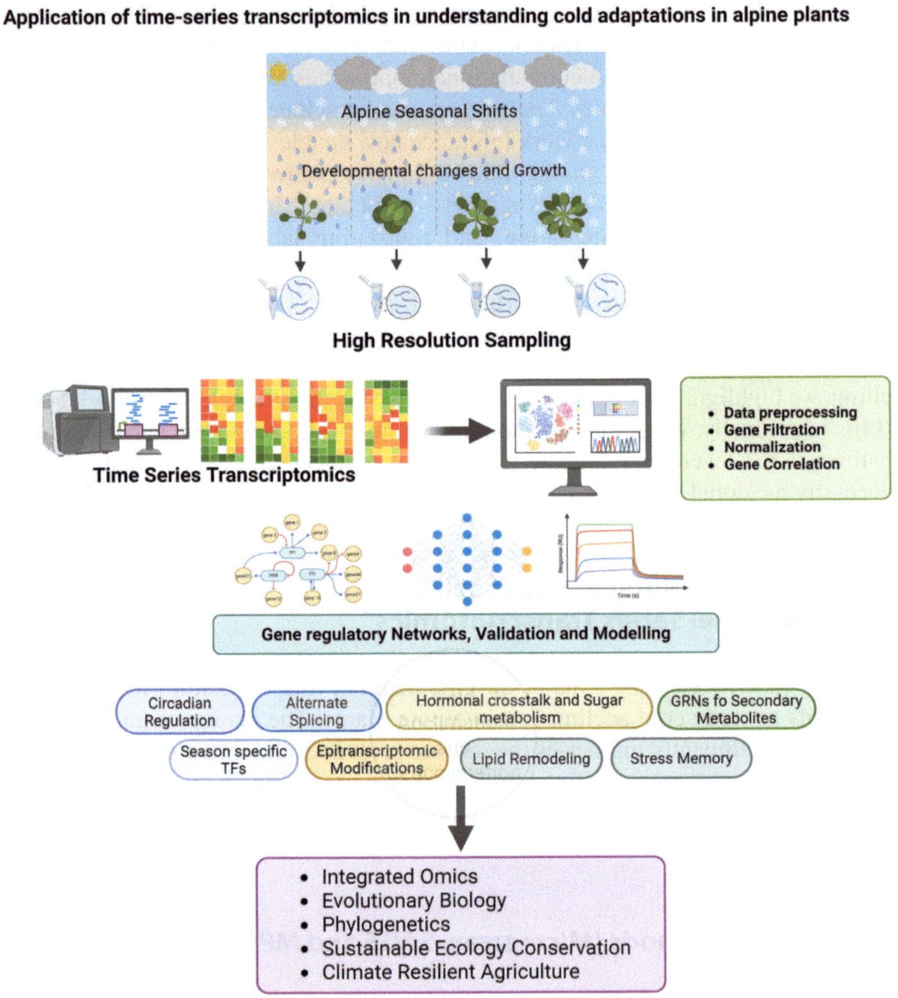

Fig. 7.1 Application of time-series transcriptomics in understanding cold adaptation in alpine plants

In parallel to microarrays, tag-based sequencing methods such as Serial Analysis of Gene Expression (SAGE) and Massively Parallel Signature Sequencing (MPSS) offered truly exploratory transcriptome profiling without the need for predesigned probes. These approaches generated short sequence tags corresponding to transcript fragments, which were quantitatively counted to assess expression levels across time points and treatment conditions. SAGE provided early temporal resolution of gene expression changes but was limited by relatively low sequencing depth and short tag length. An illustrative case is the Deep Super-SAGE analysis of cold acclimation in lentil (*Lens culinaris*), where recombinant inbred lines differing in frost tolerance were subjected to chilling treatments. Deep Super-SAGE produced 26-bp "supertags," enabling robust annotation and quantification of low-abundance and novel transcripts during cold acclimation, including candidate transcription factors and stress-related genes that would have been under-represented in microarray datasets (Barrios et al., 2017).

MPSS significantly increased both coverage and dynamic range by sequencing millions of 17–20-nt signature tags per library. Although dedicated MPSS studies focused exclusively on cold stress remain scarce, the Plant MPSS Databases for *Arabidopsis*, rice, and grape include libraries derived from multiple abiotic stress treatments—among them cold exposure—facilitating the identification and quantitative comparison of cold-induced transcripts, antisense RNAs, and unannotated intergenic signals (Meyers et al., 2004).

While the high cost and complex library preparations limited the widespread adoption of SAGE and MPSS, these methods served as critical transitional technologies. They broadened the detectable transcriptome—uncovering rare regulatory RNAs and novel cold-responsive loci—and laid the groundwork for the advent of RNA-Seq, which now dominates time-series transcriptomics through its superior dynamic range, sensitivity, and *de novo* discovery capabilities.

2.2 High-Throughput RNA Sequencing in Alpine Plants

The advent and refinement of RNA sequencing (RNA-Seq) technology have revolutionized transcriptomics by allowing comprehensive, unbiased, and quantitative assessments of gene expression in both model and non-model species. It became an important tool to study *de novo* transcriptome assembly and isoform discovery in alpine plants as most of them are non-model plants and lacks availability of high-quality genomes. RNA-Seq surpasses previous probe-dependent methods, providing superior sensitivity, dynamic range, and the ability to detect low-abundance transcripts as well as novel and alternatively spliced isoforms.

In addition, it also provides high resolution gene expression profiles to sequentially categorize activation and repression of cold acclimation networks. These phased gene expression network facilitated the reconstruction of gene co-expression modules and identification of key regulatory TFs, signaling components, and downstream effectors operating in phased responses to alpine environmental cues (Kumari et al., 2025; Wang et al., 2024). For example, Rathore et al. (2022) performed the time-series transcriptomics in *Rhododendron anthopogon* to unveil the season transcriptomic shifts. The study elucidated how winter and spring transitions were reflected in cold-responsive genes amplify during onset of freezing conditions and recede during de-acclimation, highlighting the plasticity of alpine gene regulatory networks. Similarly, Li et al. (2023) integrated multi-omics in *Rheum nobile* revealed multitude of specialized metabolic adjustments. Such as flavonoid biosynthesis in photosynthetically active bracts and enhanced antifreeze protein expression in roots, accompanied by genome duplications that underlie expanded cold tolerance gene families.

To assemble high-quality *de novo* transcriptome assembly for alpine plants is essential step because of scarcity of reference genome that usually is carried by common software platforms such as Trinity and SOAPdenovo-Trans aid assembly. However, the quality of *de novo* transcriptome assembly hugely relies on optimizations of various parameters to reduce redundancy and fragmentation while improving isoform resolution. Clearly, improvement in isoform resolution proved to be a critical factor for accurately identifying alternative splicing events that modulate stress responses (Grabherr et al., 2011). To further strengthen this, integration of long-read sequencing technologies, including Oxford Nanopore and PacBio Iso-Seq, has substantially improved transcriptome completeness and isoform-level resolution by capturing full-length transcripts in a single read. Integration proved to be valuable as it became helpful in dissecting the complex secondary metabolism and post-transcriptional modifications in cold acclimation (Wang & Qiao, 2020). In addition, current RNA-Seq protocols also have integrated parameters like strand specificity and unique molecular identifiers, improving quantification accuracy and enabling detection of antisense transcripts and noncoding RNAs.

In last two decades, development of cost-effective high-throughput workflows, combined with decreasing sequencing costs, has empowered large-scale population-level time-series studies possible. Overall, this helps in understanding of environmental and genotypic variation, essential for adaptive plasticity as well as evolutionary resilience in alpine plants.

However, with all advancements certain limitations are also associated with RNA-seq analysis framework. These challenges include; batch effects, uneven sampling intervals in alpine fieldwork, and complex data normalization in multifactorial experimental designs (Thron & Jafari, 2025). To minimize their effects on overall data, various computational tools are being utilized such as machine learning and Bayesian inference facilitating causal network inference and predictive modelling in cold acclimation studies (Greenfield et al., 2013; Kim et al., 2023).

Overall, high-throughput RNA-Seq continues to serve as a foundational platform from which increasingly sophisticated molecular models of alpine cold responses are generated (Table 7.1).

2.3 Single-Cell and Spatial Transcriptomics in Alpine Research

The transcriptome dynamics is recorded as snapshots by RNA-seq technology. But these transcriptomic snapshots individually can mask important aspects such as cellular heterogeneity and tissue-specific cold responses in alpine plants. Therefore, the further advancements such as Single-cell RNA sequencing (scRNA-Seq) and spatial transcriptomics became a transformative tool to resolve this complexity by profiling gene expression at cellular and anatomical resolution, respectively (Chen et al., 2023). Currently, these methods are providing new knowledge that is advancing our understanding of cold acclimation mechanisms at astonishing resolution particularly in alpine plants species.

Table 7.1 Time-Series transcriptomic elucidating cold adaptive responses

Plant species	Time points	Key observations	References
Saussurea involucrata	Multiple temperature/time conditions including 20°C → 4°C (24 h), 4°C (7 day), 0 °C (24 h), −2 °C (12–24 h)	Cold-responsive TFs, circadian rhythm activation, signal transduction and transport genes changed across time/temperature steps (dynamic cold-response network)	Li et al. (2017)
Elymus nutans	0, 3, 6, 12, 24 h (comparative genotypes under cold stress)	Time-resolved DEGs between tolerant and sensitive genotypes; fructan biosynthesis, dehydrin genes and early stress signaling differ across time points	Fu et al. (2016)
Malus sieversii	Time series including freezing exposure / recovery (multiple sampling during freezing)	Freezing-responsive genes and pathways identified; transcriptional switches in cold-tolerance and ice-response pathways across time	Zhou et al. (2021)
Rhododendron	Time course across cold acclimation and de-acclimation (field + lab time series; samples across CA and DA stages)	Photoprotection and light-related genes (ELIP/HY5/BBX network) strongly modulated during acclimation/deacclimation; time-dependent expression patterns reveal photoprotection contributes to freezing tolerance	Liu et al. (2022)
Rhododendron anthopogon	Seasonal / time-series sampling covering acclimation and de-acclimation stages (multiple dates across seasons)	Time-series RNA-seq revealed stages of cold acclimation and de-acclimation, with transcriptional signatures for carbohydrate metabolism, protective proteins, and phases of stress memory	Rathore et al. (2022)
Sinopodophyllum hexandrum	Rhizome harvested at day 0, 14 and 21 under two temperatures (15 °C vs 25 °C)—Time points across days	Temperature-dependent transcriptome shifts in rhizome; genes linked to podophyllotoxin biosynthesis and stress responses change over the time series	Kumari et al. (2014)
Helictotrichon virescens	Short time-course cold treatments (e.g. 0–48 h)	Crosstalk between circadian transcription factors and cold response; time-dependent TF dynamics linking clock and cold signaling	Cheng et al. (2022)
Bergenia purpurascens	Time-course cold treatments (0–48 h and sampling across treatments)	MAPK signaling and ROS-related transcriptional responses exhibit distinct temporal patterns; stress-protective gene modules activated dynamically	Zhang et al. (2023)

(continued)

Table 7.1 (continued)

Plant species	Time points	Key observations	References
Camellia sinensis	Time series during cold-acclimation treatments (multiple sampling times during CA experiments)	Cold acclimation triggers phased transcriptome changes including CBF-pathway responses, carbohydrate metabolism and protective protein induction over time	Wang et al. (2013)
Ammopiptanthus mongolicus	Time series at 2 h, 8 h, 24 h (cold and drought treatments were profiled in time course)	Comparative time-course expression under cold: rapid transcriptional responses (early signaling TFs) and later metabolic adjustments; distinct temporal profiles for cold vs drought	Wu et al. (2014)

Further, advance in practical aspects of single-cell RNA sequencing procedures such as streamlining nucleus isolation and protecting RNA quality also proved to be useful. Advances in protocols like isolating nuclei instead of digesting cell walls, researchers avoid RNA degradation that help to capture transcripts at very low concentrations. In addition, protoplast-free methods and cell-type-specific labelling further minimize preparation bias and advances allow a reliable single-cell profiling of varied plant tissues (Jean-Baptiste et al., 2019).

Although, scRNAseq research in alpine plants is still at nascent stage, studies in *Arabidopsis* reveal that different cell types respond uniquely to cold stress. In mesophyll cells, genes for photosynthesis were downregulated, while epidermal cells boost pathways inducing cuticle biosynthesis pathways to enhance surface protection. (Denyer et al., 2019; Tenorio Berrío & Dubois, 2024). These findings imply that alpine species likely show similar cell-specific responses, which are crucial for surviving rapid temperature fluctuations.

Spatial transcriptomics complements scRNA-Seq by enabling *in situ* gene expression mapping, thus retaining tissue context lost in traditional dissociation-based methods. Advanced protocols, including Slide-seq, SeqFISH, and Visium, are being adopted to map the localization of cold-responsive gene networks across anatomical landscapes (Asp et al., 2020; Ståhl et al., 2016). Similar spatial approaches uncover the orchestration of developmental transitions such as meristem dormancy, leaf senescence, and root cold sensing within their native cellular topography under natural freeze–thaw cycles (Lv et al., 2024).

Together, these approaches enable the construction of integrative, multi-scale atlases linking molecular, cellular, and tissue-level dynamics of alpine plant cold acclimation. As sequencing costs decline and high-contiguity reference genomes become increasingly available through global initiatives such as the 10KP Project (Cheng et al., 2018; Michael & VanBuren, 2020). Single-cell and spatial transcriptomics will become indispensable tools for elucidating the cellular architectures underpinning environmental resilience.

2.4 Integrative Omics

The integration of time-series transcriptomics with other omics studies like proteomics, metabolomics, and epigenomics has enabled researchers to develop a comprehensive model that provide more detailed molecular dynamics over time. This multilayer approach has become a powerful tool to decipher the integrated view of gene networks in tandem with proteins, metabolites and signaling molecules working together in response to different environmental cues. Meanwhile, alpine plants provide a dynamics platform to explore important mechanism such as active acclimations, stress resilience, and developmental changes, synchronized in highly coordinated manner.

The epigenomic assays like ATAC-seq can expose the regulatory regions of chromatin modelling during cold exposure, linking these changes to specific gene activation at different stages. The temporal methylation profiling of alpine plants can illustrate various epigenetic marks that are helpful in modulation of transcriptional plasticity and contribute to stress memory by priming gene responsiveness upon recurring cold events. The integration of all these datasets accounts for holistic understanding of transcriptional regulations controlled by multilayered chromatin landscapes and their temporal dynamics. In addition, systems biology frameworks that include epigenomic modifications help clarify how regulatory mechanisms drive phenotypic plasticity in high-altitude plants (Kumar & Rani, 2023).

Further, the integrated time series proteomics and phosphoproteomic studies revealed the post-translational modifications underlying cold acclimation. The phosphorylation patterns of signaling proteins and transcription factors corresponds dynamic temperature shifts. This aspect highlights the alteration in protein activity and stability in post-translational manner that is beyond regulatory dimension of transcriptome. For example, integrated phosphoproteomic and transcriptomic analysis in *Rhododendron chrysanthum* uncovered kinase pathways and ABA signaling regulating UV-B and cold stress tolerance, linking environmental cues to stress adaptation in alpine habitats (Sun et al., 2023).

In complementation, metabolomics helps to track the temporal accumulation of various metabolites for example in sugars (sucrose, raffinose, and trehalose) and osmo-protectants. These metabolites in turn stabilize cellular structures and mitigate oxidative damage during cold stress. Collectively this analysis could reveal if metabolic shifts are synchronized with gene expression, enabling plants to balance energy conservation with protective functions. For example, Wang et al. (2024) integrated the transcriptomic and metabolomic analysis in *Rheum pumilum,* revealing phytohormone-regulated networks. Where, ABA maintaining seed dormancy and GA and ethylene promoting rapid germination, facilitating adaptation to alpine climatic conditions.

Several studies have demonstrated the successful use of omics approaches in revealing multi-layered cold response signaling in alpine plants. For example, a transcriptomic, metabolomic, and microbiome study in *Meconopsis horridula* from the Qinghai-Tibet Plateau revealed a well-regulated upregulation of flavonoid biosynthesis and antioxidant pathways that underlie adaptation to high UV-B radiation

(Xu et al., 2025). Similarly, multi-omics research on *Rheum nobile* revealed that accumulation of protective sugars, flavonoid biosynthesis that underlying adaptation to high UV-B radiation and cold stress in alpine cold adaptation (Li et al., 2023).

In Himalayan flora, multi-omics analysis has revealed common adaptive strategies to counter the cold stress. These responses include use of heat shock proteins, anti-oxidant systems, and synthetic pathways for specialized secondary metabolites. For example, integrated analyses of transcripts, proteins, and metabolites in *Hippophae rhamnoides* and *Rheum nobile* pinpointed antioxidant, phenylpropanoid, and sugar metabolism pathways that enable cold and drought tolerance (Chen et al., 2024; Li et al., 2023).

In addition, microbiome profiles with plant metabolomics and transcriptomics reveals how microbial communities influence host plants by modulating immune responses and metabolite production. This interaction enhances alpine plant adaptation by supporting stress resilience and metabolic balance. For example, in alpine plants, *Cremanthodium napifera* changes in altitude affect both plant gene expression related to metabolite production and the composition of root-associated microbial communities These microbes can be recruited by plants to boost growth and stress tolerance, demonstrating a valuable functional partnership in harsh environments (Yang et al., 2025). This multi-omics approach deepens understanding of how microbial partners contribute to plant survival in alpine regions.

Integrative omics studies collectively reveal the complex molecular mechanisms that allow alpine plants to survive in harsh environments (Table 7.2). These insights offer valuable opportunities for conservation genomics and crop improvement to address climate challenges. They also establish a systems-level framework to guide future research focused on understanding how stress resilience is coordinated over time.

3 Key Findings from Plant Studies

Time-series transcriptomic advances have shown that cold acclimation involves not only immediate transcriptional activation but also long-term metabolic plasticity, chromatin memory, and circadian integration (Kidokoro et al., 2022). The following subsections will discuss key findings from these studies.

3.1 C-repeat Binding Factor (CBF) and Circadian Rhythms

The CBF regulon are one of the key elements of transcriptional networks activated during cold acclimation (Shi et al., 2018). In the alpine plant *Saussurea involucrata*, induction of cold-responsive genes, such as *Sik*COR413PM1, displays two dynamic phases, a priming phase followed by sustained expression that minimize metabolic costs, potentially via carbohydrate metabolism (Guo et al., 2019). Similar results were reported in a time-series study of *Meconopsis horridula*, where delayed amplification of stress-responsive genes supports extended cold tolerance (Xu et al., 2025).

Table 7.2 Transcriptomic and integrated -omics studies on alpine plants under cold stress

Plant species	Habitat/ Location	Transcriptomic approach	Key findings	References
Crucihimalaya himalaica	Qinghai-Tibet Plateau	RNA-seq	487 positively selected high-altitude adaptation genes	Qiao et al. (2016)
Rhododendron nivale subsp. *boreale*	High-altitude mountaintops	Genome + transcriptome	Auto-tetraploid haplotype showcasing mountaintop adaptations	Lyu et al. (2024)
Eutrema species	High altitudes	Comparative genomics	Comparatives demographic trajectories and evolution of R gene families (LBS-LRR, NLRs)	Guo et al. (2018)
Prunus spp.	Tibet plateau	Population Genomics + metabolomics	Positive corelation between phenylpropanoid (379 metabolites) accumulations and altitude	Wang et al. (2021)
Saussurea species	High-altitude	RNA-seq	Characterized gene involved in flavonoid biosynthesis UV protection and DNA damage	Zhang et al. (2021)
Notopterygium incisum & *N. Franchetii*	High-alpine (China endemic)	Comparative transcriptome	Polymorphic SSRs for adaptive evolution and species divergence	Jia et al. (2017)
Potentilla bifurca	High altitude	RNA-seq	DEGs for secondary metabolites production such as phenylpropanoids and flavonoids	Tang et al. (2022)
Rhodiola algida	Qinghai-Tibetan Plateau	Comparative transcriptome	Tissue-specific adaptation	Zhang et al. (2014a, b)
Picrorhiza kurroa	Himalayan region	Metabolomics + transcriptome	Differential accumulation of secondary metabolites involved in cold adaptive response	Kumari et al. (2020)
Sinopodophyllum hexandrum	High-altitude	Transcriptome + metabolome	Comparative analysis revealing upregulation of gene involved in flavonoid synthesis and light stress genes at 3300 m and 2300 m altitude	Zhao et al. (2023)

(continued)

Table 7.2 (continued)

Plant species	Habitat/ Location	Transcriptomic approach	Key findings	References
Rheum nobile	3500–4500 m (Alpine subnival)	Multi-omics	Downregulation of photosynthetic genes within bracts, upregulation of flavonol genes and antifreeze proteins and accumulation of their products in overwintering roots	Li et al. (2023)
Chorispora bungeana	Subnival alpine	Deep-sequencing Transcriptome	Chilling tolerance mechanism involved upregulation of protein phosphorylation and auto-ubiquitination	Zhao et al. (2012)
Rhododendron anthopogon	Elevation gradient	RNA-seq	Identified 200 TFs involved in regulation of metabolic pathways influenced by elevational gradient	Mangral et al. (2025)
Dactylorhiza hatagirea	IHR	RNA-seq + metabolic profiling	DEGs for Spatio-temporal expression of secondary metabolite biosynthesis	Dhiman et al. (2019)
Rhododendron arboreum	IHR	Transcriptome + metabolomics	Identified genetic and metabolic factors underlying alpine stress conditions	Choudhary et al. (2019)
Rhododendron (Hirado Hybrids)	China	RNA-seq	Cold adaptation such as reported circadian rhythms related genes involved in induction of sugar acclimations and linked acclimation involved in lowering levels of ABA and JA to decrease lignin content	Liu et al. (2022)
Rhododendron aureum	China	RNA-seq	Cold adaptation response comprising gene regulation network in hormone signal transduction, cell homeostasis, lipid metabolism and photosystem reaction subunit	Cao et al. (2022)

Circadian control was also linked to cold response as evident from gene expression of *CCA1* and *LHY* oscillate with temperature cues, shaping the timing and amplitude of stress response (Dong et al., 2011). The gated circadian control under cold responsive transcription factors such as REV4 and RVE8 in *Arabidopsis* was found linked to co-activators proteins LNK3 and LNK4 under cold stress conditions (Kidokoro et al., 2023). Omics studies also revealed histone modifications associated with of circadian control of transcription that might play a vital role in adaptive behaviours of high-altitude plants during day-night cycles (Nakamichi, 2020). This synchronization of circadian gene expression with cold induced transcription suggests its active role in efficient energy management under unpredictable temperature changes.

3.2 Specialized Metabolite Synthesis and Stress Memory

During cold acclimation, alpine and extremophile plants carefully coordinate the timing of specialized metabolite production to defend against low temperatures. After exposure to cold stress, within hours alpine plants start the biosynthesis of specialized metabolic pathways such as phenylpropanoids and flavonoids with gene activation of phenylalanine ammonia-lyase (PAL), chalcone synthase (CHS), and flavonol synthase (FLS). This leads to a burst of antioxidant molecules and signaling compounds that help limit immediate oxidative damage (Kitashova et al., 2024; Li et al., 2025). After 3 days of cold exposure secondary pathways such as terpenoid and glucosinolate pathways become active, reinforcing cell structures and facilitating inter-plant communication via compounds like linalool and nerol (He et al., 2023). It was also demonstrated using subcellular fractionation that an early shortage of flavonoids can disrupt key plastid functions such as photorespiration, carbon and nitrogen balance, and folate metabolism. These disruptions, in turn, negatively impact protein stability within the cell (Kitashova et al., 2024).

In *Tetrastigma hemsleyanum,* cold stress lasting 12–48 h induce PAL, 4-coumarate: CoA ligase (4CL), CHS, anthocyanidin reductase (ANR), FLS, and leucoanthocyanidin reductase (LAR), raising levels of catechin, epicatechin, rutin, and quercetin (Peng et al., 2019). Different species show unique responses: in *Camellia sinensis*, cold triggers coordinated carbohydrate and flavonoid metabolism through calcium signaling and can be enhanced by silicon treatment to boost flavone and flavonol synthesis (Xie et al., 2024).

These metabolic shifts are regulated by transcription factors—including MYB, WRKY, bHLH, AP2/ERF, and NAC—that balance growth and defense (Ng et al., 2018). In cold-tolerant alfalfa (*Medicago sativa*), ubiquitination-related ribosomal proteins serve as central regulators of downstream metabolite genes (Pu et al., 2024).

Importantly, specialized metabolites also contribute to stress memory, enabling plants to "remember" prior cold exposure and respond more rapidly upon subsequent stresses. Accumulation of flavonoids and related antioxidants can persist beyond initial cold periods, stabilizing cellular redox balance and acting as biochemical priming agents. Such primed states are linked to epigenetic modifications that reinforce metabolite biosynthesis pathways (Wang & Qiao, 2020; Zhang et al., 2014a, b). This metabolic memory ensures metabolic readiness and reduces lag time during recurring freeze-thaw cycles typical of alpine environments.

Time-series experiments are essential for capturing these dynamic changes and stress memory phenomena, as single time-point studies cannot reveal the full sequence of metabolic adaptations that enable plants to survive freezing temperatures.

3.3 Chromatin Modelling and Epigenetics

Time-series transcriptomic studies have shown that plants retain a molecular "memory" of cold exposure through sustained chromatin remodelling and epigenetic modifications. The long-term tracking of gene expression revealed that chromatin remodelers VIN3, SWN, and CLF remain active after cold treatment, keeping stress-response loci in a primed state (Kim & Sung, 2013). In *Arabidopsis thaliana*, efficient re-acclimation to chilling depends on reprogramming histone marks and DNA methylation patterns established during the first cold exposure, enabling plants to "recall" prior stress and respond more rapidly (Talarico et al., 2024). Genome-wide methylome profiling further confirms that cold induces dynamic DNA methylation switches at key regulatory regions, altering transcription factor binding affinity and promoter reactivation capacity for stress-responsive genes (Banerjee et al., 2017). Similar chromatin-based priming has been observed in rice and barley, indicating deep conservation of cold-stress memory mechanisms across angiosperms (Hilker & Schmülling, 2019; Kong et al., 2020). Recent spatial and single-cell transcriptomic analyses in plants have demonstrated that stress-induced gene expression patterns differ across tissues and cell types (Yu et al., 2023). For example, spatial transcriptomics of poplar stems identified separate meristematic regions, each exhibiting its own characteristic expression signature (Du et al., 2023). These observations imply that spatial context plays a crucial role in establishing epigenetic or transcriptional memory within specific tissues.

Single-cell RNA sequencing technologies provide high resolution data points that can help in understanding tissue-specific memory related to cold acclimation (Denyer et al., 2019). Spatial and time-series transcriptomics together provide a unique perspective to decipher complex process spread across time and space, controlled by interconnected genetic, metabolic, and epigenetic layers. The deciphered rhythms of plant cold stress memory could provide opportunities and ways for advanced breeding and gene-editing approaches to improve crop resilience.

4 Post Transcriptional and RNA Processing Mechanisms

The mechanism of cold acclimation in alpine plants is not only regulated by controlled gene transcription. It also involved multitude of fine-tuned post-transcriptional processes like alternative splicing, mRNA stabilization, epitranscriptomic modifications, and regulation by chromatin-associated RNA. Collectively these processes make sure of temporal and functional control of cold responsive pathways. Further, time-series transcriptomics studies have revealed that alpine species possess unique, lineage specific post transcriptional mechanisms (Rathore et al., 2022).

These discoveries highlight cold acclimation as a dynamic process, where RNA-level regulation acts as a vital bridge between early cold perception and long-term adaptive stability.

4.1 Splicing and Extended Stabilization in Alpine Species

The alternative splicing has been associated with stress responses and adaptive evolution of plants (Verta & Jacobs, 2022). However, transcriptomic analysis in alpine plants *Lepidium meyenii* under cold induction revealed this mechanism is also linked to non-sense mediated mRNA decay (Shi et al., 2019).

High-throughput transcriptomic studies have revealed that exposure to low temperatures induces extensive intron retention, exon skipping, and splice site switching (Filichkin et al., 2018). Such splicing plasticity generates functionally diverse isoforms of stress-associated genes, particularly those encoding transcription factors and signaling proteins, thereby broadening functional repertoires under cold stress (Liu et al., 2018).

In addition to splicing regulation, mRNA stability mechanisms are tightly recruited during chilling. Unlike lowland species, where these transcripts are quickly broken down, alpine plants keep cold-induced gene transcripts stable in the cytosol for extended periods. These transcripts usually are those for heat shock proteins, lipid desaturases, and enzymes involved in osmolyte production. This prolonged stability supports ongoing protein production during long chilling periods, helping prevent metabolic collapse amid daily freeze–thaw cycles.

Cold-induced activation of RNA-binding proteins (RBPs), such as glycine-rich RBPs and serine/arginine-rich splicing factors, helps maintain select pools of transcripts required for osmoprotectant synthesis and membrane remodelling (Kim et al., 2007; Staiger & Brown, 2013). Stress granules and processing bodies (P-bodies) also serve as dynamic reservoirs of transcripts by selectively stabilizing or degrading mRNAs, ensuring that metabolic resources are directed toward essential protective pathways (Kosmacz et al., 2019). Together, these fine-scale RNA modifications enable plants to balance flexibility with efficiency during rapidly changing alpine temperatures.

In, addition, transcripts of transcription factors in such as NAC (ATAF1-like) and MYB (MYB73) show unusually long persistence after being activated in high altitude plants (Diao et al., 2020). These observations suggest an existence of lineage specific transcript stabilization mechanism in high altitude plants of alpine region. Overall, these processes act like an active temporal buffering system that allow protective responses through short growing seasons marked by repeated freeze-thaw cycles.

4.2 Chromatic Remodelling and Epigenetics

Epigenetic regulation as a post-transcriptional regulation in plants usually involves the mechanism of chromatin remodelling. The recent studies, revealed cold-teriggred epigenetic remodelling that involved histone marks like H3K4me3 and H3K27ac (Faivre et al., 2024). Specifically, there histone marks are associated with activation of RNA splicing factors and RNA-binding proteins. These outcomes also signal at RNA splicing and stability occurring within a larger framework of chromatin-based control.

The existence of non-coding RNAs (ncRNAs), including small RNAs and long non-coding RNAs (lncRNAs), also play an important role in connecting chromatin states transcript processing. Often in high-altitude medicinal plants, cold-responsive lncRNAs may recruit epigenetic modifiers, reinforce transcriptional memory, and maintain stable transcriptome patterns (Tian et al., 2019). This combined regulation represents various facets of chromatin remodelling that involved splicing dynamics, accessibility of chromatin, and epigenetic flexibility. Ultimately, these facets work together to create stress memory and allow evolutionary adaptation.

Overall, these findings highlight that post-transcriptional regulation in alpine plants is a important integrated multi-layered system. Alternative splicing, transcript stability, and epigenetic mechanisms provide both immediate response and long-lasting resilience to cold stress.

5 Phytohormonal Crosstalk and Sugar Metabolism

The phytohormones and sugars are interconnected regulators of cold acclimation in alpine plants that are vital of creating delicate balance between stress responses and developmental processes. As abscisic acid (ABA) is a well-known key player in activating cold-responsive genes, recent studies also emphasize important roles for jasmonates (JAs), cytokinins (CKs), brassinosteroids (BRs), and gibberellins in coordinating complex adaptive responses (Ding et al., 2025; Kapoor et al., 2021).

A comparative transcriptomic analysis in *Arabidopsis* reveals that cytokinin oxidase/dehydrogenase (CKX) genes are upregulated during extended cold exposure, which correlates with slowed growth and the onset of temporary dormancy (Prerostova et al., 2021). Additionally, brassinosteroid signaling through BES1/BZR1 transcription factors has been associated with cold stress response by strengthening cell wall structure and optimizing photoprotective processes (Percio et al., 2025). These finding suggest that phytohormonal cross-talk coordinates the temporal gene expression that further ensures acclimations response do not compromise reproductive or survival-critical developmental pathways. Complementing this hormonal regulation, sugars act as both metabolic substrates and adaptive signals.

Metabolic profiling studies across diverse plant species reveal that cold stress significantly upregulates sugar transporters, such as SWEET family members and ERD6-like transporters, alongside cell wall invertases (CWINVs). These components drive the sequential accumulation of cryoprotective sugars, including sucrose, raffinose, and trehalose, which enhance cold tolerance (Poschet et al., 2011; Nishanth et al., 2018;). Similarly, transcriptomic and metabolomic analysis in *Saposhnikovia divaricata* an alpine plant, revealed upregulation of gens involved in production of enzymes for sugar metabolism along flavonoid/terpenoid biosynthesis and plant hormone signaling pathways (Jiang et al., 2025). These mechanisms highlight conserved cold adaptation strategies across plant species.

In addition, an adaptive significance of trehalose metabolism was also found linked to important cold response cross-talk. The homologs of trehalose-6-phosphate synthase (TPS) and trehalose-6-phosphate phosphatase (TPP) show cold-induced upregulation in Arabidopsis thaliana (Zhao et al., 2025). These metabolites were not only in stabilizing cellular structures but also in signaling through SnRK1 kinase pathways and in modulating epigenetic remodelling complexes (Fichtner et al., 2020).

In collaboration, there findings demonstrate phytohormonal cross-talk and sugar metabolism create an integrated regulatory framework. Within this scaffold, metabolites and signaling molecules work collaboratively to coordinate cold acclimation, balancing rapid protection, sustained resilience, and adaptive flexibility essential for survival in alpine environments.

6 Ecological and Evolutionary Insights

The integration of time-series transcriptomics into the study of alpine plants has significantly deepened our understanding of how molecular mechanisms interact with ecological and evolutionary processes. By shifting focus from static gene catalogs to dynamic temporal expression patterns, this approach reveals the precise timing and coordination of gene activity that underpin cold acclimation. Such temporal detail is crucial for capturing the complex interactions between environmental signals and biological responses, allowing better predictions of not only individual plant survival but also the persistence of populations and stability of ecosystems amid accelerating climate change (Miller-Rushing et al., 2010). The following subsections highlight comprehensive ecological and evolutionary insights gained through these integrative studies.

6.1 Ecological Forecasting

Time-resolved transcriptional data serve as powerful molecular phenotypes, capturing the dynamic progression of stress responses over time and across different spatial scales. Core cold-responsive modules such as CBF/DREB transcription factors, ABA-dependent signaling elements, and late embryogenesis abundant (LEA) proteins, produce identifiable expression signatures that can be incorporated into mechanistic ecological niche and phenological models (Shinozaki & Yamaguchi-Shinozaki, 2007).

By linking these transcriptomic markers with climate and geographic information, researchers improve spatial predictions of frost tolerance limits and altitudinal range shifts, essential for species restricted to narrow elevational zones like *Saussurea, Notopterygium*, and other endemic alpine plants (Körner, 2023).

Additionally, transcriptomic timing biomarkers allow for fine-scale differentiation of survival probabilities and physiological states at the microhabitat level within populations, detail that traditional ecological surveys have been unable to provide (Exposito-Alonso et al., 2018. This dynamic molecular phenotyping is vital for modelling how often and how severely freeze–thaw cycles occur, as well as anticipating phenological mismatches under climate change, which is critical for proactive conservation efforts. Combining gene expression timelines with remote sensing and local climate data offers promising avenues to develop predictive tools that emphasize temporal plasticity as a cornerstone of alpine plant resilience.

6.2 Evolutionary Plasticity

Comparative time-series transcriptomics between alpine plants and their closely related lowland counterparts reveal that successful adaptation to harsh mountain environments depends not only on genomic content but critically on the evolutionary fine-tuning of gene expression timing referred to as "temporal plasticity" (Savolainen et al., 2013). This adaptation strategy allows alpine plants to align molecular defense precisely with highly variable and often unpredictable environmental fluctuations.

For instance, trehalose-phosphate signaling components maintain prolonged stability post-cold exposure, stabilizing cellular metabolism through repeated freeze–thaw cycles (Fichtner & Lunn, 2021). Likewise, stress-responsive NAC and MYB transcription factors show extended activity, linked to improved survival under prolonged cold stress (Sharma et al., 2020). This temporal reprogramming of gene expression networks, alongside genetic structural adaptations, serves as a key mechanism for extreme-environment adaptation across species.

Importantly, these insights challenge traditional perspectives on adaptation by emphasizing temporal modulation as a novel axis of evolutionary innovation. This understanding supports the development of predictive molecular markers to gauge adaptive capacity and drive breeding programs focused on climate resilience in both wild and cultivated plant species. Moreover, evolution of transcript timing plasticity offers a mechanistic basis for rapid acclimation and transgenerational memory in alpine plant populations, highlighting its ecological and evolutionary importance.

6.3 Conservation Applications and Bioresource Management

Ecological and evolutionary insights from time-series transcriptomics are being translated into practical tools for conservation genomics and resource management in alpine ecosystems. Molecular biomarkers derived from temporal gene expression profiles offer early and sensitive indicators of a population's ability to acclimate, guiding strategies such as assisted migration, seed sourcing, restoration, and cryo-preservation amid accelerating climate change (Aitken & Whitlock, 2013). These biomarkers help conservationists prioritize populations or genotypes with dynamic transcriptomic flexibility as an indicator of heightened resilience and functional adaptability, maximizing the effectiveness of biodiversity preservation efforts (Whitehead & Crawford, 2006).

Additionally, temporal transcriptomic data are shaping modern biodiversity policy by defining "adaptive potential zones." These are landscapes where restoration or species reintroduction is most likely to succeed due to the alignment of local climatic conditions with the molecular preparedness of populations. Such zones provide a robust scientific foundation for landscape-scale management under shifting climate regimes.

In agriculture, insights from temporal transcriptomics enable the strategic design and deployment of alpine-derived regulatory elements. For example, stage-specific promoters from genes like COR15a and ZAT12 could be engineered into key crops, enhancing cold tolerance without the growth penalties often seen with constant stress gene activation (Liu et al., 2014). This precise temporal regulation optimizes energy use, stabilizes yields, and supports adaptation to increasingly erratic climate conditions.

Collectively, these advances demonstrate how incorporating the temporal dimension of gene expression into conservation genomics and bioresource management not only safeguards biodiversity but also accelerates the translation of alpine plant resilience into broader ecological and agricultural sustainability frameworks critical for the Anthropocene epoch.

7 Challenges and Future Prospects

Time-series transcriptomics has revolutionized the study of cold acclimation mechanisms in alpine plants but still faces significant challenges on multiple fronts. The harsh and unpredictable conditions of alpine habitats characterized by difficult terrain and variable weather limit the feasibility of frequent, high-resolution sampling. As a result, many datasets lack sufficient temporal detail to capture rapid, transient, cyclic, or biphasic molecular events critical to early and late phases of cold acclimation (Zuther et al., 2015).

Compounding these logistical obstacles is a notable taxonomic bias in available data. Research largely focuses on model species such as *Arabidopsis thaliana*, while comprehensive, time-resolved datasets for true alpine specialists remain limited. This gap hinders the ability to tease apart universally conserved cold response mechanisms from lineage-specific evolutionary adaptations across diverse mountain floras like those in the Himalayas, Andes, and Hengduan Mountains (Xia et al., 2022).

Technical limitations further compound these issues. Many alpine plants lack high-quality reference genomes, forcing reliance on *de novo* transcriptome assemblies, which often produce fragmented isoform annotations, diminished splice variant detection, and challenges in identifying true orthologs. Such limitations impede rigorous functional comparisons between species (Grabherr et al., 2011). Moreover, existing computational tools optimized primarily for well-controlled laboratory datasets also face difficulties analysing uneven sampling, missing replicates, and noisy field-derived data typical of alpine studies. This reduces reproducibility and limits the application of advanced predictive modelling approaches such as dynamic Bayesian networks, wavelet analysis, and machine learning frameworks (van Bergen et al., 2020).

Biological complexities further compound this: bulk tissue transcriptomics averages signals across diverse cell types, obscuring highly localized and cell-type-specific cold responses. Emerging single-cell and spatial transcriptomic technologies hold promise to address this, revealing critical heterogeneity in epidermal, vascular, and meristematic cell populations where discrete molecular circuits likely mediate adaptive responses (Denyer et al., 2019; Rich-Griffin et al., 2020). Adding to this, many cold stress experiments oversimplify the natural alpine environment by employing abrupt cold shock or static freezing regimes under chamber conditions, thus failing to capture naturalistic gradual cooling, repeated freeze–thaw cycles, and photoperiodic fluctuations integral to alpine habitats (Körner, 2023). Moreover, transcriptomic investigations often remain siloed within the RNA layer, with limited integration of complementary epigenomic, proteomic, or metabolomic data, despite strong evidence that cold acclimation is orchestrated through complex cross-omic regulatory networks (Zenda et al., 2021).

Looking ahead, numerous innovations are poised to surmount current barriers and unlock new investigative frontiers. Increasing temporal resolution will require coordinated field research networks deploying portable sequencing devices and automated sampling systems, facilitating denser, ecologically valid chronologies of transcriptomic regulation (Parker et al., 2020). Expanding the taxonomic and genomic breadth through long-read sequencing, pan-genome assemblies, and chromosome-level scaffolding will enrich reference-quality resources vital for resolving isoform diversity and characterizing noncoding RNAs implicated in cold stress (Michael & VanBuren, 2020).

The integration of multi-omics—encompassing proteomics, metabolomics, chromatin accessibility (e.g., ATAC-seq), and phosphoproteomics—into temporally resolved frameworks promises to map the complete cascade from transcriptional activation to metabolite flux, constructing comprehensive systems-level models of cold adaptation (Ortega-Recalde & Hore, 2023). Advanced computational methodologies, including machine learning algorithms, differential equation modelling, and dynamic Bayesian inference tailored for irregular, sparse field datasets, are increasingly capable of identifying causal network hubs and feedback loops that are prime targets for genetic modification or selective breeding (Greenfield et al., 2013; Marbach et al., 2012).

Single-cell and spatial transcriptomics are forecasted to revolutionize the field by uncovering fine-scale spatial and cellular heterogeneity inaccessible to bulk approaches. Successfully applied in model species, these techniques will reveal tissue- and cell-type-specific regulatory modules in mesophyll, epidermis, and meristems that mediate alpine metabolic reprogramming and cold tolerance (Rich-Griffin et al., 2020). A complementary frontier lies in elucidating epigenetic priming, stress memory, and transgenerational inheritance. Alpine plants regularly encounter cyclic cold events, and chromatin remodelling along with DNA methylation acts as an epigenetic memory system sustaining durable acclimation and potentially transmitting stress tolerance across generations (Gallusci et al., 2023). Longitudinal interdisciplinary studies integrating temporal transcriptomic data with histone modification mapping, methylome analyses, and small RNA profiling across seasonal and generational scales will deepen understanding of this critical adaptive dimension.

Progress will also depend on robust standardization in experimental design and bioinformatics. The development of universally accepted phenological benchmarks, ecologically valid stress simulation protocols, and harmonized normalization pipelines adapted for alpine plant transcriptomics is urgently needed to foster inter-study comparability, meta-analysis, and reproducibility (Kooke & Keurentjes, 2012). Sustained, large-scale ecological transcriptomic monitoring across altitudinal gradients and climatic zones, paired with environmental data collection, will establish foundational baselines required for forecasting evolutionary and phenotypic responses to ongoing climate change (López-Maury et al., 2008).

Finally, translational applications offer promising future trajectories. Precision genome editing and synthetic biology tools, including CRISPR-Cas12a, base editors, and tuneable gene switches, enable temporally precise control of cold response networks. This targeted regulation enhances protective effects against cold stress while minimizing associated fitness costs (Bandyopadhyay et al., 2020). These technologies hold great promise for advancing alpine biodiversity conservation and for developing crops with improved stress tolerance tailored to increasingly unpredictable climates.

Looking ahead, the future of alpine cold transcriptomics resides at the intersection of rapid technological innovation, ecological realism, and practical application. By overcoming current limitations and harnessing multi-omics integration, high-resolution computational modelling, and targeted gene editing, researchers can evolve fragmented molecular data into comprehensive, predictive systems-level frameworks that offer meaningful ecological and agricultural benefits.

8 Conclusion

Time-series transcriptomics has emerged as a powerful tool to unravel how alpine plants respond to cold stress by revealing the temporal layers of gene regulation essential for survival in extreme environments. By monitoring gene expression across distinct acclimation phases, this method has identified conserved stress-responsive genes, lineage-specific regulators, and circadian-linked dynamics that together shape cold tolerance. Advances in sequencing technologies, computational tools, and integrative analyses have shed light on key pathways such as CBF signaling, lipid remodelling, antioxidant defense, and ABA-mediated perception. These findings underscore that plant responses to cold are dynamic, unfolding in coordinated, time-dependent patterns. This perspective expands understanding from individual genes to complex regulatory networks involving transcriptional switches, memory effects, and feedback loops critical for resilience. Despite challenges like standardizing experimental protocols, enriching genomic resources, and enhancing cross-species analytical tools, research continues to advance. Integrating transcriptomic data with metabolomic, proteomic, and epigenetic information, alongside emerging single-cell and spatial technologies, promises to deepen insights into cold adaptation. Beyond fundamental discoveries, these advances have practical implications for agriculture, aiding in the breeding and engineering of crops with improved tolerance to temperature variability.

Time-series transcriptomics offers more than an analytical method as it provides a conceptual framework to understand how plants synchronize environmental signals with internal biological timing. Its ongoing development will enrich ecological studies of alpine plant resilience and support actionable strategies for climate-resilient agriculture.

References

Aitken, S. N., & Whitlock, M. C. (2013). Assisted gene flow to facilitate local adaptation to climate change. *Annual Review of Ecology, Evolution, and Systematics, 44*(1), 367–388.

Asp, M., Bergenstråhle, J., & Lundeberg, J. (2020). Spatially resolved transcriptomes—Next generation tools for tissue exploration. *BioEssays, 42*(10), 1900221.

Bandyopadhyay, A., Kancharla, N., Javalkote, V. S., Dasgupta, S., & Brutnell, T. P. (2020). CRISPR-Cas12a (Cpf1): A versatile tool in the plant genome editing tool box for agricultural advancement. *Frontiers in Plant Science, 11*, 584151.

Banerjee, A., Wani, S. H., & Roychoudhury, A. (2017). Epigenetic control of plant cold responses. *Frontiers in Plant Science, 8*, 1643.

Barrios, A., Caminero, C., García, P., Krezdorn, N., Hoffmeier, K., Winter, P., & Pérez de la Vega, M. (2017). Deep Super-SAGE transcriptomic analysis of cold acclimation in lentil (Lens culinaris Medik.). *BMC Plant Biology, 17*(1), 111.

Cai, J., Shen, L., Kang, H., & Xu, T. (2025). RNA modifications in plant adaptation to abiotic stresses. *Plant Communications, 6*(2), 101229.

Cao, K., Zhang, Z., Fan, H., Tan, Y., Xu, H., & Zhou, X. (2022). Comparative transcriptomic analysis reveals gene expression in response to cold stress in Rhododendron aureum Georgi. *Theoretical and Experimental Plant Physiology, 34*(3), 347–366.

Chen, C., Ge, Y., & Lu, L. (2023). Opportunities and challenges in the application of single-cell and spatial transcriptomics in plants. *Frontiers in Plant Science, 14*, 1185377.

Chen, H., Chen, X., Li, X., Lin, X., Yue, L., Liu, C., & Li, Y. (2024). Growth and physiological response of Yulu Hippophae rhamnoides to drought stress and its omics analysis. *Plant Signaling & Behavior, 19*(1), 2439256.

Cheng, S., Melkonian, M., Smith, S. A., Brockington, S., Archibald, J. M., Delaux, P. M., et al. (2018). 10KP: A phylodiverse genome sequencing plan. *GigaScience, 7*(3), giy013.

Cheng, M., Cui, K., Zheng, M., Yang, T., Zheng, J., Li, X., et al. (2022). Physiological attributes and transcriptomics analyses reveal the mechanism response of Helictotrichon virescens to low temperature stress. *Bmc Genomics, 23*(1), 280.

Choudhary, S., Thakur, S., Jaitak, V., & Bhardwaj, P. (2019). Gene and metabolite profiling reveals flowering and survival strategies in Himalayan Rhododendron arboreum. *Gene, 690*, 1–10.

Denyer, T., Ma, X., Klesen, S., Scacchi, E., Nieselt, K., & Timmermans, M. C. (2019). Spatiotemporal developmental trajectories in the Arabidopsis root revealed using high-throughput single-cell RNA sequencing. *Developmental Cell, 48*(6), 840–852.

Dhiman, N., Sharma, N. K., Thapa, P., Sharma, I., Kumar Swarnkar, M., Chawla, A., et al. (2019). De novo transcriptome provides insights into the growth behaviour and resveratrol and trans-stilbenes biosynthesis in Dactylorhiza hatagirea-An endangered alpine terrestrial orchid of western Himalaya. *Scientific Reports, 9*(1), 13133.

Diao, P., Chen, C., Zhang, Y., Meng, Q., Lv, W., & Ma, N. (2020). The role of NAC transcription factor in plant cold response. *Plant Signaling & Behavior, 15*(9), 1785668.

Ding, Y., Shi, Y., & Yang, S. (2019). Advances and challenges in uncovering cold tolerance regulatory mechanisms in plants. *New Phytologist, 222*(4), 1690–1704. https://doi.org/10.1111/nph.15696

Ding, F., Fan, X., Tian, R., Wang, M., & Sun, Z. (2025). Crosstalk of Abscisic acid with other hormones and signaling molecules in tomato cold stress tolerance. *Horticulturae, 11*(6), 647.

Dong, M. A., Farré, E. M., & Thomashow, M. F. (2011). Circadian clock-associated 1 and late elongated hypocotyl regulate expression of the C-repeat binding factor (CBF) pathway in Arabidopsis. *Proceedings of the National Academy of Sciences, 108*(17), 7241–7246.

Du, J., Wang, Y., Chen, W., Xu, M., Zhou, R., Shou, H., & Chen, J. (2023). High-resolution anatomical and spatial transcriptome analyses reveal two types of meristematic cell pools within the secondary vascular tissue of poplar stem. *Molecular Plant, 16*(5), 809–828.

Exposito-Alonso, M., Vasseur, F., Ding, W., Wang, G., Burbano, H. A., & Weigel, D. (2018). Genomic basis and evolutionary potential for extreme drought adaptation in Arabidopsis thaliana. *Nature Ecology & Evolution, 2*(2), 352–358.

Faivre, L., Kinscher, N. F., Kuhlmann, A. B., Xu, X., Kaufmann, K., & Schubert, D. (2024). Cold stress induces rapid gene-specific changes in the levels of H3K4me3 and H3K27me3 in Arabidopsis thaliana. *Frontiers in Plant Science, 15*, 1390144.

Fichtner, F., & Lunn, J. E. (2021). The role of trehalose 6-phosphate (Tre6P) in plant metabolism and development. *Annual Review of Plant Biology, 72*(1), 737–760.

Fichtner, F., Olas, J. J., Feil, R., Watanabe, M., Krause, U., Hoefgen, R., et al. (2020). Functional features of TREHALOSE-6-PHOSPHATE SYNTHASE1, an essential enzyme in Arabidopsis. *The Plant Cell, 32*(6), 1949–1972.

Filichkin, S. A., Hamilton, M., Dharmawardhana, P. D., Singh, S. K., Sullivan, C., Ben-Hur, A., et al. (2018). Abiotic stresses modulate landscape of poplar transcriptome via alternative splicing, differential intron retention, and isoform ratio switching. *Frontiers in Plant Science, 9*, 5.

Fu, J., Miao, Y., Shao, L., Hu, T., & Yang, P. (2016). De novo transcriptome sequencing and gene expression profiling of Elymus nutans under cold stress. *BMC Genomics, 17*(1), 870.

Gallusci, P., Agius, D. R., Moschou, P. N., Dobránszki, J., Kaiserli, E., & Martinelli, F. (2023). Deep inside the epigenetic memories of stressed plants. *Trends in Plant Science, 28*(2), 142–153.

Grabherr, M. G., Haas, B. J., Yassour, M., Levin, J. Z., Thompson, D. A., Amit, I., et al. (2011). Full-length transcriptome assembly from RNA-Seq data without a reference genome. *Nature Biotechnology, 29*(7), 644–652.

Greenfield, A., Hafemeister, C., & Bonneau, R. (2013). Robust data-driven incorporation of prior knowledge into the inference of dynamic regulatory networks. *Bioinformatics, 29*(8), 1060–1067.

Grene, R., Klumas, C., Suren, H., Yang, K., Collakova, E., Myers, E., et al. (2012). Mining and visualization of microarray and metabolomic data reveal extensive cell wall remodeling during winter hardening in Sitka spruce (Picea sitchensis). *Frontiers in Plant Science, 3*, 241.

Guo, X., Hu, Q., Hao, G., Wang, X., Zhang, D., Ma, T., & Liu, J. (2018). The genomes of two Eutrema species provide insight into plant adaptation to high altitudes. *DNA Research, 25*(3), 307–315.

Guo, X., Zhang, L., Dong, G., Xu, Z., Li, G., Liu, N., et al. (2019). A novel cold-regulated protein isolated from Saussurea involucrata confers cold and drought tolerance in transgenic tobacco (Nicotiana tabacum). *Plant Science, 289*, 110246.

He, J., Yao, L., Pecoraro, L., Liu, C., Wang, J., Huang, L., & Gao, W. (2023). Cold stress regulates accumulation of flavonoids and terpenoids in plants by phytohormone, transcription process, functional enzyme, and epigenetics. *Critical Reviews in Biotechnology, 43*(5), 680–697.

Hilker, M., & Schmülling, T. (2019). Stress priming, memory, and signalling in plants. *Plant, Cell & Environment, 42*(3), 753–761.

Jain, M. (2024). Gene regulatory networks in abiotic stress responses via single-cell sequencing and spatial technologies: Advances and opportunities. *Current Opinion in Plant Biology, 82*, 102662.

Jean-Baptiste, K., McFaline-Figueroa, J. L., Alexandre, C. M., Dorrity, M. W., Saunders, L., Bubb, K. L., et al. (2019). Dynamics of gene expression in single root cells of Arabidopsis thaliana. *The Plant Cell, 31*(5), 993–1011.

Jia, Y., Liu, M. L., Yue, M., Zhao, Z., Zhao, G. F., & Li, Z. H. (2017). Comparative transcriptome analysis reveals adaptive evolution of Notopterygium incisum and Notopterygium franchetii, two high-alpine herbal species endemic to China. *Molecules, 22*(7), 1158.

Jiang, M., Pan, Y., Yu, K., Ma, Y., Cui, Y., Liu, Y., et al. (2025). Metabolic profiling and gene expression analyses shed light on the cold adaptation mechanisms of *Saposhnikovia divaricata* (Turcz.) Schischk. *Scientific Reports, 15*(1), 7070.

Kapoor, B., Kumar, P., Sharma, R., & Kumar, A. (2021). Regulatory interactions in phytohormone stress signaling implying plants resistance and resilience mechanisms. *Journal of Plant Biochemistry and Biotechnology, 30*(4), 813–828.

Kidokoro, S., Shinozaki, K., & Yamaguchi-Shinozaki, K. (2022). Transcriptional regulatory network of plant cold-stress responses. *Trends in Plant Science, 27*(9), 922–935.

Kidokoro, S., Konoura, I., Soma, F., Suzuki, T., Miyakawa, T., Tanokura, M., et al. (2023). Clock-regulated coactivators selectively control gene expression in response to different temperature stress conditions in Arabidopsis. *Proceedings of the National Academy of Sciences of the United States of America, 120*(16), e2216183120.

Kim, D. H., & Sung, S. (2013). Coordination of the vernalization response through a VIN3 and FLC gene family regulatory network in Arabidopsis. *The Plant Cell, 25*(2), 454–469.

Kim, J. Y., Park, S. J., Jang, B., Jung, C. H., Ahn, S. J., Goh, C. H., et al. (2007). Functional characterization of a glycine-rich RNA-binding protein 2 in Arabidopsis thaliana under abiotic stress conditions. *The Plant Journal, 50*(3), 439–451.

Kim, T. W., Park, C. H., Hsu, C. C., Kim, Y. W., Ko, Y. W., Zhang, Z., et al. (2023). Mapping the signaling network of BIN2 kinase using TurboID-mediated biotin labeling and phosphoproteomics. *The Plant Cell, 35*(3), 975–993.

Kitashova, A., Lehmann, M., Schwenkert, S., Münch, M., Leister, D., & Nägele, T. (2024). Insights into physiological roles of flavonoids in plant cold acclimation. *The Plant Journal, 120*(5), 2269–2285.

Kong, L., Liu, Y., Wang, X., & Chang, C. (2020). Insight into the role of epigenetic processes in abiotic and biotic stress response in wheat and barley. *International Journal of Molecular Sciences, 21*(4), 1480.

Kooke, R., & Keurentjes, J. J. B. (2012). Multi-dimensional regulation of metabolic networks shaping plant development and performance. *Journal of Experimental Botany, 63*(9), 3353–3365.

Körner, C. (2023). Concepts in alpine plant ecology. *Plants, 12*(14), 2666.

Kosmacz, M., Gorka, M., Schmidt, S., Luzarowski, M., Moreno, J. C., Szlachetko, J., et al. (2019). Protein and metabolite composition of Arabidopsis stress granules. *New Phytologist, 222*(3), 1420–1433.

Kumar, M., & Rani, K. (2023). Epigenomics in stress tolerance of plants under the climate change. *Molecular Biology Reports, 50*(7), 6201–6216.

Kumari, A., Singh, H. R., Jha, A., Swarnkar, M. K., Shankar, R., & Kumar, S. (2014). Transcriptome sequencing of rhizome tissue of Sinopodophyllum hexandrum at two temperatures. *BMC Genomics, 15*(1), 871.

Kumari, M., Joshi, R., & Kumar, R. (2020). Metabolic signatures provide novel insights to Picrorhiza kurroa adaptation along the altitude in Himalayan region. *Metabolomics, 16*(7), 77.
Kumari, M., Kumar, P., Saini, V., Joshi, R., Shankar, R., & Kumar, R. (2025). Transcriptional landscape illustrates the diversified adaptation of medicinal plants to multifactorial stress combinations linked with high altitude. *Planta, 261*(5), 1–25.
Li, Y., Niu, Z., Zhu, M., Wang, Z., Xu, R., Li, M., et al. (2023). Multi-omics data provide insight into the adaptation of the glasshouse plant Rheum nobile to the alpine subnival zone. *Communications Biology, 6*(1), 906.
Li, J., Liu, H., Xia, W., Mu, J., Feng, Y., Liu, R.,... & Chen, X. (2017). De novo transcriptome sequencing and the hypothetical cold response mode of Saussurea involucrata in extreme cold environments. *International Journal of Molecular Sciences, 18*(6),1155.
Li, J., Yu, Q., Liu, C., Zhang, N., & Xu, W. (2025). Flavonoids as key players in cold tolerance: Molecular insights and applications in horticultural crops. *Horticulture Research, 12*(4), uhae366.
Liu, D., Li, W., Cheng, J., & Hou, L. (2014). Expression analysis and functional characterization of a cold-responsive gene COR15A from Arabidopsis thaliana. *Acta Physiologiae Plantarum, 36*(9), 2421–2432.
Liu, Z., Qin, J., Tian, X., Xu, S., Wang, Y., Li, H., et al. (2018). Global profiling of alternative splicing landscape responsive to drought, heat and their combination in wheat (Triticum aestivum L.). *Plant Biotechnology Journal, 16*(3), 714–726.
Liu, B., Zhao, F. M., Cao, Y., Wang, X. Y., Li, Z., Shentu, Y., et al. (2022). Photoprotection contributes to freezing tolerance as revealed by RNA-seq profiling of rhododendron leaves during cold acclimation and deacclimation over time. *Horticulture Research, 9*, uhab025.
López-Maury, L., Marguerat, S., & Bähler, J. (2008). Tuning gene expression to changing environments: From rapid responses to evolutionary adaptation. *Nature Reviews Genetics, 9*(8), 583–593.
Lv, Z., Jiang, S., Kong, S., Zhang, X., Yue, J., Zhao, W., et al. (2024). Advances in single-cell transcriptome sequencing and spatial transcriptome sequencing in plants. *Plants, 13*(12), 1679.
Lyu, Z. Y., Zhou, X. L., Wang, S. Q., Yang, G. M., Sun, W. G., Zhang, J. Y., et al. (2024). The first high-altitude autotetraploid haplotype-resolved genome assembled (Rhododendron nivale subsp. boreale) provides new insights into mountaintop adaptation. *GigaScience, 13*, giae052.
Mangral, Z. A., Islam, S. U., Majeed, A., Tariq, L., Goel, S., Barik, S. K., & Dar, T. U. H. (2025). Transcriptomic insights into elevation-dependent gene expression in rhododendron anthopogon D. Don: Implications for climate resilience. *Physiologia Plantarum, 177*(4), e70419.
Marbach, D., Costello, J. C., Küffner, R., Vega, N. M., Prill, R. J., Camacho, D. M., et al. (2012). Wisdom of crowds for robust gene network inference. *Nature Methods, 9*(8), 796–804.
Meyers, B. C., Tej, S. S., Vu, T. H., Haudenschild, C. D., Agrawal, V., Edberg, S. B., et al. (2004). The use of MPSS for whole-genome transcriptional analysis in Arabidopsis. *Genome Research, 14*(8), 1641–1653.
Michael, T. P., & VanBuren, R. (2020). Building near-complete plant genomes. *Current Opinion in Plant Biology, 54*, 26–33.
Miller-Rushing, A. J., Høye, T. T., Inouye, D. W., & Post, E. (2010). The effects of phenological mismatches on demography. *Philosophical Transactions of the Royal Society B: Biological Sciences, 365*(1555), 3177–3186.
Nakamichi, N. (2020). The transcriptional network in the Arabidopsis circadian clock system. *Genes, 11*(11), 1284.
Ng, D. W., Abeysinghe, J. K., & Kamali, M. (2018). Regulating the regulators: The control of transcription factors in plant defense signaling. *International Journal of Molecular Sciences, 19*(12), 3737.
Nishanth, M. J., Sheshadri, S. A., Rathore, S. S., Srinidhi, S., & Simon, B. (2018). Expression analysis of cell wall invertase under abiotic stress conditions influencing specialized metabolism in Catharanthus roseus. *Scientific Reports, 8*(1), 15059.

Oono, Y., Seki, M., Satou, M., Iida, K., Akiyama, K., Sakurai, T., et al. (2006). Monitoring expression profiles of Arabidopsis genes during cold acclimation and deacclimation using DNA microarrays. *Functional & Integrative Genomics, 6*(3), 212–234.

Ortega-Recalde, O., & Hore, T. A. (2023). Analytical methods and tools to study the Epigenome. In *Epigenetics in aquaculture* (pp. 149–174).

Parker, M. T., Knop, K., Sherwood, A. V., Schurch, N. J., Mackinnon, K., Gould, P. D., et al. (2020). Nanopore direct RNA sequencing maps the complexity of Arabidopsis mRNA processing and m6A modification. *eLife, 9*, e49658.

Peng, X., Wu, H., Chen, H., Zhang, Y., Qiu, D., & Zhang, Z. (2019). Transcriptome profiling reveals candidate flavonol-related genes of Tetrastigma hemsleyanum under cold stress. *BMC Genomics, 20*(1), 687.

Percio, F., Rubio, L., Amorim-Silva, V., & Botella, M. A. (2025). Crucial roles of brassinosteroids in cell wall composition and structure across species: New insights and biotechnological applications. *Plant, Cell & Environment, 48*(3), 1751–1767.

Poschet, G., Hannich, B., Raab, S., Jungkunz, I., Klemens, P. A., Krueger, S., et al. (2011). A novel Arabidopsis vacuolar glucose exporter is involved in cellular sugar homeostasis and affects the composition of seed storage compounds. *Plant Physiology, 157*(4), 1664–1676.

Prerostova, S., Černý, M., Dobrev, P. I., Motyka, V., Hluskova, L., Zupkova, B., et al. (2021). Light regulates the cytokinin-dependent cold stress responses in Arabidopsis. *Frontiers in Plant Science, 11*, 608711.

Pu, X., Fu, Y., Xu, C., Li, X., Wang, W., De, K., et al. (2024). Transcriptomic analyses provide molecular insight into the cold stress response of cold-tolerant alfalfa. *BMC Plant Biology, 24*(1), 741.

Qiao, Q., Wang, Q., Han, X., Guan, Y., Sun, H., Zhong, Y., et al. (2016). Transcriptome sequencing of Crucihimalaya himalaica (Brassicaceae) reveals how Arabidopsis close relative adapt to the Qinghai-Tibet plateau. *Scientific Reports, 6*(1), 21729.

Rathore, N., Kumar, P., Mehta, N., Swarnkar, M. K., Shankar, R., & Chawla, A. (2022). Time-series RNA-Seq transcriptome profiling reveals novel insights about cold acclimation and deacclimation processes in an evergreen shrub of high altitude. *Scientific Reports, 12*(1), 15553.

Rich-Griffin, C., Stechemesser, A., Finch, J., Lucas, E., Ott, S., & Schäfer, P. (2020). Single-cell transcriptomics: A high-resolution avenue for plant functional genomics. *Trends in Plant Science, 25*(2), 186–197.

Savolainen, O., Lascoux, M., & Merilä, J. (2013). Ecological genomics of local adaptation. *Nature Reviews Genetics, 14*(11), 807–820.

Sharma, P., Sharma, M. M. M., Patra, A., Vashisth, M., Mehta, S., Singh, B., et al. (2020). The role of key transcription factors for cold tolerance in plants. In *Transcription factors for abiotic stress tolerance in plants* (pp. 123–152). Academic.

Shi, Y., Ding, Y., & Yang, S. (2018). Molecular regulation of CBF signaling in cold acclimation. *Trends in Plant Science, 23*(7), 623–637.

Shi, Y., Su, Z., Yang, H., Wang, W., Jin, G., He, G., et al. (2019). Alternative splicing coupled to nonsense-mediated mRNA decay contributes to the high-altitude adaptation of maca (Lepidium meyenii). *Gene, 694*, 7–18.

Shinozaki, K., & Yamaguchi-Shinozaki, K. (2007). Gene networks involved in drought stress response and tolerance. *Journal of Experimental Botany, 58*(2), 221–227.

Ståhl, P. L., Salmén, F., Vickovic, S., Lundmark, A., Navarro, J. F., Magnusson, J., et al. (2016). Visualization and analysis of gene expression in tissue sections by spatial transcriptomics. *Science, 353*(6294), 78–82.

Staiger, D., & Brown, J. W. (2013). Alternative splicing at the intersection of biological timing, development, and stress responses. *The Plant Cell, 25*(10), 3640–3656.

Sun, Q., Zhou, X., Yang, L., Xu, H., & Zhou, X. (2023). Integration of phosphoproteomics and transcriptome studies reveals ABA signaling pathways regulate UV-B tolerance in Rhododendron chrysanthum leaves. *Genes, 14*(6), 1153.

Talarico, E., Zambelli, A., Araniti, F., Greco, E., Chiappetta, A., & Bruno, L. (2024). Unravelling the epigenetic code: DNA methylation in plants and its role in stress response. *Epigenomes, 8*(3), 30.

Tang, K., Zhao, L., Ren, Y., Yang, S., Zhu, J. K., & Zhao, C. (2020). The transcription factor ICE1 functions in cold stress response by binding to the promoters of CBF and COR genes. *Journal of Integrative Plant Biology, 62*(3), 258–263. https://doi.org/10.1111/jipb.12918

Tang, X., Li, J., Liu, L., Jing, H., Zuo, W., & Zeng, Y. (2022). Transcriptome analysis provides insights into Potentilla bifurca adaptation to high altitude. *Life, 12*(9), 1337.

Tenorio Berrío, R., & Dubois, M. (2024). Single-cell transcriptomics reveals heterogeneity in plant responses to the environment: A focus on biotic and abiotic interactions. *Journal of Experimental Botany, 75*(17), 5188–5203.

Thron, C., & Jafari, F. (2025). Correcting scale distortion in RNA sequencing data. *BMC Bioinformatics, 26*(1), 32.

Tian, M., Ci, D., Song, Y., & Zhang, D. (2019). Transcriptional regulation of chilling stress responsive long noncoding RNAs in Populus simonii. *Trees, 33*(3), 733–749.

van Bergen, G. H., Duenk, P., Albers, C. A., Bijma, P., Calus, M. P., Wientjes, Y. C., & Kappen, H. J. (2020). Bayesian neural networks with variable selection for prediction of genotypic values. *Genetics Selection Evolution, 52*(1), 26.

Verta, J. P., & Jacobs, A. (2022). The role of alternative splicing in adaptation and evolution. *Trends in Ecology & Evolution, 37*(4), 299–308.

Wang, L., & Qiao, H. (2020). Chromatin regulation in plant hormone and plant stress responses. *Current Opinion in Plant Biology, 57*, 164–170.

Wang, X. C., Zhao, Q. Y., Ma, C. L., Zhang, Z. H., Cao, H. L., Kong, Y. M., et al. (2013). Global transcriptome profiles of Camellia sinensis during cold acclimation. *BMC Genomics, 14*(1), 415.

Wang, X., Liu, S., Zuo, H., Zheng, W., Zhang, S., Huang, Y., et al. (2021). Genomic basis of high-altitude adaptation in Tibetan Prunus fruit trees. *Current Biology, 31*(17), 3848–3860.

Wang, A., Guo, W., Wang, S., Wang, Y., Kong, D., & Li, W. (2024). Transcriptome analysis unveiled the genetic basis of rapid seed germination strategies in alpine plant Rheum pumilum. *Scientific Reports, 14*(1), 19194.

Whitehead, A., & Crawford, D. L. (2006). Variation within and among species in gene expression: Raw material for evolution. *Molecular Ecology, 15*(5), 1197–1211.

Wu, Y., Wei, W., Pang, X., Wang, X., Zhang, H., Dong, B., et al. (2014). Comparative transcriptome profiling of a desert evergreen shrub, Ammopiptanthus mongolicus, in response to drought and cold stresses. *BMC Genomics, 15*(1), 671.

Xia, M., Liu, Y., Liu, J., Chen, D., Shi, Y., Chen, Z., et al. (2022). Out of the Himalaya-Hengduan Mountains: Phylogenomics, biogeography and diversification of Polygonatum Mill.(Asparagaceae) in the northern hemisphere. *Molecular Phylogenetics and Evolution, 169*, 107431.

Xie, X., Yang, S., Zhao, X., Shang, T., & Han, X. (2024). Mechanisms of silicon-mediated amelioration for Camellia sinensis using physiology, transcriptomics, and metabolomics under cold stress. *Beverage Plant Research, 5*(1), e003.

Xu, J., Chen, Z., Wang, F., Jia, W., & Xu, Z. (2020). Combined transcriptomic and metabolomic analyses uncover rearranged gene expression and metabolite metabolism in tobacco during cold acclimation. *Scientific Reports, 10*(1), 5242.

Xu, G., Tao, Z., & He, Y. (2022). Embryonic reactivation of FLOWERING LOCUS C by ABSCISIC ACID-INSENSITIVE 3 establishes the vernalization requirement in each Arabidopsis generation. *The Plant Cell, 34*(6), 2205–2221.

Xu, G., Guo, J., Yu, X., Zhao, N., Li, X., Yuan, T., et al. (2025). Multi-omics analysis reveals adaptive strategies of Meconopsis horridula to UV-B radiation in the Qinghai-Tibet plateau. *Plant, Cell & Environment, 48*, 8249–8263.

Yang, G., Liu, W., Mo, X., & Mei, Z. (2025). Alpine adaptive mechanism on rhizosphere microbes recruitment of Crepis napifera (Franch.) Babc. By multi-omics analysis. *Biology, 14*(4), 345.

Yu, X., Liu, Z., & Sun, X. (2023). Single-cell and spatial multi-omics in the plant sciences: Technical advances, applications, and perspectives. *Plant Communications, 4*(3), 100508.

Zenda, T., Liu, S., Dong, A., Li, J., Wang, Y., Liu, X., et al. (2021). Omics-facilitated crop improvement for climate resilience and superior nutritive value. *Frontiers in Plant Science, 12*, 774994.

Zhang, Y., Butelli, E., & Martin, C. (2014a). Engineering anthocyanin biosynthesis in plants. *Current Opinion in Plant Biology, 19*, 81–90.

Zhang, F., Gao, Q., Khan, G., Luo, K., & Chen, S. (2014b). Comparative transcriptome analysis of aboveground and underground tissues of Rhodiola algida, an important ethno-medicinal herb endemic to the Qinghai-Tibetan plateau. *Gene, 553*(2), 90–97.

Zhang, X., Sun, Y., Landis, J. B., Shen, J., Zhang, H., Kuang, T., et al. (2021). Transcriptomes of Saussurea (Asteraceae) provide insights into high-altitude adaptation. *Plants, 10*(8), 1715.

Zhang, X., Yu, F., Lyu, X., Chen, J., Zeng, H., Xu, N., et al. (2023). Transcriptome profiling of Bergenia purpurascens under cold stress. *BMC Genomics, 24*(1), 754.

Zhao, Z., Tan, L., Dang, C., Zhang, H., Wu, Q., & An, L. (2012). Deep-sequencing transcriptome analysis of chilling tolerance mechanisms of a subnival alpine plant, Chorispora bungeana. *BMC Plant Biology, 12*(1), 222.

Zhao, Q., Dong, M., Li, M., Jin, L., & Paré, P. W. (2023). Light-induced flavonoid biosynthesis in Sinopodophyllum hexandrum with high-altitude adaptation. *Plants, 12*(3), 575.

Zhao, M., Wang, L., Feng, M., Li, X., Huang, C., Lian, S., et al. (2025). Trehalose-6-phosphate synthase VvTPS10 regulates cold tolerance by participating in the CBF pathway and VvPUB19-mediated ubiquitination in grapevine. *Plant, Cell & Environment, 48*, 8086–8102.

Zhong, Y., Luo, Y., Sun, J., Qin, X., Gan, P., Zhou, Z., et al. (2024). Pan-transcriptomic analysis reveals alternative splicing control of cold tolerance in rice. *The Plant Cell, 36*(6), 2117–2139.

Zhou, P., Li, X., Liu, X., Wen, X., Zhang, Y., & Zhang, D. (2021). Transcriptome profiling of Malus sieversii under freezing stress after being cold-acclimated. *BMC Genomics, 22*(1), 681.

Zuther, E., Juszczak, I., Ping Lee, Y., Baier, M., & Hincha, D. K. (2015). Time-dependent deacclimation after cold acclimation in Arabidopsis thaliana accessions. *Scientific Reports, 5*(1), 12199.

Plant Omics and Co-expression Networks Capture Key Components of Secondary Metabolites Biosynthesis in Alpine Medicinal Herbs

8

Bhawna Kumari, Rutvik Sameer Kulkarni, Siddhant Diliprao Mahabale, Rashmi Kishor Agashe, Varun Kumar, and Rajinder Singh Chauhan

Abstract

Alpine medicinal plants grow in high-altitude conditions such as extreme cold, strong UV radiation, limited nutrients, and short growing seasons. To survive these harsh conditions, they have evolved unique physiological traits, including the production of diverse secondary metabolites with strong medicinal potential. This chapter summarizes current knowledge on alpine plant ecology, traditional uses, and phytochemical diversity, emphasizing how multi-omics approaches are advancing this field. Genomic, transcriptomic, proteomic, metabolomic, epigenomic and microbiome studies reveal how genes, enzymes and microbes work together to enhance stress tolerance and metabolite production. Integrating these datasets through systems biology and machine learning helps identify key regulatory networks and pathways for metabolic engineering and conservation.

Keywords

Alpine medicinal plants · Secondary metabolites · Genomics · Proteomics · Metabolomics · Co-expression networks

1 Introduction

Ancient systems such as Ayurveda, Traditional Chinese Medicine, Egyptian papyri, and Unani texts illustrate that plants have long served as the foundation of medical practice across civilizations. In many rural and high-altitude mountain ecosystems,

B. Kumari · R. S. Kulkarni · S. D. Mahabale · R. K. Agashe · V. Kumar · R. S. Chauhan (✉)
Center for Life Sciences, Mahindra University, Hyderabad, Telangana, India
e-mail: rajinder.chauhan@mahindrauniversity.edu.in

P. Kumar et al. (eds.), *Climate Resilience and Molecular Adaptation in Alpine Medicinal Plants*, https://doi.org/10.1007/978-981-95-6923-6_8

traditional medicine continues to serve as the primary healthcare system due to its accessibility, affordability, therapeutic efficacy, and minimal side effects (Ahmad et al., 2021). Several plant genera, including *Aconitum* (Kumar et al., 2016), *Nardostachys* (Yang et al., 2025a), *Gentiana* (Mirzaee et al., 2017), *Bergenia* (Liu et al., 2023), *Picrorhiza* (Kumar et al., 2017a), *Rhodiola* (Zhu et al., 2025), *Saussurea* (Xu et al., 2025), *Dendrobium* (Li et al., 2025a), *Valeriana* (Li et al., 2022), *Hypericum* (Shasmita et al., 2023), *Swertia* (Padhan et al., 2015) and *Podophyllum* (Kumar et al., 2015a, b), have been traditionally used to treat a wide range of ailments, including dermatological, gastrointestinal, respiratory, gynecological, and neurological disorders. These high-value species play a crucial role in sustaining the livelihoods of ruraland mountain communities in India and Nepal (Chhetri et al., 2021; Sharma & Kala, 2018). However, species such as *Podophyllum hexandrum, Swertia chirayita, Aconitum heterophyllum* and *Picrorhiza kurroa* (Kumar et al., 2015a, b, 2017a; Kumar & Van Staden, 2016; Padhan et al., 2015) have become endangered due to overharvesting and habitat degradation. Conserving these species is essential for maintaining the economic stability of mountain communities.

Alpine plants thrive at elevations reaching up to 6000 m in the Himalayas (Jandova et al., 2025), where they withstand low temperatures, drought, high UV radiation, strong winds, and nutrient-poor soils combined with short growing seasons (Binter et al., 2025). These harsh environments have driven the evolution of complex physiological and metabolic adaptations, leading to the synthesis of diverse antimicrobial, anti-inflammatory, antioxidant and neuroprotective secondary metabolites (Ashraf et al., 2024). Medicinal plants thus represent an invaluable source of bioactive compounds with significant applications in human health, pharmaceutics, functional foods (Riaz et al., 2023), agrochemicals (Kilani-Morakchi et al., 2021), cosmetics (Ochatt et al., 2022), and fragrance industries (Hazrati et al., 2024). Despite their immense potential, only a limited number of alpine species have been characterized at the molecular level. A major barrier to sustainable utilization, conservation, and discovery of new therapeutics lies in the limited understanding of the biosynthetic pathways responsible for the production of these secondary metabolites.

Recent advances in multi-omics*viz.* genomics, transcriptomics, proteomics, metabolomics, epigenomics and microbiome/meta-omics coupled with next-generation sequencing, high-resolution mass spectrometry, and machine learning approaches now enable a systems-level dissection of alpine plant biology. Genomic resources, including chromosome-scale assemblies for non-model species such as *Gentiana straminea* (Kelsang et al., 2024), *Bergenia purpurascens* (Liu et al., 2025a), *Leontopodium alpinum* (Cho et al., 2020), *Aconitum heterophyllum* (Wang et al., 2024a) and *Potentilla saundersiana* (Li et al., 2025b), reveal genome duplications and expansions in gene families associated with secondary metabolite biosynthesis. Transcriptome analyses have identified regulators of metabolite production, stress tolerance and developmental traits across species such as *Aconitum heterophyllum* (Pal et al., 2015), *Hypericum perforatum* (Selvakesavan et al., 2024), *Gentiana lutea* (Zhou et al., 2023), *Picrorhiza kurroa* (Kumari et al., 2021; Vashisht et al., 2016), *Swertia chirayita* (Pal et al., 2018), and *Dendrobium huoshanense* (Wu et al., 2023). Proteomic and metabolomic investigations in *Kobresia pygmaea*

(Li et al., 2014), *Picrorhiza kurroa* (Kumar et al., 2017b; Kumari et al., 2021), and *Taxus baccata* (Hao et al., 2017) highlight the stress-induced upregulation of protective proteins and specialized metabolic pathways.

Epigenomic and microbiome analyses further uncover DNA methylation patterns and rhizosphere symbioses, particularly with *Pseudomonas*, *Sphingomonas*, and *Bacillus* species that enhance stress resilience and secondary metabolite production (Kołtun-Jasion et al., 2025; Lv et al., 2010). Integrated multi-omics pipelines that combine correlation networks, machine learning, and pathway enrichment analyses (KEGG, GO) now facilitate the identification of key regulatory hubs for targeted metabolic engineering and conservation. Collectively, these holistic, system-level approaches pinpoint the potential of multi-omics to safeguard and sustainably utilize alpine medicinal plant biodiversity in the face of climate change.

2 Secondary Metabolites in Alpine Medicinal Herbs: Adaptation and Value

2.1 Diversity of Secondary Metabolites in Alpine Plants

Alpine medicinal plants produce a wide diversity of secondary metabolites, primarily alkaloids, terpenoids, flavonoids, phenolics, tannins, saponins, and glycosides (Table 8.1). Minor classes such as polyketides, steroids, and fatty acid derivatives are also common in specific taxa (Sekar et al., 2024). The diversity of these metabolites arises from intricate biosynthetic networks, many of which are being elucidated through genomics, transcriptomics, and metabolomics (Kumar, 2019; Zhong et al., 2025). The shikimate pathway generates phenylpropanoids, flavonoids, and tannins, essential for UV protection and antioxidant defence (Kumari et al., 2023), while the mevalonate (MVA) and methylerythritol phosphate (MEP) pathways underpin terpenoid biosynthesis (Bergman & Dudareva, 2024). Polyketide pathways also yield specialized compounds such as anthraquinones and phenylethanoid glycosides (Wang et al., 2024b).

The structural diversity of secondary metabolites in alpine plants reflect enhanced flux through stress-responsive biosynthetic pathways. This metabolic flux is regulated by multiple mechanisms, notably transcriptional control by WRKY and MYB transcription factors, which are upregulated under environmental stresses and activate genes related to secondary metabolism (Vashisht et al., 2016). For example, UV-B radiation and cold stress increase phenylalanine ammonia-lyase (PAL) activity, enhancing phenolic biosynthesis (Wang et al., 2025a). In *Rhododendron chrysanthum*, phenolic compounds such as gallocatechin, 6-methoxyflavone, kaempferol-3-O-arabinoside, naringenin chalcone, and quercetin-3-O-arabinoside increased up to 4.28-fold (Gong et al., 2023). Similarly, sesquiterpenoid content in *Artemisia brevifolia* increased by 57% (Nataraj et al., 2022), and alkaloid levels increased significantly in *Lupinus polyphyllus* (ammodendrine, lupanine; 2.86-fold) and *Hyoscyamus niger* (hyoscyamine, scopolamine; 8.06-fold) (Jan et al., 2016; Zaffar et al., 2018). Cold stress activates CBF and MYB transcription factors,

Table 8.1 Alpine plant species, their secondary metabolites, and pharmacological applications

Plant name	Family	Altitude (metres)	Secondary metabolites	Role in plants	Medicinal use	References
Rhodiola imbricata	Crassulaceae	4000–5000	Salidroside, Rosavin	UV and cold protection	Adaptogen, neuroprotective	Singamaneni et al. (2020)
Picrorhiza kurroa	Plantaginaceae	3000–5000	Picroside I, II and Kutkoside	Bitter defence pathogen protection	Hepatoprotective	Raut et al. (2023)
Swertia chirayata	Gentianaceae	1200–2100	Amarogentin, Swertiamarin	Bitter, antiherbivory, antimicrobial	Antipyretic, antidiabetic	Kumar et al. (2016)b
Podophyllum hexandrum	Berberidaceae	2700–4200	Podophyllotoxin	Antifungal, antiherbivory	Precursor to anticancer drug and effective against warts, skin diseases	Sofi et al. (2025)
Taxus wallichiana	Taxaceae	1800–3300	Paclitaxel	Antiherbivory	High fever and chronic inflammatory diseases	Juyal et al. (2014)
Nardostachys jatamansi	Caprifoliaceae	3000–5000	Jatamansone, Sesquiterpenes	Antiherbivory, antimicrobial	Anxiety, insomnia, hypotensive, anticonvulsant and antimicrobial,	Pathak and Godela (2024)
Saussurea costus	Asteraceae	2500–3800	Costunolide, Dehydrocostus	Antibacterials, antivirals, antifungals, and antifeedants	Anti-cancer, anti-inflammatory, antioxidant and anti-diabetic	Kumari et al. (2024b)
Berberis aristata	Berberidaceae	1800–3500	Berberine	Antimicrobial	Metabolic, eye, ears and dental uses	Rathi et al. (2013)
Hippophae rhamnoides	Elaeagnaceae	2500–4000	Flavonoids, Carotenoids	UV-oxidative defence	Cardiometabolic, skin	Kubczak et al. (2022)
Aconitum heterophyllum	Ranunculaceae	2500–4000	Atisine	Antiherbivory	Antidiarrhea, antipyretic	Malhotra and Sharma (2021)
Hypericum perforatum	Hypericaceae	1500–3000	Hypericin, Hyperforin	Phototoxic defence	Mild-depression	Kapoor et al. (2023)

Aconitum spicatum	Ranunculaceae	1800–4200	Atisine, Spicatine, Napelline, Aconitine, Mesaconitine, Hypaconitine	Defence against herbivores and pathogens, UV-protection	Analgesic and anti-inflammatory effects	Chapagain et al. (2019)
Allium wallichii Kunth	Amaryllidaceae	2400–4600	Organosulfur (Allicin, Diallyl sulfide, Allyl propyl disulfide, Ajoene, S-methylcysteine sulfoxide)	Defence against herbivores, insects, and soil pathogens	Cardiovascular protection, anticancer potential	Singh Rana et al. (2022)
Bergenia ciliata	Saxifragaceae	1000–3200	Bergenin, Arbutin	Defence against fungal and bacterial pathogens, protection from herbivores, and oxidative stress	Anti-urolithic (stone dissolving), diuretic	Liu et al. (2023)
Cinnamomum tamala	Lauraceae	450–2000	Cinnamaldehyde, Eugenol, Cinnamyl acetate, Linalool, Safrole, β-caryophyllene	Defence against herbivores, fungi, and insects; contribute to fragrance that attracts pollinators	Treatment against cancer, cardiac diseases, diabetes, anxiety, depression, ulcer, GI diseases	Tiwari and Talreja (2020)
Corydalis govaniana	Papaveraceae	3000–4800	Corydaline, Tetrahydropalmatine, Corydalmine, Canadine, Berberine	Stabilize membranes under cold and drought stress	Treating syphilis, scrofulla, cutaneous infection, diarrhoea	Muhammad et al. (2015)
Dactylorhiza hatagirea	Orchidaceae	2800–3960	Dactylorhins A–E, Dactyloses A–B	Store energy in tubers, help survive alpine frost, drought, and nutrient-poor soils	Curing disorders of the circulatory, respiratory, nervous, digestive, skeletal, and reproductive systems, besides boosting the immune system to fight infectious diseases	Sharma et al. (2024)

(continued)

Table 8.1 (continued)

Plant name	Family	Altitude (metres)	Secondary metabolites	Role in plants	Medicinal use	References
Delphinium himalayae Munz	Ranunculaceae	3000–4500	Delphinine, Lycoctonine	Highly toxic protect against grazing herbivores (yaks, goats, sheep) and insect pests	Joint pain, inflammation, and wounds	Parajuli et al. (2021)
Dioscorea deltoidea	Dioscoreaceae	450–3100	Diosgenin	Bitter, toxic to many herbivores and soil pathogens, protect rhizomes	Gastrointestinal and urogenital disorders, diarrhoea, irritability, abdominal pain, wounds, intestinal worms, anaemia, and with ophthalmic conditions	Semwal et al. (2021)
Ephedra gerardiana	Ephedraceae	2400–4500	Ephedrine, Pseudoephedrine, Norephedrine, Norpseudoephedrine	Defence compounds, bitter taste deter herbivores and pests; also regulate osmotic stress in dry habitats	Used for asthma, bronchitis, cough, cold, fever	Choudhary et al. (2021)
Fritillaria cirrhosa	Liliaceae	3000–4600	Edpetiline, Peimine, Peiminine, Imperialine, Verticine, Fritimine	Bitter and toxic strong defence against herbivores and grazing animals	Anti-asthmatic, antioxidant, antitumor, and anti-inflammatory functions	Wang et al. (2025b)
Hedychium spicatum	Zingiberaceae	1500–2100	Curzerene, Coronarin, Curdione, Linderazulene	Protect rhizomes from microbial infection, herbivores, and soil-borne pathogens; also attract pollinators	Bronchitis, asthma, hiccups, digestive disorders	Rawat et al. (2021)

Heracleum candicans	Apiaceae	2200–3800	Psoralen, Bergapten, Imperatorin, Xanthotoxin, Isopimpinellin	UV-activated toxins that deter grazing animals (cattle, sheep, yak) and insects; strong antimicrobial activity	Skin diseases, ulcers, and rheumatism	Sher et al. (2021)
Nardostachys jatamansi	Caprifoliaceae	3200–5000	Jatamansone (Valeranone), Nardostachone, Jatamansinol, Nardosinone	Defence against soil microbes and herbivores; root aroma also prevents insect and fungal attack	Neuroprotective, anxiety, insomnia, hypotensive, anticonvulsant antimicrobialand anxiolytic	Pathak and Godela (2024)
Paris polyphylla	Melanthiaceae	2000–3000	Polyphyllin I–VIII	Deter grazing herbivores, antifungal & antibacterial	Anticancer, anti-inflammatory andanalgesic	Thapa et al. (2022)
Potentilla lineata	Rosaceae	1600–4800	Quercetin, Kaempferol, Rutin, Isoquercitrin	Protect leaves from UV-B damage	Diarrhea, dysentery, and bleeding disorders	Tomczyk and Latté (2009)
Rheum acuminatum	Polygonaceae	3300–4200	Emodin, Rhein, Chrysophanol, Aloe-emodin, Physcion	Act as phytoalexins, providing defence against fungal/bacterial pathogens; bitter/toxic compounds deter grazing animals	Laxative, stomach-ache, liver tonic, anti-inflammatory and blood purifier	Rokaya et al. (2012)
Rheum australe	Polygonaceae	3200–4200	Emodin, Rhein, Chrysophanol, Aloe-emodin, Physcion			
Rhododendron anthopogon	Ericaceae	3000–4800	Monoterpenes (α-pinene, β-pinene, Limonene, Borneol, 1,8-Cineole, Camphene, Myrcene)	Provide UV-B protection at high altitudes, pigmentation for pollination, and ROS scavenging during cold/drought stress	Cough, cold, headache, asthma, and digestive ailments	Mangral et al. (2025)

(continued)

Table 8.1 (continued)

Plant name	Family	Altitude (metres)	Secondary metabolites	Role in plants	Medicinal use	References
Rubia manjith	Rubiaceae	1200–2100	Manjistin, Purpurin, Munjistin, Alizarin, Rubiadin, Nordamnacanthal	Red pigment in roots, protecting against herbivores and root pathogens	Blood purifier, anti-inflammatory, detoxifier, and in skin diseases, gynaecological problems, liver ailments	Pradhan et al. (2021)
Valeriana hardwickei	Caprifoliaceae	1200–4000	Valtrate, Didrovaltrate, Acevaltrate	Volatile defence molecules, feeding deterrents to insects and grazing animals	Insomnia, hysteria, epilepsy, headaches, heart palpitations, rheumatism, and digestive disorders	Bashir et al. (2011)
Zanthoxylum armatum	Rutaceae	1100–2500	Armatamide, Nitidine, Chelerythrine, Magnoflorine, Berberine	Toxic to insects, grazing animals, and pathogenic microbes; strong chemical defence	Used as a natural toothbrush; fruits and bark relieve toothache, gum problems, and bad breath	Shah et al. (2025)

redirecting metabolic flux from amino acids such as phenylalanine, tyrosine, and tryptophan toward secondary metabolism, resulting in a 2.67-fold increase in phenolic and flavonoid content (Tang et al., 2025). Drought similarly promotes flavonoid accumulation by upregulating key enzymes such as chalcone synthase (CHS) and flavanone-3-hydroxylase (F3H), leading to increased production of osmoprotectants and antioxidants including flavonoids, proline, and phenolic acids (Cai et al., 2023).

Exposure to abiotic stress enhances the diversity and accumulation of secondary metabolites in alpine medicinal plants, thereby enriching their pharmacological potential. For example, alkaloids such as reserpine from *Rauwolfia serpentina* exhibit antihypertensive properties (Lobay, 2015), while aconitine and atisine from *Aconitum* act as anti-inflammatory, analgesic, and antiarrhythmic agents (Zheng et al., 2024). Terpenoids such as artemisinin from *Artemisia annua* possesses antimalarial activity (Yin et al., 2025), and α-pinene and limonene from *Juniperus communis* exhibit diuretic, antiseptic, and carminative properties (Raina et al., 2019). Phenolic compounds, including salidroside from *Rhodiola imbricata* (neuroprotective and cardioprotective), hypericin and hyperforin from *Hypericum perforatum* (antidepressant and anti-anxiety), exhibit diverse therapeutic roles (Kapoor et al., 2018; Macar & Macar, 2024). Similarly, glycosides such as amarogentinfrom *Gentiana kurroo*, picrosides from *Picrorhiza kurroa*, and swertiamarin and mangiferin from *Swertia chirayita* possess hepatoprotective and antidiabetic properties (Kumar et al., 2017a; Padhan et al., 2015; Skinder et al., 2017). Essential oils from *Valeriana jatamansi* (valepotriates; sedative), *Origanum vulgare* (carvacrol, thymol; antimicrobial), and *Thymus linearis* (thymol, borneol; antioxidant, anti-inflammatory) further contribute to their pharmacological importance (Rathore et al., 2023). Quinones and xanthones such as chrysophanol and emodinin *Rheum emodi* show potent laxative activity (Pandith et al., 2014). Altogether, alpine herbs represent a reservoir of bioactive secondary metabolites with significant pharmacological and nutraceutical applications.

2.2 Role of Secondary Metabolites in Alpine Plant Adaptation

Plants face diverse biotic and abiotic stresses and adapt to these challenges by producing secondary metabolites. These compounds act as natural defences by deterring herbivores, inhibiting pathogens, and mitigating oxidative damage. Environmental factors such as drought, salinity, extreme temperatures, cold, heavy metals, wind, and UV-B radiation strongly influence secondary metabolite biosynthesis, making them key drivers of plant resilience (Rabeh et al., 2025). These factors directly or indirectly shape plant diversity and community composition.

Along altitude gradients, around 3.4% of species are trees, 14.9% are shrubs and 81.7% are herbs. Herbs generally possess richer secondary metabolite profiles than trees or shrubs, which enables them to survive at higher elevations (up to 4900 m), whereas shrubs typically grow up to 4700 m and trees up to 3900 m (Sekar et al., 2024; Tiwari et al., 2023). The high abundance and diversity of secondary metabolites in herbs not only facilitate survival under cold stress but also contribute to their medicinal and ecological importance.

Multiple secondary metabolites in alpine herbs confer tolerance to extreme conditions. For example, phenolics in *Rhodiola rosea* and *Gentiana lutea*, and lignin in *Artemisia glacialis* and *Primula auricula* increase under cold stress, strengthening cell walls to prevent dehydration and ice-induced damage (Ahlawat et al., 2024). Flavonoids in *Leontopodium alpinum* and *Thymus serpyllum*act as antioxidants that stabilize cell membranes and proteins, while terpenoids in *Salvia pratensis* and *Achillea millefolium* maintain membrane fluidity and protect chloroplasts (Jangpangi et al., 2025). Carotenoids in *Saxifraga oppositifolia* and *Campanula alpina* serve as precursors of signalling molecules involved in stress responses (Jangpangi et al., 2025). Osmoprotectants in *Silene acaulis* and *Dryas octopetala* help maintain osmatic balance and reduce ice crystal formation during freezing.

High UV radiation represents another major stress in alpine environments. Flavonoids (flavanols, anthocyanins, and flavones) in *Leontopodium alpinum*, *Gentiana lutea*, and *Primula auricula* act as potent UV-B absorbers, protecting against cellular damage (Xue et al., 2023). Phenolic acids in *Artemisia glacialis* and *Thymus serpyllum*, and tannins in *Dryas octopetala* and *Silene acaulis*, absorb UV-light and neutralize ROS generated under UV stress (Oluwole et al., 2022; Takshak & Agrawal, 2019; Wang et al., 2025a; Xue et al., 2023).

Additionally, herbivory and pathogen attacks increase with altitude due to food scarcity. Alpine plants mitigate these threats by producing bitter and toxic secondary metabolites in their leaves, which act as natural insecticides and deterrents. Alkaloids such as atisine and indole alkaloids in *Aconitum napellus* and *Delphinium elatum* are highly toxic to herbivores (Divekar et al., 2022). Terpenoids in *Artemisia glacialis* and *Salvia officinalis* function as insecticidal and antifungal agents (Tholl, 2015). Saponins in *Primula veris* and *Gentiana lutea*, and glycosides in *Digitalis purpurea* and *Veratrum album*, disrupt pathogen membranes and reduce insect feeding due to their bitterness (Shakeel et al., 2025). Phenolic compounds such as tannins in *Dryas octopetala* and *Leontopodium alpinum* bind to proteins, rendering leaves indigestible (Sun & Shahrajabian, 2023), while quinones in *Thymus serpyllum* and *Saxifraga oppositifolia* exhibit strong antifungal and antibacterial activity (Goura et al., 2025; Iyer, 2024).

Overall, secondary metabolites play crucial roles in plant adaptation to multiple alpine stresses, including hypoxia (due to reduced atmospheric pressure limiting respiration and energy metabolism), strong winds (mechanical stress), short growing seasons (driving early flowering and clonal reproduction), water stress (caused by rapid snowmelt and low retention), nutrient limitation, and high soil salinity (arising from mineral-rich soil sand evaporation) (Kumari et al., 2024a; Pandey et al., 2021; Sumner et al., 2022; Wang et al., 2024c).

3 Plant Omics Approaches for Studying Secondary Metabolites Biosynthesis

Plant omics, the integration of genomics, transcriptomics, proteomics, and metabolomics, has emerged as a powerful framework for elucidating the biosynthesis and regulation of secondary metabolites (Fig. 8.1). Genomic and transcriptomic analyses reveal biosynthetic gene clusters, pathway-specific transcription factors, and regulatory networks, while proteomics provides insights into enzymes and post-translational modifications driving metabolite production. Metabolomics offers direct quantification and profiling of secondary metabolites across developmental and environmental conditions. Integrating these omics layers enables systems-level understanding of plant secondary metabolism, facilitating pathway engineering, metabolic flux optimization, and the discovery of novel bioactive compounds (Miao et al., 2023; Zhang et al., 2022).

3.1 Genomics

Advances in sequencing and genome annotation have revolutionized research on plant secondary metabolism by enabling identification of biosynthetic gene clusters and key enzyme-coding genes such as polyketide synthases, terpene synthases, and cytochrome P450s. Functional annotation through conserved domains, sequence homology, and pathway databases facilitates predictive gene function, while large-scale sequencing continues to uncover new genes and pathways, expanding the repertoire of bioactive compounds relevant to agriculture, nutrition, and pharmaceuticals (Hu et al., 2023).

Genome annotation also promote integrative omics approaches that connect genotype to phenotype by aligning transcriptomic and proteomic data to genomic loci, elucidating regulatory networks, and revealing lineage-specific metabolic innovations. Such insights illuminate how plants evolve new secondary metabolites to

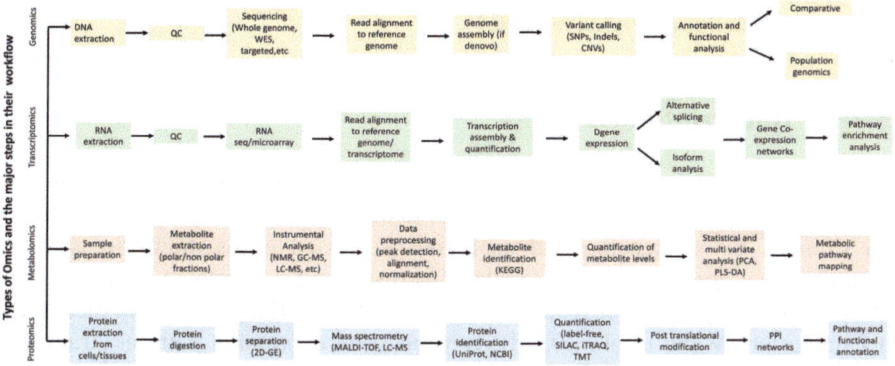

Fig. 8.1 Schematic representation of different types of omics and steps in their workflow

support ecological diversification. For example, complete chloroplast genome sequencing of *Fritillaria cirrhosa* and related species revealed a conserved circular quadripartite structure and highly variable loci useful for species identification and phylogenetic studies, which increases understanding of evolutionary relationships within the genus (Chen et al., 2019).

Despite these advances, sequencing non-model alpine plants remains challenging due to large, repeat-rich, and polyploid genomes, limited reference data, and difficulties in obtaining high-quality DNA from remote habitats. Their complex biochemical profiles further complicate genome assembly compared to model species (Zhang et al., 2024). However, recent breakthroughs in long-read sequencing (PacBio, Oxford Nanopore), Hi-C scaffolding, and optical mapping now enable highly contiguous genome assemblies even for polyploid or repeat-rich genomes. Combined with refined annotation pipelines, these technologies are accelerating the functional characterization of novel genes and unlocking unique genetic resources with potential applications in biotechnology, ecology, and stress biology (Higashi & Saito, 2013; Zhang et al., 2025).

Notable recent efforts include the *Picrorhiza kurroa* draft genome (~1.7 Gb, 12,924 scaffolds), assembled using Illumina and PacBio data, which identified 24,798 protein-coding and 9789 non-coding genes alongside 710 temperature-responsive miRNAs, providing a foundation for genetic improvement and conservation of this alpine species (Sharma et al., 2021a). Similarly, the chromosome-level genome of *Bergenia purpurascens* (650.70 Mb; N50 = 34.19 Mb scaffolds) annotated 45,841 protein-coding and 10,797 non-coding RNAs, offering key insights into bergenin biosynthesis and alpine adaptation (Liu et al., 2025a). Transcriptomic and organellar genome studies further enrich alpine plant genomics: the *Nardostachys jatamansi* novo transcriptome (48,411 unigenes) identified core biosynthetic genes and SSR markers (Dhiman et al., 2020), while mitochondrial genomes of *Gentiana* and *Phlomoides* species revealed structural diversity, codon bias, and potential molecular markers for population genetics (Ala et al., 2023; Liu et al., 2025b). The complete chloroplast genome of *Saussurea involucrata* (152,490 bp; 37.7% GC) and the chromosome-level assembly of *Roscoea debilis* (1.6 Gb; 98.7% completeness) further contribute to understanding the evolutionary and ecological significance of alpine flora (Peng et al., 2024).

3.2 Transcriptomics

Transcriptomics provides a powerful support to decipher how plants regulate secondary metabolite biosynthesis in response to developmental cues and environmental stresses. By sequencing and analyzing expressed genes, transcript scan be identified that are up- or down-regulated across tissues, developmental stages, or stress conditions, thereby revealing the temporal and spatial dynamics of metabolite accumulation (Wang et al., 2024d). Beyond gene expression patterns, transcriptomic analyses illuminate the regulatory machinery controlling metabolite production. Differential expression profiling enables the identification of transcription factors, transporters, and signalling components co-expressed with biosynthetic enzymes, offering insights into pathway activation and metabolite transport (Ahmad et al., 2022).

Differential gene expression (DGE) analysis remains a robust approach for identifying genes significantly modulated under specific conditions and for elucidating the molecular basis of secondary metabolism (Sharma et al., 2021b). This method is particularly useful in non-model species where biosynthetic pathways are poorly annotated, as it can decipher novel enzyme-coding genes involved in specialized metabolism (Mala et al., 2021). For example, expression profiling has revealed that *MYB* and *bHLH* transcription factors are key regulators of flavonoid and alkaloid biosynthetic pathways (Cao et al., 2020). Moreover, in *Pinellia ternata*, a genome-wide survey identified 188 MYB genes and pinpointed *PtMYB77*, a heat-responsive 1R-MYB that localizes to the nucleus, directly binds promoters of heat-responsive genes and, when overexpressed in Arabidopsis, enhances thermotolerance by activating heat shock factors and boosting antioxidant defenses (Wang et al., 2025c). An Integrating DGE data with co-expression networks and pathway analyses enables the identification of both structural and regulatory genes, advancing our understanding of secondary metabolic networks and facilitating the identification of molecular targets for metabolic engineering and crop improvement (Liu et al., 2019).

Recent studies highlight the versatility of transcriptomics in diverse plant systems. De novo transcriptome assembly of *Dactylorhiza hatagirea* identified genes involved in resveratrol and *trans*-stilbene biosynthesis, with 817 unigenes mapped to glycolysis, sucrose metabolism, and phenylpropanoid pathways, providing insights into stress adaptation and conservation genetics (Dhiman et al., 2019). In *Bergenia purpurascens*, characterization of the *WRKY* gene family revealed that *BpWRKY13* positively regulates cold-responsive genes, with all BpWRKY proteins sharing conserved motifs and were nucleus-localized indicative of evolutionary conservation (Chen et al., 2025). Similarly, transcriptomic analysis of *Lamiophlomis rotata* under low nitrogen stress showed altered nitrogen metabolism, monoterpenoid biosynthesis, and photosynthesis, suggesting metabolic diversion for nitrogen conservation (Zhong et al., 2025). In *Paeonia rockii*, 28,155 DEGs and 66,563 unigenes were identified, implicating *MYB*, *WD*, *RING1*, and *LRR*gene families in carpel development and yield traits (Liu et al., 2019). Comparative transcriptomics in *Orinus* species revealed phylogenetic proximity and genomic adaptations with alpine environments (Su et al., 2022), while RNA-seq analysis of *Kingdonia uniflora* across altitudes identified tissue-specific expression of light-response and root development genes, providing insights into high-altitude adaptation mechanisms (Nong et al., 2023).

3.3 Proteomics

Proteomics provide direct insights into the proteins catalyzing and regulating secondary metabolite biosynthesis by enabling large-scale identification and quantification of enzymes and regulators. Unlike transcriptomics, which predicts protein function from mRNA levels, proteomics reveals actual protein abundance, modifications, and activity states (Vogel & Marcotte, 2012). Advanced techniques such as liquid chromatography coupled with tandem mass spectrometry (LC-MS/MS) allow high-throughput detection of pathway-specific enzymes, including synthases, transferases, cytochrome P450s, and transporters involved in metabolite trafficking.

Quantitative proteomic approaches such as label-free quantification, SILAC, and iTRAQ facilitate comparative analyses across tissues, developmental stages, and environmental conditions, linking protein dynamics with metabolite accumulation. These strategies not only validate candidate genes identified through genomics and transcriptomics but also reveal post-translational modifications and protein–protein interactions that fine-tune pathway regulation (Lao et al., 2014; Zhou et al., 2017). By integrating protein abundance and post-translational modifications data, proteomics elucidate how biosynthetic enzymes and regulatory proteins are regulated beyond transcriptional control, providing a comprehensive understanding of pathway flux, metabolite diversity, and adaptive metabolic responses. Such insights are crucial for metabolic engineering, as they indicate not only which proteins are present, but also how their activities are regulated in plants (Lao et al., 2014).

3.4 Metabolomics

Metabolomic analysis is a powerful approach for the identification and quantification of the diverse secondary metabolites produced by alpine plants to withstand extreme environments such as high UV radiation, low temperatures, and nutrient-poor soils. Advanced analytical platforms, including liquid chromatography-mass spectrometry (LC-MS), gas chromatography-mass spectrometry (GC-MS), and nuclear magnetic resonance (NMR) spectroscopy, enable high-throughput characterization of major metabolite classes such as alkaloids, flavonoids, terpenoids, and phenolic compounds. By comparing metabolite profiles across tissues, developmental stages, and stress conditions, metabolomics help link specific metabolite patterns to adaptive traits and supports the discovery of new bioactive molecules relevant to pharmaceuticals and functional foods (Saito et al., 2008; Yang et al., 2025b).

Recent studies demonstrate how metabolomics, integrated with transcriptomics, can elucidate regulatory mechanisms underlying stress adaptations in alpine species. For example, 6-benzylaminopurine (6-BA) treatment triggered seed dormancy release in *Polygonatum cyrtonema* by altering genes involved in hormone signalling, starch metabolism, and phenylpropanoid biosynthesis, accompanied by changes in flavonoids, amino acids, and sugar accumulation (Zhang et al., 2022). Similarly, comparative transcriptome analysis of *Rheum australe* identified differentially expressed genes associated with secondary metabolism and stress signalling, providing insights into its medicinal properties (Mala et al., 2021). In *Rhodiola linearifolia*, growth at higher altitude (2687 m) significantly enhanced the accumulation of antioxidant metabolites, emphasizing altitude-associated metabolic adaptations that contribute to stress tolerance and medicinal value (Terletskaya et al.,

2024). Moreover, the chromosome-level genome assembly of *Gynostemma pentaphyllum* enabled reconstruction of gypenoside biosynthetic pathway and identification of key UDP-glucuronosyltransferases, providing a foundation for metabolic engineering (Huang et al., 2021).

Targeted metabolomics allows precise quantification of known stress-associated metabolites. For example, flavonoids involved in UV protection, alkaloids mediating herbivore defence, and terpenoids contributing to cold tolerance (He et al., 2022). In contrast, untargeted metabolomics provides an unbiased assessment of diverse metabolites (metabolome), enabling the discovery of novel stress-induced compounds and metabolic intermediates (Tang et al., 2025). Together, these approaches offer an integrated understanding of how secondary metabolism promote alpine plant adaptation and support both fundamental research on stress resilience and applied efforts in natural product discovery and crop improvement (Zhang et al., 2022).

4 Gene Co-expression Network (GCN) Analysis in the Context of Secondary Metabolite Biosynthesis

4.1 Principles of Gene Co-expression Network Construction

Gene co-expression refers to the correlated expression of genes across tissues, developmental stages, or environmental conditions. Such correlations often indicate shared regulatory mechanisms or participation in related biological pathways (Fig. 8.2). The underlying assumption is that genes involved in common processes are co-regulated and therefore display similar expression profiles. By analysing large-scale transcriptomic datasets, co-expression networks can be constructed in which clusters of genes form distinct modules based on shared expression patterns. These modules allow functional inference of uncharacterized genes through their association with well-annotated genes, a concept commonly referred as "guilt by

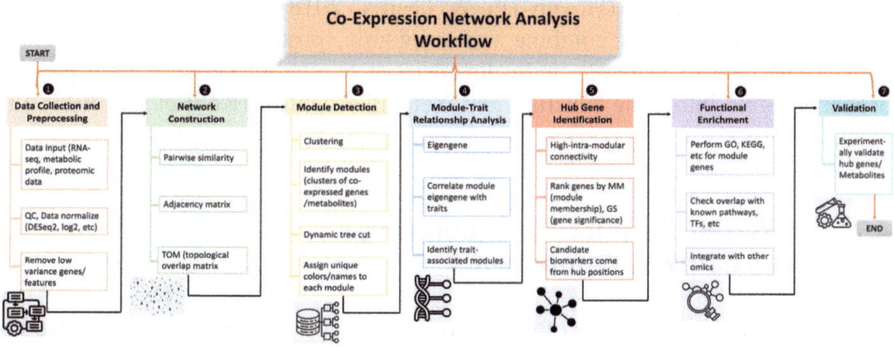

Fig. 8.2 Overview of data processing and network construction in co-expression analysis

association" (Langfelder & Horvath, 2008).Co-expression analysis has been particularly effective in secondary metabolism, where it has facilitated the identification of pathway enzyme-encoding genes and regulatory transcription factors, thereby offering targets for functional validation and metabolic engineering (Langfelder & Horvath, 2008).

For example, transcriptome-wide co-expression analysis in ginseng identified 40 gene modules associated with environmental variables and saponin accumulation. Of these, 18 modules correlated with ecological factors and 22 with saponin biosynthesis. The study proposed and validated a regulatory network model linking environmental cues to ginsenoside biosynthesis, demonstrating how network approaches can guide metabolic pathway regulation (Wang et al., 2023). Integrating transcriptomic data with proteomic and metabolomic profiles further enhances network robustness by linking gene expression to protein abundance and metabolites accumulation, enabling both global and context-specific interpretations of gene function (Higashi & Saito, 2013).

Several algorithms are used to buildgene co-expression networks. Simple approaches rely on correlation coefficients (e.g., Pearson or Spearman) to measure pairwise similarity among genes. However, more advanced techniques such as Weighted Gene Co-expression Network Analysis (WGCNA) generate scale-free networks by weighing gene pairs according to correlation strength, and identify modules of strongly co-expressed genes (Langfelder & Horvath, 2008). Mutual information-based algorithms, such as ARACNE (Algorithm for the Reconstruction of Accurate Cellular Networks), capture non-linear relationships, while machine learning and Bayesian network models infer directional interactions (Jiang et al., 2023; Margolin et al., 2006).

Multiple computational platforms support co-expression network construction and visualization. WGCNA (R package) is one of the most widely used tools for building weighted networks and identifying co-expression modules (Saito et al., 2008), while Cytoscape provides a flexible platform for network visualizing with functional enrichmentplugins such as ClueGO (Su et al., 2014). Tools like ARACNE and GENIE3 facilitate regulatory network inference, and web-based tools including ATTED-II and CoExpNetViz offer pre-computed co-expression networks for numerous plant species (Huynh-Thu & Geurts, 2018). Increasingly, integrative multi-omics frameworks combine transcriptomic, proteomic, metabolomic, and epigenomic data to generate multi-dimensional regulatory networks that provide deeper insights into gene function and metabolic control (Kohl et al., 2011; Liu et al., 2019).

4.2 Inferring Regulatory Relationships and Pathway Components

Gene co-expression networks offer a powerful strategy predicting genes involved in secondary metabolite biosynthesis, based on the principle that genes participating in the same metabolic pathway often exhibit coordinated expression patterns (Sharma et al., 2021b). In these networks, genes encoding characterized pathway enzymes act as "anchor" nodes, while strongly co-expressed genes are considered likely contributors to the same biosynthetic route. This strategy is particularly useful for identifying missing steps in incomplete pathways and uncovering species-specific genes associated with unique metabolite profiles (Langfelder & Horvath, 2008).

The topology of co-expression networks also provides insights into pathway regulation by highlighting patterns of connectivity among genes. Highly connected "hub" nodes often represent key regulatory elements, such as transcription factors or signalling molecules, that coordinate the expression of multiple enzymesin a pathway (Hu et al., 2023). In contrast, peripheral nodes with lower connectivity may encode specialized enzymes responsible for diversifying metabolic structures (Langfelder & Horvath, 2008). Network parameters include node degree, clustering coefficients, and centrality facilitate prioritization of candidate genes for functional verification, enabling a systems-level understanding of secondary metabolism (Langfelder & Horvath, 2008; Liu et al., 2019).

Coupling transcriptomic data with metabolite profiling further enhances the identification of candidate genes by correlating gene expression modules with metabolite abundance patterns (Higashi & Saito, 2013; Saito et al., 2008). For example, in *Arundina graminifolia*, integrated transcriptome-metabolome analysis identified candidate genes associated with flavonoid, phenylpropanoid, and bibenzyl biosynthesis, with expression patterns corresponding to tissue-specific metabolite accumulation. The perennial flowering and high accumulation of antioxidative and antitumor bibenzyls reveal the medical value of this species, highlighting how co-expression analysis can guide the discovery of key biosynthetic and regulatory genes (Ahmad et al., 2022).

5 Integrating Plant Omics and Co-expression Networks for Deeper Insights

5.1 Multi-omics Integration Strategies

Integrating multi-omics datasets such as transcriptomics, proteomics, and metabolomics, with gene co-expression network analysis significantly enhances the ability to reveal regulatory interactions in secondary metabolism (Fig. 8.2). Transcriptomics provides temporal and special gene expression profiles, metabolomics reflects metabolite abundance, and proteomics reveals enzyme abundance and modification that drive biosynthetic reactions (Higashi & Saito, 2013). When combined, these datasets enable co-expression networks to be mapped not only to genes but also to metabolic profiles and enzyme activities, resulting in more accurate identification of candidate genes, pathway enzymes, and regulatory factors. Such multi-omics integration strengthens functional predictions, fills gaps in biosynthetic pathways, and provides a systems-level understanding of how secondary metabolites are controlled under different conditions, ultimately advancing pathway discovery and metabolic engineering in model and non-model plant species (Miao et al., 2023).

Linking metabolite profiles to gene co-expression networks is particularly effective for predicting gene function. For example, a gene that co-expresses with known pathway genes and whose expression correlates with the accumulation of a specific alkaloid or flavonoid is strongly supported as part of that pathway (Hua et al., 2022; Jiang et al., 2023). This approach distinguishes true pathway genes from false positives, and highlights environment- or tissue-specific metabolic regulation.

Incorporating proteomic data further refines these networks by providing evidence of enzyme abundance and activity rather than transcript level alone. Studies in *Uncaria rhynchophylla* showed that changes in abundance of key enzymes, such as strictosidine synthase, linked to indole alkaloid accumulation, indicating their regulatory roles (Jiang et al., 2023). Similarly, integrating proteomic and transcriptomic analyses in *Arabidopsis thaliana* reveal discrepancies between mRNA and protein levels, highlighting post-transcriptional controls that influence metabolite production (Lao et al., 2014; Zhou et al., 2017).

Proteomics also captures post-translational modifications, such as phosphorylation or glycosylation, which modulate enzyme stability, activity, and subcellular localization in pathways including phenylpropanoid and flavonoid biosynthesis (Joshi & Joshi, 2025). Such integration not only improves pathway reconstruction but also reveals possible regulatory processes and points of intervention for metabolic engineering (Zhang et al., 2022; Miao et al., 2023; Langfelder & Horvath, 2008). Figure 8.2 workflow includes 7 stages and subsequent steps in each segment. The schematic representation of stages is: 1. Data collection and preprocessing 2. Network construction 3. Module detection 4. Module-trait relationship analysis 5. Hub gene identification 6. Functional enrichment 7. Validation.

5.2 Case Studies: Examples of Successful Integration in Alpine Medicinal Herbs

In *Dioscorea nipponica*, dioscin biosynthesis was elucidated from cholesterol and C-24 alkylsterols, identifying these sterols as key intermediates. Organ-specific transcript profiling revealed strong root-based expression of cytochrome P450s and glycosyltransferases, particularly after methyl jasmonate (MeJA) treatment, highlighting coordinated transcriptional regulation of dioscin production (Sun et al., 2017). Similarly, metabolite and transcript co-expression analyses in *Uncaria rhynchophylla* identified higher monoterpene indole alkaloid (UR-MIA) levels in hook and branch tissues compared to leaves, alongside 79 metabolite features, 30 of which showed differential accumulation. Overexpression of two strictosidine synthases (*UrSTR1/5*) enhanced pathway flux, and co-expression network analysis connected 58 pathway genes with 12 regulatory hubs, providing key candidates for reconstituting UR-MIA biosynthesis in heterologous systems (Jiang et al., 2023).

In *Picrorhiza kurroa*, large-scale transcriptome sequencing across tissues identified 47,726–66,979 transcripts, enabling the construction of gene co-expression networks underlying the biosynthesis of hepatoprotective iridoid glycosides picroside-I and picroside-II. These networks, comprising ~2810 nodes and 15,626 edges, indicated complex transcriptional coordination of metabolite production (Sharma et al., 2021b). Similarly, integrative transcriptome-metabolite analysis in *Paris Polyphylla* mapped the polyphyllin biosynthetic pathways and characterized PpUGT73C1 as a key glycosyltransferase acting on diosgenin and pennogenin (Hua et al., 2022). In *Panax ginseng*, 32 trihelix transcription factorswith 218 alternative transcripts were identified and classified into five subgroups. Their expression displayed tissue-specific, age-dependent, and MeJA-responsive patterns, and co-expression network analyses suggested synergistic regulatory roles (Hu et al., 2023). Furthermore, WGCNA-based analysis of *Arabidopsis* somatic embryogenesis identified 26 gene modules and 260 hub genes involved in developmental transitions, offering deep insights into molecular control of embryogenic reprogramming (De Silva et al., 2022).

5.3 Identifying Key Regulatory Genes and Transcription Factors

Network-based analyses are particularly effective for identifying master regulators by evaluating connectivity patterns within biosynthetic gene modules. For example, co-expression analysis in *Uncaria rhynchophylla* revealed correlations between monoterpene indole alkaloid (MIA)accumulation and the expression of candidate biosynthetic genes, enabling the identification of transcription factor hubs that modulate MIA biosynthesis (Jiang et al., 2023). Similarly, in *Arabidopsis*, R2R3-MYBs transcription factors (*MYB28*, *MYB29*) and a flavin-monooxygenase gene were linked to aliphatic glucosinolate and cellulose biosynthesis, with mutant validation emphasizing the utility of co expression networks for functional genomics (Saito et al., 2008).

Genome-wide analyses also highlight how gene family diversification shapes stress resilience and metabolic specialization. In three *Notopterygium* species, 63 LEA genes were classified into six subfamilies, with gene duplication and negative selection contributing to expansion. Variation in cis-regulatory elements and stress-responsive expression patterns supports key roles of LEA genes in heat and drought tolerance in alpine environments (Wu et al., 2025). Similarly, OSCA gene family in *N. franchetii*, a Qinghai-Tibetan Plateau medicinal plant, comprises 29 genes across 11 chromosomes, many showing root-specific and time-dependent upregulation under drought, providing insight into stress adaptation in alpine medicinal plants (Zhang et al., 2025).

6 Challenges and Future Perspectives

6.1 Limitations of Current Plant Omics and Network Approaches

Although plant omics and network-based strategies have significantly improved our understanding of gene-protein-metabolite interactions, several limitations still persist. These include challenges in integrating heterogeneous multi-omics datasets (genomics, transcriptomics, proteomics, metabolomics) due to differences in data quality, analytical platforms, and normalization procedures (Yang et al., 2025b). Network construction is further constrained by incomplete or uneven data coverage, leading to missing nodes or artificial connections. Additionally, dynamic and context-dependent interactions such as those varying across developmental stages, tissues, or environmental conditions, are difficult to capture in static network models. Functional validation of predicted interactions remains limited, as experimental confirmation is resource-intensive. Moreover, the lack of standardized, plant-specific computational pipelines limits reproducibility and scalability, restricting the translation of network understanding into crop improvement and applied plant research (Wu et al., 2025).

Research on non-model alpine plants is further complicated by both biological and logistical constraints. Unlike well-established model species, many alpine plants lack comprehensive genomic, transcriptomic, and metabolomic resources, which limits functional annotation and systems-level interpretation of their adaptive traits (Peng et al., 2024). Field studies are challenging in alpine plants due to extreme environmental conditions, limited accessibility, and short growing seasons, making sample collection and long-term monitoring difficult. Maintaining alpine species under controlled growth conditions is also challenging, as many fail to survive outside their natural habitats (Terletskaya et al., 2024). Additionally, their ecological specialization and genetic variation across small, isolated populations complicate comparative analyses and reduce reproducibility. Collectively, these challenges restrict the scientific progress and the integration of alpine species into broader plantomics and network biology frameworks (Salick et al., 2009).

6.2 Future Directions and Opportunities

There is a pressing need for multi-omics studies across diverse alpine medicinal herbs to better elucidate their complex biology, adaptative strategies, and therapeutic potential. At present, such investigations are largely limited to a few model species, leaving many ecologically and pharmacologically valuable alpine plants underexplored (Vafaee et al., 2025). Integrating genomics, transcriptomics, proteomics, and metabolomics can uncover biosynthetic pathways, stress-response mechanisms, and the unique adaptations associated with high-elevation environments, facilitating the discovery of new bioactive compounds. Increasingly powerful long-read sequencing platforms, such as PacBio and Oxford Nanopore, further enhance these efforts by enabling full-length isoform resolution, detection of structural variants, and characterization of alternative splicing, capabilities especially useful for non-model alpine species (Wang et al., 2024a). Functional genomics tools, including CRISPR/Cas-mediated gene editing, RNA interference, and gene overexpression, are equally essential for validating candidate genes identified through transcriptomic and network analyses. These approaches allow direct investigation of stress tolerance pathways, secondary metabolite biosynthesis, and altitude-specific adaptation.

7 Conclusion

Integrating co-expression network analysis with multi-omics data has become a powerful approach for deciphering complex biosynthetic pathways of alpine medicinal plants. Genomic, transcriptomic, proteomic, and metabolomic datasets provide complementary layers of information that, when combined, enable the identification of candidate genes, regulatory hubs, and biosynthetic bottlenecks responsible for the production of bioactive compounds. Co-expression network analysis further strengthens these insights by revealing functionally related gene modules and their dynamic interactions under specific environmental or developmental conditions.

Together, these integrated approaches enhance our understanding of the molecular basis underlying metabolite diversity in alpine plants and open new areas for functional validation, metabolic engineering, and the sustainable utilization of high-value medicinal species.

References

Ahlawat, Y. K., Singh, M., Manorama, K., & Others. (2024). Plant phenolics: Neglected secondary metabolites in plant stress tolerance. *Brazilian Journal of Botany, 47*, 703–721. https://doi.org/10.1007/s40415-023-00949-x

Ahmad, K., Ahmad, M., Huber, F. K., & Weckerle, C. S. (2021). Traditional medicinal knowledge and practices among the tribal communities of Thakht-e-Sulaiman Hills, Pakistan. *BMC Complementary Medicine and Therapies, 21*(1), 230. https://doi.org/10.1186/s12906-021-03403-1

Ahmad, S., Gao, J., Wei, Y., Chuqiao, L., Zhu, G., & Yang, F. (2022). The transcriptome profiling of flavonoids and Bibenzyls reveals medicinal importance of rare orchid *Arundina graminifolia. Frontiers in Plant Science, 13*. https://doi.org/10.3389/fpls.2022.923000

Ala, K. G., Zhao, Z., Ni, L., & Wang, Z. (2023). Comparative analysis of mitochondrial genomes of two alpine medicinal plants of *Gentiana* (Gentianaceae) edited by X. J. Min. *PLoS One, 18*(1), e0281134. https://doi.org/10.1371/journal.pone.0281134

Ashraf, M. V., Khan, S., Misri, S., Gaira, K. S., Rawat, S., Rawat, B., Khan, M. A. H., Shah, A. A., Asgher, M., & Ahmad, S. (2024). High-altitude medicinal plants as promising source of phytochemical antioxidants to combat lifestyle-associated oxidative stress-induced disorders. *Pharmaceuticals, 17*(8), 975. https://doi.org/10.3390/ph17080975

Bashir, S., Memon, R., & Gilani, A. H. (2011). Antispasmodic and antidiarrheal activities of *Valeriana hardwickii* Wall. Rhizome are putatively mediated through Calcium Channel

Blockade. *Evidence-based Complementary and Alternative Medicine, 2011*(1), 304960. https://doi.org/10.1155/2011/304960

Bergman, M. E., & Dudareva, N. (2024). Plant specialized metabolism: Diversity of terpene synthases and their products. *Current Opinion in Plant Biology, 81*, 102607. https://doi.org/10.1016/j.pbi.2024.102607

Binter, J., Macek, M., & Dolezal, J. (2025). Integrating morphological, anatomical, and physiological traits to explain elevational distributions in Himalayan steppe and alpine plants. *Journal of Integrative Plant Biology, 67*(10), 2643–2657. https://doi.org/10.1111/jipb.13971

Cai, C., Shen, J., Chen, L., & Li, J. (2023). Content, composition, and biosynthesis of anthocyanin in *Fragaria* species: A review. *HortScience, 58*(9), 988–995. https://doi.org/10.21273/HORTSCI17207-23

Cao, Y., Li, K., Li, Y., Zhao, X., & Wang, L. (2020). MYB transcription factors as regulators of secondary metabolism in plants. *Biology, 9*(3), 61. https://doi.org/10.3390/biology9030061

Chapagain, D. J., Meilby, H., & Ghimire, S. K. (2019). Plant density and life history traits of *Aconitum spicatum* in North-central Nepal: Effects of elevation and anthropogenic disturbances. *PeerJ, 7*, e7574. https://doi.org/10.7717/peerj.7574

Chen, Q., Wu, X., & Zhang, D. (2019). Phylogenetic analysis of *Fritillaria cirrhosa* D. Don and its closely related species based on complete chloroplast genomes. *PeerJ, 7*. https://doi.org/10.7717/peerj.7480

Chen, J., Zeng, H., Yan, F., Jiang, Z., Chen, J., Wang, W., & Zhu, Q. (2025). Identification of the WRKY gene family in *Bergenia Purpurascens* and functional analysis of BpWRKY13 under cold stress. *Plant Physiology and Biochemistry, 223*, 109832. https://doi.org/10.1016/j.plaphy.2025.109832

Chhetri, V. T., Shrestha, S., Thapa, S., & Timilsina, S. (2021). Status and role of Medicinal and Aromatic Plants (MAPs) in Nepalese livelihood. *International Journal of Environment, 10*(1), 112–136. https://doi.org/10.3126/ije.v10i1.38405

Cho, W. K., Kim, H.-I., Kim, S.-Y., Seo, H. H., Song, J., Kim, J., Shin, D. S., Jo, Y., Choi, H., Lee, J. H., & Moh, S. H. (2020). Anti-aging effects of *Leontopodium alpinum* (Edelweiss) callus culture extract through transcriptome profiling. *Genes, 11*(2), 230. https://doi.org/10.3390/genes11020230b

Choudhary, S., Chaudhary, G., & Kaurav, H. (2021). Medicinal importance of *Ephedra gerardiana* in ayurveda and modern sciences: A review. *Asian Journal of Pharmacy and Pharmacology, 7*(3), 110–117. https://doi.org/10.31024/ajpp.2021.7.3.1

De Silva, K. K., Dunwell, J. M., & Wickramasuriya, A. M. (2022). Weighted Gene Correlation Network Analysis (WGCNA) of Arabidopsis Somatic Embryogenesis (SE) and identification of key gene modules to uncover SE-associated hub genes. *International Journal of Genomics, 2022*, 1–24. https://doi.org/10.1155/2022/7471063

Dhiman, N., Sharma, N. K., Thapa, P., Sharma, I., Kumar Swarnkar, M., Chawla, A., Shankar, R., & Bhattacharya, A. (2019). De novo transcriptome provides insights into the growth behaviour and resveratrol and trans-stilbenes biosynthesis in *Dactylorhiza hatagirea*—An endangered alpine terrestrial orchid of western Himalaya. *Scientific Reports, 9*(1), 13133. https://doi.org/10.1038/s41598-019-49446-w

Dhiman, N., Kumar, A., Kumar, D., & Bhattacharya, A. (2020). De novo transcriptome analysis of the critically endangered alpine Himalayan herb *Nardostachys Jatamansi* reveals the biosynthesis pathway genes of tissue-specific secondary metabolites. *Scientific Reports, 10*(1), 17186. https://doi.org/10.1038/s41598-020-74049-1

Divekar, P. A., Narayana, S., Divekar, B. A., Kumar, R., Gadratagi, B. G., Ray, A., Singh, A. K., Rani, V., Singh, V., Singh, A. K., Kumar, A., Singh, R. P., Meena, R. S., & Behera, T. K. (2022). Plant secondary metabolites as defense tools against herbivores for sustainable crop protection. *International Journal of Molecular Sciences, 23*(5), 2690. https://doi.org/10.3390/ijms23052690

Gong, F., Yu, W., Zeng, Q., Dong, J., Cao, K., Xu, H., & Zhou, X. (2023). *Rhododendron chrysanthum*'s primary metabolites are converted to phenolics more quickly when exposed to UV-B radiation. *Biomolecules, 13*(12), 1700. https://doi.org/10.3390/biom13121700

Goura, K., Legrifi, I., Kallali, N., et al. (2025). Beyond survival: The role of secondary metabolites in plant defense mechanisms. *Journal of Crop Health, 77*, 121. https://doi.org/10.1007/s10343-025-01183-z

Hao, J., Guo, H., Shi, X., Wang, Y., Wan, Q., Song, Y.-B., Zhang, L., Dong, M., & Shen, C. (2017). Comparative proteomic analyses of two Taxus species (*Taxus × media and Taxus mairei*) reveals variations in the metabolisms associated with paclitaxel and other metabolites. *Plant and Cell Physiology, 58*(11), 1878–1890. https://doi.org/10.1093/pcp/pcx128

Hazrati, S., Mousavi, Z., & Nicola, S. (2024). Harvest time optimization for medicinal and aromatic plant secondary metabolites. *Plant Physiology and Biochemistry, 212*, 108735. https://doi.org/10.1016/j.plaphy.2024.108735

He, J., Yao, L., Pecoraro, L., Liu, C., Wang, J., Huang, L., & Gao, W. (2022). Cold stress regulates accumulation of flavonoids and terpenoids in plants by phytohormone, transcription process, functional enzyme, and epigenetics. *Critical Reviews in Biotechnology, 43*(5), 680–697. https://doi.org/10.1080/07388551.2022.2053056

Higashi, Y., & Saito, K. (2013). Network analysis for gene discovery in plant-specialized metabolism. *Plant, Cell & Environment, 36*(9), 1597–1606. https://doi.org/10.1111/pce.12069

Hu, J., Liu, T., Huo, H., Liu, S., Liu, M., Liu, C., Zhao, M., Wang, K., Wang, Y., & Zhang, M. (2023). Genome-wide characterization, evolutionary analysis, and expression pattern analysis of the trihelix transcription factor family and gene expression analysis under MeJa treatment in *Panax ginseng. BMC Plant Biology, 23*(1), 376. https://doi.org/10.1186/s12870-023-04390-w

Hua, X., Song, W., Wang, K., Yin, X., Hao, C., Duan, B., Zhichao, X., Tongbing, S., & Xue, Z. (2022). Effective prediction of biosynthetic pathway genes involved in bioactive polyphyllins in *Paris polyphylla. Communications Biology, 5*(1), 50. https://doi.org/10.1038/s42003-022-03000-z

Huang, D., Ming, R., Xu, S., Wang, J., Yao, S., Li, L., Huang, R., & Tan, Y. (2021). Chromosome-level genome assembly of Gynostemma pentaphyllum provides insights into gypenoside biosynthesis. DNA Research, 28(5):dsab018. https://doi.org/10.1093/dnares/dsab018

Huynh-Thu, V., & Geurts, P. (2018). dynGENIE3: Dynamical GENIE3 for the inference of gene networks from time series expression data. *Scientific Reports, 8*, 3384. https://doi.org/10.1038/s41598-018-21715-0

Iyer, R. V. (2024). Role of plant secondary metabolites in biotic interactions and defense. *International Journal of Biology Sciences, 6*(1), 172–175. https://doi.org/10.33545/26649926.2024.v6.i1c.428

Jan, S., Kamili, A. N., Parray, J. A., Bedi, Y. S., & Ahmad, P. (2016). Microclimatic variation in UV perception and related disparity in tropane and quinolizidine alkaloid composition of *Atropa acuminata, Lupinus polyphyllus* and *Hyoscyamus niger. Journal of Photochemistry and Photobiology B: Biology, 161*, 230–235. https://doi.org/10.1016/j.jphotobiol.2016.04.028

Jandova, V., Altman, J., Sehadova, H., Macek, M., Fibich, P., Ruka, A. T., & Dolezal, J. (2025). Climate warming promotes growth in Himalayan alpine cushion plants but threatens survival through increased extreme snowfall. *New Phytologist, 247*(1), 115–127. https://doi.org/10.1111/nph.70206

Jangpangi, D., Patni, B., Chandola, V., & Chandra, S. (2025). Medicinal plants in a changing climate: Understanding the links between environmental stress and secondary metabolite synthesis. *Frontiers in Plant Science, 16*, 1587337. https://doi.org/10.3389/fpls.2025.1587337

Jiang, C., Yu, J., Fei, X., Pan, X., Zhu, N., Lin, C., Zhou, D., Zhu, H., Qi, Y., & Wu, Z. (2023). Gene coexpression networks allow the discovery of two strictosidine synthases underlying monoterpene indole alkaloid biosynthesis in *Uncaria rhynchophylla. International Journal of Biological Macromolecules, 226*, 1360–1373. https://doi.org/10.1016/j.ijbiomac.2022.11.249

Joshi, S., & Joshi, R. (2025). Quantifying redox Signalling regulatory transcriptional dynamics in *Nardostachys Jatamansi* under abiotic stress response. *Biochimica et Biophysica Acta – General Subjects, 1869*(6), 130788. https://doi.org/10.1016/j.bbagen.2025.130788

Juyal, D., Thawani, V., Thaledi, S., & Joshi, M. (2014). Ethnomedical properties of *Taxus Wallichiana* Zucc. (Himalayan Yew). *Journal of Traditional and Complementary Medicine, 4*(3), 159–161. https://doi.org/10.4103/2225-4110.136544

Kapoor, S., Raghuvanshi, R., Bhardwaj, P., Sood, H., Saxena, S., & Chaurasia, O. P. (2018). Influence of light quality on growth, secondary metabolites production and antioxidant activity in callus culture of *Rhodiola imbricata* Edgew. *Journal of Photochemistry and Photobiology B: Biology, 183*, 258–265. https://doi.org/10.1016/j.jphotobiol.2018.04.018

Kapoor, S., Chandel, R., Kaur, R., Kumar, S., Kumar, R., Janghu, S., Kaur, A., & Kumar, V. (2023). The flower of *Hypericum perforatum* L.: A traditional source of bioactives for new food and pharmaceutical applications. *Biochemical Systematics and Ecology, 110*, 104702. https://doi.org/10.1016/j.bse.2023.104702

Kelsang, G. A., Ni, L., & Zhao, Z. (2024). Insights from the first chromosome-level genome assembly of the alpine gentian *Gentiana straminea* Maxim. *DNA Research, 31*(5), dsae022. https://doi.org/10.1093/dnares/dsae022

Kilani-Morakchi, S., Morakchi-Goudjil, H., & Sifi, K. (2021). Azadirachtin-based insecticide: Overview, risk assessments, and future directions. *Frontiers in Agronomy, 3*, 676208. https://doi.org/10.3389/fagro.2021.676208

Kohl, M., Wiese, S., & Warscheid, B. (2011). Cytoscape: Software for visualization and analysis of biological networks. In M. Hamacher, M. Eisenacher, & C. Stephan (Eds.), *Data mining in proteomics* (Vol. 696, pp. 291–303). Humana Press. https://doi.org/10.1007/978-1-60761-987-1_18

Kołtun-Jasion, M., Czerwiec, K., Parzonko, A., Bakiera, A., Ożarowski, M., & Kiss, A. K. (2025). Comprehensive profiling of *Rhodiola rosea* roots and corresponding products: Phytochemical insights and modulation of neuroinflammation in BV2 microglial cell model. *Frontiers in Pharmacology, 16*, 1608767. https://doi.org/10.3389/fphar.2025.1608767

Kubczak, M., Khassenova, A. B., Skalski, B., Michlewska, S., Wielanek, M., Skłodowska, M., Aralbayeva, A. N., Nabiyeva, Z. S., Murzakhmetova, M. K., Zamaraeva, M., Bryszewska, M., & Ionov, M. (2022). *Hippophae rhamnoides* L. leaf and twig extracts as rich sources of nutrients and bioactive compounds with antioxidant activity. *Scientific Reports, 12*(1), 1095. https://doi.org/10.1038/s41598-022-05104-2

Kumar, V. (2019). OMICS-based approaches for elucidation of Picrosides biosynthesis in *Picrorhiza Kurroa*. In R. Banerjee, G. V. Kumar, & S. P. J. Kumar (Eds.), *OMICS-based approaches in plant biotechnology* (1st ed., pp. 145–166). Wiley. https://doi.org/10.1002/9781119509967.ch8

Kumar, V., & Van Staden, J. (2016). A review of *Swertia chirayita* (Gentianaceae) as a traditional medicinal plant. *Frontiers in Pharmacology, 6*. https://doi.org/10.3389/fphar.2015.00308

Kumar, P., Pal, T., Sharma, N., Kumar, V., Sood, H., & Chauhan, R. S. (2015a). Expression analysis of biosynthetic pathway genes vis-à-vis podophyllotoxin content in *Podophyllum hexandrum* Royle. *Protoplasma, 252*(5), 1253–1262. https://doi.org/10.1007/s00709-015-0757-x

Kumar, V., Sharma, N., Shitiz, K., Singh, T. R., Tandon, C., Sood, H., & Chauhan, R. S. (2015b). An insight into conflux of metabolic traffic leading to picroside-I biosynthesis by tracking molecular time course changes in a medicinal herb, *Picrorhiza kurroa*. *Plant Cell, Tissue and Organ Culture, 123*(2), 435–441. https://doi.org/10.1007/s11240-015-0839-7

Kumar, V., Malhotra, N., Pal, T., & Chauhan, R. S. (2016). Molecular dissection of pathway components unravelatisine biosynthesis in a non-toxic Aconitum species, *A. heterophyllum* Wall. *3 Biotech, 6*(1), 106. https://doi.org/10.1007/s13205-016-0417-7

Kumar, V., Chauhan, R. S., & Tandon, C. (2017a). Biosynthesis and therapeutic implications of iridoid glycosides from *Picrorhiza* genus: The road ahead. *Journal of Plant Biochemistry and Biotechnology, 26*(1), 1–13. https://doi.org/10.1007/s13562-016-0364-8

Kumar, V., Bansal, A., & Chauhan, R. S. (2017b). Modular design of Picroside-II biosynthesis deciphered through NGS transcriptomes and metabolic intermediates analysis in naturally variant Chemotypes of a medicinal herb, *Picrorhiza kurroa*. *Frontiers in Plant Science, 8*, 564. https://doi.org/10.3389/fpls.2017.00564

Kumari, M., Pradhan, U. K., Joshi, R., Punia, A., Shankar, R., & Kumar, R. (2021). In-depth assembly of organ and development dissected *Picrorhiza kurroa* proteome map using mass spectrometry. *BMC Plant Biology, 21*(1), 604. https://doi.org/10.1186/s12870-021-03394-8

Kumari, A., Kumar, V., Ovadia, R., & Oren-Shamir, M. (2023). Phenylalanine in motion: A tale of an essential molecule with many faces. *Biotechnology Advances, 68*, 108246. https://doi.org/10.1016/j.biotechadv.2023.108246

Kumari, S., Nazir, F., Maheshwari, C., Kaur, H., Gupta, R., Siddique, K. H. M., & Khan, M. I. R. (2024a). Plant hormones and secondary metabolites under environmental stresses: Enlightening defense molecules. *Plant Physiology and Biochemistry, 206*, 108238. https://doi.org/10.1016/j.plaphy.2023.108238

Kumari, R., Negi, M., Thakur, P., Mahajan, H., Raina, K., Sharma, R., Singh, R., Anand, V., Ming, L. C., Goh, K. W., Calina, D., Sharifi-Rad, J., & Chaudhary, A. (2024b). *Saussureacostus* (Falc.) Lipsch.: A comprehensive review of its pharmacology, phytochemicals, ethnobotanical uses, and therapeutic potential. *Naunyn-Schmiedeberg's. Archives of Pharmacology, 397*(3), 1505–1524. https://doi.org/10.1007/s00210-023-02694-0

Langfelder, P., & Horvath, S. (2008). WGCNA: An R package for weighted correlation network analysis. *BMC Bioinformatics, 9*(1), 559. https://doi.org/10.1186/1471-2105-9-559

Lao, Y., Wang, X., Naihan, X., Zhang, H., & Hongxi, X. (2014). Application of proteomics to determine the mechanism of action of traditional Chinese medicine remedies. *Journal of Ethnopharmacology, 155*(1), 1–8. https://doi.org/10.1016/j.jep.2014.05.022

Li, X., Yang, Y., Ma, L., Sun, X., Yang, S., Kong, X., Hu, X., & Yang, Y. (2014). Comparative proteomics analyses of *Kobresia pygmaea* adaptation to environment along an elevational gradient on the Central Tibetan Plateau. *PLoS One, 9*(6), e98410. https://doi.org/10.1371/journal.pone.0098410

Li, J., Li, X., Wang, C., Zhang, M., Ye, M., & Wang, Q. (2022). The potential of Valeriana as a traditional Chinese medicine: Traditional clinical applications, bioactivities, and phytochemistry. *Frontiers in Pharmacology, 13*, 973138. https://doi.org/10.3389/fphar.2022.973138

Li, Z., Li, J., Shu, Z., Xu, M., Zhang, Y., Gu, J., Chen, J., Li, X., & Wang, M. (2025a). Comparative metabolomic analysis provides insights into the metabolite profiles of wild and cultivated *Dendrobium flexicaule*. *BMC Plant Biology, 25*(1), 217. https://doi.org/10.1186/s12870-025-06054-3

Li, X., Xu, H., Yu, J., Han, Y., Han, S., Niu, Y., & Zhang, F. (2025b). Comparative and phylogenetic analysis of *Potentilla* and *Dasiphora* (Rosaceae) based on plastid genome. *BMC Plant Biology, 25*(1), 176. https://doi.org/10.1186/s12870-025-06186-6

Liu, N., Cheng, F., Zhong, Y., & Guo, X. (2019). Comparative transcriptome and Coexpression network analysis of carpel quantitative variation in *Paeonia Rockii*. *BMC Genomics, 20*(1), 683. https://doi.org/10.1186/s12864-019-6036-z

Liu, Y., An, Z., & He, Y. (2023). The traditional uses, phytochemistry, pharmacology and toxicology of *Bergenia purparescens*: A review comments and suggestions. *Heliyon, 9*(11), e22249. https://doi.org/10.1016/j.heliyon.2023.e22249

Liu, X.-D., Twyford, A. D., Hollingsworth, P. M., Luo, Y.-H., Ming, S.-P., Yang, B.-Y., Gao, L.-M., & Tan, S.-L. (2025a). A chromosome-level genome assembly of the alpine medicinal plant *Bergenia Purpurascens* (Saxifragaceae). *Scientific Data, 12*(1), 121. https://doi.org/10.1038/s41597-025-04473-3

Liu, H., Tian, Z., Danzin, T., Tan, X., Wang, J., La, Q., & Li, W. (2025b). Characterization and comparative analysis of the complete mitochondrial genome of *Phlomoides rotata*, a traditional Tibetan medicinal plant. *BMC Genomics, 26*(1), 727. https://doi.org/10.1186/s12864-025-11871-9

Lobay, D. (2015). *Rauwolfia* in the treatment of hypertension. *Integrative Medicine (Encinitas, Calif.), 14*(3), 40–46.

Lv, Y., Zhang, F., Chen, J., Cui, J., Xing, Y., Li, X., & Guo, S. (2010). Diversity and antimicrobial activity of endophytic Fungi associated with the alpine plant *Saussurea involucrata*. *Biological and Pharmaceutical Bulletin, 33*(8), 1300–1306. https://doi.org/10.1248/bpb.33.1300

Macar, T. K., & Macar, O. (2024). A study on the effect of *Hypericum perforatum* L. extract on vanadium toxicity in *Allium cepa* L. *Scientific Reports, 14*, 28486. https://doi.org/10.1038/s41598-024-79535-4

Mala, D., Awasthi, S., Sharma, N. K., Swarnkar, M. K., Shankar, R., & Kumar, S. (2021). Comparative transcriptome analysis of *Rheum Australe*, an endangered medicinal herb, growing in its natural habitat and those grown in controlled growth chambers. *Scientific Reports, 11*(1), 3702. https://doi.org/10.1038/s41598-020-79020-8

Malhotra, N., & Sharma, S. (2021). *Aconitum heterophyllum*. In *Himalayan medicinal plants* (pp. 5–25). Elsevier. https://doi.org/10.1016/B978-0-12-823151-7.00015-5

Mangral, Z. A., Islam, S. U., Majeed, A., Tariq, L., Goel, S., Barik, S. K., & Dar, T. U. H. (2025). Transcriptomic insights into elevation-dependent gene expression in *Rhododendron anthopogon* D.Don: Implications for climate resilience. *Physiologia Plantarum, 177*(4), e70419. https://doi.org/10.1111/ppl.70419

Margolin, A. A., Nemenman, I., Basso, K., Wiggins, C., Stolovitzky, G., Dalla Favera, R., & Califano, A. (2006). ARACNE: An algorithm for the reconstruction of gene regulatory networks in a mammalian cellular context. *BMC Bioinformatics, 7*(S1), S7. https://doi.org/10.1186/1471-2105-7-S1-S7

Miao, B.-B., Dong, W., Yi-Xin, G., Han, Z.-F., Luo, X., Ke, C.-H., & You, W.-W. (2023). OmicsSuite: A customized and pipelined suite for analysis and visualization of multi-omics big data. *Horticulture Research, 10*(11). https://doi.org/10.1093/hr/uhad195

Mirzaee, F., Hosseini, A., Jouybari, H. B., Davoodi, A., & Azadbakht, M. (2017). Medicinal, biological and phytochemical properties of Gentiana species. *Journal of Traditional and Complementary Medicine, 7*(4), 400–408. https://doi.org/10.1016/j.jtcme.2016.12.013

Muhammad, N., Lal Shrestha, R., Adhikari, A., Wadood, A., Khan, H., Khan, A. Z., Maione, F., Mascolo, N., & De Feo, V. (2015). First evidence of the analgesic activity of govaniadine, an alkaloid isolated from *Corydalis govaniana* Wall. *Natural Product Research, 29*(5), 430–437. https://doi.org/10.1080/14786419.2014.951933

Nataraj, N., Hussain, M., Ibrahim, M. H., Hausmann, A. E., Rao, S., Kaur, S., Khazir, J., Mir, B. A., Olsson, S. B., et al. (2022). Effect of altitude on volatile organic and phenolic compounds of *Artemisia brevifolia* Wall. ex DC. from the Western Himalayas. *Frontiers in Ecology and Evolution, 10*, Article 864728. https://doi.org/10.3389/fevo.2022.864728

Nong, M.-L., Luo, X.-H., Zhu, L.-X., Zhang, Y.-N., Dun, X.-Y., & Huang, L. (2023). Insights into the adaptation to high altitudes from transcriptome profiling: A case study of an endangered species, *Kingdonia uniflora*. *Genes, 14*(6), 1291. https://doi.org/10.3390/genes14061291

Ochatt, S., Alan, A. R., Bhattacharya, A., Hano, C., Kiselev, K. V., Marconi, P. L., Otoni, W. C., Park, S. Y., Tang, K. X., & Weathers, P. J. (2022). Secondary metabolites: A boon from plants, the best chemist in nature: Preface from the editors. *Plant Cell, Tissue and Organ Culture, 149*(1–2), 1–6. https://doi.org/10.1007/s11240-022-02289-2

Oluwole, O., Fernando, W. B., Lumanlan, J., Ademuyiwa, O., & Jayasena, V. (2022). Role of phenolic acid, tannins, stilbenes, lignans and flavonoids in human health – A review. *International Journal of Food Science & Technology, 57*(10), 6326–6335. https://doi.org/10.1111/ijfs.15936

Padhan, J. K., Kumar, V., Sood, H., Singh, T. R., & Chauhan, R. S. (2015). Contents of therapeutic metabolites in *Swertia chirayita* correlate with the expression profiles of multiple genes in corresponding biosynthesis pathways. *Phytochemistry, 116*, 38–47. https://doi.org/10.1016/j.phytochem.2015.05.007

Pal, T., Malhotra, N., Chanumolu, S. K., & Chauhan, R. S. (2015). Next-generation sequencing (NGS) transcriptomes reveal association of multiple genes and pathways contributing to secondary metabolites accumulation in tuberous roots of *Aconitum heterophyllum* Wall. *Planta, 242*(1), 239–258. https://doi.org/10.1007/s00425-015-2304-6

Pal, T., Padhan, J. K., Kumar, P., Sood, H., & Chauhan, R. S. (2018). Comparative transcriptomics uncovers differences in photoautotrophic versus photoheterotrophic modes of nutrition in relation to secondary metabolites biosynthesis in *Swertia chirayita*. *Molecular Biology Reports, 45*(2), 77–98. https://doi.org/10.1007/s11033-017-4135-y

Pandey, V., Tiwari, D. C., Dhyani, V., Bhatt, I. D., Rawal, R. S., & Nandi, S. K. (2021). Physiological and metabolic changes in two Himalayan medicinal herbs under drought, heat and combined stresses. *Physiology and Molecular Biology of Plants, 27*(7), 1523–1538. https://doi.org/10.1007/s12298-021-01027-w

Pandith, S. A., Hussain, A., Bhat, W. W., Dhar, N., Qazi, A. K., Rana, S., Razdan, S., Wani, T. A., Shah, M. A., Bedi, Y. S., Hamid, A., & Lattoo, S. K. (2014). Evaluation of anthraquinones from Himalayan rhubarb (*Rheum emodi* Wall. Ex Meissn.) as antiproliferative agents. *South African Journal of Botany, 95*, 1–8. https://doi.org/10.1016/j.sajb.2014.07.012

Parajuli, R., Timilsina, B., Jan, H. A., Sher, H., Ur Rahman, I., Hussain, W., Kunwar, R. M., Bussmann, R. W., & Paniagua-Zambrana, N. Y. (2021). *Delphinium brunonianum* Royle *Delphinium himalayae* Munz *Delphinium roylei* Munz *Delphinium vestitum* Wall. Ex Royle Ranunculaceae. In R. M. Kunwar, H. Sher, & R. W. Bussmann (Eds.), *Ethnobotany of the Himalayas* (pp. 1–12). Springer International Publishing. https://doi.org/10.1007/978-3-030-45597-2_79-1

Pathak, S., & Godela, R. (2024). *Nardostachys jatamansi*: Phytochemistry, ethnomedicinal uses, and pharmacological activities: A comprehensive review. *Fitoterapia, 172*, 105764. https://doi.org/10.1016/j.fitote.2023.105764

Peng, X.-C., Huang, A.-D., Wang, W.-J., Li, G.-S. X., & Zhao, J.-L. (2024). Draft genome of *Roscoea Debilis*, the first genome in the alpine ginger *Roscoea* (Zingiberaceae). *BMC Genomic Data, 25*(1), 77. https://doi.org/10.1186/s12863-024-01261-7

Pradhan, D. K., Ulak, S., Charmakar, S., Kunwar, R. M., Bussmann, R. W., & Paniagua-Zambrana, N. Y. (2021). *Rubia manjith* Roxb. Ex Fleming *Rubia tinctorium* L. Rubiaceae. In R. M. Kunwar, H. Sher, & R. W. Bussmann (Eds.), *Ethnobotany of the Himalayas* (pp. 1–8). Springer International Publishing. https://doi.org/10.1007/978-3-030-45597-2_207-1

Rabeh, K., Hnini, M., & Oubohssaine, M. (2025). A comprehensive review of transcription factor-mediated regulation of secondary metabolites in plants under environmental stress. *Stress Biology, 5, 15.* https://doi.org/10.1007/s44154-024-00201-w

Raina, R., Verma, P. K., Peshin, R., & Kour, H. (2019). Potential of *Juniperus communis* L as a nutraceutical in human and veterinary medicine. *Heliyon, 5*(8). https://doi.org/10.1016/j.heliyon.2019.e02376

Rathi, B., Sahu, J., Koul, S., & Kosha, R. (2013). Detailed pharmacognostical studies on *Berberis aristata* DC plant. *Ancient Science of Life, 32*(4), 234. https://doi.org/10.4103/0257-7941.131981

Rathore, S., Mukhia, S., Kumar, R., & Kumar, R. (2023). Essential oil composition and antimicrobial potential of aromatic plants grown in the mid-hill conditions of the Western Himalayas. *Scientific Reports, 13*(1), 4878. https://doi.org/10.1038/s41598-023-31875-3

Raut, A., Dhami-Shah, H., Phadke, A., Shindikar, A., Udipi, S., Joshi, J., Vaidya, R., & Vaidya, A. D. B. (2023). *Picrorhiza kurroa*, Royle ex Benth:Traditional uses, phytopharmacology, and translational potential in therapy of fatty liver disease. *Journal of Ayurveda and Integrative Medicine, 14*(1), 100558. https://doi.org/10.1016/j.jaim.2022.100558

Rawat, A., Prakash, O., Kumar, R., Arya, S., & Srivastava, R. M. (2021). *Hedychium spicatum* Sm.: Chemical composition with biological activities of Methanolic and Ethylacetate Oleoresins from rhizomes. *Journal of Biologically Active Products from Nature, 11*(3), 269–288. https://doi.org/10.1080/22311866.2021.1923572

Riaz, M., Khalid, R., Afzal, M., Anjum, F., Fatima, H., Zia, S., Rasool, G., Egbuna, C., Mtewa, A. G., Uche, C. Z., & Aslam, M. A. (2023). Phytobioactive compounds as therapeutic agents for human diseases: A review. *Food Science & Nutrition, 11*(6), 2500–2529. https://doi.org/10.1002/fsn3.3308

Rokaya, M. B., Maršík, P., & Münzbergová, Z. (2012). Active constituents in *Rheum acuminatum* and *Rheum australe* (Polygonaceae) roots: A variation between cultivated and naturally growing plants. *Biochemical Systematics and Ecology, 41*, 83–90. https://doi.org/10.1016/j.bse.2011.11.004

Saito, K., Hirai, M. Y., & Yonekura-Sakakibara, K. (2008). Decoding genes with coexpression networks and metabolomics – 'Majority report by precogs'. *Trends in Plant Science, 13*(1), 36–43. https://doi.org/10.1016/j.tplants.2007.10.006

Salick, J., Fang, Z., & Byg, A. (2009). Eastern Himalayan alpine plant ecology, Tibetan ethnobotany, and climate change. *Global Environmental Change, 19*(2), 147–155. https://doi.org/10.1016/j.gloenvcha.2009.01.008

Sekar, K. C., Thapliyal, N., Pandey, A., Joshi, B., Mukherjee, S., Bhojak, P., Bisht, M., Bhatt, D., Singh, S., & Bahukhandi, A. (2024). Plant species diversity and density patterns along altitude gradient covering high-altitude alpine regions of west Himalaya, India. *Geology, Ecology, and Landscapes, 8*(4), 559–573. https://doi.org/10.1080/24749508.2022.2163606

Selvakesavan, R. K., Nuc, M., Kolarčik, V., Krajewski, P., & Franklin, G. (2024). Single molecule real-time sequencing data sets of *Hypericum perforatum* L. plantlets and cell suspension cultures. *Scientific Data, 11*(1), 42. https://doi.org/10.1038/s41597-023-02878-6

Semwal, P., Painuli, S., & Cruz-Martins, N. (2021). *Dioscorea deltoidea* wall. Ex Griseb: A review of traditional uses, bioactive compounds and biological activities. *Food Bioscience, 41*, 100969. https://doi.org/10.1016/j.fbio.2021.100969

Shah, S. S., Ahmed, S., Zhou, B., & Shi, L. (2025). A review on pharmacological activities and phytochemical constituents of *Zanthoxylum armatum* DC. *Natural Product Research, 39*(11), 3240–3259. https://doi.org/10.1080/14786419.2024.2409984

Shakeel, A., Noor, J. J., Jan, U., Gul, A., Handoo, Z., & Ashraf, N. (2025). Saponins, the unexplored secondary metabolites in plant defense: Opportunities in integrated pest management. *Plants, 14*(6), 861. https://doi.org/10.3390/plants14060861

Sharma, N., & Kala, C. P. (2018). Harvesting and management of medicinal and aromatic plants in the Himalaya. *Journal of Applied Research on Medicinal and Aromatic Plants, 8*, 1–9. https://doi.org/10.1016/j.jarmap.2017.09.003

Sharma, T., Sharma, N. K., Kumar, P., Panzade, G., Rana, T., Swarnkar, M. K., Singh, A. K., Singh, D., Shankar, R., & Kumar, S. (2021a). The first draft genome of *Picrorhiza kurrooa*, an endangered medicinal herb from Himalayas. *Scientific Reports, 11*(1), 14944. https://doi.org/10.1038/s41598-021-93495-z

Sharma, A., Bhattacharyya, D., Sharma, S., & Chauhan, R. S. (2021b). Transcriptome profiling reveals key hub genes in co-expression networks involved in iridoid glycosides biosynthetic machinery in *Picrorhiza kurroa*. *Genomics, 113*(5), 3381–3394. https://doi.org/10.1016/j.ygeno.2021.07.024

Sharma, S., Kumar, V., Seth, C. A., Sourirajan, A., El-Shazly, M., & Dev, K. (2024). A comprehensive review on the phytochemistry, pharmacological properties, and in vitro propagation of an endemic medicinal orchid, *Dactylorhiza hatagirea*. *Naunyn-Schmiedeberg's Archives of Pharmacology, 397*(5), 2621–2635. https://doi.org/10.1007/s00210-023-02827-5

Shasmita, Behera, S., Mishra, P., Samal, M., Mohapatra, D., Monalisa, K., & Naik, S. K. (2023). Recent advances in tissue culture and secondary metabolite production in *Hypericum perforatum* L. *Plant Cell, Tissue and Organ Culture, 154*(1), 13–28. https://doi.org/10.1007/s11240-023-02525-3

Sher, H., Kunwar, R. M., Bussmann, R. W., & Paniagua-Zambrana, N. Y. (2021). *Heracleum candicans* Wall. Ex DC. Apiaceae. In I. R. Kunwar, H. Sher, & R. W. Bussmann (Eds.), *Ethnobotany of the Himalayas* (pp. 1–10). Springer International Publishing. https://doi.org/10.1007/978-3-030-45597-2_115-1

Singamaneni, V., Sharma, U., Lone, B., & Gupta, P. (2020). Phytochemistry and pharmacological activities of *Rhodiola imbricata* Edgew., a high value medicinal herb of Cold Desert Himalaya. In B. Singh (Ed.), *Botanical leads for drug discovery* (pp. 159–178). Springer. https://doi.org/10.1007/978-981-15-5917-4_8

Singh Rana, V., Sethiya, N. K., Duseja, M., Gupta, R., Bisht, D., & Gangal, A. (2022). Ethnomedicinal and traditional application of *Allium wallichii*Kunth (Himalayan Onion): An unexplored and underutilized nutraceutical plant foods from Himalayan regions. *Ethnobotany Research and Applications, 24*. https://doi.org/10.32859/era.24.15.1-16

Skinder, B. M., Ganai, B. A., & Wani, A. H. (2017). Scientific study of *Gentiana kurroo* Royle. *Medicines, 4*(4), 74. https://doi.org/10.3390/medicines4040074

Sofi, M. A., Sofi, M. A., Nanda, A., Nayak, B. K., Othman, Z., & Sadikan, M. Z. (2025). Exploring the therapeutic potential of *Podophyllum hexandrum* root extract: Chemical composition, antimicrobial efficacy, and antioxidant and anticancer activities. *Scientifica, 2025*(1), 5100547. https://doi.org/10.1155/sci5/5100547

Su, G., Morris, J. H., Demchak, B., & Bader, G. D. (2014). Biological network exploration with Cytoscape 3. *Current Protocols in Bioinformatics, 47*(1), 8.13.1–8.13.24. https://doi.org/10.1002/0471250953.bi0813s47

Su, X., Tao Liu, Y., Liu, P., Harris, A., & Chen, J. Y. (2022). Adaptive radiation in *Orinus*, an endemic alpine grass of the Qinghai-Tibet plateau, based on comparative transcriptomic analysis. *Journal of Plant Physiology, 277*, 153786. https://doi.org/10.1016/j.jplph.2022.153786

Sumner, E. E., Williamson, V. G., Gleadow, R. M., Wevill, T., & Venn, S. E. (2022). Acclimation to water stress improves tolerance to heat and freezing in a common alpine grass. *Oecologia, 199*(4), 831–843. https://doi.org/10.1007/s00442-022-05245-1

Sun, W., & Shahrajabian, M. H. (2023). Therapeutic potential of phenolic compounds in medicinal plants—Natural health products for human health. *Molecules, 28*(4), 1845. https://doi.org/10.3390/molecules28041845

Sun, W., Wang, B., Yang, J., Wang, W., Liu, A., Leng, L., Xiang, L., Song, C., & Chen, S. (2017). Weighted gene co-expression network analysis of the dioscin rich medicinal plant *Dioscorea nipponica*. *Frontiers in Plant Science, 8*, 789. https://doi.org/10.3389/fpls.2017.00789

Takshak, S., & Agrawal, S. B. (2019). Defense potential of secondary metabolites in medicinal plants under UV-B stress. *Journal of Photochemistry and Photobiology B: Biology, 193*, 51–88. https://doi.org/10.1016/j.jphotobiol.2019.02.002

Tang, W., Li, Z., Xu, Z., et al. (2025). Transcriptomic and metabolomic analysis reveal the cold tolerance mechanism of common beans under cold stress. *BMC Plant Biology, 25*, 340. https://doi.org/10.1186/s12870-025-06333-z

Terletskaya, N. V., Erbay, M., Mamirova, A., Ashimuly, K., Korbozova, N. K., Zorbekova, A. N., Kudrina, N. O., & Hoffmann, M. H. (2024). Altitude-dependent Morphophysiological, anatomical, and Metabolomic adaptations in *Rhodiola Linearifolia* Boriss. *Plants, 13*(19), 2698. https://doi.org/10.3390/plants13192698

Thapa, C. B., Paudel, M. R., Bhattarai, H. D., Pant, K. K., Devkota, H. P., Adhikari, Y. P., & Pant, B. (2022). Bioactive secondary metabolites in *Paris polyphylla* Sm. and their biological activities: A review. *Heliyon, 8*(2), e08982. https://doi.org/10.1016/j.heliyon.2022.e08982

Tholl, D. (2015). Biosynthesis and biological functions of terpenoids in plants. In J. Schrader & J. Bohlmann (Eds.), *Biotechnology of isoprenoids. Advances in biochemical engineering/biotechnology* (Vol. 148). Springer. https://doi.org/10.1007/10_2014_295

Tiwari, S., & Talreja, S. (2020). Importance of *Cinnamomum tamala* in the treatment of various diseases. *Pharmacognosy Journal, 12*(6s), 1792–1796. https://doi.org/10.5530/pj.2020.12.241

Tiwari, D., Kewlani, P., Gaira, K. S., Bhatt, I. D., Sundriyal, R. C., & Pande, V. (2023). Predicting phytochemical diversity of medicinal and aromatic plants (MAPs) across eco-climatic zones and elevation in Uttarakhand using Generalized Additive Model. *Scientific Reports, 13*(1), 10888. https://doi.org/10.1038/s41598-023-37495-1

Tomczyk, M., & Latté, K. P. (2009). Potentilla—A review of its phytochemical and pharmacological profile. *Journal of Ethnopharmacology, 122*(2), 184–204. https://doi.org/10.1016/j.jep.2008.12.022

Vafaee, Y., Wallace, L., Khoddamzadeh, A., & Nalousi, A. M. (2025). Multiomics approaches for conservation of endangered orchids: Advances, applications, and future prospects. In S. Gantait & P. Chmielarz (Eds.), *Conservation of plant genetic resources* (pp. 299–350). Springer Nature. https://doi.org/10.1007/978-981-96-8331-4_10

Vashisht, I., Pal, T., Sood, H., & Chauhan, R. S. (2016). Comparative transcriptome analysis in different tissues of a medicinal herb, *Picrorhiza kurroa* pinpoints transcription factors regulating picrosides biosynthesis. *Molecular Biology Reports, 43*(12), 1395–1409. https://doi.org/10.1007/s11033-016-4073-0

Vogel, C., & Marcotte, E. M. (2012). Insights into the regulation of protein abundance from proteomic and transcriptomic analyses. *Nature Reviews Genetics, 13*(4), 227–232. https://doi.org/10.1038/nrg3185

Wang, Z., Chen, Z., Tang, Y., Zhang, M., & Huang, M. (2023). Regulation of transcriptome networks that mediate ginsenoside biosynthesis by essential ecological factors. *PLoS One, 18*(8), e0290163. https://doi.org/10.1371/journal.pone.0290163

Wang, T., Rangji, C., Liu, W., Ma, J., Zhou, R., Leng, L., & Zhang, Y. (2024a). Multi-omics on traditional medicinal plant of the genus Aconitum: Current progress and prospect. *Molecules, 30*(1), 118. https://doi.org/10.3390/molecules30010118

Wang, X., Kong, D.-K., Zhang, H.-R., & Zou, Y. (2024b). Discovery of a polyketide carboxylate phytotoxin from a polyketide glycoside hybrid by β-glucosidase mediated ester bond hydrolysis. *Chemical Science, 15*(41), 17183–17192. https://doi.org/10.1039/D4SC05256K

Wang, Y., Suonan, J., Liu, K., Gao, Y., Zhu, S., Liu, Q., & Zhao, N. (2024c). The third dimension of alpine plant leaf traits is related to cold-tolerance. *Plant Diversity, 47*(1), 159–165. https://doi.org/10.1016/j.pld.2024.10.001

Wang, A., Guo, W., Wang, S., Wang, Y., Kong, D., & Li, W. (2024d). Transcriptome analysis unveiled the genetic basis of rapid seed germination strategies in alpine plant *Rheum pumilum*. *Scientific Reports, 14*(1), 19194. https://doi.org/10.1038/s41598-024-70320-x

Wang, G., Sun, X., Li, Y., Wang, Y., & Jin, C. (2025a). The role of UV-B radiation in modulating secondary metabolite biosynthesis and regulatory mechanisms in medicinal plants. *BioResources, 20*(2). https://doi.org/10.15376/biores.20.2.Wang

Wang, Y., Liu, J., Zhang, E., Yang, Y., Lu, Q., Zhu, Z., & Li, R. (2025b). Metabolite profiling and anti-inflammatory activities of *Fritillaria cirrhosa* D. Don bulbs derived from tissue culture. *Molecules, 30*(3), 623. https://doi.org/10.3390/molecules30030623

Wang, R., Li, Z., Liu, X., Singh, B., Chauhan, R. S., Duan, Y., Xue, J., Kumar, V., & Xue, T. (2025c). A novel transcription factor *PtMYB77* from *Pinellia ternata* enhances heat tolerance in Arabidopsis by inducing early expression of heat shock factor genes. *Industrial Crops and Products, 227*, 120791. https://doi.org/10.1016/j.indcrop.2025.120791

Wu, L., Meng, F., Su, X., Chen, N., Peng, D., & Xing, S. (2023). Transcriptomic responses to cold stress in *Dendrobium huoshanense* C.Z. Tang et S.J. Cheng. *Physiology and Molecular Biology of Plants, 29*(11), 1633–1646. https://doi.org/10.1007/s12298-023-01385-7

Wu, X., He, X., Wang, X., Liu, P., Ai, S., Liu, X., Li, Z., & Wang, X. (2025). Genome-wide identification, phylogenetic evolution, and abiotic stress response analyses of the late embryogenesis abundant gene family in the alpine cold-tolerant medicinal *Notopterygium* species. *International Journal of Molecular Sciences, 26*(2), 519. https://doi.org/10.3390/ijms26020519

Xu, Y., Cao, Y., Tie, F., Kong, X., Liu, Y., Zhang, Y., Guan, W., Hu, N., Wang, H., Qin, X., Wu, Z., & Fu, C. (2025). Generation of suspension cell cultures with high syringin content and anti-inflammatory activity through overexpressing glycotransferase SiUGT72BZ2 in *Saussurea involucrata*. *Plant Biotechnology Journal, 23*(5), 1713–1724. https://doi.org/10.1111/pbi.70001

Xue, J.-S., Qiu, S., Jia, X.-L., Shen, S.-Y., Shen, C.-W., Wang, S., Xu, P., Tong, Q., Lou, Y.-X., Yang, N.-Y., Cao, J.-G., Hu, J.-F., Shen, H., Zhu, R.-L., Murray, J. D., Chen, W.-S., & Yang, Z.-N. (2023). Stepwise changes in flavonoids in spores/pollen contributed to terrestrial adaptation of plants. *Plant Physiology, 193*(1), 627–642. https://doi.org/10.1093/plphys/kiad313

Yang, X., Liang, Y., Liu, P., Song, L., & Gao, J. (2025a). Extraction, structural characteristics, and antioxidant activity of a glucose-dominated polysaccharide, NJPW, purified from *Nardostachys jatamansi* (D.Don) DC. *ACS Omega, 10*(28), 30100–30115. https://doi.org/10.1021/acsomega.5c00085

Yang, G., Liu, W., Mo, X., & Mei, Z. (2025b). Alpine adaptive mechanism on rhizosphere microbes recruitment of *Crepisnapifera* (Franch.) Babc. By multi-omics analysis. *Biology, 14*(4), 345. https://doi.org/10.3390/biology14040345

Yin, Q., Xiang, L., Han, X., Zhang, Y., Lyu, R., Yuan, L., & Chen, S. (2025). The evolutionary advantage of artemisinin production by *Artemisia annua*. *Trends in Plant Science, 30*(2), 213–226. https://doi.org/10.1016/j.tplants.2024.09.006

Zaffar, S., Khan, I., Zaffar, A., & Kamili, A. N. (2018). Altitudinal Variation in Alkaloid Composition of *Hyoscyamus niger*: A study with reference to Kashmir region of Himalayas. *International Journal of Advances in Scientific Research, 4*(3), 13. https://doi.org/10.7439/ijasr.v4i3.4728

Zhang, W., Xia, L., Peng, F., Song, C., Manzoor, M. A., Cai, Y., & Jin, Q. (2022). Transcriptomics and metabolomics changes triggered by exogenous 6-Benzylaminopurine in relieving epicotyl dormancy of *Polygonatum cyrtonema* Hua seeds. *Frontiers in Plant Science, 13*, 961899. https://doi.org/10.3389/fpls.2022.961899

Zhang, J., Dong, K.-L., Ren, M.-Z., Wang, Z.-W., Li, J.-H., Sun, W.-J., Zhao, X., Xin-Xing, F., Ye, J.-F., Liu, B., Zhang, D.-M., Wang, M.-Z., Zeng, G., Niu, Y.-T., Li-Min, L., Jun-Xia,

S., Liu, Z.-J., Soltis, P. S., Soltis, D. E., & Chen, Z.-D. (2024). Coping with alpine habitats: Genomic insights into the adaptation strategies of *Triplostegia glandulifera* (Caprifoliaceae). *Horticulture Research, 11*(5). https://doi.org/10.1093/hr/uhae077

Zhang, Q.-Y., He, X.-J., Xie, Y.-Z., Zhou, L.-P., Meng, X., Kang, J., Luo, C.-Y., Wang, Y.-N., Li, Z.-H., & Guan, T.-X. (2025). Genome-wide identification, phylogeny, and abiotic stress response analysis of OSCA family genes in the alpine medicinal herb *Notopterygium franchetii*. *International Journal of Molecular Sciences, 26*(11), 5043. https://doi.org/10.3390/ijms26115043

Zheng, J., Jiang, H., Yan, Y., & Yin, T. (2024). Overview of the chemistry and biological activities of natural atisine-type diterpenoid alkaloids. *RSC Advances, 14*(32), 22882–22893. https://doi.org/10.1039/D4RA03305A

Zhong, L., Li, T., Zhang, J., Zhang, J., Zhang, J., Li, L., Chen, G., Zhong, S., & Rui, G. (2025). Integrated physiological, biochemical and transcriptomic analyses elucidate the response of alpine plant *Lamiophlomis rotata* (Benth.) Kudo to low-nitrogen stress. *BMC Plant Biology, 25*(1), 941. https://doi.org/10.1186/s12870-025-06953-5

Zhou, X., Chen, S., Hui, W., Yang, Y., & Hongwei, X. (2017). Biochemical and proteomics analyses of antioxidant enzymes reveal the potential stress tolerance in *Rhododendron chrysanthum* Pall. *Biology Direct, 12*(1), 10. https://doi.org/10.1186/s13062-017-0181-6

Zhou, T., Zhao, Y., Zhou, L., Chen, X., Jia, Y., & Bai, G. (2023). Comparative transcriptome analyses of three Gentiana species provides signals for the molecular footprints of selection effects and the phylogenetic relationships. *Molecular Genetics and Genomics, 298*(2), 399–411. https://doi.org/10.1007/s00438-022-01991-2

Zhu, R., Fang, C., Zhang, S., Han, Z., Zhu, G., Cai, S., Zheng, C., Tang, Y., & Wang, Y. (2025). Comprehensive review on *Rhodiola crenulata*: Ethnopharmacology, Phytochemistry, pharmacological properties and clinical applications. *Chinese Journal of Integrative Medicine, 31*(8), 752–759. https://doi.org/10.1007/s11655-025-4010-y

Molecular Phylogeny and Phylogeography of Alpine Medicinal Plants

9

Shubham Sharma

Abstract

Alpine medicinal plants underpin mountain ethnopharmacopoeias and provide model systems for studying adaptation to extreme environments. This chapter synthesizes current advances in molecular phylogeny and phylogeography of alpine medicinal plants, with an emphasis on widely used markers (nuclear internal transcribed spacer, chloroplast regions, complete plastid genomes), next-generation sequencing approaches (restriction-site associated DNA sequencing, genome skimming), and integrative analyses that couple genetic variation with ecological and biogeographic data. Using representative taxa—including *Rhodiola* (Crassulaceae), *Saussurea* and *Artemisia* (Asteraceae), and *Gentiana* (Gentianaceae)—we highlight patterns of rapid diversification, glacial refugia, postglacial range dynamics, and phylogenetic–taxonomic resolutions relevant to conservation and sustainable use. We also outline a transparent workflow from field sampling and laboratory pipelines to computational inference (maximum likelihood, Bayesian frameworks) and phylogeographic modeling (coalescent analysis, haplotype networks, species distribution modeling). The synthesis frames alpine diversity in a comparative context across the Himalayas, the European Alps, the Andes, and the Tibetan Plateau, setting the stage for multi-omics and conservation genomics in later sections. Findings inform taxonomy, prioritization of evolutionarily significant units, and climate-resilient management strategies.

S. Sharma (✉)
Institute of Nano Science and Technology (INST),
Sector-81, Knowledge City, Sahibzada Ajit Singh Nagar, Mohali, Punjab, India
e-mail: shubham.sharma@alumni.iitgn.ac.in

P. Kumar et al. (eds.), *Climate Resilience and Molecular Adaptation in Alpine Medicinal Plants*, https://doi.org/10.1007/978-981-95-6923-6_9

Keywords

Alpine medicinal plants · Molecular phylogeny · Phylogeography · Plastid genomes · ITS · RAD-seq · Genome skimming · Climate change · Himalayan flora · Conservation genomics

1 Introduction

Alpine ecosystems are among the most challenging environments on Earth. Found above the treeline in mountain systems worldwide, they are characterized by extreme abiotic stressors such as low temperatures, short growing seasons, intense ultraviolet (UV) radiation, hypoxia, and strong diurnal fluctuations in climate. Despite these harsh conditions, alpine zones host a remarkable diversity of plant species, many of which have developed unique adaptations at the physiological, biochemical, and molecular levels. These adaptations not only allow survival but also underpin the production of secondary metabolites with important medicinal properties. Alpine medicinal plants therefore occupy a dual role: they are integral to local ethnopharmacopoeias and livelihoods, while also serving as global sources of phytochemicals of pharmacological interest.

The Himalayas, European Alps, Andes, Tibetan Plateau, and other high mountain systems harbor numerous medicinal taxa. *Rhodiola* species provide adaptogenic compounds that modulate stress responses; *Saussurea* species are used in traditional Tibetan medicine for their anti-inflammatory properties; *Artemisia* taxa include the lowland *A. annua*, source of artemisinin, but high-elevation relatives also yield valuable terpenoids; and *Gentiana* species are used for their bitter iridoid glycosides with hepatoprotective effects. Beyond these well-known genera, alpine floras include *Delphinium*, *Meconopsis*, *Primula*, and many other medicinally important taxa with unresolved systematics (Kala (2005)).

Conservation of alpine medicinal plants faces severe challenges. Many species are overharvested, both for local use and for global trade, while climate change is driving upward shifts in species distributions and altering competitive dynamics. Mountain building and Quaternary climatic oscillations have left complex signatures on their evolutionary histories, making phylogenetic and phylogeographic analyses indispensable for conservation planning.

Molecular phylogeny provides a framework to resolve taxonomic ambiguities, test hypotheses about adaptive radiations, and identify cryptic lineages. Phylogeography, by integrating molecular data with geography, uncovers patterns of genetic structure, historical demography, and postglacial migration. Together, they reveal how evolutionary history, geography, and climate have shaped present-day alpine diversity. These approaches also underpin modern concepts of conservation biology such as delineation of evolutionarily significant units (ESUs).

A transparent workflow (Fig. 9.1) links field sampling, herbarium voucher verification, DNA extraction, sequencing, bioinformatics pipelines, and phylogenetic/phylogeographic analyses. Molecular markers—ranging from nuclear ribosomal

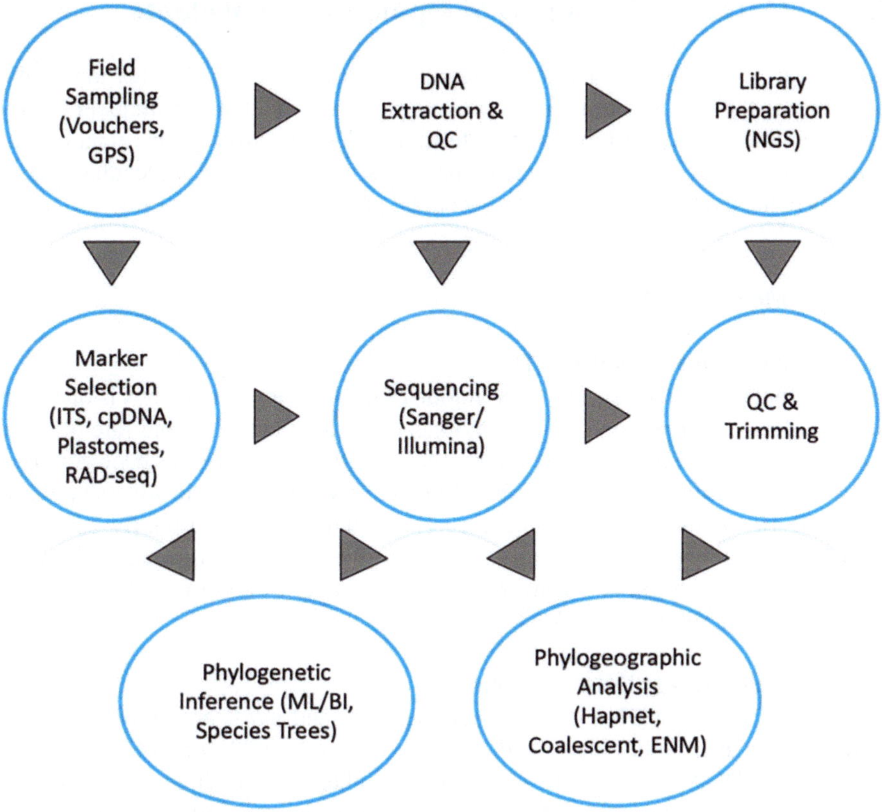

Fig. 9.1 Workflow: molecular phylogeny & phylogeography

ITS and chloroplast DNA regions to complete plastomes and genome-wide SNPs—have been extensively applied to alpine plants. Next-generation sequencing (NGS) technologies, particularly restriction-site associated DNA sequencing (RAD-seq), genotyping-by-sequencing (GBS), and genome skimming, now permit resolution of complex relationships even in recently radiated clades.

In this chapter, Sects. 1, 2 and 3 provide the foundations. Section 2 reviews molecular markers and analytical methods, with case studies from representative alpine medicinal genera. Section 3 focuses on phylogeographic insights, integrating demographic and ecological modeling, and comparing mountain systems worldwide. Subsequent Sects. 4 and 5 will expand to integrative frameworks and emerging approaches.

2 Molecular Phylogeny of Alpine Medicinal Plants

Molecular phylogeny reconstructs relationships among taxa using heritable DNA sequences. Alpine medicinal plants present particular challenges: frequent convergent evolution (e.g., cushion or rosette morphologies), rapid diversification, and hybridization obscure morphological characters. Consequently, molecular datasets have become essential to reconstruct accurate phylogenetic hypotheses.

2.1 Molecular Markers

2.1.1 Nuclear Ribosomal ITS

The internal transcribed spacer (ITS) region of nuclear ribosomal DNA is one of the most widely used markers in plant phylogenetics. Its relatively rapid rate of evolution makes it useful for resolving interspecific relationships. ITS has been employed to study *Rhodiola* complexes across Eurasia, revealing cryptic diversity within *R. rosea*. In *Gentiana*, ITS has resolved clades that correspond poorly with traditional morphology, highlighting taxonomic over-splitting. Limitations include paralogy, pseudogenes, and concerted evolution, necessitating corroboration with plastid and genome-wide markers.

2.1.2 Chloroplast DNA Regions

Chloroplast DNA (cpDNA) regions such as *rbcL* and *matK* are global DNA barcodes, while intergenic spacers like *trnH–psbA* and *rpl32–trnL* offer additional variation. cpDNA markers are uniparentally inherited, yielding clear haplotype structures but failing to capture biparental inheritance and hybridization. In *Saussurea*, *trnH–psbA* has revealed east–west divergence across the Himalayas. In *Gentiana rigescens*, cpDNA haplogroups are partitioned by river barriers in the Sino-Himalayan region.

2.1.3 Complete Plastid Genomes

With genome skimming, entire plastid genomes (plastomes) can be recovered efficiently. Plastomes provide thousands of informative sites and enable partitioned analyses of coding vs. noncoding regions. In *Rhodiola*, plastome phylogenies resolved the placement of cryptic species. In *Saussurea*, plastome data clarified subgeneric relationships. Complete plastomes are now a standard in alpine phylogenomics.

2.1.4 Genome-Wide SNPs from RAD-Seq and GBS

RAD-seq and GBS generate thousands of SNPs from across the genome. These data permit coalescent-based species tree inference and population-level studies. In *Gentiana*, RAD-seq revealed strong population structure consistent with sky-island dynamics. In *Saussurea*, RAD-seq showed introgression and secondary contact zones in sympatric taxa. Proper bioinformatics is crucial to avoid spurious results due to paralogy or missing data.

2.1.5 Whole Genomes and Transcriptomes

With decreasing sequencing costs, whole-genome sequencing (WGS) and transcriptome sequencing (RNA-seq) are increasingly used. In alpine *Primula*, transcriptomic analyses have identified candidate genes for cold adaptation. While expensive, these datasets support integrative analyses linking genotype to phenotype. Table 9.1 summarizes the strengths and limitations of key marker classes.

2.2 Analytical Methods

Phylogenetic inference traditionally relied on maximum parsimony and distance methods but now is dominated by model-based approaches such as maximum likelihood (ML) and Bayesian inference (BI). Programs like RAxML, IQ-TREE, and MrBayes are standard. Coalescent-based species tree methods (e.g., *ASTRAL, SVDquartets*) are essential when incomplete lineage sorting is expected. Molecular dating using relaxed-clock models (BEAST) provides temporal frameworks to link diversification with geological and climatic events. Species delimitation methods (*BPP, STACEY*) integrate genetic data to detect cryptic taxa.

Table 9.1 Comparative overview of molecular markers for phylogeny and phylogeography in alpine plants

Marker class	Example Loci	Advantages	Limitations	Typical application	Reference(s)
Nuclear ribosomal DNA	ITS	High variability; widely used	Paralogs, pseudogenes	Species-level phylogeny	Baldwin et al. (1995)
cpDNA barcodes	rbcL, matK, trnH–psbA	Universally amplifiable; standard DNA barcodes	Uniparental inheritance; low variation in some taxa	DNA barcoding; haplotype analysis	Kress and Erickson (2007)
Complete plastomes	Whole cpDNA	Thousands of sites; backbone resolution	Maternal inheritance; expensive (but decreasing costs)	Deep phylogenies; molecular dating	Straub et al. (2012), Wicke et al. (2021)
Reduced-representation NGS	RAD-seq, GBS	Thousands of SNPs; genome-wide	Bioinformatic complexity; missing data	Population genomics; coalescent species trees	Davey and Blaxter (2010), Excoffier and Heckel (2006)
Whole-genome sequencing	Nuclear genomes	High resolution; structural variants	Cost-intensive; large data	Pangenomics; adaptation genomics	Favre et al. (2015)
Transcriptomics	RNA-seq	Expression + adaptive loci	Tissue-specific; no non-expressed genes	Stress adaptation; functional genomics	Liu et al. (2014)

2.3 Case Studies in Representative Lineages

Rhodiola **(Crassulaceae)** ITS and plastome data have revealed repeated colonization of alpine zones and cryptic species within *R. rosea*. Plastome phylogenies suggest diversification pulses during Quaternary glacial cycles.

Saussurea **(Asteraceae)** One of the most species-rich alpine genera, *Saussurea* has diversified rapidly on the Qinghai–Tibetan Plateau. Plastome and RAD-seq data confirm rapid radiations, with hybridization zones at elevational ecotones.

Gentiana **(Gentianaceae)** Discordance between plastid and nuclear datasets reflects incomplete lineage sorting and introgression. RAD-seq reconciles these signals. Plastome phylogenies indicate deep divergences coinciding with Himalayan uplift.

Artemisia **(Asteraceae)** A widespread genus across Eurasia, *Artemisia* lineages adapted to alpine environments exhibit genetic continuity along steppe corridors. Phylogenomic data reveal convergent adaptation to cold and drought.

Primula*, *Meconopsis*, *Delphinium Other medicinal alpine taxa show similar patterns: rapid radiations, introgression, and adaptation to high-elevation niches. Table 9.2 summarizes representative alpine medicinal taxa with major phylogenetic insights.

3 Phylogeography of Alpine Medicinal Plants

Phylogeography connects genealogical lineages to geography. For alpine taxa, phylogeography reveals how Pleistocene glaciations, refugial dynamics, and topographic barriers shaped genetic diversity. (Avise (2000))

3.1 Markers for Phylogeography

cpDNA haplotypes track seed-mediated gene flow, while nuclear microsatellites and genome-wide SNPs reveal biparental structure. Approximate Bayesian Computation (ABC) and coalescent simulations test demographic scenarios.

Table 9.2 Representative alpine medicinal plants with major phylogenetic/phylogeographic insights

Genus/Species	Family	Distribution	Molecular approach	Key findings	References
Rhodiola rosea complex	Crassulaceae	Circumboreal (Eurasia, North America)	ITS + plastomes	Multiple cryptic lineages; Eurasian refugia; late Pleistocene expansion	Zhang et al. (2014)
Saussurea obvallata	Asteraceae	Himalayas (sky-islands)	cpDNA + RAD-seq	Fine-scale structuring; limited gene flow; sky-island dynamics	Wang et al. (2019)
Gentiana rigescens	Gentianaceae	Sino-Himalayas	cpDNA + SNPs	River-barrier structuring; secondary contact in mid-elevation corridors	Chen et al. (2020)
Artemisia alpine taxa	Asteraceae	Eurasia (steppe–alpine transitions)	RAD-seq	Steppe corridors support continuity; deserts act as barriers	Liu et al. (2021)
Primula spp.	Primulaceae	Himalayas, Alps	Transcriptomes + cpDNA	Cold adaptation genes; refugial persistence in valleys	Liu et al. (2014)
Meconopsis spp.	Papaveraceae	Eastern Himalayas	Plastomes	Radiation linked to orogeny and climate oscillations	Hundsdoerfer et al. (2022)

3.2 Models and Frameworks

Genealogy-based haplotype networks (TCS, PopART) show lineage relationships. Coalescent methods (BEAST, IMa2) infer divergence times and migration. Ecological niche modeling (MaxEnt) reconstructs potential distributions during glacial cycles, allowing tests of refugial persistence vs. recolonization.

3.3 Regional Comparisons

- **Himalayas & Tibetan Plateau:** Deep structure across river gorges and rain shadows. Eastern Himalayas often harbor refugial lineages.
- **European Alps:** Southern refugia with postglacial northward expansions.
- **Andes:** Latitudinal clines and sky-island divergence.
- **Caucasus:** Long-term isolated haplotypes, limited connectivity.
- **Rockies & Japanese Alps:** Postglacial recolonization, reduced diversity compared to Eurasia.

3.4 Representative Species

Rhodiola rosea shows Eurasian refugia and late Pleistocene expansions. *Saussurea obvallata* exhibits fine-scale structuring in Himalayan sky-islands. *Gentiana rigescens* demonstrates river-barrier partitioning, with nuclear SNPs indicating secondary contact. *Artemisia* lineages illustrate east–west continuity with desert interruptions. Table 9.3 summarizes alpine diversity hotspots and Himalayan detail.

Table 9.3 Regional phylogeographic patterns of alpine medicinal plants

Mountain system	Dominant processes	Representative taxa	Refugial patterns	Postglacial dynamics	References
Himalayas & Tibetan plateau	Deep phylogeographic structure across valleys and rain shadows	Saussurea, Rhodiola, Gentiana	Refugia in river gorges and eastern Himalayas	Upslope tracking; secondary contact	Wang et al. (2019), Chen et al. (2020)
European Alps	Southern refugia, northward expansion	Primula, Gentiana	Refugia in southern/ eastern Alps	Northward colonization	Schönswetter et al. (2005)
Andes	Latitudinal clines; sky-islands	Valeriana, Gentianella	Refugia in inter-Andean valleys	Stepwise upslope tracking	Muellner-Riehl (2019)
Rockies	Postglacial recolonization	Artemisia, erigeron	Refugia south of ice sheets	Expansion into alpine meadows	Hewitt (2000)
Japanese Alps	Climatic refugia in isolated mountains	Saussurea, Gentiana	Localized refugia	Rapid postglacial spread	Liu et al. (2014)

4 Integrative Framework: Linking Molecular Phylogeny, Phylogeography, and Ecology

Molecular phylogeny and phylogeography are powerful in isolation, but their true strength emerges when combined with ecological and environmental data. Alpine medicinal plants provide ideal systems to integrate these approaches because they inhabit ecologically complex regions where topography, climate, and human use intersect. An integrative framework allows us to connect evolutionary processes across scales: from genes to populations, species, and ecosystems.

4.1 From Phylogeny to Conservation Units

Phylogenetic trees inform taxonomic revisions and delineate evolutionarily significant units (ESUs). For alpine medicinal plants, this is critical because morphological convergence often masks genetic distinctiveness. For example, *Rhodiola rosea* was long treated as a single widespread species; phylogenomic analyses revealed multiple cryptic lineages with distinct evolutionary histories. Recognizing such lineages as ESUs ensures that conservation efforts maintain genetic diversity rather than collapsing distinct taxa into a single unit.

4.2 Landscape Genomics

Landscape genomics integrates genetic data with spatial environmental variables to identify loci under selection. In *Saussurea* populations across the Himalayas, RAD-seq data correlated with temperature and precipitation gradients reveal candidate genes for cold and drought tolerance. Linking these adaptive loci to environmental niches allows prediction of which populations are most vulnerable to climate change.

4.3 Species Distribution Modeling and Genomic Data

Species distribution models (SDMs), such as those generated with MaxEnt, can be combined with genomic data to project the future of alpine medicinal plants. For instance, SDMs for *Gentiana rigescens* indicate contraction of suitable habitats under warming scenarios. Coupled with genomic scans, we can identify whether populations at the trailing edge harbor unique alleles or adaptive variants. These populations may be conservation priorities even if they occur in marginal habitats.

4.4 Multi-omics Integration

Genomics, transcriptomics, metabolomics, and epigenomics together provide a holistic view of adaptation. Transcriptomic analyses of *Primula* under cold stress reveal upregulation of genes involved in antioxidant pathways. Metabolomic profiling of *Rhodiola* species shows variation in salidroside content linked to genetic lineages. Epigenomic studies suggest DNA methylation as a mechanism for rapid acclimatization to changing climates. Integrating these omics layers enables mechanistic understanding of trait evolution.

4.5 Human Dimensions and Ethnobotany

Phylogeny and phylogeography can also be integrated with ethnobotanical knowledge. Local communities often distinguish plant varieties based on subtle morphological or chemical cues. Molecular data sometimes corroborate these distinctions, revealing cryptic species recognized by traditional knowledge systems. For conservation planning, aligning scientific and indigenous perspectives enhances both legitimacy and efficacy.

5 Emerging Approaches and Future Directions

Alpine medicinal plants face accelerating threats from climate change, overharvesting, and habitat disturbance. To address these challenges, research is moving toward new molecular technologies and applied frameworks that transcend traditional phylogeny and phylogeography.

5.1 High-Quality Reference Genomes and Pangenomics

Advances in long-read sequencing (PacBio HiFi, Oxford Nanopore) are enabling chromosome-scale assemblies for alpine taxa. Reference genomes provide insights into genome structure, transposable elements, and adaptive gene families. For genera such as *Saussurea* and *Gentiana*, pangenomic approaches—incorporating genomes from multiple populations—reveal structural variants associated with altitude adaptation.

5.2 Epigenetics and Phenotypic Plasticity

Epigenetic modifications, including DNA methylation and histone modification, mediate rapid responses to environmental stress. Studies in *Primula* suggest methylation changes under cold stress that revert in warmer conditions, highlighting reversible plasticity. Understanding such epigenetic regulation may inform assisted adaptation strategies.

Fig. 10.1 A conceptual model illustrating how alpine plants utilize epigenetic mechanisms to cope with environmental stress

2 Fundamentals of Epigenetic Mechanisms in Plants

Epigenetics, at its core, is the study of heritable changes in how genes are expressed that ariseexcluding any alteration to the fundamental DNA sequence (Dar et al., 2022; Saeed et al., 2022). The term was initially conceptualized by Conrad Waddington, who defined epigenetics as the intricate interactions between genes and their products that ultimately shape the observable characteristics, or phenotype, of an organism (Ahtisham & Obaid, 2024). A key feature of epigenetic modifications is their potential to be transmitted through both mitotic cell divisions, which occur during growth and development, and meiotic divisions, which lead to the formation of reproductive cells (Iwasaki & Paszkowski, 2014). This heritability allows for a form of cellular memory, where past environmental experiences can influence future responses (Iwasaki & Paszkowski, 2014). For plants, which are sessile organisms unable to escape adverse conditions, these epigenetic mechanisms are not merely important but essential for their proper function and adaptation to the ever-changing environment (Dar et al., 2022). The capacity for epigenetic changes to be passed down through generations is particularly significant for alpine plants, as it allows them to not only respond to immediate stress but also potentially prepare for future similar stresses or even confer a degree of preparedness to their offspring. This form of anticipatory adaptation could be crucial in an environment characterized by predictable seasonal changes and the possibility of recurring extreme events.

The major epigenetic mechanisms that govern gene expression in plants include DNA methylation, histone modifications, and the action of non-coding RNAs (Table 10.1). DNA methylation adds a methyl group to cytosine residues within the DNA molecule (Miryeganeh, 2021). This modification can silence genes by

Table 10.1 Major epigenetic mechanisms and their roles in plant stress response.

Epigenetic mechanism	Description	Key enzymes/ factors	General effect on gene expression	Examples of role in stress response
DNA methylation	Addition of a methyl group to cytosine bases in DNA	DNA methyltransferases (DNMTs), RNA-directed DNA methylation (RdDM) pathway components	Typically, transcriptional silencing	Involved in responses to heat, cold, drought, salinity, and pathogen attacks (Miryeganeh, 2021; Ahtisham & Obaid, 2024)
Histone modifications	Chemical alterations to histone proteins (e.g., acetylation, methylation)	Histone acetyltransferases (HATs), Histone deacetylases (HDACs), Histone methyltransferases (HMTs), Histone demethylases (HDMs)	Can lead to activation or repression	Regulates gene expression in response to various abiotic and biotic stresses, including heat, cold, drought, salinity, and pathogen attacks (Nunez-Vazquez et al., 2022; Sharma et al., 2023)
Non-coding RNAs	RNA molecules that do not code for proteins (e.g., miRNAs, siRNAs, lncRNAs)	Dicer-like (DCL) enzymes, Argonaute (AGO) proteins, RNA polymerases IV and V	Can lead to transcriptional or post-transcriptional silencing or activation	Involved in responses to heat, cold, drought, salinity, and pathogen attacks (Gelaw & Sanan-Mishra, 2021; Yadav et al., 2024)

preventing transcription factors, the proteins that start gene expression, from accessing the DNA (Dar et al., 2022). A critical pathway in plants for establishing DNA methylation is RNA-directed DNA methylation (RdDM), where non-coding RNAs direct methylation machinery to precise genomic loci (Ahtisham & Obaid, 2024). For plants, DNA methylation happens in three sequence contexts: CG, CHG, and CHH, each potentially having distinct regulatory roles (Deng et al., 2018). The complexity of DNA methylation, with its various contexts and the RdDM pathway, suggests a highly nuanced regulatory role in alpine plants, potentially enabling exact regulation of how genes are expressed as a response to particular environmental signals.

Histone modifications represent another crucial layer of epigenetic regulation (Ramirez-Prado et al., 2018; Kang et al., 2022). Histones are proteins that serve as spools for DNA to wind around, creating a complex known as chromatin, the structural framework of chromosomes. These histone proteins are chemically altered in various ways, including acetylation, methylation, phosphorylation, ubiquitination, and sumoylation (Miryeganeh, 2021). The structure of chromatin is influenced by these modifications, making genes either more or less accessible to the transcriptional machinery (Bashir et al., 2025). Histone acetylation generally leads to a more open chromatin structure, thereby enhancing gene transcription. In contrast, histone methylation can have varied effects on transcription, either activating or repressing it, based on which amino acid residue is modified (Dar et al., 2022). Key regulatory protein complexes, like the Polycomb group (PcG) and trithorax group (TrxG) proteins, exert their control over gene expression through the methylation of specific lysine residues on histones (Dar et al., 2022). Furthermore, the incorporation of histone variants, which are different forms of histone proteins, can also influence chromatin conformation and gene expression (Miryeganeh, 2021). The diversity of histone modifications and the involvement of these regulatory protein complexes point to a sophisticated system for regulating how genes are expressed in alpine plants, potentially allowing for rapid and reversible responses to fluctuating conditions.

Non-coding RNAs (ncRNAs) represent a diverse class of RNA molecules that do not code for proteins but play significant regulatory roles in plants. These include small non-coding RNAs (sncRNAs), like microRNAs (miRNAs) and small interfering RNAs (siRNAs), as well as long non-coding RNAs (lncRNAs) (Miryeganeh, 2021). miRNAs and siRNAs, usually 20–24 nucleotides in length, can silence gene expression either by degrading messenger RNAs (mRNAs) after transcription (post-transcriptional gene silencing) or by inducing chromatin modifications that prevent gene transcription (transcriptional gene silencing) (Bashir et al., 2025). lncRNAs are non-coding RNA molecules, typically over 200 nucleotides in length, that regulate gene expression by interacting with DNA and proteins (Malhotra & Sharma, 2021). They can be transcribed from various regions of the genome and often complement functional DNA regions, allowing them to target specific genes for regulation (Dar et al., 2022). The involvement of both small and long non-coding RNAs adds another layer of complexity to the epigenetic regulation in alpine plants, potentially allowing for coordinated control of multiple genes and pathways involved in stress response.

It is important to note that these different epigenetic mechanisms do not operate in isolation but rather exhibit significant interplay and crosstalk. For example, DNA methylation can influence the patterns of histone modifications and recruit protein complexes that further modify chromatin (Abdulraheem et al., 2024; Talarico et al., 2024). Conversely, non-coding RNAs, particularly siRNAs, can direct DNA methylation to specific genomic regions through the RdDM pathway (Ahtisham & Obaid, 2024). Furthermore, miRNAs have been shown to interact with epigenetic pathways, influencing chromatin remodeling and thereby affecting gene expression (Kaya et al., 2024). This interconnectedness suggests that the epigenetic response in alpine plants is not a collection of independent events but rather a highly coordinated regulatory network. Changes in one epigenetic mark can trigger a cascade of effects on others, leading to complex and finely tuned responses to the multitude of environmental stresses encountered in alpine habitats. Understanding these intricate interactions is vital for an inclusive comprehension of how alpine plants regulate their gene expression in the face of adversity.

3 Alpine-Specific Environmental Stresses and Their Impact on Plants

Alpine environments impose a unique and severe set of abiotic stresses on the plants that inhabit them. Among the most significant is temperature stress. Alpine plants are frequently subjected to extreme cold, with temperatures often dropping below freezing, particularly during the night and outside the insulating snow cover (Körner & Hiltbrunner, 2021; Satyakam et al., 2022; Streb et al., 2025). The growing season at high altitudes is typically short, limiting the time available for plants to complete their life cycle, including development and reproduction. Furthermore, alpine plants often experience dramatic diurnal temperature fluctuations, with leaf temperatures potentially varying by as much as 30 °C between night and day (Streb et al., 2025). Such rapid and large temperature changes can significantly impact the rates of biochemical reactions within the plant.

Radiation stress is another major challenge for alpine flora. High altitudes are characterized by increased levels of both UV-B and visible light intensity. This excess light energy can overwhelm the photosynthetic machinery and lead to the production of toxic reactive oxygen species (ROS), which can damage cellular components (Streb et al., 2025). Water stress is also a significant concern in alpine regions. While precipitation may be abundant, it often falls as snow, which can be unavailable to plants during the early growing season or may melt too quickly, leading to periods of drought (Körner & Hiltbrunner, 2021). Additionally, the thin soils and steep slopes common in alpine areas can limit water retention. Strong winds, prevalent at high altitudes, can further exacerbate water stress by increasing the rate of transpiration and leading to desiccation.

Beyond these primary stresses, alpine plants often face nutrient limitation due to the poorly developed and often leached soils found in these environments (Streb et al., 2025). Mechanical stress from wind and snow can also cause physical damage to plant tissues. Moreover, the partial pressure of gases, including CO_2, is lower at high altitudes, which can directly affect the efficiency of photosynthesis (Muhammad et al., 2021). The unique combination and intensity of these abiotic stresses in alpine environments likely necessitate specialized and integrated stress response mechanisms in plants compared to those in less extreme habitats.

The impact of these stresses on alpine plants is profound, affecting fundamental physiological processes and ultimately influencing their survival and distribution. The stresses can directly modulate and disturb photosynthesis, the very process that fuels progressive change in plants (Muhammad et al., 2021). Excess light energy, as mentioned, can trigger the concentration of damaging ROS. The significant temperature fluctuations can cause biochemical reactions within the plant to vary dramatically, potentially disrupting metabolic pathways (Streb et al., 2025). Water scarcity, or drought, poses a direct threat to the biochemical and physiological activities essential for plant life (Kaya et al., 2024). In a broader context, stress, regardless of its specific nature, can induce physiological or even pathological changes within the plant (Kovalchuk, 2023). Notably, when plants are subjected to multiple stressors simultaneously, even if each individual stressor is relatively mild, the combined effect can significantly and negatively impact their growth and survival (Shriti et al., 2024). Therefore, understanding how epigenetic mechanisms enable alpine plants to mitigate these multifaceted impacts is crucial for appreciating their adaptation to these challenging high-altitude ecosystems (Kurpisz & Pawłowski, 2022; Jiang & Wang, 2024).

4 Role of Epigenetic Mechanisms in Response to Environmental Stresses in Alpine Plants

In their challenging high-altitude habitats, alpine plants are subjected to a unique combination of severe abiotic and biotic stresses, including temperature extremes, high radiation, and herbivory. To survive and thrive in these conditions, plants have developed highly flexible and responsive regulatory mechanisms. These dynamic molecular mechanisms provide a critical layer of adaptive plasticity, allowing for swift and flexible adjustments to environmental cues (Table 10.2).

4.1 Response to Abiotic Stresses

4.1.1 Response to Cold Stress

Alpine plants frequently encounter temperatures that fluctuate between chilling (0–18 °C) and freezing (<0 °C), presenting distinct types of cold stress (Song et al., 2015). Research on the alpine plant *Chorisporabungeana* have indicated exposure to both chilling and freezing conditions leads to alterations in cytosine methylation patterns (Song et al., 2015). Notably, these changes in DNA methylation occur rapidly following exposure to cold and appear to be distinct for chilling versus freezing temperatures, evolving gradually over time. This suggests that *C. bungeana* employs DNA methylation as a rapid and flexible mechanism to respond to different types of cold stress in its alpine environment. In some plant species, cold stress has been observed to induce hypomethylation at specific genomic locations (Iwasaki & Paszkowski, 2014). The ability to rapidly modify DNA methylation in response to varying degrees of cold stress underscores the importance of this epigenetic mark in the immediate adaptation of alpine plants to their fluctuating thermal environment (Satyakam et al., 2022).

Histone modifications are also instrumental in the acclimation and tolerance of plants, including those in alpine regions, to cold stress. In some species, cold stress has been shown to promote an increase in histone H3 lysine 4 methylation (H3K4me) and H3 acetylation, modifications generally associated with gene activation (Iwasaki & Paszkowski, 2014). Histone deacetylases (HDACs), a class of enzymes that

Table 10.2 Alpine-specific environmental stresses and their potential epigenetic responses.

Alpine stress	Impact on plants	Examples of associated epigenetic modifications in alpine plants
Temperature extremes (cold &frost)	Metabolic disruption, damage to cellular structures, limited growing season	DNA methylation changes in *Chorisporabungeana* (Song et al., 2015), histone modifications in cold acclimation (Ramakrishnan et al., 2022), regulation by miRNAs and lncRNAs (Huo et al., 2022)
High UV radiation &light intensity	Formation of reactive oxygen species (ROS), damage to DNA and proteins	DNA methylation changes in response to UV-B (Quan et al., 2021), histone acetylation for DNA repair (Boycheva et al., 2024), regulation by miRNAs (Xu et al., 2024)
Water stress (drought &desiccation)	Impaired physiological processes, reduced growth and photosynthesis	DNA methylation changes contributing to drought stress memory (Zi et al., 2024), histone modification dynamics (Kaya et al., 2024), role of siRNAs and lncRNAs (Gelaw & Sanan-Mishra, 2021)
Herbivory	Tissue damage, loss of resources, reduced reproductive success	DNA methylation changes in response to defense hormones (Latzel et al., 2012), histone modifications regulating defense gene expression (Bashir et al., 2025), involvement of miRNAs and lncRNAs (Yadav et al., 2024)
Pathogen attacks	Disease development, tissue damage, impaired growth	DNA methylation changes in defense responses (Bashir et al., 2025), histone modifications and chromatin remodeling (Kang et al., 2022), role of miRNAs, siRNAs, and lncRNAs (Miryeganeh, 2021)

catalyse the deacetylation of histones, are also involved in influencing the expression patterns of stress-related genes during exposure to low temperatures. The chromatin remodeler PICKLE (PKL) has been identified as a crucial factor in cold stress responses in plants (Ashapkin et al., 2020). Furthermore, proteins like HOS15 and the histone deacetylase HD2C are involved in the regulation of cold acclimation through histone modification pathways (Ashapkin et al., 2020). These findings indicate that histone modifications are integral to the longer-term process of cold acclimation in plants, likely including alpine species, by modulating the gene expression that confer tolerance to low temperatures.

Non-coding RNAs, both small and long, are also critical components of the regulatory network that enables plants to cope with cold stress. Small non-coding RNAs (sncRNAs), which include microRNAs (miRNAs) and small interfering RNAs (siRNAs), are crucial regulators of various cold response processes in plants (Huo et al., 2022). Long non-coding RNAs (lncRNAs) have been shown to play a role in how plants respond and acclimate to cold temperatures (Liu et al., 2022). For instance, a novel cold-induced lncRNA named CIL1 facilitates cold stress tolerance in *Arabidopsis* by regulating the gene network associatedwith stress responses (Liu et al., 2022). The involvement of both sncRNAs and lncRNAs highlights the complexity of the regulatory mechanisms that alpine plants utilize to withstand cold stress, encompassing control transcriptionally and post-transcriptionally.

4.1.2 Response to Drought Stress

Water availability in alpine environments can be unpredictable, with periods of drought occurring due to various factors. In response to such conditions, plants employ epigenetic mechanisms, including DNA methylation, to trigger changes in gene expression that ultimately improve their adaptation (Zi et al., 2024). Drought stress has been shown to induce epigenetic markers, with DNA methylation playing a crucial role in this response (Zi et al., 2024). Notably, drought stress can modulate DNA methylation in plants, and these changes are believed to contribute to the phenomenon of drought stress memory, allowing plants to respond more effectively to subsequent drought episodes (Rao et al., 2024). DNA methylation is recognized as a key mechanism in plants' adaptation to various environmental changes, including water deficit (Deng et al., 2018). In some plant species, drought stress has been observed to lead to an increase in methylation levels. This suggests that DNA methylation is a vital component of how alpine plants respond to drought, potentially by regulating genes involved in water conservation, osmotic adjustment, and overall stress tolerance, and by contributing to the plant memory for future water scarcity.

Histone modifications are also essential in regulating the genes and hormones that are critical for a plant's adaptation to drought stress (Kaya et al., 2024). When confronted with non-biological stressors, including drought, plants often exhibit global changes in histone acetylation (Nunez-Vazquez et al., 2022). The dynamic balance between histone acetyltransferases (HATs), which add acetyl groups, and histone deacetylases (HDACs), which remove them, maintains histone acetylation levels. This balance regulates the accessibility of chromatin to transcriptional machinery, thereby influencing gene expression in response to drought stress (Li et al., 2021; Nunez-Vazquez et al., 2022; Sharma et al., 2023). Histone variants can influence the abiotic stress response, such as drought, by modifying chromatin conformation (Nunez-Vazquez et al., 2022). These dynamic changes in histone modifications enable alpine plants to rapidly adjust their gene expression in response to fluctuations in water availability, influencing the activity of genes that are key to their survival under drought conditions.

Non-coding RNAs (ncRNAs) have gained recognition as significant controllers in modulating how plants respond to drought and other abiotic stresses (Gelaw & Sanan-Mishra, 2021). In the realm of plant genetics, the response to drought has been linked to the activities of both small interfering RNAs (siRNAs) and long non-coding RNAs (lncRNAs) (Gelaw & Sanan-Mishra, 2021; Yang et al., 2023). MicroRNAs (miRNAs) also hold a significant role in drought response, primarily by controlling the expression of genes crucial for stress tolerance (Kaya et al., 2024). These non-coding RNAs, along with other epigenetic mechanisms like DNA methylation and histone modifications, form an intricate regulatory system. This system is instrumental in how plants, particularly those in alpine environments, sense and respond to drought at a molecular level. The coordinated action of these epigenetic mechanisms allows for a fine-tuned and robust response to the challenges posed by limited water availability in the alpine environment (Kurpisz & Pawłowski, 2022; Jiang et al., 2024).

4.1.3 Response to Radiation Stress (UV-B and High Light)

Alpine environments are characterized by high levels of UV-B radiation, which can induce significant stress in plants. UV-B radiation has been observed to induce modifications in the DNA methylation patterns of plants (Quan et al., 2021). Similarly, exposure to high light stress can lead to differential methylation of genes that are involved in plant growth responses (Gao et al., 2014; Nair et al., 2025). Moreover, UV-B radiation can alter the total levels of DNA methylation and the epigenetic profiles in plants (Quan et al., 2021; Shi & Liu, 2021). These observations indicate that DNA methylation is involved in the response of alpine plants to intense UV-B radiation. This may be accomplished through the regulation of genes related to DNA repair and the synthesis of photoprotective compounds.

Histone modifications are involved in how plants respond to radiation stress. Key enzymes called histone acetyltransferases (HATs) regulate chromatin structure and gene expression, which helps with the essential DNA repair mechanisms plants need when exposed to UV radiation (Boycheva et al., 2024). Histone acetylation leads to a more relaxed chromatin state, which facilitates the transcriptional activation of genes necessary for DNA repair processes (Yusuf et al., 2025). Furthermore, UV radiation can act as a signal that triggers photomorphogenesis, a process that affects gene expression related to plant development and enhances the activation of DNA repair mechanisms under light conditions (Boycheva et al., 2024). These observations indicate that histone modifications, particularly acetylation, are vital for enabling alpine plants to repair the DNA damage caused by excessive UV radiation with a rise in accessibility of the damaged DNA to the repair machinery.

Non-coding RNAs, especially microRNAs (miRNAs), are known to be involved in the plant abiotic stress response, which likely includes the high light conditions prevalent in alpine regions (Contreras-Cubas et al., 2012). Specific miRNAs have been shown to have their expression levels altered (either increased or decreased) when exposed to UV-B radiation in various plant species, thereby implying their role in regulating target genes that are part of the plant defense mechanism to this stress (Xu et al., 2024). This indicates that non-coding RNAs, particularly miRNAs, may regulate gene expression in alpine plants that are exposed to high light and UV-B radiation, allowing them to fine-tune their protective responses and minimize the damage caused by excessive radiation.

4.2 Response to Biotic Stresses

4.2.1 Response to Herbivory

Herbivory poses a significant biotic stress to plants, including those in alpine environments. Epigenetic mechanisms, such as DNA methylation, have been shown to play a role in mediating plant responses to herbivore attacks. Studies have indicated that variation in DNA methylation can lead to heritable differences in how plants respond to defense hormones like jasmonic acid and salicylic acid, which are crucial in inducing defenses against herbivores (Latzel et al., 2012). In some plant species, herbivore damage has been observed to alter the global levels of DNA methylation (Bashir et al., 2025; Zhang et al., 2025). Furthermore, DNA methylation may influence the expression of genes that are involved in both direct (e.g., production of toxins) and indirect (e.g., release of volatile compounds to attract predators) defense mechanisms against herbivores. However, research on alpine *Arabidopsis lyrata* populations suggests that DNA methylation at genic regions might be largely insensitive to environmental changes, while transposable elements (TEs) showed environmental effects on methylation (Hämälä et al., 2022). This indicates that the role of DNA methylation in herbivore response might vary depending on the plant species and even the specific population within a species. Nevertheless, it appears to be involved in regulating defense-related genes and potentially contributing to heritable variation in herbivore resistance in some alpine plants.

Histone modifications are also critical in orchestrating plant responses to herbivory. Modifications like acetylation and methylation of histones can quickly regulate gene expression in response to herbivore attacks. These changes impact chromatin structure, thereby influencing gene activity (Troyee et al., 2022). For example, anescalation in histone acetylation within the promoter regions of defense genes can enhance the transcription of genes that produce compounds that deter herbivores. Additionally, specific histone methylation marks have been linked to the priming of plant immune responses, enabling plants to react more quickly and effectively when they encounter herbivores (Bashir et al., 2025). These findings emphasize the importance of histone modifications in allowing alpine plants to swiftly activate or repress defense-related genes upon herbivore attack, facilitating a rapid and tailored response.

Non-coding RNAs, such as small RNAs like microRNAs (miRNAs) and small interfering RNAs (siRNAs), are crucial in how plants defend themselves against herbivores. These small RNAs work by silencing genes from pathogens and also by suppressing certain plant genes that might increase susceptibility. Additionally, they help regulate the signaling pathways that are part of the plant's defense response to herbivores. Long non-coding RNAs (lncRNAs) also have a key role in regulating plant biological processes, including how plants respond to stresses caused by living organisms, like herbivory (Sharma et al., 2022; Yadav et al., 2024). This suggests that non-coding RNAs contribute to a nuanced defense response in alpine plants against herbivores by regulating gene expression at multiple levels, optimizing the plant's ability to defend itself while minimizing any potential self-damage.

4.2.2 Response to Pathogen Attacks

Plants in alpine environments are also susceptible to attacks from various pathogens. DNA methylation, a key epigenetic mechanism, is critical for how plants respond to invasions by pathogens (Sun et al., 2022; Qiao et al., 2025). Modifications in DNA methylation patterns can either suppress genes that increase a plant's vulnerability to pathogens or activate defense-related genes, which helps the plant prepare for an effective defense (Bashir et al., 2025). As an example, the hypermethylation of gene promoter regions can lead to gene silencing, which prevents pathogens from exploiting the plant's cellular functions. Conversely, the demethylation of promoters of defense genes can enhance their expression, strengthening the plant's defense arsenal against pathogens. Interestingly, pre-treating plants with non-pathogenic strains has been shown to induce hypomethylation patterns that can be remembered by the plant, leading to a more effective response against subsequent pathogen infections (Kramer et al., 2023; Mierziak & Wojtasik, 2024). Furthermore, a decrease in DNA methylation within gene bodies has been associated with the hyperactivation of defense response genes during pathogen infection (Lee et al., 2023). These findings underscore that DNA methylation is a critical component of the alpine plant immune system, enabling rapid reprogramming of gene expression to combat pathogen attacks and potentially facilitating a form of immune memory.

The regulation of gene expression during a pathogenic attack is heavily dependent on histone modifications and chromatin remodeling. By altering the structure of chromatin, these mechanisms can modulate the accessibility of specific genes to transcriptional machinery. A prime example is the upregulation of genes that produce antimicrobial compounds, which can be achieved through an increase in histone acetylation at the loci of these defense genes. Additionally, chromatin remodelers, which are protein complexes that can change the organization of chromatin, are also involved in regulating plant immunity during fungal infections (Ding & Wang, 2015; Mierziak & Wojtasik, 2024). The ability of alpine plants to rapidly and effectively activate defense genes following pathogen detection is facilitated by processes that modify the accessibility of their genome to the transcriptional machinery.

5 Epigenetic Memory: Long-Term Adaptation to Alpine Stress

Plants possess the remarkable ability to retain a memory of past stressful experiences and utilize this memory to enhance their survival when they encounter similar conditions again (Kinoshita & Seki, 2014). This phenomenon of stress memory is particularly crucial in fluctuating environments such as alpine regions, where plants may face recurring challenges like extreme temperatures, drought, or pathogen attacks (Ahtisham & Obaid, 2024). Epigenetic modifications are key regulators in the establishment and maintenance of stress memory in plants(Kinoshita & Seki, 2014; Iwasaki & Paszkowski, 2014).

Evidence suggests that the memory of stress responses can be inherited across plant generations in alpine species (Ahtisham & Obaid, 2024). This transgenerational epigenetic inheritance can contribute to the evolutionary adaptation of plant populations to their environment over time (de Carvalho M, 2023). Stress memory in plants can manifest in different forms, including somatic memory within an individual plant, intergenerational memory affecting the directly exposed generation and their immediate offspring, and transgenerational memory that persists across multiple generations (Ramakrishnan et al., 2022). The potential for stress-induced epigenetic modifications to be passed down to future generations provides a mechanism for rapid adaptation in alpine plant populations, allowing them to become better suited to their challenging environment in a relatively short timeframe.

Epigenetic memory is established and maintained through crucial mechanisms like alterations in DNA methylation and histone modifications (Ramakrishnan et al., 2022). Particular histone modifications have been shown to be involved in transmitting stress memory across generations. As a relatively stable modification, DNA methylation is also believed to be essential for creating and maintaining drought stress memory in plants (Deng et al., 2018). Both of these epigenetic mechanisms appear to be vital for alpine plants to recall previous stresses, allowing them to respond more effectively to future exposures.

Transgenerational epigenetic inheritance has significant implications for the adaptation of alpine plant species. By providing a mechanism for stress memory to be passed on to offspring, it enables these offspring to respond more effectively to potential future stresses. This form of inheritance contributes to stress memory and helps plants to respond better to recurring or long-term environmental challenges (Thiebaut et al., 2019; Rajpal et al., 2022). In the context of the predictable seasonal changes and the potential increase in extreme weather events due to climate change in alpine regions, transgenerational epigenetic inheritance could be a key factor in the survival and adaptation of these plant species.

6 Epigenetic Variation and Adaptation in Alpine Plant Populations

Epigenetic variability is increasingly recognized as an important factor underlying environmental adaptation in plants, including those inhabiting alpine regions. Environmental changes can induce epigenetic modifications, either independently of or in conjunction with genetic variation, contributing to the overall phenotypic plasticity of plant populations (Rajpal et al., 2022). Studies on the alpine plant species, *Ranunculuskuepferi*, reveal distinct methylation patterns that vary among different cytotypes (diploid and tetraploid) and reproductive strategies (sexual, mixed, or apomictic). Research indicates these patterns correlate with environmental factors and geographical distribution (Schinkel et al., 2020). Similarly, research on *Deyeuxia angustifolia* in the alpine tundra of the Changbai Mountains in China has revealed epigenetic variation that may provide a basis for adaptation, particularly in cases where genetic variation is low (Ni et al., 2021). These findings suggest that alpine plant populations exhibit significant epigenetic variation, which likely plays a crucial role in their ability to adapt to the diverse microhabitats and environmental conditions found across alpine landscapes.

Epigenetics also appears to be instrumental in facilitating rapid adaptation to local environmental conditions along the altitudinal gradients that are characteristic of alpine regions (Zhang et al., 2023). Epigenetic processes can alter plant phenotypes, leading to improved adaptation to divergent environments, including the varying conditions encountered at different elevations in alpine areas (Zhang et al., 2025). For example, an investigation into the alpine plant *Wahlenbergiaceracea* revealed that variations in DNA methylation are linked to adaptive plasticity in its temperature response to a warmer climate (Nicotra et al., 2015). Furthermore, the observation of epigenetic and phenotypic divergence between native and invasive plant populations in different environments suggests that epigenetics may contribute to the colonization of new habitats (Campoy et al., 2022). This indicates that epigenetic variation allows alpine plants to exhibit adaptive plasticity, enabling them to rapidly adjust to the specific environmental conditions at different altitudes and potentially facilitating their ability to colonize new or changing habitats.

The adaptation of alpine plants is likely a result of the complex interaction between their underlying genetic makeup and the epigenetic modifications that allow them to respond and adjust to their environment (de Carvalho M, 2023). Genetic variation lays the groundwork for adaptation, but epigenetic modifications provide a more adaptable and rapid way to control how genes are expressed in direct response to environmental signals (de Carvalho M, 2023). The interplay between these two levels of regulation ultimately determines how effectively a plant can adapt to the environmental challenges it faces. In some instances, stress-induced epigenetic changes have been observed to occur even in the absence of significant genetic diversity, highlighting the critical role that epigenetics can play in adaptation (Wu & Fan, 2025).

Several case studies of alpine plant species provide compelling evidence for epigenetic adaptation (Table 10.3). The rapid and distinct DNA methylation responses to cold stress observed in *Chorisporabungeana* illustrate how epigenetics can mediate adaptation to temperature extremes (Song et al., 2015). The increased DNA methylation variation associated with adaptive plasticity in response to warming in *Wahlenbergiaceracea* highlights the role of epigenetics in adapting to changing climate conditions (Nicotra et al., 2015). The correlations between methylation patterns and environmental gradients in *Ranunculuskuepferi* suggest that epigenetics is involved in adaptation to different ecological niches within the alpine zone (Schinkel et al., 2020). Finally, the epigenetic diversity found in *Deyeuxia angustifolia* in the alpine tundra indicates that epigenetics can provide a basis for adaptation even when genetic variation is limited (Ni et al., 2021). These case studies provide concrete examples of how epigenetic mechanisms, particularly DNA methylation, are involved in the adaptation of specific alpine plant species to the unique challenges of their high-altitude environment.

7 Future Perspectives and Research Directions

Although our understanding of epigenetic regulation in alpine plant stress responses has advanced considerably, certain questions and limitations have yet to be addressed. The precise mechanisms underlying epigenetic inheritance without alterations to the DNA sequence are still not fully elucidated (Meera et al., 2025). Furthermore, translating research findings in plant epigenetics into practical applications, such as crop improvement, faces several limitations and challenges (Abdulraheem et al., 2024). The inherent complexity of epigenetic interactions and

Table 10.3 Case studies of epigenetic adaptation in alpine plant species

Plant species	Type of stress studied	Key epigenetic modifications observed	Main findings related to adaptation	Reference
Chorisporabungeana	Cold stress (chilling &freezing)	Alterations in cytosine methylation	Rapid and distinct DNA methylation changes in response to different types of cold stress, suggesting a crucial role in immediate adaptation	Song et al. (2015)
Wahlenbergiaceracea	Warming (temperature response)	Increased variation in DNA methylation	Adaptive plasticity in temperature response to warming is associated with increased DNA methylation variation	Nicotra et al. (2015)
Ranunculuskuepferi	Environmental gradients (elevation, temperature)	Variation in methylation patterns related to ploidy level and reproductive strategy	Methylation patterns correlated with elevation and temperature gradients, suggesting adaptation to different ecological niches	Schinkel et al. (2020)
Deyeuxia angustifolia	Alpine tundra conditions	Epigenetic variation and DNA methylation levels	Epigenetic variation may provide a basis for adaptation in the case of low genetic variation in the alpine tundra	Ni et al. (2021)

the variability in stress responses across different plant species also pose ongoing challenges to researchers (de Carvalho M, 2023). Additionally, the relationship between epigenetics and broader ecological phenomena, such as plant responses to climate change or the invasion of new habitats, is still not completely understood (Campoy et al., 2022).

Future research should prioritize several key areas to advance our understanding. One important direction is to investigate the combined effects of multiple stressors on the alpine plant epigenome (Jing et al., 2024). Alpine plants in their natural environment rarely encounter a single stressor in isolation; instead, they are typically exposed to a complex interplay of factors such as temperature extremes, high radiation, and water limitation. Given that future climate scenarios predict an increased frequency of multiple stressful environments for plants (Jing et al., 2024), research that focuses on how alpine plants integrate their epigenetic responses to these interacting stresses is crucial for predicting their resilience (Shriti et al., 2024).

Another promising avenue for future research involves exploring the role of specific epigenetic marks in the adaptation of alpine plants to the diverse microhabitats found within alpine zones (Streb et al., 2025). Alpine environments are often highly heterogeneous, with significant variations in conditions like temperature, moisture, and solar exposure over relatively small spatial scales. Investigating how particular epigenetic tags, like specific histone modifications or DNA methylation patterns at defined genomic regions, contribute to local adaptation within these microhabitats would provide valuable insights into the fine-tuning of plant responses to their immediate surroundings.

A critical need for future research is the development of tools and techniques specifically tailored for studying epigenetics in non-model alpine species (Abdulraheem et al., 2024). While much of our current understanding of plant epigenetics comes from studies on model organisms like *Arabidopsis*, the vast diversity of alpine flora likely harbors a wide array of unique epigenetic mechanisms and adaptations. Expanding research efforts to encompass a broader range of non-model alpine species through the application of advanced genomic and epigenomic technologies will be essential for a more comprehensive understanding of how these plants thrive in their challenging environment.

Furthermore, future research should focus on clarifying the broader ecological and evolutionary importance of epigenetic variation within alpine plant communities (Abdulraheem et al., 2024; Wu & Fan, 2025). This includes investigating the heritability and stability of stress-induced epigenetic changes in natural alpine populations over multiple generations, as well as exploring how epigenetic variation contributes to ecological dynamics, species interactions, and the long-term evolutionary trajectories of these plant communities. Finally, the knowledge gained from studying epigenetics in alpine plants has the potential for practical applications in the conservation and management of these unique ecosystems. Epigenetic markers could potentially be used to assess the health and resilience of alpine plant populations, and understanding the epigenetic mechanisms of stress tolerance could inform conservation strategies in the face of rapid environmental change (Akhter et al., 2021). Epigenetic approaches might even be developed to enhance stress tolerance in vulnerable alpine species, contributing to their long-term survival in a changing world (Ahtisham & Obaid, 2024).

8 Conclusions

The evidence presented underscores the critical role of epigenetic mechanisms in enabling alpine plants to not only survive but also thrive in their challenging high-altitude environments. The unique combination of abiotic stresses, including extreme temperature fluctuations, intense radiation, and potential water scarcity, necessitates highly responsive and adaptable regulatory systems. The ability of alpine plants to adjust their gene expression in a dynamic environment is facilitated by epigenetic mechanisms, including DNA methylation, histone modifications, and non-coding RNAs. Furthermore, the capacity for epigenetic memory and transgenerational inheritance suggests a mechanism for rapid adaptation at the population level, potentially allowing alpine flora to cope with the ongoing and future impacts of climate change. While significant progress has been made, future research focusing on the interplay of multiple stressors, adaptation to microhabitats, the study of non-model alpine species, and the ecological and evolutionary significance of epigenetic variation will be crucial for a more complete understanding of these remarkable biological systems. The knowledge gained has the potential to inform conservation and management strategies aimed at preserving the unique biodiversity of alpine ecosystems.

References

Abdulraheem, M. I., Xiong, Y., Moshood, A. Y., Cadenas-Pliego, G., Zhang, H., & Hu, J. (2024). Mechanisms of plant epigenetic regulation in response to plant Stress: Recent discoveries and implications. *Plants, 13*(2), 163.

Ahtisham, M., & Obaid, Z. (2024). Harnessing epigenetic mechanisms for crop resilience: A comprehensive review of plant responses to biotic and abiotic stresses. *Premier Journal of Plant Biology, 1*, 100007.

Akhter, Z., Bi, Z., Ali, K., Sun, C., Fiaz, S., Haider, F. U., & Bai, J. (2021). In response to abiotic stress, DNA methylation confers epiGenetic changes in plants. *Plants (Basel), 10*(6), 1096. https://doi.org/10.3390/plants10061096. PMID: 34070712; PMCID: PMC8227271.

Ashapkin, V. V., Kutueva, L. I., Aleksandrushkina, N. I., & Vanyushin, B. F. (2020). Epigenetic mechanisms of plant adaptation to biotic and abiotic stresses. *International Journal of Molecular Sciences, 21*(20), 7457.

Bashir, H., Samanhudi, P., & Setyawati, A. (2025). Decoding the mechanisms of plant epigenetic adaptations to stress. *Reviews in Agricultural Science, 13*(2), 1–19.

Boycheva, I., Bonchev, G., Manova, V., Stoilov, L., & Vassileva, V. (2024). How histone acetyltransferases shape plant photomorphogenesis and UV response. *International Journal of Molecular Sciences, 25*(14), 7851.

Campoy, J. G., Sobral, M., Carro, B., Lema, M., Barreiro, R., & Retuerto, R. (2022). Epigenetic and phenotypic responses to experimental climate change of native and invasive *Carpobrotus edulis*. *Frontiers in Plant Science, 13*, 888391.

Contreras-Cubas, C., Palomar, M., Arteaga-Vázquez, M., Reyes, J. L., & Covarrubias, A. A. (2012). Non-coding RNAs in the plant response to abiotic stress. *Planta, 236*(4), 943–958. https://doi.org/10.1007/s00425-012-1693-z. Epub 2012 Jul 4. PMID: 22761008.

Dar, F. A., Mushtaq, N. U., Saleem, S., Rehman, R. U., Dar, T. U. H., & Hakeem, K. R. (2022). Role of epigenetics in modulating phenotypic plasticity against abiotic stresses in plants. *International Journal of Genomics, 2022*, 1092894. https://doi.org/10.1155/2022/1092894. PMID: 35747076; PMCID: PMC9213152.

Deng, J., Kou, S., Zhang, C., Zou, Q., Li, P., et al. (2018). DNA methylation and plant stress responses. *Journal of Plant Physiology & Pathology, 6*, 4.

Ding, B., & Wang, G. L. (2015). Chromatin versus pathogens: The function of epigenetics in plant immunity. *Frontiers in Plant Science, 6*, 675. https://doi.org/10.3389/fpls.2015.00675. PMID: 26388882; PMCID: PMC4557108.

Gao, G., Li, J., Li, H., Li, F., Xu, K., Yan, G., Chen, B., Qiao, J., & Wu, X. (2014). Comparison of the heat stress induced variations in DNA methylation between heat-tolerant and heat-sensitive rapeseed seedlings. *Breeding Science, 64*, 125–133.

Gelaw, T. A., & Sanan-Mishra, N. (2021). Non-coding RNAs in response to drought stress. *International Journal of Molecular Sciences, 22*(22), 12519. https://doi.org/10.3390/ijms222212519. PMID: 34830399; PMCID: PMC8621352.

H. de Carvalho, M. (2023). Understanding the role of epigenetics in plant stress responses. *Journal of Agricultural Science and Botany, 8*(4), 254.

Hämälä, T., Ning, W., Kuittinen, H., Aryamanesh, N., & Savolainen, O. (2022). Environmental response in gene expression and DNA methylation reveals factors influencing the adaptive potential of *Arabidopsis lyrata*. *eLife, 11*, e83115. https://doi.org/10.7554/eLife.83115. PMID: 36306157; PMCID: PMC9616567.

Huo, C., Zhang, B., & Wang, R. (2022). Research progress on plant noncoding RNAs in response to low-temperature stress. *Plant Signaling & Behavior, 17*(1), 2004035. https://doi.org/10.1080/15592324.2021.2004035. Epub 2021 Dec 19. PMID: 34927551; PMCID: PMC8932918.

Iwasaki, M., & Paszkowski, J. (2014). Epigenetic memory in plants. *The EMBO Journal, 33*, 1987–1998.

Jiang, X. L., & Wang, W. F. (2024). Genetic face-lifting: Applications and prospects of epigenetic modifications in stress response of trees. *Plant Gene and Trait, 15*(2), 52–61.

Jing, Z., Liu, N., Zhang, Z., & Hou, X. (2024). Research progress on plant responses to stress combinations in the context of climate change. *Plants, 13*(4), 469.

Kang, H., Fan, T., Wu, J., Zhu, Y., & Shen, W.-H. (2022). Histone modification and chromatin remodeling in plant response to pathogens. *Frontiers in Plant Science, 13*, 986940.

Kaya, C., Uğurlar, F., & Adamakis, I.-D. S. (2024). Epigenetic modifications of hormonal signaling pathways in plant drought response and tolerance for sustainable food security. *International Journal of Molecular Sciences, 25*(15), 8229.

Kinoshita, T., & Seki, M. (2014). Epigenetic memory for stress response and adaptation in plants. *Plant & Cell Physiology, 55*(11), 1859–1863. https://doi.org/10.1093/pcp/pcu125. Epub 2014 Oct 8. PMID: 25298421.

Körner, C., & Hiltbrunner, E. (2021). Why is the alpine flora comparatively robust against climatic warming? *Diversity, 13*(8), 383.

Kovalchuk, I. (2023). Role of epigenetic factors in response to stress and establishment of somatic memory of stress exposure in plants. *Plants, 12*(21), 3667.

Kramer, H. M., Cook, D. E., Seidl, M. F., & Thomma, B. P. H. J. (2023). Epigenetic regulation of nuclear processes in fungal plant pathogens. *PLoS Pathogens, 19*(8), e1011525. https://doi.org/10.1371/journal.ppat.1011525. PMID: 37535497; PMCID: PMC10399791.

Kurpisz, B., & Pawłowski, T. A. (2022). Epigenetic mechanisms of tree responses to climatic changes. *International Journal of Molecular Sciences, 23*(21), 13412.

Latzel, V., Zhang, Y., Karlsson Moritz, K., Fischer, M., & Bossdorf, O. (2012). Epigenetic variation in plant responses to defence hormones. *Annals of Botany, 110*(7), 1423–1428. https://doi.org/10.1093/aob/mcs088. Epub 2012 Apr 26. PMID: 22543179; PMCID: PMC3489142.

Lee, S., Choi, J., Park, J., et al. (2023). DDM1-mediated gene body DNA methylation is associated with inducible activation of defense-related genes in *Arabidopsis*. *Genome Biology, 24*, 106.

Li, S., He, X., Gao, Y., Zhou, C., Chiang, V. L., & Li, W. (2021). Histone acetylation changes in plant response to drought stress. *Genes, 12*(9), 1409. https://doi.org/10.3390/genes12091409. PMID: 34573391; PMCID: PMC8468061.

Liu, G., Liu, F., Wang, Y., & Liu, X. (2022). A novel long noncoding RNA CIL1 enhances cold stress tolerance in *Arabidopsis*. *Plant Science, 323*, 111370.

Malhotra, N., & Sharma, S. (2021). Genome-wide analysis of long noncoding RNAs in Sorghum and their roles in development and stress. In S. K. Upadhyay (Ed.), *Long noncoding RNAs in plants* (pp. 75–91). Academic.

Meera, S. P., & Kumar, A. (2025). Role of epigenetics in mangroves: Recent progress and future perspectives. *Plant & Cell Physiology, 66*(5), 674–686. https://doi.org/10.1093/pcp/pcaf017. PMID: 39916474.

Mierziak, J., & Wojtasik, W. (2024). Epigenetic weapons of plants against fungal pathogens. *BMC Plant Biology, 24*, 175.

Miryeganeh, M. (2021). Plants' epigenetic mechanisms and abiotic stress. *Genes, 12*(8), 1106. https://doi.org/10.3390/genes12081106. PMID: 34440280; PMCID: PMC8394019.

Muhammad, I., Shalmani, A., Ali, M., Yang, Q. H., Ahmad, H., & Li, F. B. (2021). Mechanisms regulating the dynamics of photosynthesis under abiotic stresses. *Frontiers in Plant Science, 11*, 615942.

Nair, A. U., Kundariya, H. S., Samantaray, D., Dopp, I. J., Allu, A. D., & Mackenzie, S. A. (2025). Short-term high light stress analysis through differential methylation identifies root architecture and cell size responses. *Plant Cell Environment, 48*(5), 3269–3280. https://doi.org/10.1111/pce.15325. Epub 2024 Dec 25. PMID: 39722567; PMCID: PMC11963490.

Ni, B., You, J., Li, J., Du, Y., Zhao, W., & Chen, X. (2021). Genetic and epigenetic changes during the upward expansion of *Deyeuxia angustifolia* Kom. In the alpine tundra of the Changbai Mountains, China. *Plants, 10*(2), 291.

Nicotra, A. B., Segal, D. L., Hoyle, G. L., Schrey, A. W., Verhoeven, K. J., & Richards, C. L. (2015). Adaptive plasticity and epigenetic variation in response to warming in an Alpine plant. *Ecology and Evolution, 5*(3), 634–647. https://doi.org/10.1002/ece3.1329. Epub 2015 Jan 13. PMID: 25691987; PMCID: PMC4328768.

Nunez-Vazquez, R., Desvoyes, B., & Gutierrez, C. (2022). Histone variants and modifications during abiotic stress response. *Frontiers in Plant Science, 13*, 984702.

Qiao, S., Song, W., Hu, W., Wang, F., Liao, A., Tan, W., & Yang, S. (2025). The role of plant DNA methylation in development, stress response, and crop breeding. *Agronomy, 15*(1), 94.

Quan, J., Latzel, V., Tie, D., Zhang, Y., Münzbergová, Z., Chai, Y., Liu, X., & Yue, M. (2021). Ultraviolet B radiation triggers DNA methylation change and affects foraging behavior of the clonal plant *Glechoma longituba*. *Frontiers in Plant Science, 12*, 633982.

Rajpal, V. R., Rathore, P., Mehta, S., Wadhwa, N., Yadav, P., Berry, E., Goel, S., Bhat, V., & Raina, S. N. (2022). Epigenetic variation: A major player in facilitating plant fitness under changing environmental conditions. *Frontiers in Cell and Developmental Biology, 10*, 1020958.

Ramakrishnan, M., Zhang, Z., Mullasseri, S., Kalendar, R., Ahmad, Z., Sharma, A., Liu, G., Zhou, M., & Wei, Q. (2022). Epigenetic stress memory: A new approach to study cold and heat stress responses in plants. *Frontiers in Plant Science, 13*, 1075279.

Ramirez-Prado, J. S., Piquerez, S. J. M., Bendahmane, A., Hirt, H., Raynaud, C., & Benhamed, M. (2018). Modify the histone to win the battle: Chromatin dynamics in plant–pathogen interactions. *Frontiers in Plant Science, 9*, 355.

Rao, X., Yang, S., Lü, S., & Yang, P. (2024). DNA methylation dynamics in response to drought stress in crops. *Plants (Basel), 13*(14), 1977. https://doi.org/10.3390/plants13141977. PMID: 39065503; PMCID: PMC11280950.

Saeed, F., Chaudhry, U. K., Bakhsh, A., Raza, A., Saeed, Y., Bohra, A., & Varshney, R. K. (2022). Moving beyond DNA sequence to improve plant stress responses. *Frontiers in Genetics, 13*, 874648.

Satyakam, Zinta, G., Singh, R. K., & Kumar, R. (2022). Cold adaptation strategies in plants—An emerging role of epigenetics and antifreeze proteins to engineer cold resilient plants. *Frontiers in Genetics, 13*, 909007.

Schinkel, C. C. F., Syngelaki, E., Kirchheimer, B., Dullinger, S., Klatt, S., & Hörandl, E. (2020). Epigenetic patterns and geographical parthenogenesis in the alpine plant species *Ranunculus kuepferi* (Ranunculaceae). *International Journal of Molecular Sciences, 21*(9), 3318.

Sharma, Y., Sharma, A., Madhu, Shumayla, Singh, K., & Upadhyay, S. K. (2022). Long non-coding RNAs as emerging regulators of pathogen response in plants. *Non-coding RNA, 8*(1), 4.

Sharma, M., Sidhu, A. K., Samota, M. K., Gupta, M., Koli, P., & Choudhary, M. (2023). Post-translational modifications in histones and their role in abiotic stress tolerance in plants. *Proteomes, 11*(4), 38. https://doi.org/10.3390/proteomes11040038. PMID: 38133152; PMCID: PMC10747722.

Shi, C., & Liu, H. (2021). How plants protect themselves from ultraviolet-B radiation stress. *Plant Physiology, 187*(3), 1096–1103. https://doi.org/10.1093/plphys/kiab245. PMID: 34734275; PMCID: PMC8566272.

Shilpa, T. R., & Prasad, P. (2024). Epigenetic regulation of abiotic stress responses in plants. *Biochimica et Biophysica Acta (BBA)-General Subjects, 1868*(9), 130661.

Shriti, S., Bhar, A., & Roy, A. (2024). Unveiling the role of epigenetic mechanisms and redox signaling in alleviating multiple abiotic stress in plants. *Frontiers in Plant Science, 15*, 1456414.

Song, Y., Liu, L., Feng, Y., Wei, Y., Yue, X., He, W., Zhang, H., & An, L. (2015). Chilling- and freezing-induced alterations in cytosine methylation and its association with the cold tolerance of an alpine subnival plant, Chorispora bungeana. *PLoS One, 10*(8), e0135485. https://doi.org/10.1371/journal.pone.0135485. PMID: 26270551; PMCID: PMC4535906.

Streb, P., Cornic, G., & Bligny, R. (2025). How do plants cope with alpine stress? Encyclopedia of the environment. https://www.encyclopedie-environnement.org/en/life/how-do-plants-cope-with-alpine-stress-2/. Accessed 23 June 2025.

Sun, M., Yang, Z., Liu, L., & Duan, L. (2022). DNA methylation in plant responses and adaption to abiotic stresses. *International Journal of Molecular Sciences, 23*(13), 6910. https://doi.org/10.3390/ijms23136910. PMID: 35805917; PMCID: PMC9266845.

Talarico, E., Zambelli, A., Araniti, F., Greco, E., Chiappetta, A., & Bruno, L. (2024). Unravelling the epigenetic code: DNA methylation in plants and its role in stress response. *Epigenomes, 8*(3), 30.

Thiebaut, F., Hemerly, A. S., & Ferreira, P. C. G. (2019). A role for epigenetic regulation in the adaptation and stress responses of non-model plants. *Frontiers in Plant Science, 10*, 246.

Troyee, A. N., Medrano, M., Müller, C., & Alonso, C. (2022). Variation in DNA methylation and response to short-term herbivory in *Thlaspi arvense*. *Flora, 293*, 152106.

White, F. J., Rosbakh, S., Orsenigo, S., & Mondoni, A. (2023). Fit for the future? Alpine plant responses to climatic sress over two decades of seed bank storage. *Biological Conservation, 286*, 110267.

Wu, W., & Fan, G. (2025). The role of epigenetics in plant pathogens interactions under the changing environments: A systematic review. *Plant Stress, 15*, 100753.

Xu, C., Huang, X., Ma, N., Liu, Y., Xu, A., Zhang, X., Li, D., Li, Y., Zhang, W., & Wang, K. (2024). MicroRNA164 affects plant responses to UV radiation in perennial ryegrass. *Plants (Basel), 13*(9), 1242. https://doi.org/10.3390/plants13091242. PMID: 38732457; PMCID: PMC11085334.

Yadav, A., Mathan, J., Dubey, A. K., & Singh, A. (2024). The emerging role of non-coding RNAs (ncRNAs) in plant growth, development, and stress response signaling. *Noncoding RNA, 10*(1), 13. https://doi.org/10.3390/ncrna10010013. PMID: 38392968; PMCID: PMC10893181.

Yang, H., Cui, Y., Feng, Y., Hu, Y., Liu, L., & Duan, L. (2023). Long non-coding RNAs of Plants in response to abiotic stresses and their regulating roles in promoting environmental adaption. *Cells, 12*(5), 729.

Yusuf, A., Wakaya, K., Sakamoto, T., Uemura, T., Okamura, K., Ramadan, A., Nozawa, A., Suzuki, T., Inui, Y., Matsunaga, S., Sawasaki, T., & Arimura, G. I. (2025). Histone modification-dependent transcriptional regulation of defence genes in early response of Arabidopsis to Spodoptera litura attack. *Plant Cell Environment, 48*(5), 3257–3268. https://doi.org/10.1111/pce.15345. Epub 2024 Dec 25. PMID: 39722556; PMCID: PMC11963488.

Zhang, X., Guo, R., Shen, R., Landis, J. B., Jiang, Q., Liu, F., Wang, H., & Yao, X. (2023). The genomic and epigenetic footprint of local adaptation to variable climates in kiwifruit. *Horticulture Research, 10*(4), uhad031. https://doi.org/10.1093/hr/uhad031. PMID: 37799629; PMCID: PMC10548413.

Zhang, K.-L., Leng, Y.-N., Hao, R.-R., Zhang, W.-Y., Li, H.-F., Chen, M.-X., & Zhu, F.-Y. (2024). Adaptation of high-altitude plants to harsh environments: Application of phenotypic-variation-related methods and multi-omics techniques. *International Journal of Molecular Sciences, 25*(23), 12666.

Zhang, H., Zhang, P., Niu, Y., et al. (2025). Genetic basis of camouflage in an alpine plant and its long-term co-evolution with an insect herbivore. *Nature Ecology & Evolution, 9*, 628–638.

Zi, N., Ren, W., Guo, H., Yuan, F., Liu, Y., & Fry, E. (2024). DNA methylation participates in drought stress memory and response to drought in *Medicago ruthenica*. *Genes, 15*(10), 1286.

11 Regulation of Anti-oxidative and Oxidative Enzymes in Alpine Medicinal Plants Under Stress Conditions

Roop Lal, Palak Thakur, Ruchika Kumari, and Ashun Chaudhary

Abstract

Alpine plants are constantly subjected to extreme conditions like intense ultraviolet radiation, low temperature, and violent wind that contribute to the excessive production of reactive oxygen species (ROS) and resulting oxidative stress. To combat this stress, they have therefore evolved a robust antioxidant defense system with enzymatic and non-enzymatic components. Among enzymatic antioxidants, superoxide dismutase (SOD) reduces superoxide radicals to hydrogen peroxide, which is then detoxified by catalase (CAT) and ascorbate peroxidase (APX). APX, operating in the ascorbate–glutathione cycle, uses ascorbic acid as a substrate, whereas glutathione reductase (GR) sustains redox homeostasis through the regeneration of reduced glutathione (GSH). Guaiacol peroxidase (GPX) and various other peroxidases help in the elimination of peroxides during stress by catalysing the ROS. Non-enzymatic antioxidants such as ascorbic acid, glutathione, carotenoids, flavonoids, and osmolytes are also responsible for ROS scavenging and membrane stabilization. Furthermore, modulation in expression of transcription factors like DREB, MYB, NAC, and WRKY suggesting dynamic genetic regulation under stress conditions. These responses not only provide survival under oxidative stress but also ensure maintenance of medicinal property in alpine plants. Comprehension of these processes provides promising lines of approach for improving abiotic stress tolerance in plants *via* conservation practices and biotechnological strategies.

R. Lal
Department of Botany, Government College Chowari, Chamba, Himachal Pradesh, India

P. Thakur · R. Kumari (✉) · A. Chaudhary (✉)
Department of Plant Sciences, School of Life Sciences, Central University of Himachal Pradesh, Shahpur Campus, Kangra, Himachal Pradesh, India

P. Kumar et al. (eds.), *Climate Resilience and Molecular Adaptation in Alpine Medicinal Plants*, https://doi.org/10.1007/978-981-95-6923-6_11

Keywords

Alpine plants · Oxidative stress · Enzymatic antioxidants · Non-enzymatic antioxidants · ROS · Redox homeostasis · Stress-responsive genes · Abiotic stress tolerance

1 Introduction

Alpine medicinal plants serve a crucial function by bridging contemporary pharmaceutical ingenuity, folk therapeutic traditions, and biodiversity conservation. Traditional medicine systems based on medicinal plants serve at least 70% of the global population, which have been highly valued for millennia by rural and indigenous individuals, particularly from developing countries (Jeelani et al., 2018; Mehta et al., 2021). The valuable knowledge regarding identification and harvesting of these medicinal plants passed on to next generation, which also enriches the cultural heritage (Liu et al., 2009). Current statistics showed that global trade in medicinal plants was more than USD 215.4 billion in the year 2024 and is predicted to increase steadily to USD 403.5 billion by year 2033 with annual rate of 7.22% (Market Data Forecast, 2025). Indigenous knowledge systems approximate that Asia contains a medicinal flora of more than 38,660 species found across the continent (Astutik et al., 2019). After China, India stands second with respect to production of medicinal herbs (Zamani et al., 2025). Large number of irreplaceable medicinal plants (1748 species) described from Indian Himalayan regions with varied therapeutic value (Mehta et al., 2021). Nonetheless, the long-term sustainability of these precious resources is seriously jeopardized by increased demand, non-sustainable harvesting, and environmental degradation triggered through climate change, causing the intensification of stress conditions (Shrestha et al., 2022; Mykhailenko et al., 2025). Thus, restoration of medicinal plants particularly in alpine ecosystems needs an integrated approach that reconciles environmental conservation and sustainable utilization. Some alpine medicinal plants possess great symbolic and traditional importance (Cheng et al., 2022a). In Himachal Pradesh, the Dhauladhar mountain range is characterized by temperate, sub-alpine, and alpine forests, which support a diverse range of vegetation. Prominent tree species in the region include *Cedrus deodara* (Roxb. ex D.Don) G.Don, *Aesculus indica* (Wall. ex Cambess.) Hook., *Pyrus pashia* Buch.-Ham. ex D.Don *Picea smithiana* (Wall.) Boiss., and *Rhododendron arboreum* Sm. In addition, the area hosts various important herbaceous and medicinal plants such as *Aconitum heterophyllum* Wall., *Rheum australe* D.Don, *Polygonatum verticillatum* (L.) All., *Angelica glauca* Edgew. and *Ligusticopsis wallichiana* (DC.) Pimenov & Kljuykov (Kumari et al., 2025; Thakur et al., 2025; Thakur et al., 2024). The availability of a variety of bioactive chemicals with high therapeutic potential makes them commonly used herbs in simple yet potent forms, including plant extracts and decoctions (Joshi et al., 2016;

Ashraf et al., 2024; Thapliyal et al., 2024). Numerous valuable species present in alpine floras with broad therapeutic utility such as hepatoprotective, anti-inflammatory, antioxidant, anticancer, hypoglycemic, antidiabetic, and immunomodulatory activities that have attracted the interest of the pharmaceutical industry for using these in a variety of diseases like cancer, heart diseases, diabetes and other related infections (Joshi et al., 2016; Zhu et al., 2022; Ashraf et al., 2024).

Alpine medicinal plants are exposed to a range of stresses, which are mainly characterized by temperature fluctuations, strong solar radiations, and limited water conditions. At these extreme conditions, overproduction of reactive oxygen species (ROS) occurs, disturbing the normal growth, photosynthesis, and other critical metabolic processes. These ROS are crucial signalling molecules at low concentration that trigger defense reactions, although they possess harmful potential at high concentration due to their ability to degrade the structural and functional aspects of cellular macromolecules like proteins, DNA, and lipids. To achieve metabolic balance and ensure survival under harsh conditions, alpine plants developed intricate adaptive morphological, physiological, and biochemical mechanisms that constitute effective antioxidant systems (Wu et al., 2024). This adaptive response is primarily governed by stress-induced regulation of anti-oxidative and oxidative enzymes.

Oxidative stress occurs when the production of ROS surpasses the scavenging potential by the antioxidant defense system of a plant, within its cells (Dumanović et al., 2021). In that respect, ROS can act against DNA, lipids, and proteins indiscriminately, leading to cell lysis, and programmed cell death mediated through membrane leakage (Van Breusegem & Dat, 2006; Mansoor et al., 2022). To overcome this stress, plants employ a coordinated defense system comprising of endogenous antioxidants. This coordinated defense system collectively scavenges and detoxify ROS, hence enhancing tolerance towards harsh stress conditions (Hasanuzzaman et al., 2019).

The innate defence mechanisms enhance the robustness of plants by affecting the development and accumulation of secondary metabolites, thus enhancing their effectiveness. These systems are essential in maintaining plant health under alpine conditions. Under stress, elevated level of antioxidants indicates the equilibrium between biochemical regulation and prevention of damage. Environmental stressors, although imparting harmful effects, also play a part in synthesizing secondary metabolites, thereby rendering them more useful in therapeutic intervention. Both the deleterious and protective effects of ROS highlight the intricate interrelationship between oxidative stress, enzymatic control, and adaptive responses in plants (Ashraf et al., 2024). Deciphering the defense mechanism of anti-oxidative and oxidative enzymes in alpine medicinal plants is important to preserve them and to utilize them as a pharmacological agent in various stress conditions. This chapter investigates the biochemical and molecular processes involved in antioxidant defense system during stress conditions in alpine plants, providing insights into how these systems may enhance ecological sustainability and new therapeutic approaches.

2 Oxidative Stress Players that Generate ROS and Oxidative Factors

The normal biochemical reactions constantly generate ROS as by-products. Precisely, oxidative enzymes like NADPH oxidases, xanthine oxidases, and peroxidases contribute to generating ROS under stressful conditions. These enzymes are integral to several physiological processes, including growth and stress response. Understanding the role of these enzymes is crucial to understanding how these plants adapted to harsh conditions while also maintaining their medicinal attributes. Among these, NADPH oxidases are the major contributors of ROS and are dynamically involved in redox signalling, connected with growth, development, and stress responses (Hu et al., 2020; Petry & Görlach, 2021). Phosphorylation and ubiquitination of such enzymes at post-translational level regulate their function for quick response towards stressful conditions (Zhang et al., 2025a). Xanthine oxidases are the other oxidative enzymes responsible for ROS synthesis by catalysing the oxidation of xanthine. They are commonly linked with purine metabolism and can modulate oxidative stress levels in plants (Maraldi, 2013). Peroxidases are a class of oxidative enzymes that catalyse ROS biosynthesis, through oxidation of compounds like indole acetic acid and salicylic acid. Together, these enzymes play a role in oxidative burst and have the ability to initiate intracellular signalling pathways, critical for growth and defense induction (Kawano, 2003; Kumari et al., 2024). The functions of these enzymes have been established and recorded in different species, but their specificity towards imparting resilience and enhancing medicinal properties in alpine plants is not yet fully understood.

During stress conditions, oxidative enzymes trigger signalling pathways and thus mediate acute stress response. The functional role of NADPH oxidase and its specific molecular regulation in imparting adaptive response has been reported earlier in medicinal plants like *Opisthopappus taihangensis* (Y.Ling) C.Shih (Ushio-Fukai, 2001; Ye et al., 2024). Particular data on the regulation of Xanthine oxidase enzyme in alpine plants is not available; however, its application towards ROS generation indicates a possible similar mode of action as that of NADPH oxidase. Additionally, peroxide oxidases prevent oxidative damage by detoxifying ROS like hydrogen peroxide, thus reducing ROS accumulation.

Conversely, whereas these enzymes play crucial roles in stress acclimation, overproduction of ROS may result in oxidative damage, thus requiring proper balanced regulation of processes in alpine medicinal plants.

2.1 Oxidative Factors under Normal and Stressed Conditions

Oxidative factors involve various ROS, generated during redox reactions under normal and stressed conditions. These factors include singlet oxygen [$^1O_2$], superoxide anion [$O_2^{\bullet -}$], hydrogen peroxide [H_2O_2], and hydroxyl radical [$OH^{\bullet}$] (Apel & Hirt, 2004; Dickinson & Chang, 2011). The ROS are mainly produced at various sites within the cell, like chloroplasts, mitochondria, peroxisomes, endoplasmic reticulum, plasma membrane, cell wall, and apoplast (Asada, 2006; Das & Roychoudhury, 2014). However, under stressful conditions, their production increased in these locations, and extra ROS production takes place in the cytosol, endoplasmic reticulum, and apoplast (Das & Roychoudhury, 2014; Czarnocka & Karpiński, 2018).

Out of these, $^1O_2$ is mainly generated in chloroplasts, in the antenna system, during photosynthetic light reaction, after interaction of triplet chlorophyll with oxygen. Under stress conditions, poor availability of intracellular CO_2 as well as stomatal closure favours the generation of $^1O_2$ (Fig. 11.1). However, $O_2^{\bullet-}$ is generated in chloroplasts, mitochondria, and peroxisomes, during oxygen reduction through the electron transport chain or the Ca^{2+} mediated action of NADPH oxidase within the plasma membrane (Asada, 2006; Kumar et al., 2012;). It is also generated in the cytosol and endoplasmic reticulum through metabolic activities. It is moderately reactive but transforms into more reactive and toxic forms such as $^1O_2$ and $OH^{\bullet}$ radical (Halliwell, 2006). Likewise, H_2O_2 arises when there is hydronation as well as univalent reduction of $O_2^{\bullet-}$ takes place. It is also mildly reactive and generated after enzymatic dismutation of $O_2^{\bullet-}$ in the chloroplast, mitochondria, peroxisomes, ER, cytoplasm, and apoplast, both under normal and stress conditions. Conditions like stomatal closure and low intracellular CO_2 concentrations coupled with the

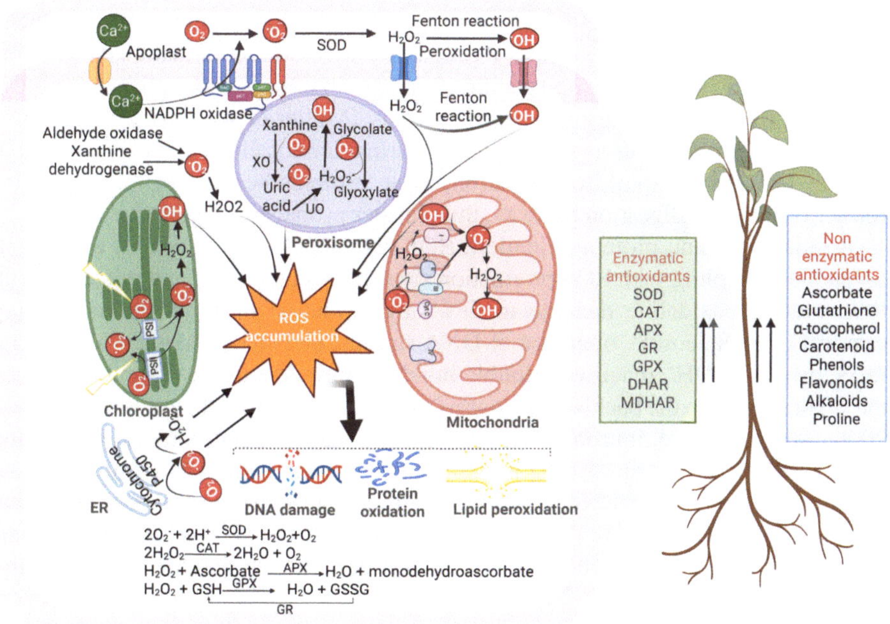

Fig. 11.1 Oxidative stress and its regulation through the antioxidant defense system in alpine medicinal plants. An illustration showing the production of ROS in response to oxidative stress in alpine plants and the response of medicinal plants against oxidative stress through increased concentration or upregulation of enzymatic and non-enzymatic antioxidants

action of oxidative enzymes favour its high generation or overproduction under stress (Noctor et al., 2002; Gill & Tuteja, 2010; Mhamdi & Van Breusegem, 2018).

Among ROS, $OH^{\bullet}$ is the most reactive and toxic factor, generated *via* the Fenton reaction involving H_2O_2 and $O_2^{\bullet -}$ through Fe^{2+} or Cu^+ (Richards et al., 2014; Treml & Šmejkal, 2016). Its scavenging mechanism is not strong enough, therefore leading to its excess accumulation and ultimately cellular death (Pinto et al., 2003).

2.2 Effect of ROS on Cellular Components

ROS under low concentration acts as a signalling molecule that mediates physiological processes and facilitates stress response. Conversely, at high concentration, it becomes highly reactive and evokes oxidative modifications of cellular macromolecules [lipids, proteins, and DNA] (Farmer & Mueller, 2013; Choudhury et al., 2017; Czarnocka & Karpiński, 2018; Sharma et al., 2021) (Fig. 11.1). During stress conditions, when the level of ROS surpasses recovery, a chain reaction of lipid peroxidation (LPO) occurs, which increases oxidative stress and further induces damage to other lipids and proteins. ROS predominantly, $^1O_2$ and $OH^{\bullet}$ radical, act on polyunsaturated fatty acids by interacting with their methylene group and lead to the formation of conjugated diene, lipid peroxyl radical, and hydroperoxides (Smirnoff, 2000). Overall, LPO increases fluidity of the membrane, permeability and physical integrity, damage and inactivates membrane receptors, enzymes, and channels. Along with LPO, ROS also causes direct as well as indirect oxidation of proteins through modifications like site-specific alteration in amino acid, enhanced proteolytic susceptibility, and varying enzyme activity (Møller et al., 2007). Among proteins, specifically sulphur and thiol group bearing amino acids are most susceptible to ROS ($^1O_2$ and $OH^{\bullet}$). After oxidation, these proteins become better targets for further proteolytic digestion by ROS, thus further propagating oxidative damage.

Besides lipid and protein oxidation, ROS also attacks DNA located within cell organelles, coupling with ROS production sites in the chloroplast and mitochondria. ROS mediate oxidative damage to DNA by oxidation of sugar, alteration and removal of a nucleotide, breakage of DNA strands, and cross-linking with protein. ROS, primarily $OH^{\bullet}$, react with double bonds of nitrogenous bases, and also harm the sugar residue in the backbone through the removal of a hydrogen atom (Halliwell, 2006). Upon removal, it produces a sugar radical that further causes a single-strand DNA break (Evans et al., 2004). It also produces irreversible DNA-protein cross-links when it reacts with DNA or with proteins and subsequently proves to be lethal for the plant cell, and it is inevitable to quench ROS. The survival of alpine plants depends on maintaining redox balance between the production and scavenging of ROS.

3 Responses of Alpine Plants to Oxidative Stress

Survival of alpine medicinal plants in severe stressful environments is due to adaptation, primarily at physiological and molecular levels. Different adaptation mechanisms are adapted by these plants, such as the regulation of secondary metabolites and antioxidants to protect against oxidative stress, produced at high altitudes. Some medicinal plants adapt by storing secondary metabolites as well as by increasing antioxidant enzyme activity, while others generally depend on the storage of secondary metabolites or increased enzyme activity. The regulation of increased activity of antioxidant enzymes and accumulation of secondary metabolites in alpine medicinal plants counteract the ROS-induced oxidative damage and boost antioxidant defense systems (Ibrahim et al., 2022).

3.1 Antioxidants Defense System

The antioxidant defense system comprises enzymatic and non-enzymatic antioxidant machinery that assist in reducing damage caused by oxidative stress.

3.1.1 Enzymatic Antioxidants

Superoxide dismutase (SOD), catalase (CAT), ascorbate peroxidase (APX), glutathione reductase (GR), peroxidase (POD), monodehydroascorbate reductase (MDHAR), dehydroascorbate reductase (DHAR), Guaiacol peroxidase (GPX) are some of the enzymes present in subcellular compartment and constitute antioxidant enzymatic machinery. Superoxide dismutase (SOD), a member of metalloenzymes family, exist in three different isoforms, in different sites namely, cytosol, peroxisomes, mitochondria, and chloroplasts (Raina et al., 2023; Mittler, 2002). It catalyse the elimination of $O_2^{\bullet -}$, by dismutating free radicals into less reactive H_2O_2 and O_2 in normal conditions. Additionally, it also removes the possibility of the formation of $OH^{\bullet}$ through the Haber-Weiss reaction. Under stress conditions, it constitutes the first line of defense against oxidative damage from stressors. The different isoforms get upregulated differentially to counteract the elevated ROS level (Boguszewska et al., 2010). Similarly, energy-efficient enzyme CAT catalyses the conversion of H_2O_2 into H_2O and O_2. Mostly found in peroxisomes, but also located in chloroplasts, cytosol, and mitochondria, and considered to have a high turnover rate (Mhamdi et al., 2010). Under stress conditions, higher energy requirement is met by increased catabolism, leads to generation of H_2O_2.

APX performs a similar function to catalase in the chloroplast and cytosol. Through the use of ascorbic acid as a reducing agent, it converts H_2O_2 into H_2O and MDHA. Five different isoforms of APX have been discovered in various cellular compartments such as thylakoid, stroma, mitochondria, peroxisomes, and cytosol. It has a greater preference for H_2O_2 than CAT, diversely distributed, and is more effective scavenger during stress (Sharma & Dubey, 2004).

GR regenerate reduced glutathione (GSH) from Glutathione disulfide (GSSG). GSH is further utilized to recycle ascorbic acid from MDHA and DHA, hence transformed into an oxidized form. It is primarily located in the chloroplast but is also present in trace amounts in mitochondria and cytosol. Under normal conditions, it maintains GSH/GSSG balance. However, under stress conditions, it sustains a higher cellular ratio of GSH/GSSG (Das & Roychoudhury, 2014).

MDHAR and DHAR recycle ascorbic acid from short-lived MDHA and DHA. They also occur in the peroxisomes, mitochondria, cytosol, and chloroplasts (Mittler, 2002). Under stress conditions, these uphold the redox state of the cell by maintaining the ascorbic acid content of the cell (Chen & Gallie, 2006). GPX functions under normal and stressful situations to eliminate excess H_2O_2. It plays a significant role in lignin synthesis to mitigate the biotic stress. It employs aromatic compounds like pyragallol and guaiacol as electron donors to remove H_2O_2 actively from intracellular, extracellular, and cell wall regions (Asada, 1999).

3.1.2 Non-enzymatic Antioxidants

These also constitute the antioxidant machinery and include ascorbic acid, reduced glutathione (GSH), α-tocopherol, carotenoids, phenolics, flavonoids, and proline. These have a dual role in protection from stressful conditions as well as growth and development induction by maintaining processes right from cell division till cell death (de Pinto & De Gara, 2004). Ascorbic acid is the most ubiquitously occurring and widely investigated antioxidant. It behaves like an electron donor in both enzymatic and non-enzymatic reactions. The majority of ascorbic acid is produced in the mitochondria as a product of the Smirnoff-Wheeler pathway. Its pool is sequestered in cytosol and apoplast, thereby forming a primary defense against ROS (Barnes et al., 2002). It is synthesized as a by-product from oxidation of MDHA and DHA. Under normal conditions, it acts as a metabolic regulator of key physiological processes and functions as a cofactor to ensure the quenching of excitation energy (Smirnoff, 2000). However, upon reaction with ROS, it restores α-tocopherol from tocopheroxyl radical, protects the activities of metal-binding enzymes, and inhibits photo-oxidation, thus preventing oxidative damage (Shao et al., 2005). Glutathione is another non-enzymatic antioxidant, occurring in large amounts in cytosol, chloroplasts, mitochondria, peroxisomes, endoplasmic reticulum, apoplast, and vacuoles. It is actively involved in controlling several processes, including division, growth, differentiation, senescence, and cell death, metabolite conjugation, protein synthesis, nucleotide synthesis, phytochelatin synthesis, and stress-inducible gene expression (Mullineaux & Rausch, 2005). Strong reducing ability makes it a versatile antioxidant. It scavenges all classes of ROS and confers protection to biomolecules and forms GSSG as by-products. It regenerates ascorbic acid and produces GSSG. GSSG is further reduced to GSH, with the help of the enzyme GR, thereby replenishing the pool of GSH.

Being a member of the lipophilic antioxidants family, α-tocopherol is very effective in scavenging ROS and lipid radicals and thus makes them irreplaceable guardians of membranes (Kiffin et al., 2006). It is exclusively formed by photosynthetic tissues and possesses maximum antioxidant activity. During oxidative stress, they shield the membrane of chloroplasts, interact with O_2, thereby safeguarding the PSII. Simultaneously, α-tocopherol terminates the propagation during the LPO cycle, interacts with lipid radicals, and gets oxidized to TOH•, which is then recycled by GSH and ascorbic acid (Igamberdiev et al., 2004). Carotenoids are other lipophilic antioxidants, which are found in plastids. They protect the photosynthetic machinery through the xanthophyll cycle both under normal and stressed conditions. Flavonoids like flavonols, flavones, isoflavones, and anthocyanins are commonly found in the leaves and floral organs. They play varied functions in imparting colours to flowers and fruits, pollen germination, and protection against pathogens. Under stress, flavonoids have been regarded as an alternative ROS defensive system in plants when damage is associated with photosynthetic machinery (Fini et al., 2011). They also play a function in neutralizing $^1O_2$ and mitigating the injuries inflicted on the chloroplast membrane (Agati et al., 2012).

Osmolyte proline is most commonly employed as a non-enzymatic antioxidant to check the damaging activities of various ROS. Proline is formed from glutamic acid. It is an effective scavenger of OH• radical and $^1O_2$ and can check the damage caused by LPO. Under stress, proline accumulates in plants to very high levels (Verbruggen & Hermans, 2008).

4 Survival Strategies Adopted by Alpine Medicinal Plants under Stress Conditions

Energy dissipation mechanism (avoidance) and robust antioxidant machinery in addition to secondary metabolite accumulation (endurance) are two survival strategies adopted by medicinal plants to live in adverse alpine conditions. Nevertheless, the regulation of the endurance mechanism heavily promotes the survival of alpine species under stress. Alpine medicinal plants have developed strong antioxidant systems to survive oxidative stress caused by a harsh environment. *Sibbaldianthe bifurca* (L.) Kurtto & T.Erikss. [Syn. *Potentilla bifurca* L.], *Rheum australe* D.Don, and *Persicaria vivipara* (L.) Ronse Decr. [Syn. *Polygonum viviparum* L.] are capable of surviving at high photon flux, extreme temperature changes, and low moisture levels with growth adaptation potential (Tang et al., 2022; Mala et al., 2021; Wang et al., 2009).

Physiological, biochemical, and molecular plasticity confer adaptation capacity to the Himalayan medicinal herbs, such as *Caragana jubata* (Pall.) Poir. and *Lepidium latifolium* L. to grow in high altitude and extreme conditions (Bhardwaj et al., 2013; Bhat et al., 2016). Overall, the stressors in alpine ecosystems overproduce the ROS, hence requiring an effective regulation between the generation and scavenging of ROS, which can be realized through the regulation of endogenous antioxidants. Most of the species mainly utilize the upregulation of enzymatic and non-enzymatic antioxidants under stress conditions.

4.1 Enzymatic Antioxidants Response

Regulation of enzymatic antioxidants is crucial for the survival and retaining therapeutic potential of most medicinal plants in alpine habitats. At higher altitude, medicinal plant like *Coleus forskohlii* Anon. demonstrated higher activity of SOD, CAT, POD, APX, GR (Rana & Saklani, 2019). Upregulation of CAT and APX act along with downregulation of polyphenol oxidase, protects these plants from extreme stress conditions (Ibrahim et al., 2022). The induction of anti-oxidative enzymes such as SOD, POD, APX, CAT, can reduce oxidative stress efficiently, hence inhibiting the oxidation of membrane during cold stress and light stress (Cheng et al., 2022b; Kumar et al., 2021a; Yao et al., 2008). Compilation of medicinal alpine plants with stress-survival response behaviour is presented in Table 11.1. Notable examples includes *Herpetospermum pedunculosum* (Ser.) C.B.Clarke, *Primula minima* L., *Potentilla saundersiana* Royle, *Nardostachys jatamansi* (D.Don) DC., *S. bifurca, L. latifolium, and P. vivipara,* exhibiting considerable rise of total and specific enzymatic antioxidants under stressed conditions (Table 11.1).

4.2 Non-enzymatic Antioxidants Response

The non-enzymatic antioxidants also get accumulated during survival under stress. Flavonoids, phenolics, anthocyanins and certain compounds like carotenoids, α-tocopherol, and ascorbic acid, scavenge ROS (Blokhina et al., 2003; Landi et al., 2015). At higher altitudes, *Soldanella* leaves possess higher amount of glutathione and α-tocopherol (Wildi & Lütz, 1996). Many plants in alpine grazing areas were reported to have high level of α-tocopherol in the Norwegian region (Sickel et al., 2012). Carotenoids, phenols and flavonoids were increased at high altitude (Ibrahim et al., 2022). Flavonoids neutralize ROS through direct scavenging, metal chelation, enzyme inhibition, and antioxidant enzyme activation. Greater accumulations of flavonoids at greater elevations were established in medicinal plants like *H. pendunculosum and Picrorhiza kurroa* Royle ex. Benth. (Zhao et al., 2019; Kumari et al., 2020). Metabolic profiling of Tibetan medicinal plant, *Phlomoides rotata* (Benth. ex Hook.f.) Mathiesen under varied elevations (4300–5000) and slope orientations in Bundanla Mountain, showed altitude-specific production and accumulation of flavonoid and slope-specific accumulation of different metabolites. Altitude triggered enhanced production of flavonoids is the adaptive process to UV mediated stress (Wang et al., 2025). Metabolomics of *Rheum tanguticum* Maxim. ex Balf. grown under different altitudinal gradient in Qinghai-Tibet Plateau, indicate higher concentration of flavonoid at higher altitudes (Wang et al., 2024). A strong relationship between phenolic content and antioxidant potential with altitude has been noted in species like *Scrophularia striata* Boiss. (Zargoosh et al., 2019). Increased regulation of phenolic content was found in the medicinal species with increasing altitude in Sinai Peninsula (Ibrahim et al., 2022). Altitude-based prominent accumulation of proline content was observed in four medicinal plants, while assessing the adaptive response of these to high altitude oxidative stress (Ibrahim et al., 2022). Abiotic

Table 11.1 Altitudinal distribution and stress-survival response behaviour of key alpine medicinal plants along with relevant references

Plant species	Family	Distribution range	Elevation (m asl)	Stress-survival response behaviour	References
Sibbaldianthe bifurca (L.) Kurtto & T.Erikss. [Syn. *Potentilla bifurca* L.	Rosaceae	Qinghai-Tibet plateau	3215	Upregulation of ubiquitin-conjugating enzyme, and anti-oxidative enzymes POX and SOD	Tang et al. (2022)
Lepidium latifolium L.	Brassicaceae	Ladakh region of Western Himalayas	2500–4500	Higher induction of anti-oxidative enzymes	Bhat et al. (2016)
Potentilla saundersiana Royle	Rosaceae	Northwestern Tibetan plateau	4350–5200	Increased activity of antioxidant enzymes and metabolites	Ma et al. (2015)
Nardostachys jatamansi (D. Don) DC.	Caprifoliaceae	Himalayan region	3000–5000	Enhanced activity of antioxidant enzymes, secondary metabolite and total sugar content	Joshi and Joshi (2025)
Persicaria vivipara (L.) Ronse Decr. [Syn. *Polygonum viviparum* L.]	Polygonaceae	Tianshan Mountain in China	3900	Increased concentration of anti-oxidative enzymes- SOD with new isoforms, DHAR, and GR; elevation of non-enzymatic antioxidants- ASC, GSH, ASC/DHA, GSH/[GSH + GSSG] ratio	Wang et al. (2009)

(continued)

Table 11.1 (continued)

Plant species	Family	Distribution range	Elevation (m asl)	Stress-survival response behaviour	References
Herpetospermum pedunculosum (Ser.) C.B. Clarke	Cucurbitaceae	Tibetan Plateu	2300–2500	Increased activities of anti-oxidative enzymes-SOD, APX, GR, MDHAR, DHAR; elevation in non-enzymatic antioxidants-glutathione, total ascorbate content and the ASA/DHA ratio	Zhao et al. (2019)
Primula minima L.	Primulaceae	Eastern Alps	1500–3000	Increased activity of anti-oxidative enzymes-CAT, APX, GR and glucose-6-phosphate dehydrogenase	Roach et al. (2023)
Picrorhiza kurroa Royle ex Benth.	Scrophulariacea	Tungnath, Uttrakhand, India	3600	Increased concentration of non-enzymatic antioxidants; enhanced production of total protein, soluble sugars and proline content	Agrawal et al. (2023) and Agrawal and Saklani (2023)
Persicaria vivipara (L.) Ronse Decr. [Syn. *Polygonum viviparum* L.]	Polygonaceae	Qinghai-Tibet plateau	4200–4300	Accumulation of flavonoids, soluble sugar, amino acids and lipids; upregulation of special metabolites	Wei et al. (2024)
Rheum australe D. Don	Polygonaceae	Indian Himalaya	3000–4200	Up-regulation of antioxidants, chaperones, osmo-protectants, and secondary metabolites	Mala et al. (2021)

Espeletia schultzii Wedd.	Asteraceae	Andean region	3100–4200	Accumulation of total soluble sugars; enhanced activity of RUBISCO	Castrillo (2006)
Persicaria alpina Gross. [Syn. *Polygonum alpinum* All.]; *Galium aparine* L.; *Onobrychis dealbata* Stocks	Polygonaceae, Rubiaceae Fabaceae	Hunza Valley	3000–3500	Increased endogenous content of ABA, sugar, proline, and proteins	Bano et al. (2009)
Rhodiola kirilowii (Regel) Maxim. [Syn. *Rhodiola linearifolia* Boriss.]	Crassulaceae	Northeastern Tianshan Mountain	2500–3100	Increased concentration of secondary metabolites, antioxidants at higher altitudes	Terletskaya et al. (2024)
Cotoneaster orbicularis Schltdl.; *Crataegus sinaica* Boiss.; *Echinops spinosissimus* subsp. *Spinosissimus* Turra.; and *Tanacetum sinaicum* (Fresen.) Delile ex K. Bremer & Humphries	Rosaceae, Asteraceae	Saint Katherine protectorate (SKP) in the south part of Sinai peninsula	1500–2250	Stress based differential activity of SOD, CAT, APX, and POX in studied species at different altitudes	Ibrahim et al. (2022)

stresses elevate the proline content in plant which acts like metabolic signal and osmo-protectant (Fu et al., 2018). Proline is the potent regulatory molecule which regulates oxidative stress by elevating PSII activities, regulating cytosol acidity, NAD^+/NADH ratio, and lowering the LPO (Kavi Kishor et al., 2022).

The increased elevation has been reported to increase the Trolox-equivalent antioxidant capacity, total phenols, and tannins, as adaptative measure against oxidative stress. Furthermore, increased protein and fatty acid contents also support high metabolic activity in alpine regions (Cui et al., 2016). Also, most endemic species elevate their soluble sugar, protein and proline content at higher altitudes (Castrillo, 2006; Bano et al., 2009; Hashim et al., 2020; Terletskaya et al., 2024). Many common alpine medicinal plants such as *P. saundersiana*, *P. kurroa*, *P. vivipara*, *R. australe* were reported to upregulate metabolite expression and their accumulation (Table 11.1). Specific secondary metabolites were reported to be upregulated upon exposure to stress. Tang et al. (2022) noted that upregulation of various differentially expressed genes in *S. bifurca* at high altitudes were involved with biosysnthesis of secondary metabolites. The research indicates that their accumulation was associated with *S. bifurca* adaptation at alpine condition. Likewise, notable upregulation of Tricetin was noted with an increase in altitude in the case of *H. penduncu-losum* (Zhao et al., 2019). Secondary metabolites, aside from enzymatic antioxidants, are crucial in oxidative stress alleviation and adaptation to alpine stresses. These compounds often surpass the functional contribution of primary metabolites under stress. Decreased levels of malondialdehyde (MDA), an indicator of LPO, in alpine plants indicate effective oxidative stress response. Various metabolites such as flavonoids, phenylpropanoids, fatty acids, terpenoids, alkaloids, etc. have been reported as adaptive molecules in the case of *Saussurea obvallata* (DC.) Edgew. (Zhang et al., 2025b).

4.3 From Genes to Enzyme: the Inner Working of Plant Stress Response

Stress conditions in the alpine region demand the induction and activation of strong antioxidative machinery to scavenge excessive ROS and to uphold redox balance. Sensing and perception of stress-induced signals generate and accumulate molecules like ROS, Ca^{2+} ions, and phytohormones that act as secondary messengers and further transduce stress signals through protein kinase, phosphatase, mitogen-activated protein kinase, and Ca^{2+} dependent protein kinase. These messengers activate the particular transcription factor, like WRKY, MYB, DREB, NAC, and bZIP, etc. These transcription factors bind to the cis-regulatory elements on antioxidative gene promoters/stress-responsive genes. The transcriptional and post-transcriptional regulation leads to the formation of functional proteins. These functional proteins may further behave like transcription factors or signal transducer molecules. Post-translational modifications (PTMs) like phosphorylation, S-nitrosylation, oxidation, glutathionylation, and ubiquitination further activate or enhance the expression of stress response factors that include enzymatic and non-enzymatic antioxidants as well as secondary metabolites (Fig. 11.2).

Potential role of transcription factors like DREB1/CBF in inducing SOD, CAT, APX, and LEA protein genes expression and differential role of WRKY factor in regulating responsive genes expression against different abiotic stresses have been observed (Akhtar et al., 2012; Chen et al., 2012; Shi et al., 2018; Li et al., 2020). In the stress response under low temperature, regulation of several transcription factors such as AP2/ERF, TRAF, bHLH, bZIP/GRAS, AUX/IAA, and GARP-G2 takes place, which were also identified in stress-tolerant plants at low altitudinal areas (Ding et al., 2020). Induction of CBF, LEA proteins, and thaumatin-like proteins in *Prunus mira* Koehne demonstrates the versatility of alpine stress response network (Cao et al., 2022). DREB transcription factors, especially SmDREB1, also induce gene expression of antioxidants (SOD, POD, and CAT) and optimize photosynthetic performance with restricted water supplies (Li et al., 2017).

In *Sophoria moorcroftiana* (Benth.) Benth. ex Baker, activation and upregulation of several drought resistance genes like transcription factors (AP2/ERF, MYB, NAC, WRKY, Zn-finger), plant hormone regulator genes (responsive and

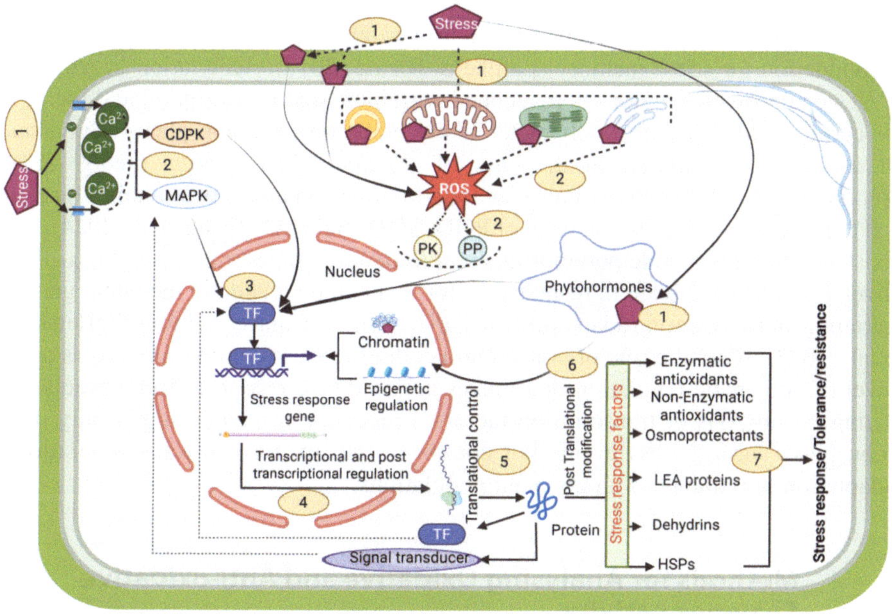

Fig. 11.2 Stress-response mechanism in alpine medicinal plants: (1) Signal sensing and perception, (2) Signal transduction by secondary messengers (Ca^{2+} ions, ROS, and phytohormones), (3) Activation of transcription factors (WRKY, MYB, DREB, NAC, bZIP), (4) Epigenetic (histone acetylation, deacetylation, ubiquitinization), transcriptional and post-transcriptional regulation (co-transcription, splicing, capping, tailing), (5) Translational control and post-translational modifications (Ubiquitinization, S-nitrosylation), (6) Enhanced production of stress response factors [Enzymatic antioxidants, non-enzymatic antioxidants, osmo-protectants, LEA proteins, dehydrins, heat shock proteins (HSPs)] (7) Stress response and tolerance

auxin-induced protein), and stress-responsive genes have been reported (Li et al., 2015). Zhou et al. (2016) have reported enhanced stress resistance through activation of stress-responsive gene SpCIPK26 in alpine medicinal grass, *Stipa purpurea* Griseb., through the ABA-dependent pathway. Activation of the calcium-signalling pathway through protein kinase represents the best adaptive mechanisms against salt stress by various taxa (Luan, 2009; Li et al., 2016). Integrated action of signalling with hormonal pathways has been demonstrated to enhance anti-oxidative activity and supporting cellular ion homeostasis (Sun et al., 2025). These transcription factors induce stress tolerance through the activation or repression of the target proteins (Hirayama & Shinozaki, 2010). Ma et al. (2015) reported the enhanced expression of APX and SOD in alpine medicinal plant *P. saundersiana* with increasing altitude. Beyond transcriptional regulation, PTMs provide a rapid mechanism for adjusting the activity, stability, cellular localization, and interactions of antioxidant enzymes. For example S-nitrosylation has been proven to control the redox-sensitive enzyme GR for the regeneration of GSH during oxidative stress.

Apart from genetic regulation, epigenetic mechanisms, including DNA methylation, histone modifications, and chromatin remodelling, contributed significantly to the long-term regulation in alpine medicinal plants. These modifications can silence or activate gene expression. Sometimes, during extended stress conditions, the induced epigenetic marks can be carried forward through cell divisions and provide a foundation for stress memory (Lämke & Bäurle, 2017). Acetylation of APX and CAT gene promoter regions is associated with increased transcription and memory of earlier cold exposure, a character likely common to alpine plants (Satyakam et al., 2022). Similarly, in the case of *Rhododendron aureum* Georgi [Syn. *Rhododendron chrysanthum* Pall.], acetylation of telomeric repeat binding protein led to increase the expression level of POD, SOD, and CAT (Gong et al., 2024). The existence of high genetic polymorphism in *Rhodiola* species, indicating high allelic diversity of SOD and auxin response factor that can be related to the latent genetic variation and their strong adaptability to harsh environmental conditions (Terletskaya et al., 2023). Also, the role of non-protein coding RNA is coming forward as a key level of epigenetic regulation that makes antioxidative responses more precise by acting on mRNAs or transcription factors. Epigenetic mechanisms not only offer robustness during stress events but could also be involved in trans-generational adaptation of indigenous alpine plant populations.

5 Methods for Analysing Oxidative and Anti-oxidative Enzymes Activity

To comprehensively evaluate the differential plant response and anti-oxidative defense mechanisms, multiple biochemical, molecular, and omics-based methods are used together under stress conditions. Enzyme activity assays form the baseline for monitoring the functionality of anti-oxidative enzymes. These assays are generally spectrophotometric, based on substrate conversion or cofactor oxidation/

reduction, and give direct estimation of the enzymatic activity of medicinal plants under conditions of stress (Panda, 2012).

In addition to enzymatic assays, gene expression through qRT-PCR enables quantitative estimation of antioxidant-related genes and hence governs transcriptional regulation upon stress (Zhang et al., 2024). Concurrently, proteomic techniques such as SDS-PAGE and immunoblotting facilitate the identification and quantification of several defensive and stress-tolerant proteins with differential expression, as well as post-translational modulation, as a multiple adaptive approach in alpine medicinal plants (Ma et al., 2015). Additionally, metabolomic profiling via GC-MS, LC-MS, NMR, HPLC, and CE-MS facilitates the detection of alterations in secondary metabolites, providing information regarding metabolic reorganization under oxidative stress (Kumar et al., 2021b). In addition, localization and detection of ROS are realized through histochemical stains like DAB (3,3′-diaminobenzidine) for hydrogen peroxide and NBT (nitro blue tetrazolium) for superoxide, or through fluorophore-based probes like dihydrofluoroscein diacetate and dihydroethidium, which can be observed through fluorescence microscopy or measured through fluorometers and flow cytometry (Akter et al., 2021). Collectively, these combined approaches yield a multifaceted picture of oxidative signalling, enzymatic responses, and molecular regulation of plants under environmental stress factors, especially high altitude alpine ecosystems.

6 Implications for Conservation and Utilization

The greater stress tolerance of alpine medicinal plants is not only crucial to their survival in severe high altitude conditions but also bears deep significance for their conservation and sustainable use. Their intrinsic capacity to modulate oxidative and anti-oxidative enzyme systems makes them resistant to extreme abiotic factors that are predicted to be strengthened through on-going climate change. Conservation of these species is important, as they contain diverse phytochemicals with strong therapeutic potential that might be lost due to changes in the stress conditions, leading to changes in their metabolic profiles. An elucidation of the enzymatic and molecular mechanisms underlying their stress responses provides important inputs to *ex-situ* conservation, restoration ecology, and the creation of climate-resilient cultivars through biotechnology. Biotechnological advancements and genetic engineering may utilize these specific stress-regulatory genes with significant functions, *e.g.*, encoding SOD, APX, LEA proteins, and DREB transcription factors, to enhance abiotic stress resistance in plant species. In addition, it is crucial to preserve the medicinal activity of these plants under changing environmental conditions since stress-induced modifications in secondary metabolite production may affect their pharmacological potential and utility. Thus, integrative strategies that integrate ecological, biochemical, and molecular practices are crucial for preserving alpine plant diversity and realizing their potential in climate adaptive natural product development.

7 Future Perspectives and Conclusion

Insight into the regulation of oxidative and anti-oxidative enzymes in alpine medicinal plants under stress conditions is invaluable to their specialized stress-resilience mechanisms. Future studies must focus on integrative omics platforms to decipher gene-environment interactions and epigenetic adjustments underlying these mechanisms. Biotechnological strategies can be utilized to enhance stress tolerance and phytochemical consistency in these plants, particularly against the changing era of climate change. Conservation efforts must equally be focused on safeguarding their indigenous habitats, as they require sustainable cultivation. In conclusion, a better understanding of these biochemical defense systems not only promotes ecological and medicinal science but also provide lead to maintaining biodiversity and traditional therapeutic resources in alpine ecosystems.

Acknowledgemnets We are grateful to the University Grant Commission (UGC), New Delhi, India, for providing financial support to Ruchika Kumari (No. 231610273039, Ref. No: NSFDC/E-81088) and Central University of Himachal Pradesh for providing necessary facilities.

Competing Interests The authors declare no conflict of interest.

References

Agati, G., Azzarello, E., Pollastri, S., & Tattini, M. (2012). Flavonoids as antioxidants in plants: Location and functional significance. *Plant Science, 196*, 67–76.

Agrawal, S., & Saklani, P. (2023). Physiological responses of the leaves of a high-altitude plant *Picrorhiza kurroa* to cold stress. *bioRxiv*, [preprint]. Available at: https://doi.org/10.1101/2023.05. Accessed 27 June 2025.

Agrawal, S., Bhandari, N., & Saklani, P. (2023). Effect of altitude on the non-enzymatic antioxidant potential of the leaves of a high-altitude medicinal plant, Picrorhiza kurroa. *Biochemical and Cellular Archives, 23*(2), 1085–1090.

Akhtar, M., Jaiswal, A., Taj, G., Jaiswal, J. P., Qureshi, M. I., & Singh, N. K. (2012). DREB1/CBF transcription factors: Their structure, function and role in abiotic stress tolerance in plants. *Journal of Genetics, 91*, 385–395.

Akter, S., Khan, M. S., Smith, E. N., & Flashman, E. (2021). Measuring ROS and redox markers in plant cells. *RSC Chemical Biology, 2*(5), 1384–1401.

Apel, K., & Hirt, H. (2004). Reactive oxygen species: Metabolism, oxidative stress, and signal transduction. *Annual Review of Plant Biology, 55*(1), 373–399.

Asada, K. (1999). The water–water cycle in chloroplasts: Scavenging of active oxygens and dissipation of excess photons. *Annual Review of Plant Biology, 50*(1), 601–639.

Asada, K. (2006). Production and scavenging of reactive oxygen species in chloroplasts and their functions. *Plant Physiology, 141*(2), 391–396.

Ashraf, M. V., Khan, S., Misri, S., Gaira, K. S., Rawat, S., Rawat, B., Khan, M. H., Shah, A. A., Asgher, M., & Ahmad, S. (2024). High-altitude medicinal plants as promising source of phytochemical antioxidants to combat lifestyle-associated oxidative stress-induced disorders. *Pharmaceuticals, 17*(8), 975.

Astutik, S., Pretzsch, J., & Ndzifon Kimengsi, J. (2019). Asian medicinal plants' production and utilization potentials: A review. *Sustainability, 11*(19), 5483.

Bano, A., Rehman, A., & Winiger, M. (2009). Altitudinal variation in the content of protein, proline, sugar and abscisic acid (ABA) in the alpine herbs from Hunza valley, Pakistan. *Pakistan Journal of Botany, 41*(4), 1593–1602.

Barnes, J., Zheng, Y., & Lyons, T. (2002). Plant resistance to ozone: The role of ascorbate. In *Air pollution and plant biotechnology: Prospects for phytomonitoring and phytoremediation* (pp. 235–252). Springer.

Bhardwaj, P. K., Kapoor, R., Mala, D., Bhagwat, G., Acharya, V., Singh, A. K., Vats, S. K., Ahuja, P. S., & Kumar, S. (2013). Braving the attitude of altitude: *Caragana jubata* at work in cold desert of Himalaya. *Scientific Reports, 3*(1), 1022.

Bhat, H. A., Kaur, T., Bhat, R., & Vyas, D. (2016). Physiological and biochemical plasticity of *Lepidium latifolium* as 'sleeper weed' in Western Himalayas. *Physiologia Plantarum, 156*(3), 278–293.

Blokhina, O., Virolainen, E., & Fagerstedt, K. V. (2003). Antioxidants, oxidative damage and oxygen deprivation stress: A review. *Annals of Botany, 91*(2), 179–194.

Boguszewska, D., Grudkowska, M., & Zagdańska, B. (2010). Drought-responsive antioxidant enzymes in potato (*Solanum tuberosum* L.). *Potato Research, 53*, 373–382.

Cao, K., Peng, Z., Zhao, X., Li, Y., Liu, K., Arus, P., Fang, W., Chen, C., Wang, X., Wu, J., & Fei, Z. (2022). Chromosome-level genome assemblies of four wild peach species provide insights into genome evolution and genetic basis of stress resistance. *BMC Biology, 20*(1), 139.

Castrillo, M. (2006). Photosynthesis in three altitudinal populations of the Andean plant *Espeletia schultzii* (Compositae). *Revista de Biología Tropical, 54*(4), 1143–1149.

Chen, Z., & Gallie, D. R. (2006). Dehydroascorbate reductase affects leaf growth, development, and function. *Plant Physiology, 142*(2), 775–787.

Chen, L., Song, Y., Li, S., Zhang, L., Zou, C., & Yu, D. (2012). The role of WRKY transcription factors in plant abiotic stresses. *Biochimica et Biophysica Acta (BBA) – Gene Regulatory Mechanisms, 1819*(2), 120–128.

Cheng, Z., Hu, X., Lu, X., Fang, Q., Meng, Y., & Long, C. (2022a). Medicinal plants and fungi traditionally used by Dulong people in Northwest Yunnan, China. *Frontiers in Pharmacology, 13*, 895129.

Cheng, M., Cui, K., Zheng, M., Yang, T., Zheng, J., Li, X., Luo, X., Zhou, Y., Zhang, R., Yan, D., & Yao, M. (2022b). Physiological attributes and transcriptomics analyses reveal the mechanism response of Helictotrichon virescens to low temperature stress. *BMC Genomics, 23*(1), 280.

Choudhury, F. K., Rivero, R. M., Blumwald, E., & Mittler, R. (2017). Reactive oxygen species, abiotic stress and stress combination. *The Plant Journal, 90*(5), 856–867.

Cui, G., Wei, X., Degen, A. A., Wei, X., Zhou, J., Ding, L., Shang, Z., Liu, S., & Long, R. (2016). Trolox-equivalent antioxidant capacity and composition of five alpine plant species growing at different elevations on the Qinghai–Tibetan Plateau. *Plant Ecology & Diversity, 9*(4), 387–396.

Czarnocka, W., & Karpiński, S. (2018). Friend or foe? Reactive oxygen species production, scavenging and signaling in plant response to environmental stresses. *Free Radical Biology and Medicine, 122*, 4–20.

Das, K., & Roychoudhury, A. (2014). Reactive oxygen species (ROS) and response of antioxidants as ROS-scavengers during environmental stress in plants. *Frontiers in Environmental Science, 2*, 53.

de Pinto, M. C., & De Gara, L. (2004). Changes in the ascorbate metabolism of apoplastic and symplastic spaces are associated with cell differentiation. *Journal of Experimental Botany, 55*(408), 2559–2569.

Dickinson, B. C., & Chang, C. J. (2011). Chemistry and biology of reactive oxygen species in signaling or stress responses. *Nature Chemical Biology, 7*(8), 504–511.

Ding, Y., Shi, Y., & Yang, S. (2020). Molecular regulation of plant responses to environmental temperatures. *Molecular Plant, 13*(4), 544–564.

Dumanović, J., Nepovimova, E., Natić, M., Kuča, K., & Jaćević, V. (2021). The significance of reactive oxygen species and antioxidant defense system in plants: A concise overview. *Frontiers in Plant Science, 11*, 552969.

Evans, M. D., Dizdaroglu, M., & Cooke, M. S. (2004). Oxidative DNA damage and disease: Induction, repair and significance. *Mutation Research/Reviews in Mutation Research, 567*(1), 1–61.
Farmer, E. E., & Mueller, M. J. (2013). ROS-mediated lipid peroxidation and RES-activated signaling. *Annual Review of Plant Biology, 64*(1), 429–450.
Fini, A., Brunetti, C., Di Ferdinando, M., Ferrini, F., & Tattini, M. (2011). Stress-induced flavonoid biosynthesis and the antioxidant machinery of plants. *Plant Signaling & Behavior, 6*(5), 709–711.
Fu, Y., Ma, H., Chen, S., Gu, T., & Gong, J. (2018). Control of proline accumulation under drought via a novel pathway comprising the histone methylase CAU1 and the transcription factor ANAC055. *Journal of Experimental Botany, 69*(3), 579–588.
Gill, S. S., & Tuteja, N. (2010). Reactive oxygen species and antioxidant machinery in abiotic stress tolerance in crop plants. *Plant Physiology and Biochemistry, 48*(12), 909–930.
Gong, F., Meng, J., Xu, H., & Zhou, X. (2024). The molecular mechanism regulating flavonoid production in *Rhododendron chrysanthum* Pall. against UV-B damage is mediated by RcTRP5. *International Journal of Molecular Sciences, 25*(24), 13383.
Halliwell, B. (2006). Reactive species and antioxidants. Redox biology is a fundamental theme of aerobic life. *Plant Physiology, 141*(2), 312–322.
Hasanuzzaman, M., Bhuyan, M. B., Anee, T. I., Parvin, K., Nahar, K., Mahmud, J. A., & Fujita, M. (2019). Regulation of ascorbate-glutathione pathway in mitigating oxidative damage in plants under abiotic stress. *Antioxidants, 8*(9), 384.
Hashim, A. M., Alharbi, B. M., Abdulmajeed, A. M., Elkelish, A., Hozzein, W. N., & Hassan, H. M. (2020). Oxidative stress responses of some endemic plants to high altitudes by intensifying antioxidants and secondary metabolites content. *Plants, 9*(7), 869.
Hirayama, T., & Shinozaki, K. (2010). Research on plant abiotic stress responses in the post-genome era: Past, present and future. *The Plant Journal, 61*(6), 1041–1052.
Hu, C. H., Wang, P. Q., Zhang, P. P., Nie, X. M., Li, B. B., Tai, L., et al. (2020). NADPH oxidases: The vital performers and center hubs during plant growth and signaling. *Cells, 9*(2), 437.
Ibrahim, I. A., Jabbour, A. A., Abdulmajeed, A. M., Elhady, M. E., Almaroai, Y. A., & Hashim, A. M. (2022). Adaptive responses of four medicinal plants to high altitude oxidative stresses through the regulation of antioxidants and secondary metabolites. *Agronomy, 12*(12), 3032.
Igamberdiev, A. U., Seregelyes, C., Manac'h, N., & Hill, R. D. (2004). NADH-dependent metabolism of nitric oxide in alfalfa root cultures expressing barley hemoglobin. *Planta, 219*, 95–102.
Jeelani, S. M., Rather, G. A., Sharma, A., & Lattoo, S. K. (2018). In perspective: Potential medicinal plant resources of Kashmir Himalayas, their domestication and cultivation for commercial exploitation. *Journal of Applied Research on Medicinal and Aromatic Plants, 8*, 10–25.
Joshi, S., & Joshi, R. (2025). Quantifying redox signalling regulatory transcriptional dynamics in *Nardostachys jatamansi* under abiotic stress response. *Biochimica et Biophysica Acta (BBA) – General Subjects, 1869*(6), 130788.
Joshi, R. K., Satyal, P., & Setzer, W. N. (2016). Himalayan aromatic medicinal plants: A review of their ethnopharmacology, volatile phytochemistry, and biological activities. *Medicine, 3*(1), 6.
Kavi Kishor, P. B., Suravajhala, P., Rathnagiri, P., & Sreenivasulu, N. (2022). Intriguing role of proline in redox potential conferring high temperature stress tolerance. *Frontiers in Plant Science, 13*, 867531.
Kawano, T. (2003). Roles of the reactive oxygen species-generating peroxidase reactions in plant defense and growth induction. *Plant Cell Reports, 21*, 829–837.
Kiffin, R., Bandyopadhyay, U., & Cuervo, A. M. (2006). Oxidative stress and autophagy. *Antioxidants & Redox Signaling, 8*(1–2), 152–162.
Kumar, A., Dutt, S., Bagler, G., Ahuja, P. S., & Kumar, S. (2012). Engineering a thermo-stable superoxide dismutase functional at sub-zero to >50 °C, which also tolerates autoclaving. *Scientific Reports, 2*(1), 387.
Kumar, A., Guleria, S., Ghosh, D., Dogra, V., & Kumar, S. (2021a). Managing reactive oxygen species—Some learnings from high altitude extremophytes. *Environmental and Experimental Botany, 189*, 104525.

Kumar, M., Kumar Patel, M., Kumar, N., Bajpai, A. B., & Siddique, K. H. (2021b). Metabolomics and molecular approaches reveal drought stress tolerance in plants. *International Journal of Molecular Sciences, 22*(17), 9108.

Kumari, M., Joshi, R., & Kumar, R. (2020). Metabolic signatures provide novel insights to *Picrorhiza kurroa* adaptation along the altitude in Himalayan region. *Metabolomics, 16*, 1.

Kumari, R., Chopra, S., Thakur, N., Rana, M., Thakur, P., Raina, K., Anand, V., Sharma, R., & Chaudhary, A. (2024). An insight into quantitative, qualitative, and analytical methods for the measurement of antioxidant activity through various assays. *Vegetos, 38*, 1–11.

Kumari, R., Thakur, P., Sharma, V., & Chaudhary, A. (2025). Exploring two medicinally important *Selinum* species: bridging traditional knowledge with modern pharmacological insights. Drug Metabolism and Personalized Therapy, 40(4), 199–215.

Lämke, J., & Bäurle, I. (2017). Epigenetic and chromatin-based mechanisms in environmental stress adaptation and stress memory in plants. *Genome Biology, 18*, 124.

Landi, M., Tattini, M., & Gould, K. S. (2015). Multiple functional roles of anthocyanins in plant-environment interactions. *Environmental and Experimental Botany, 119*, 4–17.

Li, H., Yao, W., Fu, Y., Li, S., & Guo, Q. (2015). De novo assembly and discovery of genes that are involved in drought tolerance in Tibetan *Sophora moorcroftiana. PLoS One, 10*(1), e111054.

Li, J., Jiang, M. M., Ren, L., Liu, Y., & Chen, H. Y. (2016). Identification and characterization of CBL and CIPK gene families in eggplant (*Solanum melongena* L.). *Molecular Genetics and Genomics, 291*, 1769–1781.

Li, H., Zhang, Y., Guo, Q., & Yao, W. (2017). Molecular characterisation of a DREB gene from *Sophora moorcroftiana*, an endemic species of plateau. *Protoplasma, 254*, 1735–1741.

Li, W., Pang, S., Lu, Z., & Jin, B. (2020). Function and mechanism of WRKY transcription factors in abiotic stress responses of plants. *Plants, 9*(11), 1515.

Liu, Y., Dao, Z., Yang, C., Liu, Y., & Long, C. (2009). Medicinal plants used by Tibetans in Shangri-la, Yunnan, China. *Journal of Ethnobiology and Ethnomedicine, 5*, 1.

Luan, S. (2009). The CBL–CIPK network in plant calcium signaling. *Trends in Plant Science, 14*(1), 37–42.

Ma, L., Sun, X., Kong, X., Galvan, J. V., Li, X., Yang, S., et al. (2015). Physiological, biochemical and proteomics analysis reveals the adaptation strategies of the alpine plant *Potentilla saundersiana* at altitude gradient of the Northwestern Tibetan Plateau. *Journal of Proteomics, 112*, 63–82.

Mala, D., Awasthi, S., Sharma, N. K., Swarnkar, M. K., Shankar, R., & Kumar, S. (2021). Comparative transcriptome analysis of *Rheum australe*, an endangered medicinal herb, growing in its natural habitat and those grown in controlled growth chambers. *Scientific Reports, 11*(1), 3702.

Mansoor, S., Ali Wani, O., Lone, J. K., Manhas, S., Kour, N., Alam, P., et al. (2022). Reactive oxygen species in plants: From source to sink. *Antioxidants, 11*(2), 225.

Maraldi, T. (2013). Natural compounds as modulators of NADPH oxidases. *Oxidative Medicine and Cellular Longevity, 2013*, 271602.

Market Data Forecast. (2025). *Herbal medicine market – Size, share, trends, COVID-19 impact & forecast (2024–2029)*. Available at: https://www.marketdataforecast.com/market-reports/herbal-medicine-market. Accessed: 06 June 2025.

Mehta, P., Bisht, K., & Sekar, K. C. (2021). Diversity of threatened medicinal plants of Indian Himalayan region. *Plant Biosystems – An International Journal Dealing with all Aspects of Plant Biology, 155*(6), 1121–1132.

Mhamdi, A., & Van Breusegem, F. (2018). Reactive oxygen species in plant development. *Development (Cambridge, England), 145*(15), dev164376.

Mhamdi, A., Queval, G., Chaouch, S., Vanderauwera, S., Van Breusegem, F., & Noctor, G. (2010). Catalase function in plants: A focus on Arabidopsis mutants as stress-mimic models. *Journal of Experimental Botany, 61*(15), 4197–4220.

Mittler, R. (2002). Oxidative stress, antioxidants and stress tolerance. *Trends in Plant Science, 7*(9), 405–410.

Møller, I. M., Jensen, P. E., & Hansson, A. (2007). Oxidative modifications to cellular components in plants. *Annual Review of Plant Biology, 58*(1), 459–481.

Mullineaux, P. M., & Rausch, T. (2005). Glutathione, photosynthesis and the redox regulation of stress-responsive gene expression. *Photosynthesis Research, 86*, 459–474.

Mykhailenko, O., Jalil, B., McGaw, L. J., Echeverría, J., Takubessi, M., & Heinrich, M. (2025). Climate change and the sustainable use of medicinal plants: A call for "new" research strategies. *Frontiers in Pharmacology, 15*, 1496792.

Noctor, G., Veljovic-Jovanovic, S., Driscoll, S., Novitskaya, L., & Foyer, C. H. (2002). Drought and oxidative load in the leaves of C3 plants: A predominant role for photorespiration? *Annals of Botany, 89*(7), 841–850.

Panda, S. K. (2012). Assay guided comparison for enzymatic and non-enzymatic antioxidant activities with special reference to medicinal plants. *Antioxidant Enzyme, 14*(1), 382–400.

Petry, A., & Görlach, A. (2021). NADPH-Oxidasen und HIF: zentrale Elemente der Redoxhomöostase. *BIOspektrum, 27*, 142–145.

Pinto, E., Sigaud-Kutner, T. C., Leitao, M. A., Okamoto, O. K., Morse, D., & Colepicolo, P. (2003). Heavy metal–induced oxidative stress in algae. *Journal of Phycology, 39*(6), 1008–1018.

Raina, K., Kumari, R., Thakur, P., Sharma, R., Singh, R., Thakur, A., Anand, V., Sharma, R., & Chaudhary, A. (2023). Mechanistic role and potential of Ayurvedic herbs as anti-aging therapies. Drug Metabolism and Personalized Therapy, 38(3), 211–226. https://doi.org/10.1515/dmpt-2023-0024

Rana, P. S., & Saklani, P. (2019). Analyzing effect of altitudinal variation in enzymatic antioxidants of *Coleus forskohlii* from Uttarakhand, India. *bioRxiv*, 662528.

Richards, S. L., Laohavisit, A., Mortimer, J. C., Shabala, L., Swarbreck, S. M., Shabala, S., & Davies, J. M. (2014). Annexin 1 regulates the H_2O_2-induced calcium signature in *Arabidopsis thaliana* roots. *The Plant Journal, 77*(1), 136–145.

Roach, T., Neuner, G., Kranner, I., & Buchner, O. (2023). Heat acclimation under drought stress induces antioxidant enzyme activity in the alpine plant *Primula minima*. *Antioxidants, 12*(5), 1093.

Satyakam, Z., Singh, G., Singh, R. K., & Kumar, R. (2022). Cold adaptation strategies in plants—An emerging role of epigenetics and antifreeze proteins to engineer cold resilient plants. *Frontiers in Genetics, 13*, 909007.

Shao, H. B., Liang, Z. S., Shao, M. A., & Sun, Q. (2005). Dynamic changes of anti-oxidative enzymes of 10 wheat genotypes at soil water deficits. *Colloids and Surfaces B: Biointerfaces, 42*(3–4), 187–195.

Sharma, P., & Dubey, R. S. (2004). Ascorbate peroxidase from rice seedlings: Properties of enzyme isoforms, effects of stresses and protective roles of osmolytes. *Plant Science, 167*(3), 541–550.

Sharma, P., Jha, A. B., Dubey, R. S., & Pessarakli, M. (2021). Reactive oxygen species generation, hazards, and defense mechanisms in plants under environmental (abiotic and biotic) stress conditions. In *Handbook of plant and crop physiology* (pp. 617–658). CRC Press.

Shi, Y., Ding, Y., & Yang, S. (2018). Molecular regulation of CBF signaling in cold acclimation. *Trends in Plant Science, 23*(7), 623–637.

Shrestha, U. B., Lamsal, P., Ghimire, S. K., Shrestha, B. B., Dhakal, S., Shrestha, S., & Atreya, K. (2022). Climate change-induced distributional change of medicinal and aromatic plants in the Nepal Himalaya. *Ecology and Evolution, 12*(8), e9204.

Sickel, H., Bilger, W., & Ohlson, M. (2012). High levels of α-tocopherol in Norwegian alpine grazing plants. *Journal of Agricultural and Food Chemistry, 60*(31), 7573–7580.

Smirnoff, N. (2000). Ascorbic acid: Metabolism and functions of a multi-facetted molecule. *Current Opinion in Plant Biology, 3*(3), 229–235.

Sun, P., Hao, R., Fan, F., Wang, Y., & Zhu, F. (2025). Adaptation of high-altitude plants to plateau abiotic stresses: A case study of the Qinghai-Tibet plateau. *International Journal of Molecular Sciences, 26*(5), 2292.

Tang, X., Li, J., Liu, L., Jing, H., Zuo, W., & Zeng, Y. (2022). Transcriptome analysis provides insights into *Potentilla bifurca* adaptation to high altitude. *Life, 12*(9), 1337.

Terletskaya, N. V., Turzhanova, A. S., Khapilina, O. N., Zhumagul, M. Z., Meduntseva, N. D., Kudrina, N. O., & Kalendar, R. (2023). Genetic diversity in natural populations of *Rhodiola* species of different adaptation strategies. *Genes, 14*(4), 794.

Terletskaya, N. V., Erbay, M., Mamirova, A., Ashimuly, K., Korbozova, N. K., Zorbekova, A. N., & Hoffmann, M. H. (2024). Altitude-dependent morphophysiological, anatomical, and metabolomic adaptations in *Rhodiola linearifolia* Boriss. *Plants, 13*(19), 2698.

Thakur, A., Kumari, R., Kumar, A., & Chaudhary, A. (2024). Quantitative ethnobotanical study of medicinal plants used by native people of selected areas of Chhota Bhangal, Himachal Pradesh. *Ethnobotany Research and Applications, 28*, 1–34.

Thakur, P., Kumari, R., Kumar, A., Bhatia, A., Sharma, U., & Chaudhary, A. (2025). Uncovering the Mechanisms of Angelica glauca Edgew. In Breast Cancer: A Combined In Vitro and In Silico Approach. Chemistry & Biodiversity, e202402554.

Thapliyal, N., Bhojak, P., Arya, D., & Sekar, K. C. (2024). Ethnobotanical assessment of medicinal plants used in alpine regions of Uttarakhand, West Himalaya. *Ethnobotany Research and Applications, 27*, 1–24.

Treml, J., & Šmejkal, K. (2016). Flavonoids as potent scavengers of hydroxyl radicals. *Comprehensive Reviews in Food Science and Food Safety, 15*(4), 720–738.

Ushio-Fukai, M. (2001). Die NADPH-Oxidase im Gefäßsystem. In G. Kojda (Ed.), *Pentaerithrityltetranitrat*. Heidelberg. https://doi.org/10.1007/978-3-642-87803-9_6

Van Breusegem, F., & Dat, J. F. (2006). Reactive oxygen species in plant cell death. *Plant Physiology, 141*(2), 384–390.

Verbruggen, N., & Hermans, C. (2008). Proline accumulation in plants: A review. *Amino Acids, 35*, 753–759.

Wang, Y., He, W., Huang, H., An, L., Wang, D., & Zhang, F. (2009). Antioxidative responses to different altitudes in leaves of alpine plant *Polygonum viviparum* in summer. *Acta Physiologiae Plantarum, 31*, 839–848.

Wang, L., Zhao, S., Li, J., & Zhou, G. (2024). Metabolomics analysis reveals the metabolite profiles of *Rheum tanguticum* grown under different altitudinal gradients. *BMC Plant Biology, 24*(1), 226.

Wang, L., Wang, H., Chen, J., Lamu, Y., Qi, X., Lei, L., & Tso, S. (2025). Analysis of metabolic differences in Tibetan medicinal plant *Phlomoides rotata* leaves in different habitats based on non-targeted metabolomics. *Frontiers in Plant Science, 16*, 1503218.

Wei, N., Ci, D., Dai, Y., et al. (2024). Comparative metabolomics revealed the effect of high altitude on the metabolic pattern of *Polygonum viviparum* L. *Preprint*, Research Square, [online] Available at: https://doi.org/10.21203/rs.3.rs-4156978/v1. Accessed 27 June 2025.

Wildi, B., & Lütz, C. (1996). Antioxidant composition of selected high alpine plant species from different altitudes. *Plant, Cell & Environment, 19*(2), 138–146.

Wu, X., Zhong, L., Chen, G., Zhong, S., & Gu, R. (2024). Morphological and physiological plasticity of alpine medicinal plants along an elevational gradient. *Journal of Applied Research on Medicinal and Aromatic Plants, 44*, 100613.

Yao, Y., Yang, Y., Li, Y., & Lutts, S. (2008). Intraspecific responses of *Fagopyrum esculentum* to enhanced ultraviolet B radiation. *Plant Growth Regulation, 56*, 297–306.

Ye, H., Liu, H., Han, M., Zhang, N., Feng, X., Gao, T., & Wang, Y. (2024). The genome-wide identification, characterization, and expression profiles of the NADPH oxidase (NOX) gene family under drought and salt stress in *Opisthopappus taihangensis* (Asteraceae). *Agronomy, 14*(4), 653.

Zamani, S., Fathi, M., Ebadi, M. T., & Máthé, Á. (2025). Global trade of medicinal and aromatic plants: A review. *Journal of Agriculture and Food Research, 21*, 101910.

Zargoosh, Z., Ghavam, M., Bacchetta, G., & Tavili, A. (2019). Effects of ecological factors on the antioxidant potential and total phenol content of *Scrophularia striata* Boiss. *Scientific Reports, 9*(1), 16021.

Zhang, X., Zhang, Y., Li, M., Jia, H., Wei, F., Xia, Z., & Wang, Z. (2024). Overexpression of the WRKY transcription factor gene *NtWRKY65* enhances salt tolerance in tobacco (*Nicotiana tabacum*). *BMC Plant Biology, 24*(1), 326.

Zhang, X., Zhang, D., Zhong, C., Li, W., Dinesh-Kumar, S. P., & Zhang, Y. (2025a). Orchestrating ROS regulation: Coordinated post-translational modification switches in NADPH oxidases. *New Phytologist, 245*(2), 510–522.

Zhang, S., Wang, S., Wang, S., Su, H., & De, J. (2025b). Genetic diversity and metabolic profile of Tibetan medicinal plant *Saussurea obvallata*. *Genes, 16*(5), 593.

Zhao, Y., Xu, F., Liu, J., Guan, F., Quan, H., & Meng, F. (2019). The adaptation strategies of *Herpetospermum pedunculosum* (Ser.) Baill at altitude gradient of the Tibetan plateau by physiological and metabolomic methods. *BMC Genomics, 20*, 1–15.

Zhou, Y., Sun, X., Yang, Y., Li, X., Cheng, Y., & Yang, Y. (2016). Expression of *Stipa purpurea SpCIPK26* in *Arabidopsis thaliana* enhances salt and drought tolerance and regulates abscisic acid signaling. *International Journal of Molecular Sciences, 17*(6), 966.

Zhu, H., Liu, C., & Qian, H. (2022). Pharmaceutical potential of high-altitude plants for fatigue-related disorders: A review. *Plants, 11*(15), 2004.

Advancements in Tissue Culture Techniques for Alpine Medicinal Plants

12

Shagun Sharma, Deepika, Pankaj Kumar, Hemant Sood, Ashish R. Warghat, and Rohit Sharma

Abstract

Alpine plants represent a unique group of high-altitude flora that thrive under extreme environmental conditions categorized by low temperature, high ultraviolet radiation, reduced atmospheric pressure and short growing seasons. These adaptive species produce a diverse array of bioactive secondary metabolites with significant pharmaceutical value, supporting traditional and modern medicine systems. However, habitat loss, climate change and unsustainable harvesting have placed many Himalayan alpine species under severe threat, making conservation and sustainable utilization an urgent priority. Conventional propagation approaches are often constrained by seed dormancy, low germination rates and recalcitrant tissues, necessitating alternative biotechnological interventions. In vitro techniques including micropropagation, callus-mediated regeneration,

Authors Shagun Sharma and Pankaj Kumar have contributed equally to this work and share first authorship.

S. Sharma · Deepika · P. Kumar (✉)
Department of Biotechnology, College of Horticulture, Dr. Yashwant Singh Parmar University of Horticulture and Forestry, Solan, Himachal Pradesh, India

H. Sood
Department of Biotechnology & Bioinformatics, Jaypee University of Information Technology, Solan, Himachal Pradesh, India

A. R. Warghat
Biotechnology Division, CSIR-Institute of Himalayan Bioresource Technology, Palampur, Himachal Pradesh, India

Academy of Scientific and Innovative Research, Ghaziabad, India

R. Sharma
Department of Forest Products, College of Forestry, Dr. Yashwant Singh Parmar University of Horticulture and Forestry, Solan, Himachal Pradesh, India

P. Kumar et al. (eds.), *Climate Resilience and Molecular Adaptation in Alpine Medicinal Plants*, https://doi.org/10.1007/978-981-95-6923-6_12

somatic-embryogenesis, cryopreservation and production of synthetic seeds, have emerged as reliable strategies for mass multiplication, germplasm conservation and metabolite enhancement of critically endangered alpine taxa. This chapter includes advancements made in key Himalayan alpine plants such as *Aconitum* spp., *Angelica* spp., *Arnebia* spp., *Fritillaria* spp., *Picrorhiza* spp., *Podophyllum* spp., *Rhodiola* spp., *Saussurea* spp., *Swertia* spp., *Trillium* spp. and *Valeriana* spp. The evaluation of tissue culture studies highlights optimization of explant choice, hormonal combinations, regeneration pathways and conservation protocols. The compiled knowledge not only underscores the potential of plant tissue culture for overcoming recalcitrance and ensuring ex situ conservation but also provides a framework for future research on biotechnological interventions to safeguard alpine biodiversity and promote sustainable utilization of these valuable resources.

Keywords

Alpine · Conservation · Cryopreservation · Micropropagation · Synthetic seed

1 Introduction

The Himalayan alpine region, lying above the tree line and just below the permanent snow, represents one of the most unique and fragile ecosystems on Earth (Leng et al., 2023). Within this zone, the Indian Himalayan Region (IHR) supports an extraordinary diversity of flora, remarkable for its richness, high endemism and ecological significance (Salaria et al., 2025). Typically found between 3800 and 5000 m above mean sea level, alpine plants have evolved remarkable adaptations such as dwarf growth, small thick leaves, protective trichomes and synthesis of secondary metabolites that enable them to withstand harsh conditions (Sharma et al., 2025a). These conditions involve exposure of plants to extreme abiotic stressors like intense irradiance, hypoxic conditions, temperature fluctuations, reduced atmospheric pressure, nutrient poor soils and restricted water availability (Sharma et al., 2025b). To endure such stresses, alpine plants have evolved distinctive physiological and biochemical mechanisms, including efficient carbon assimilation, reduced photoinhibition, synthesis of UV-absorbing compounds, antifreeze lipids, detoxification of reactive oxygen species and regulation of membrane fluidity (Pandey et al., 2019; Sharma et al., 2025b). These mechanisms collectively ensure survival under conditions of cold stress, scarce carbon dioxide and elevated ultraviolet exposure (Kumari et al., 2025). Ecologically, the Himalaya host extraordinary alpine biodiversity, with 762 recorded taxa of which 373 (48.95%) are endemic. Endemic species frequently dominate as indicator taxa, making alpine habitats irreplaceable reservoirs of genetic diversity (Kumar & Khanduri, 2024). Local and indigenous communities have traditionally relied on alpine plants for a wide range of purposes including, medicinal remedies, livestock grazing, timber, dyes and spiritual practices. More than half of the alpine flora is directly used for medicinal, grazing and spiritual purposes (Chaudhary et al., 2017). Their pharmacological limit is remarkable: alpine species often synthesize rare and potent specialized metabolites such as alkaloids, flavonoids, terpenoids and glycosides (Singh & Singh, 2022). These compounds confer ecological resilience while offering therapeutic potential against gastrointestinal disorders, respiratory ailments, wounds, inflammation, cardiovascular diseases, fever, hypertension and stress-related conditions (Banik et al., 2020; Ashraf et al., 2024). Traditional systems of medicine (TSM), including Ayurveda, Unani, Siddha, Chinese and Tibetan medicine, have relied on these species for centuries (Pan et al., 2014; Adhikari & Paul, 2018). Modern pharmacological research has validated many of these applications, identifying alpine plants as valuable reservoirs for herbal medicine, nutraceuticals and drug discovery (Chaachouay & Zidane, 2024).

Although due to their ecological and medicinal value, alpine plants face multiple threats. Anthropogenic pressures, overharvesting and environmental change have accelerated their decline (Fig. 12.1) (Singh et al., 2021). Rapid extinction of these species not only disrupts fragile alpine ecosystems but also undermines the livelihoods and healthcare systems of local communities who depend on them (Chauhan et al., 2015; Ganie et al., 2019). Furthermore, climate warming ($\approx$+1.5 °C) and increased precipitation are driving species upslope, where many rare and endemic taxa risk being displaced by lower-elevation invaders (Fig. 12.1) (Liang et al., 2016). The decline of alpine plants therefore threatens biodiversity, ecosystem amenities, cultural heritage and future drug discovery potential. Conservation of alpine medicinal plants is thus of urgent importance (Grabherr, 2009). Preserving their wild populations safeguards unique genetic resources, stabilizes mountain ecosystems and supports climate-resilient livelihoods. Among the modern strategies, plant tissue culture has emerged as powerful tool for the conservation and propagation of endangered medicinal taxa (Oseni et al., 2018). It offers precise control over physical and chemical conditions, enabling rapid mass-propagation, year-round production of secondary metabolites and long-term germplasm conservation in limited space (Sharma & Thokchom, 2014; Sood, 2020). For species that are recalcitrant to conventional propagation methods, in vitro approaches have shown particular promise to regenerate these species (Hussain et al., 2012). Recent advancements in in vitro tissue culture have been recognized as vital strategies for safeguarding endangered alpine flora, offering reliable methods for ex situ conservation and sustainable utilization (Fig. 12.1) (Chandran et al., 2023; Majumder et al., 2023). Given these challenges and opportunities, this chapter compiles and critically examines advancements in tissue culture-based approaches for conserving endangered alpine medicinal plants. Particular emphasis is placed on protocols for callus induction, regeneration, organogenesis, somatic embryogenesis, cryopreservation and synthetic seed technology developed across different species. The chapter compiles Himalayan alpine plants; *Aconitum* spp., *Angelica* spp., *Arnebia* spp., *Fritillaria* spp., *Picrorhiza* spp., *Podophyllum* spp., *Rhodiola* spp., *Saussurea* spp., *Swertia* spp., *Trillium* spp. and *Valeriana* spp., highlighting progress, limitations and future prospects.

Fig. 12.1 Illustration depicting anthropogenic threats and biotechnological interventions for the conservation of alpine medicinal plants. (Icons used in the figure were sourced from Flaticon)

1.1 *Aconitum* spp.

Aconitum species, commonly known as monkshood or aconite, are perennial herbs in the Ranunculaceae family, thriving in alpine and subalpine meadows of the Himalayas and other high-altitude regions (Kumar et al., 2024a). These plants feature striking hooded flowers in shades of blue, purple or white, with tuberous roots adapted to cold, moist soils at elevations of 2000–4500 m (Kakkar et al., 2023). Medicinally, *Aconitum* species contain potent alkaloids like aconitine, used in TSM such as Ayurveda and Traditional Chinese Medicine (TCM) for treating pain, inflammation and cardiac issues (Talreja & Tiwari, 2023). Due to overharvesting for pharmaceutical and export purposes, habitat loss from climate change and grazing, many species like *A. heterophyllum* are vulnerable and critically endangered as per International Union for Conservation of Nature (IUCN) listings, prompting urgent conservation through in vitro propagation (Wani et al., 2022; Kakkar et al., 2023).

The initial work on *Aconitum heterophyllum* was carried out by Giri et al. (1993), who reported somatic-embryogenesis from callus-derived from in vitro raised leaf and petiole explants. Building on this, Giri et al. (1997) successfully established hairy root cultures of *A. heterophyllum* by using *Agrobacterium rhizogenes* strains (*A4, LBA 9360, LBA 9402*). The transformed roots exhibited a significantly higher alkaloid content (2.96%), nearly four times greater than that of non-transformed roots (0.79%) and thin layer chromatography confirmed the presence of hetidine, atisine and heteratisine. In *A. balfourii*, Pandey et al. (2004) described a detailed 5 step regeneration protocol involving (i) axillary bud culture and sprouting, (ii) callus induction using leaf segments, (iii) shoot initiation using callus, (iv) shoot multiplication (v) rooting and hardening. This was further refined by Pandey et al.

(2004), who developed a standardized regeneration protocol from leaf sections derived from in vitro raised axillary buds. Callus induction was optimized on Murashige and Skoog (MS) medium (Murashige & Skoog, 1962) with NAA (Naphthalene Acetic Acid) (5.5 mg/L), and BAP (6-Benzylaminopurine) (1.0 mg/L), shoot induction achieved on MS medium with BAP (2.0 mg/L) and NAA (0.3 mg/L) and maximum shoot proliferation was obtained on a lower concentration of BAP (0.25 mg/L). Rooting was most successful on MS medium supplemented using IBA (Indole-3-Butyric Acid) (2.5 mg/L) and hardened plants were successfully transferred to the green house conditions. Jabeen et al. (2006) established a callus-based protocol for shoot proliferation in *A. heterophyllum*. Friable callus formation was obtained on MS medium augmented with NAA (0.6 mg/L), although long-term maintenance was hindered by leaching of phenolic compounds.

Mahajan et al. (2015) later optimized hormonal combinations for mass shoot multiplication in *A. heterophyllum*, while Belwal et al. (2016) further improved the regeneration protocol by reducing growth regulator concentrations and successfully regenerating plantlets from shoot tips. Rana and Sreenivasulu (2013) addressed the problem of seed dormancy in *A. heterophyllum*. They found that ethanol (80%) treatment enhanced seed germination and proteomic analysis of treated versus untreated seeds revealed 40 differentially expressed proteins using two-dimensional (2-D) electrophoresis, providing insight into the mechanisms of dormancy breaking. More recently, Hadi et al. (2023) optimized in vitro germination and regeneration in *A. violaceum*. Germination reached 51.84% on MS medium containing TDZ (Thidiazuron) (1.0 mg/L), and sucrose (30 g/L). Callus induction from nodal explants achieved 58.88% on MS medium with 2,4-D (2,4-Dichlorophenoxyacetic acid) (1.0 mg/L) and Kn (kinetin) (0.5 mg/L). Further, multiple shoot proliferation was obtained with a 70.66% culture response and an average of five shoots/explant on MS medium supplemented with kinetin (0.1 mg/L) and BAP (0.5 mg/L). Rooting efficiency reached about 73% upon sub-culturing shoots on MS medium containing Kn (0.1 mg/L), and IAA (Indole-3-Acetic Acid) (0.5 mg/L). Synthetic seed technology was also explored, with complete bead formation achieved using 3.0% sodium alginate and 100 mM $CaCl_2$ alginate-encapsulated micro-shoots showed 44% germination on MS medium enriched with IAA (0.5 mg/L) and Kn (1.0 mg/L). Concusively, significant progress has been made in emerging in vitro plant regeneration and propagation protocols for various *Aconitum* species. The studies highlighted show a clear progression, from early work on *A. heterophyllum* to more recent advancements in *A. violaceum*, addressing key challenges like seed dormancy and optimizing hormonal combinations for enhanced growth and regeneration (Table 12.1). These cumulative efforts have led to a greater understanding of the factors influencing plant tissue culture in *Aconitum*, paving the way for more efficient conservation and mass production of these species.

1.2 *Angelica glauca*

Angelica species, part of the Apiaceae family, are robust biennial or perennial herbs found in alpine and subalpine zones of the Himalayas (Khoja et al., 2024). They grow to 1–3 m tall with compound leaves and umbels of white or greenish flowers, preferring moist meadows and damp forests at 2000–4000 m elevation (Sharma & Sharma, 2024). In traditional medicine, species like *A. archangelica* and *A. glauca* are valued for their roots and seeds containing coumarins and essential oils, used to treat digestive issues, menstrual disorders and respiratory ailments in Ayurveda, TCM and European herbalism (Sarker & Nahar, 2004). Overharvesting, habitat fragmentation due to agriculture and climate-induced shifts have led to endangered status for some species, such as *A. glauca*, classified as vulnerable, necessitating cultivation and protected reserves for conservation (Jan & Abbas, 2018).

Pandey et al. (2011) described a reproducible protocol in *A. glauca* for somatic-embryogenesis and plant regeneration using epicotyls, hypocotyls, and cotyledonary nodes as explants. Callus induction frequency reached 95.80% when epicotyl explants were cultured on MS medium enriched with 2,4-D (0.9 mg/L). The highest inducted embryogenic callus was obtained from epicotyls cultured on MS medium containing BAP (0.5 mg/L) and NAA (0.4 mg/L). Plant regeneration was achieved when somatic embryos were transferred to half strength MS medium without growth regulators, with a survival rate of 85% after acclimatization. Furthermore, subculturing primary embryos on MS medium with BAP (0.5 mg/L) and NAA (0.4 mg/L) induced secondary embryos, thereby providing a rapid approach for large-scale propagation. Huang et al. (2015) and Zhang et al. (2019), working with *Angelica sinensis*, further confirmed that callus growth rate is strongly influenced by medium composition, pH, sucrose concentration, inoculum density and photoperiod, emphasizing the importance of fine-tuned culture conditions across *Angelica* species. Recent work by Huang et al. (2023) highlighted the influence of inoculum density, temperature and photoperiod on growth and metabolite accumulation in *A. glauca*. Maximum growth rate, accumulation of bioactive metabolites (ferulic acid, soluble sugars, flavonoids and phenolics) and antioxidant activity were observed at 9–12 days when cultures were initiated with an inoculum density of 40 g/L, incubated at 22 °C, with a 16/8 h light and dark cycle. Callus induction rates varied with hormonal combinations, aseptic seedling roots responded best (99%) to IBA (2.0 mg/L), and Kn (0.2 mg/L), while field seedling roots showed the highest induction (97%) on 2,4-D (1.5 mg/L) + NAA (0.4 mg/L) + BAP (0.4 mg/L).

Report on somatic embryogenesis and artificial seed production in *A. glauca* was provided by Bisht et al. (2015). Using mature leaf explant derived from newly sprouted rhizomes, 86% callus proliferation was achieved on MS medium containing 2,4-D (3.0 μM). Subculturing calli into media containing NAA and BA, either alone or in combination, promoted differentiation into globular structures. A combination of NAA (2.0 μM) and BAP (2.0 μM) produced an average of 2.2 shoots/culture within 8 weeks. For acclimatization, plantlets transferred to a sand:soil:peat moss (1:1:1) mixture exhibited a survival rate of 75%. Additionally, somatic embryos encapsulated in sodium alginate (~3.0%) and $Ca(NO_3)_2$ (calcium nitrate

100 μM) for 30 min successfully formed synthetic seeds, demonstrating the potential of this technology for germplasm conservation. Conclusively, the tissue culture advancements for *Angelica glauca* have established effective protocols for its micropropagation and conservation (Table 12.1). The successful implementation of techniques like secondary somatic embryogenesis and synthetic seed technology provides powerful tools for large-scale propagation and safeguarding its genetic diversity, offering a promising solution to mitigate its endangered status.

1.3 *Arnebia* spp.

Arnebia species, from the Boraginaceae family, are perennial herbs native to alpine and subalpine regions of the Himalayas and Central Asia, growing at 3000–4500 m in rocky slopes and sandy soils (Ambrish & Srivastava, 2014). Known for their bright red roots and tubular flowers, they adapt to arid, cold environments with hairy foliage to reduce water loss. Medicinally, species like *A. benthamii* and *A. euchroma* yield shikonin derivatives used in Ayurveda and TCM for wound healing, antimicrobial effects and anti-inflammatory treatments, with roots prized for dyes and ointments (Dulta et al., 2023). Intensive harvesting for pharmaceuticals and cosmetics, combined with habitat degradation and climate change, has rendered many critically endangered, such as *A. benthamii*, urging in vitro propagation and sustainable farming initiatives.

Studies by Manjkhola et al. (2005) demonstrated the simultaneous occurrence of direct regenration and somatic embryogenesis in *Arnebia euchroma* from leaf derived callus, a finding not observed in root-derived callus which only produced roots. The most successful organogenesis, yielding an average of 12.2 shoots/culture, was achieved with 1.0 mM IBA and 2.5 mM BAP. For somatic embryogenesis, maximum number of embryos (16.3/culture) was induced with a different balance of IBA and BAP 2.5 mM each. The shoots were subjected to rooting efficiently in half strength MS medium containing 2.0 mM IBA and encapsulated embryos showed a promising 60.6% germination rate. The plantlets were successfully transferred to a soil mixture with a 72% survival rate. For *A. benthamii*, Quadri et al. (2012) developed a propagation system from shoot tip explants, finding that a half strength MS medium with a specific concentration of BAP and IBA resulted in maximum shoot multiplication. After inducing rooting with various auxins, 80% of plantlets were effectively acclimatized. Another significant protocol was established by Jiang et al. (2005), who used the cytokinin TDZ to induce shoot organogenesis in *A. euchroma* from cotyledon and hypocotyl explants, leading to a high yield of shoots. This research also found that adding polyvinylpyrrolidone to the medium was effective in reducing the browning of the plant's basal portion. Pal and Chaudhury (2010) achieved direct plant regeneration from shoot tips and nodal segments, which led to the formation of multiple shoot clusters and also in vitro flowering when the plantlets were maintained for an extended period. Also, Malik et al. (2016) developed a highly efficient and economical method using liquid culture media for *A. euchroma*. They observed that a liquid shake flask culture with TDZ

Table 12.1 Plant tissue culture interventions for propagation and conservation of Himalayan alpine medicinal plants

Sr. No.	Plant species	Explant used	Technique	Key outcome	Reference
1	*Aconitum* spp.	Leaf, petiole, nodal explants, seeds, nodal explants	Somatic embryogenesis, callus culture, hairy roots, seed dormancy breaking, germination, callus induction	Somatic embryos induction and hairy roots with high alkaloids. Regeneration protocol development with shoot regeneration, rooting, synthetic seed production	Giri et al. (1993, 1997), Pandey et al. (2004), and Jabeen et al. (2006)
2	*Angelica* spp.	Nodal segments, petioles	Callus induction, shoot regeneration	High frequency shoot induction protocol on MS medium containing BAP and rooting optimized with IBA	Huang et al. (2015), and Zhang et al. (2019)
3	*Arnebia* spp.	Leaf, petiole explants	Somatic embryogenesis, synthetic seeds	Somatic embryos induced from callus; encapsulated beads showed successful regeneration potential on MS medium	Guo-feng et al. (1992), Manjkhola et al. (2005), and Pal and Chaudhury (2010)
4	*Fritillaria* spp.	Bulb scales, nodal explants	Bulblet production, callus culture, cryopreservation	Efficient callus induction from bulb scales explants were achieved. LC/MS based metabolite quantification was done and potential for ex situ conservation induced using cryopreservation	Joshi et al. (2007), Kumar et al. (2020), and Tao et al. (2024)
5	*Picrorhiza* spp.	Rhizome, leaf explants	Shoot organogenesis, rooting, picrocide biosynthesis, synthetic seeds	Direct shoots from rhizome explants were achieved; Picrocide accumulation of *in vitro* regenerated plantlets, synthetic seed technology explored	Sharma and Sharma (2003), Mahajan et al. (2016), and Sharma et al. (2016a, b)

(continued)

Table 12.1 (continued)

Sr. No.	Plant species	Explant used	Technique	Key outcome	Reference
6	*Podophyllum* spp.	Rhizome buds, seeds	Callus induction, somatic embryogenesis, specialized metabolite	Embryogenic callus development achieved with in vitro propagation supports conservation efforts	Nadeem et al. (2000), Chattopadhyay et al. (2002), and Sultan et al. (2006)
7	*Rhodiola* spp.	Shoot tips, leaf explants, callus	Organogenesis, cryopreservation, conservation	Shoot multiplication achieved; cryopreservation protocols standardized for callus for conservation	Zych et al. (2005), Jian-feng et al. (2007), and Matvieieva et al. (2023)
8	*Saussurea* spp.	Leaf, petiole explants, bud, apical shoot	Callus induction, somatic embryogenesis, synthetic seed	Somatic embryos reported; production of syringin and rutin and synthetic seed production	Kuo et al. (2015), Sharma et al. (2019), and Mehreen et al. (2020)
9	*Swertia* spp.	Nodal, leaf explants, seedlings	Direct and indirect organogenesis, shoot multiplication	In vitro flowering was achieved; axillary multiplication was done and rooting induced using NAA and IBA	Joshi et al. (2007), Chaudhuri et al. (2008), and Sharma et al. (2014)
10	*Trillium* spp.	Mini rhizome, nodal explants	Callus induction, shoot organogenesis	Protocols for shoot regeneration from rhizomes and nodal segments were successfully established	Gagliardo et al. (2012), Chauhan et al. (2020), and Chandola et al. (2023)
11	*Valeriana* spp.	Nodal, leaf, rhizome explants	Synthetic seeds, callus induction, organogenesis, cryopreservation	Somatic embryogenesis and shoot organogenesis achieved, root induction and enhancement of bioactive compounds, synthetic seed production using calcium chloride beads	Kaur et al. (1999), Das et al. (2013), and Chen et al. (2014)

led to a remarkable 14-fold increase in shoot numbers, proving that a periodic transfer from this liquid medium to an agar-based medium could ensure sustained and healthy shoot production.

Only study that has explored the cryopreservation in *Arnebia euchroma* to preserve its genetic material was done by Guo-feng et al. (1992). They tested two cryoprotectants: (1) 10% dimethyl sulfoxide (DMSO) with 0.5 mol/L sorbitol and (2) 7.5% DMSO, 5.0% glycerin and 5.0% sucrose. Callus samples cultured for 10 and 15 days were subjected to a slow-freezing method, cooled at rate of 1.0 °C/min to −40 °C, held for 2 h and then submerged in liquid nitrogen (−196 °C). The samples were later thawed in the water bath at 40 °C. The study found that calli subcultured for 10 days showed normal growth after cryopreservation using the second cryoprotectant mixture and the slow-freezing method. While the growth rate was initially retarded, the ability to restore normal growth indicated this as an effective cryopreservation method. In conclusion, the collective research on *Arnebia* species demonstrates that a variety of in vitro propagation methods, from somatic embryogenesis to liquid culture systems can effectively aid in their conservation (Table 12.1). These studies have not only established efficient protocols for micropropagation but also successfully integrated cryopreservation techniques, offering a robust strategy for the long term preservation of this critically endangered alpine plant.

1.4 *Fritillaria* spp.

Fritillaria species, commonly known as kakoli, are perennial herbaceous plants belonging to the Liliaceae family, thriving in alpine and subalpine zones of the Himalayas, Alps and other high-altitude regions (Kumar et al., 2020). These plants, characterized by their nodding, bell-shaped flowers with bulbous underground structures, that can stay in dormant stage for 8–9 months (Gochar et al., 2024; Kumar et al., 2021). Species such as *F. roylei*, *F. cirrhosa* and *F. meleagris* are particularly valued for their medicinal properties, containing alkaloids like peimine, peimisine and sipeimine (Sharma et al., 2023). Many species, such as *F. cirrhosa*, are listed as vulnerable or endangered on regional conservation lists (Fig. 12.1). Initial work on tissue culture propagation of *Fritillaria* began with several species: *F. thunbergii* (Sun et al., 1977), *F. pallidiflora* (Hao et al., 1982), *F. ussuriensis* (Zhou, 1983), *F. meleagris* (Kukulezanka et al., 1989), *F. taipaiensis* (Liu et al., 1996), *F. unibracteata* (Gao et al., 1999), *F. cirrhosa* (Chen et al., 2000), *F. hupehensis* (Shiau et al., 2000), *F. roylei* (Joshi et al., 2007), *F. alburyana* and *F. whittallii* (Ozcan et al., 2007).

Joshi et al. (2007) regenerated bulblets using bulb scales as explants in *F. roylei*, with white bulblets structure emerging within 3 weeks and reaching 4–5 mm in 7–8 weeks. Combination of 0.4 mg/L NAA and 1.0 mg/L Kn gave the highest regeneration (95.8%) and produced 10.1 bulblets/explant. Similarly, Petric et al. (2013) investigated dormancy in in vitro–regenerated *F. meleagris* bulblets. MS medium supplemented 1.0 mg/L TDZ, casein hydrolysate and proline promoted bulblet formation, with the best results after 8 weeks at 4 °C however, at 15 °C, morphogenetic potential was highest (10.3 bulblets). Further, acclimatization was achieved with a survival rate of 46.8%. On the other hand, Carasso and Mucciarelli (2014) rescued immature zygotic embryos of *F. tubiformis*, showing callus formation on MS medium with 2.0 mg/L BAP and 0.5 mg/L NAA. Cold stratification (4.0 °C, dark) improved germination and rooting was achieved on half-strength MS with 1.0 mg/L IBA. For bulblet and callus induction in *F. aurea* Kizil et al. (2016) confirmed successful results using bulb scales, with MS medium having 0.05 mg/L TDZ + 0.1 mg/L NAA or 0.1 mg/L TDZ + 0.1 mg/L NAA. Rooting success reached 95% on 1.0 mg/L IBA and 100% on 0.5 mg/L IBA. Sadr et al. (2018) demonstrated callus induction and its regeneration in *F. raddeana* from petal and leaf explants. MS medium with 0.5 mg/L IBA and 2.0 mg/L TDZ resulted in the highest callus regeneration (82.66%) and indirect regeneration (36.66%). Kumar et al. (2020) enhanced callus induction and in vitro plant regeneration in *F. roylei* from bulb scales. Maximum callus formation (88.89%) occurred on MS with 0.5 mg/L TDZ and 2.0 mg/L picloram, while highest regeneration (77.78%) occurred at MS medium with KIN 1.0 mg/L + NAA 0.5 mg/L. Phytochemical analysis also confirmed phenolic and flavonoid accumulation in in vitro tissues. In a different approach, Chang et al. (2020) regenerated *F. cirrhosa* bulblets from seeds. Germination rates varied (36.9–91.0%), with maximum germination (91.0%) on MS medium with 1.0 mg/L BAP + 0.4 mg/L NAA. About 37.5% of seedlings developed bulblets and 19.2% showed callus induction. Transfer to half strength MS medium with sucrose and gellan gum enhanced survival to 91.4%.

Cryopreservation techniques were advanced by Seydi et al. (2020) developed micropropagation and cryopreservation protocols for *F. imperialis*. The best growth (3.5 leaves, 5.7 roots) occurred with 0.5 mg/L KIN +1.0 mg/L NAA. Encapsulation-dehydration before liquid nitrogen storage improved survival to 70.5%. He et al. (2023) vitrified dormant buds of *F. przewalskii* after cold acclimation, preculture in sucrose medium and dehydration, achieving 93% recovery. Regenerated bulblets formed within 5 weeks without callus formation. Recent studies have continued to refine these techniques in which, refinements by Gochar et al. (2024) demonstrated efficient bulb multiplication on Gamborg's B_5 medium supplemented with 1.0 mg/L NAA, BAP, Kn each and 20.0 mg/L Spermidine, resulted 10–12 bulblets per explant. Their subsequent phytochemical analysis detected four key steroidal alkaloids: solanidine, peimisine, petisidine and fritilline-A. Furthermore, elicitation treatments significantly enhanced accumulation, achieving a 2.6 fold increase in solanidine and up to a 1.4 fold increase in the other three alkaloids. Also, Zhang et al. (2024) showed that embryogenic callus proliferation in *F. cirrhosa* was strongly influenced by hormones and temperature. The best response (4.11 fold proliferation) was with NAA (0.2 mg/L) + BAP (1.0 mg/L) + Gibberellic acid (GA_3) (1.0 mg/L) at 20 °C. Tao et al. (2024) further established an efficient regeneration and artificial seed system in *F. cirrhosa*. Callus induction (84.93%) occurred on MS medium containing 1.0 mg/L picloram. Differentiation into bulblets was optimal on MS with 1.0 mg/L BAP + 0.2 mg/L NAA. Regenerated bulblets were encapsulated in MS with sucrose, NAA, BAP and sodium alginate, achieving up to 81.67%

germination with fungicide carbendazim and activated carbon. Concluding with, a substantial progress has been made in developing micropropagation and somatic embryogenesis protocols for *Fritillaria* spp., along with initial success in synthetic seed production and cryopreservation (Table 12.1 and Fig. 12.1). These advances highlight tissue culture as a promising tool for conserving this RET genus, though refinement of protocols.

1.5 *Picrorhiza* spp.

Picrorhiza species, in the Plantaginaceae family, are small perennial herbs endemic to the alpine Himalayas, growing at 3000–5000 m in rocky crevices and meadows (Bakane et al., 2023). With rhizomes and small, white or purple flowers with corolla, they are adapted to extreme cold and short growing seasons (Mondal et al., 2013). Medicinally, *P. kurroa* and *P. scrophulariiflora* contain iridoid glycosides like kutkin, used in Ayurveda and TCM for liver disorders, fever and digestive issues with hepatoprotective and anti-inflammatory properties (Kumar et al., 2017). Overexploitation for pharmaceuticals, habitat loss from climate change and low regeneration rates have led to critically endangered status per IUCN, prompting tissue culture and sustainable cultivation for conservation (Fig. 12.1).

Early studies on *Picrorhiza kurroa* focused on callus induction and in vitro regeneration using different types of explants. Lal et al. (1988) and Upadhyay et al. (1989) described that shoot tips cultured on MS medium containing Kn (3.0–5.0 mg/L), stem cuttings at BAP (0.11–2.25 mg/L) either alone or in combination with IAA or GA_3 and runners with axillary shoots on BAP (0.23 mg/L) could be successfully induced to regenerate shoots. Later, Lal and Ahuja (1996) achieved callus induction from nodal cuttings and leaf explants on MS medium supplemented with 0.5–2.0 mg/L 2,4-D, with proliferation being optimal on 4.0 mg/L NAA and 1.0 mg/L Kn. Shoot organogenesis required sequential media transfer, where bud primordia were first induced on MS with 0.25 mg/L BAP and later converted into shoots on MS with 0.12 mg/L BAP under reduced nitrogen supply, followed by efficient rooting on MS with 0.2 mg/L NAA. Chandra et al. (2006) confirmed that BAP and GA_3 combinations enhanced regeneration efficiency in shoot tip and runner cultures. Sood and Chauhan (2009) refined callus initiation studies, showing root segments as the best explant source (70% callusing), followed by leaf discs (56.3%), and nodal segments (38.3%), with shoot regeneration being highest from root- and leaf-derived calli on MS with BAP (2.0 mg/L) and Kn (3.0 mg/L). Jan et al. (2010) observed that MS medium containing 2,4-D and BAP 0.25 mg/L each, supported both callus induction and direct shoot regeneration, while also reporting indirect and direct adventitious shoot formation, elongation and multiplication with NAA (0.2–1.0 mg/L) alone or in combination with IAA and IBA. Mahajan et al. (2016) then developed a highly efficient mass multiplication protocol where MS medium containing Kn (0.5 mg/L), BAP (1.0 mg/L), and GA_3 (1.0 mg/L) resulted highest shoot regeneration potential, while callus on MS containing BAP and IAA generated both shoots and roots. Gupta et al. (2016) reported prolific shoot regeneration from leaf derived calli (38 shoots/explant) and stem-derived callus (36 shoots/explant) on MS enriched with BAP and Kn.

Alongside regeneration, cryopreservation and synthetic seed technology were explored as strategies for conservation. Sharma and Sharma (2003) successfully cryopreserved shoot tips of *P. kurroa* through vitrification, where 4 weeks of cold hardening at 4.0 °C significantly improved post-thaw survival (70%) and regeneration (35%). Later, Mishra et al. (2011) achieved efficient synthetic seed development using alginate encapsulated micro-shoots, reporting 89.33% regeneration after 3 months of storage. Rooting was successfully induced in half-strength MS medium supplemented with 0.18 mg/L NAA and the acclimatized plantlets showed a survival rate of 95% under greenhouse conditions. Biochemical investigations added another dimension by linking morphogenesis with metabolite biosynthesis. Sood and Chauhan (2010) first explored that picroside-I (P-I) was biosynthesized in shoot cultures, while picroside-II (P-II) was absent in in vitro grown plants. Kawoosa et al. (2010) demonstrated that P-I and P-II production was strongly influenced by environmental conditions, with higher picroside content observed at 15 °C under light compared to 25 °C in darkness. Sharma et al. (2016a, b) tracked P-I biosynthesis across morphogenetic stages, revealing a reduction during de-differentiation in callus induction and callus biomass, but a progressive increase during redifferentiation in shoot primordia, multiple number shoots and fully developed stages, thereby identifying shoots as the primary site of P-I biosynthesis in *P. kurroa*. The extensive research on *P. kurroa* has established a robust foundation for its propagation and conservation through tissue culture. The successful development of high-frequency regeneration protocols, combined with effective cryopreservation and synthetic seed technologies, provides a comprehensive approach to safeguard this valuable medicinal plant (Table 12.1 and Fig. 12.1). Furthermore, studies linking in vitro morphogenesis with picroside biosynthesis offer a unique opportunity to enhance the production of its key medicinal compounds.

1.6 *Podophyllum* spp.

Podophyllum species, from the Berberidaceae family, are perennial herbs in alpine and temperate forests of the Himalayas, at 2000–4000 m (Jagdish et al., 2018). It has large umbrella like leaves and solitary flowers of white/pink/red depending upon species (www.planthuntersfairs.co.uk/newsletters). Medicinally, *P. hexandrum* (Himalayan Mayapple) contains podophyllotoxin, a precursor for anti-cancer drugs like etoposide, used traditionally for warts and tumors (Butt & Amjad, 2015). Overharvesting of rhizomes, habitat destruction from logging and slow growth have classified *P. hexandrum* as endangered by IUCN, with Convention on International Trade in Endangered Species of Wild Fauna and Flora (CITES) restrictions, necessitating micropropagation and protected cultivation (Khan et al., 2024).

Early efforts to propagate *P. hexandrum* were described by Nadeem et al. (2000), who used rhizome segments treated with NAA and IBA for rooting and demonstrated the role of GA_3 (250.0 µM) and BAP (250.0 µM) in enhancing germination.

Addition of BAP (0.23 mg/L) and IAA (0.17–0.70 mg/L) into MS medium enabled shoot multiplication from zygotic embryos within 4–5 weeks, while somatic embryogenesis was attained on MS fortified with BAP (0.1 mg/L) and NAA (1.0 mg/L). Later, Sultan et al. (2006) standardized a protocol for callus-mediated shoot multiplication using root-derived callus initiated on Gamborg's B_5 (Gamborg et al., 1968) medium with 2,4-D (0.5–1.5 mg/L) and BAP (0.2–1.0 mg/L). Shoot proliferation occurred on MS supplemented with BAP (0.5 mg/L) and IAA (1.0 mg/L), while rooting was improved with NAA and activated charcoal. Further improvement came from Sultan et al. (2009), who achieved rapid embryo germination on MS medium enriched with BAP (1.0–4.0 mg/L) and IAA (0.5–2.0 mg/L), leading to efficient plantlet development. Direct organogenesis was later established by Chakraborty et al. (2010) from rhizome explants, where MS medium enriched with IAA (2.0 mg/L) induced the maximum multiple shoot formation, followed by NAA (0.5 mg/L) and BAP (2.5 mg/L). Efficient root initiation was obtained in ½ MS liquid medium supplemented with IBA (100 μM).

Synthetic seed technology was explored in *P. peltatum* by Lata et al. (2010), who encapsulated buds in alginate beads, enabling prolonged storage under low temperature (10 °C) and osmotic agents such as sorbitol and mannitol, achieving up to 100% survival after 8 months. This laid the foundation for alpine *P. hexandrum*. Recently, Khan et al. (2025) developed an advanced slow-growth storage protocol for *P. hexandrum*, where rhizome explants and calli were maintained at 5.0 °C under sorbitol, mannitol, spermidine and calcium pantothenate supplementation. Their study reported up to 100% survival, callus recovery and regeneration, demonstrating a reliable long-term conservation method while maintaining genetic stability. Biochemical investigations were initiated by Chattopadhyay et al. (2002), who studied podophyllotoxin synthesis in cell suspension cultures. They noted medium browning, clumping and pH decline that reduced biomass and metabolite production, but inclusion of polyvinylpyrrolidone (10 mg/L) significantly improved growth and podophyllotoxin yield. Subsequently, Sultan et al. (2009) demonstrated that in vitro raised plants retained podophyllotoxin (5.97%) and its β-D glycoside (5.72%), confirming biosynthetic potential of cultured tissues. HPLC (High Performance Liquid Chromatography) analyses by Chakraborty et al. (2010) detected podophyllotoxin in leaf-derived callus cultures, reinforcing the role of differentiated tissues in metabolite accumulation. Rajesh et al. (2014) further optimized adventitious root cultures by manipulating nutrient sources, reporting maximum biomass in ½ strength MS with 2.0% sucrose and IBA (3.0 mg/L), while podophyllotoxin accumulation peaked under 6.0% sucrose and medium pH 6.0. Also, recently HPLC profiling of callus from in vivo and in vitro explants confirmed notable podophyllotoxin content, with the highest yields from root-derived callus, particularly from high-altitude populations of Kinnaur and Lahaul-Spiti, highlighting the influence of source material and culture conditions on metabolite production. In summary, tissue culture studies on *Podophyllum* have enabled regeneration from multiple explants and facilitated production of key lignans (Table 12.1). However, protocol standardization for mass-scale conservation and cryogenic storage is still needed.

1.7 *Rhodiola* spp.

Rhodiola species, in the Crassulaceae family, are succulent perennials in alpine and subarctic regions of the Himalayas, subarctic, at 2000–5000 m (Beaulieu et al., 2016; Yang et al., 2023). With fleshy leaves and yellow or red flowers, they adapt to cold, rocky soils via stress-resistant compounds. Medicinally, *R. rosea* and *R. imbricata* contain rosavins and salidroside, used as adaptogens in Indian System of Medicine (ISM), TCM and Russian folk remedies for fatigue, stress and altitude sickness. Overharvesting for supplements, habitat loss and climate warming have led to critically endangered status for some, like *R. rosea*, requiring CITES monitoring and cultivation programs (Fig. 12.1) (Aneva et al., 2021; Butenkova et al., 2024).

Early progress on related species was reported by Zych et al. (2005), who encapsulated differentiated callus and lateral buds of *Rhodiola kirilowii* in calcium alginate beads, achieving successful shoot and seedling growth on MS medium. This opened prospects for similar synthetic seed technology in *R. imbricata*. Subsequently, Tasheva and Kosturkova (2010) developed efficient micropropagation protocol for *R. rosea*, while Zhao et al. (2012) improved temporary immersion bioreactors for *R. crenulata*, which reduced hyperhydration and enhanced shoot multiplication. Comparable approaches were later adopted for *R. imbricata*, where suspension cultures on MS medium supplemented with BAP (1.0 mg/L) and IBA (2.0 mg/L) supported shoot regeneration (Pundir et al., 2019). Work specifically on *R. imbricata* was advanced by Sharma (2016), who achieved effective shoot and root induction on MS medium containing BAP (2.0 mg/L) and IBA (4.0 mg/L). Bhardwaj et al. (2018) further optimized morphogenesis, reporting that NAA (10–15 μM) induced rhizogenous callus, while shoot regeneration acheived highest on NAA (5.0 μM) with BAP (2.5 μM) or NAA (1.0 μM) with BAP (5.0 μM), yielding 38.88% and 37.49% shoot formation, respectively. Rooting was most effective under low NAA (0.5 μM), producing well-developed plantlets with over 80% survival during acclimatization. In parallel, Pundir et al. (2019) standardized MS medium containing BAP (1.0 mg/L) with IBA (2.0 mg/L) as optimal for multiple shoot culture, whereas BAP (2.0 mg/L) and IBA (4.0 mg/L) supported both shoot and root induction. Further insights from *R. rosea* expanded the methodological framework. Erst and Yakubov (2019) demonstrated that medium containing BAP (1.0 mg/L) and NAA (0.5 mg/L) induced the largest shoot number, though at reduced height, while control MS medium favored taller shoots. Their study also documented in vitro flowering at later passages, highlighting physiological plasticity in tissue culture. More recently, Matvieieva et al. (2023) showed that MS medium with BAP (2.5 mg/L) and 2,4-D (1.0 mg/L) enabled 100% regeneration in *R. rosea*, whereas altered auxin-cytokinin ratios reduced regeneration frequency, underscoring the importance of fine hormonal balance in optimizing propagation protocols.

Cryogenic approaches were pioneered in *R. sachalinensis* by Jian-feng et al. (2007), who vitrified callus tissues after sugar preconditioning and PVS2 treatment, achieving 78.24% survival upon recovery, with healthy regrowth and vigor. Although synthetic seed encapsulation was demonstrated in *R. kirilowii* (Zych et al., 2005), such technology remains to be fully applied in *R. imbricata*.

Nonetheless, the success of related Rhodiola species provides a foundation for future conservation strategies. In *R. imbricata*, Bhardwaj et al. (2018) conducted the first metabolic evaluation, reporting a significant upregulation of cinnamyl alcohol levels, 49.6% in in vitro roots and 30.6% in in vitro shoots compared to their mother plant counterparts. This highlights the potential of in vitro cultures for enhancing bioactive compounds. Similarly, Pundir et al. (2019) emphasized the use of callus cultures, particularly with TDZ (1.0 mg/L), as a viable substitute for direct explants in both regeneration and metabolite production, paving the way for biotechnological exploitation of this high-value alpine species. The collective research on *R. imbricata* and its related species has established a strong foundation for its conservation and biotechnological exploitation (Fig. 12.1). The development of specialized protocols for shoot proliferation, root induction and metabolic enhancement, along with the successful application of cryopreservation in related species, offers a comprehensive strategy for the sustainable propagation and future commercial use of this valuable alpine medicinal plant (Table 12.1).

1.8 *Saussurea* spp.

Saussurea species, from the Asteraceae family, are perennial herbs in alpine meadows and rocky slopes of the Himalayas and Central Asia, at 3000–5000 m (Butola & Samant, 2010). With thistle-like flowers and woolly leaves, they withstand extreme cold and UV radiation (Yang & Sun, 2009). They support pollinators and prevent erosion in fragile high-altitude habitats. Medicinally, *S. costus* (Kuth) yields costunolide from roots, used in Ayurveda for anti-inflammatory, antimicrobial and digestive treatments (Semwal et al., 2020). Overexploitation for perfumes and medicine, habitat degradation and climate change have made *S. costus* as critically endangered status, demanding tissue culture and sustainable farming for preservation (Vishvamitera et al., 2024).

The first attempt at in vitro regeneration of *Saussurea lappa* was initiated by Arora and Bhojwani (1989), who tested multiple seedling derived explants. Leaf explants showed maximum shoot regeneration, followed by cotyledons whereas roots and hypocotyls lacked morphogenetic potential. Shoot multiplication (3.5-fold every 3 weeks) was achieved on MS medium containing BAP and GA_3, while efficient rooting (90%) occurred on MS with 0.1 mg/L NAA. Remarkably, shoot cultures stored in dark at 5.0 °C for 12 months without subculturing retained 100% viability, suggesting a simple means of medium-term germplasm storage. Johnson et al. (1997) demonstrated rapid micropropagation of *S. lappa* using shoot tips (0.5–1.0 cm) from two-week-old seedlings. MS medium with TDZ (1.0 mg/L) yielded callus-free multiple shoots, while liquid MS medium having TDZ (0.05 mg/L) promoted better multiplication compared to agar-solidified medium. Rooting obtained on MS medium containing NAA (0.2 mg/L) and hardened plantlets survived well in soil. Work on other species soon followed. Joshi and Dhar (2003) reported successful micropropagation of *S. obvallata* from epicotyl explants. MS medium augmented with Kn (0.2 mg/L) and NAA (0.05 mg/L) regenerated up to 5.0 shoots/explant within 75 days. Rooting was 100% on half-strength MS with IBA (0.5 mg/L) and 66.7% of acclimatized plants survived under field conditions. Similarly, Chen and Li (2005) described in vitro regeneration of *S. laniceps*, while Guo et al. (2007) established efficient shoot organogenesis in *S. involucrata* leaf explants, achieving 66% regeneration frequency and 87% rooting on IAA (0.44 mg/L). Gailite et al. (2010) initiated tissue cultures of *S. esthonica* from seeds, observing best shoot proliferation on MS supplemented with BAP (0.5–1.0 mg/L). Kinetin promoted shoot elongation and root formation but was not effective for multiplication. About 30% rooting was recorded. Guo et al. (2013) later refined protocols for *S. involucrata*, while Kuo et al. (2015) contributed additional improvements to its organogenesis. Warghat et al. (2016) standardized regeneration conditions for *Saussurea costus*, demonstrating that 2,4-D (3.0 mg/L) and Kn (5.0 mg/L) were the most suitable for callus initiation across diverse explant types (leaf, epicotyl, hypocotyl, cotyledonary root and stem), with stem and root showing earliest responses. Sharma et al. (2019) developed a micro-propagation protocol for *S. costus* using shoot tip explants. MS medium fortified with NAA (0.5 mg/L) and TDZ (0.25 mg/L) yielded highest shoot regeneration, while multiplication was enhanced on MS with BAP (1.0 mg/L), Kn (0.5 mg/L) and GA_3 (0.5 mg/L). Rooting was recorded on MS with IBA (0.5 mg/L) and plantlets acclimatized successfully *ex vitro*. Similarly to this, Zaib-Un-Nisa et al. (2019) optimized *S. lappa* micropropagation, reporting 90% shoot induction on MS with BAP (1.0 mg/L), while rooting was most effective with IBA (1.0 mg/L). Callus induction was best from root and seedling explants on MS with 2,4-D (1.0 mg/L) + Kn (0.25 mg/L). Sharma et al. (2019) further improved *S. costus* protocols, achieving 73.33% regenerated shoots and an average of 11.4 shoots/explant on MS with TDZ (0.25 mg/L) + NAA (0.5 mg/L). Rooting (77.78%) was highest with IBA (0.5 mg/L), producing vigorous plants. Khan et al. (2021) reported indirect organogenesis and phytochemical profiling of *S. costus*. Callus was induced from seeds, leaves, petioles and roots with maximum shoot induction (82%) on MS medium supplemented with BAP (2.0 mg/L), NAA (1.0 mg/L) and GA_3 (0.25 mg/L). More recently, Mehreen et al. (2020) studied in vitro propagation of *S. heteromalla*, broadening the scope of *Saussurea* tissue culture applications.

Guo et al. (2013) devised a vitrification protocol for *S. involucrata* callus. Callus preconditioned at 4.0 °C for 12 days was treated with PVS2 (60% for 40 min; 100% for 60 min) before liquid nitrogen storage. Survival reached 75%, regrowth 60% and over 90% of regenerated shoots established in complete plantlets, with 81% surviving acclimatization. Although artificial seed technology has not been explicitly applied to *Saussurea*, the success of callus storage and vitrification in *S. involucrata* suggests its potential for germplasm conservation in other threatened species like *S. costus* and *S. obvallata*. Bhandari et al. (2013) explored in vitro conservation of *S. costus*, employing multiple explants for mass propagation. Maximum shoot induction occurred on MS with BAP (2.0 mg/L), while half-strength MS with NAA (1.0 mg/L) was best for rooting. Field transfer was successful, supporting ex situ conservation efforts. Warghat et al. (2016) and Khan et al. (2021) linked regeneration studies with phytochemical profiling, showing that optimized culture conditions not only supported propagation but also maintained the medicinal metabolite profile of *S. costus*. Similarly, Mehreen et al. (2020) combined in vitro regeneration with phytochemical studies in *S. heteromalla*, highlighting the dual application of micropropagation for conservation and bioactive compound production. Collectively, research on *Saussurea* demonstrates efficient regeneration pathways through callus induction, shoot organogenesis and synthetic seed techniques (Table 12.1). While significant strides have been achieved, large-scale propagation and long-term genetic stability need further exploration for effective conservation.

1.9 *Swertia* spp.

Swertia species, in the Gentianaceae family, are annual or perennial herbs in alpine and subalpine Himalayas, at 2000–5000 m in meadows and forests (Samaddar et al., 2014). They contribute to biodiversity as indicators of pristine habitats. Medicinally, *S. chirayita* contains swertiamarin and amarogentin, used in Ayurveda for fever, diabetes and liver ailments as a bitter tonic (Verma et al., 2024). Overharvesting, habitat loss and low germination have classified *S. chirayita* as critically endangered status by IUCN, requiring micropropagation and protected areas for conservation (Fig. 12.1) (Thapa et al., 2024).

Chaudhuri et al. (2007) developed a shoot multiplication protocol for *Swertia chirayita* using nodal segments. Half-strength MS medium supplemented with BAP (0.1 mg/L) and 6-furfurylaminopurine (Kinetin) (1.0 mg/L) supported regeneration, while the maximum number of shoots was attained on MS medium with KNO_3 (10 mM) and casein hydrolysate (75 mg/L). The regenerated plants were effectively transferred to field for viable seed production. Joshi and Dhawan (2007a, b) developed an efficient micropropagation system using 4 week old seedling-derived nodal explants. Shoot multiplication was achieved on MS medium containing BAP (0.9 mg/L) and 2-iP (N6-(2-Isopentenyl)adenine) (0.3 mg/L). For rooting, NAA (0.2 mg/L) with activated charcoal (500 mg/L) was effective. Indirect organogenesis and somatic embryogenesis in *S. chirayita* were later reported by Chaudhuri et al. (2008, 2009). Leaf and nodal explants, as well as internodes from in vitro grown seedlings, were used to induce embryogenic callus, offering a potential route for large-scale propagation. Balaraju et al. (2009) optimized in vitro regeneration of *S. chirata* utilizing shoot tips from in vitro seedlings. MS medium fortified with BAP (1.0 mg/L) and Kn (0.1 mg/L) produced the maximum shoot number (42.16 ± 1.05)/explant. Elongation was enhanced on MS with GA_3 (0.1 mg/L), followed by transfer to rooting medium. Wang et al. (2009) confirmed that auxins, either alone or in combination, were effective in rooting of *S. chirayita* microshoots, complementing earlier reports. Shailja et al. (2017) developed a reproducible protocol for *S. chirayita* propagation using leaf explants. Maximum callus induction and regeneration occurred on MS medium containing BAP (3.0 mg/L) with NAA (0.1 mg/L). Shoot multiplication was highest with BAP (2.5 mg/L) + Kn (0.1 mg/L). Singh et al. (2021) reviewed explant choices (seeds, nodal segments, axillary buds and leaves) from both field raised and in vitro regenerated plantlets of *Swertia* spp.

They highlighted that BAP alone or in combination with Kn, 2-iP and IAA, IBA, NAA in lower concentrations was optimal for shoot multiplication, while indirect organogenesis remained a reliable approach for mass multiplication.

Joshi and Dhawan (2007a, b) achieved rooting on MS medium containing NAA (0.2 mg/L) + activated charcoal, demonstrating efficient conversion of shoots into plantlets. Balaraju et al. (2009) recorded maximum rooting frequency (22.48 ± 1.08) on half-strength MS with NAA (0.1 mg/L). Hardened plantlets acclimatized successfully under controlled indoor conditions. Shailja et al. (2017) reported that ½ strength MS medium enriched with NAA (0.1 mg/L) and AC (400 mg/L) yielded the highest rooting frequency, improving the survival of tissue-cultured plants. Chaudhuri et al. (2008, 2009) demonstrated somatic embryogenesis in *S. chirayita* using nodal segments, leaf explants, internodes and seedlings, showing its capability for large scale propagation. Sharma et al. (2014) documented in vitro flowering in *S. chirayita*, marking a unique finding in the genus. Sharma et al. (2016a, b) further established auxin-mediated rooting efficiencies in *S. chirayita*, consolidating earlier work on rooting responses. Overall, these studies confirm that *Swertia* spp. can be effectively propagated using nodal and leaf explants with cytokinin-rich media, while auxins facilitate rooting. Somatic embryogenesis, in vitro flowering and optimized callus culture further enhance its potential for conservation and commercial use. Overall, studies on *Swertia* species have contributed standardized methods for shoot regeneration and metabolite production under in vitro conditions (Table 12.1 and Fig. 12.1). Advances in cryopreservation and encapsulated seed techniques also broaden conservation opportunities, but further integration with field based restoration is necessary.

1.10 *Trillium govanianum*

Trillium species, from the Melanthiaceae family, are perennial herbs in alpine and temperate forests of the Himalayas and North America, at 2000–4000 m (Kanaujia et al., 2023). Medicinally, *T. govanianum* (Himalayan Trillium) contains steroidal saponins used traditionally for pain and inflammation, with potential anti-cancer properties (Verma et al., 2021). Overharvesting of rhizomes and habitat loss have made many, like *T. govanianum*, critically endangered per IUCN, urging propagation and conservation gardens (Kanaujia et al., 2023).

Early studies by Pence and Soukup (1993) demonstrated the successful in vitro regeneration of mini-rhizomes in *T. grandiflorum* and *T. erectum* using stem and leaf as explant. Surface sterilization was achieved with 70% ethanol and 0.1% sodium hypochlorite, after which explants were cultured on full- and half-strength MS medium enriched with various cytokinins (BAP, Kn, zeatin, 2-iP) and different auxins (NAA, IAA, 2,4-D). Both species produced mini-rhizomes capable of regenerating shoots and roots. Interestingly, *T. erectum* exhibited rhizome formation even without growth regulators, with stem explants responding more effectively than leaf tissues. This work provided an alternative and faster propagation method compared to traditional approaches. Further progress was reported by Jun (1998) in *T.*

tsehonoskii, where explants from different organs and developmental stages were cultured using combinations of BAP, IAA, NAA and zeatin. Among these, the combination of BAP and IAA yielded the best response for callus induction and sprout differentiation. Conservation-oriented tissue culture research was carried out on *T. persistens* and *T. reliquum*, two threatened species, by Gagliardo et al. (2012). Their study emphasized the role of in vitro methods in safeguarding species at risk from overharvesting and habitat degradation. Protocols for establishing sterile cultures, root initiation and optimized media formulations were also developed by them.

More recent work on *T. govanianum* by Kumar et al. (2024b) has focused on identifying suitable explants and refining sterilization protocols. Survival rates of 86.67–93.33% were achieved in seed, root, rhizome disc and rhizome bud explants when treated with carbendazim and $HgCl_2$, though phytotoxic effects were noted in leaf and stem explants. Among all explants, rhizome buds proved to be most effective, showing 96.67% shoot regeneration using MS medium containing BAP (0.5 mg/L), kinetin (0.5 mg/L) and GA_3 (0.5 mg/L). Chlorosis observed during shoot proliferation was corrected using a modified MS medium enriched with ammonium sulphate, calcium chloride and Fe-EDDHA. Enhanced production of lateral buds and mini-rhizomes was obtained on MS medium supplemented with 6.0% sucrose, 0.5 mg/L BAP, 0.5 mg/L kinetin, 0.5 mg/L GA_3 and casein hydrolysate, achieving up to 93.33% bud proliferation. Rhizome buds have consistently emerged as the most reliable explants for regeneration across *Trillium* species (Chauhan et al., 2020; Chandola et al., 2023). Literature evidence highlights the critical role of cytokinins, as their type and concentration strongly influence shoot induction through interactions with endogenous hormone levels in explants. The tissue culture research on *T. govanianum* and related species has successfully established efficient in vitro approaches for their propagation and preservation. The development of protocols for mini-rhizome production and high frequency shoot proliferation from reliable explants like rhizome buds offers a viable alternative to traditional, slower propagation methods (Table 12.1). These advancements provide a crucial tool for protecting this endangered plant from overexploitation and habitat loss.

1.11 *Valeriana* spp.

Valeriana species, in the Caprifoliaceae family, are perennial herbs in alpine meadows and forests of the Himalayas and Europe, at 2000–4000 m (Joshi et al., 2016). Medicinally, *V. jatamansi* contains valerenic acid, used in Ayurveda and TCM as a sedative for anxiety, insomnia and nervous disorders. Overexploitation for essential oils, habitat degradation and climate change have led to critically endangered status for *V. jatamansi* as per IUCN, necessitating in vitro culture and sustainable harvesting (Fig. 12.1) (Dhiman et al., 2023).

Rodriguez (1997) reported micropropagation of *V. edulis* where seedlings produced adventitious shoots on MS medium containing BAP (5.0 mg/L) and IAA (0.2 mg/L). Rooting occurred on MS medium containing IAA (0.4 mg/L) with

2.0 mg/L sucrose. Kaur et al. (1999) described mass propagation in *V. jatamansi* using shoot bud explants. MS medium enriched with BAP (1.0 mg/L) yielded 100% regeneration, with average of 5.1 ± 0.2 shoots/explant, 7.9 ± 2.4 leaves and 6.4 ± 0.2 cm shoot length. Combinations of BAP with IAA or NAA also supported regeneration. Gao and Bjork (2000) regenerated plants from shoot tips and axillary buds of *V. officinalis* cultured on MS enriched with BAP (5.0 mg/L) and IAA (1.0 mg/L). Regenerated plants displayed variability in height (59 cm) and root biomass (295 g). Abdi and Khosh-Khui (2007) established direct embryogenesis in *V. officinalis* using leaf explants. Maximum shoot regeneration (74.1 ± 2.1%) was attained on MS containing BAP (1.0 mg/L). 2iP (6.0 mg/L) + IAA (3.0 mg/L) yielded 93.3 ± 2.2% of multiplied shoot culture, while rooting on MS + NAA (1.0 mg/L) produced 100% efficiency with 34.7 mm root length. Das et al. (2013) reported callus induction in *V. jatamansi*, showing rhizome explants as most responsive. MS medium containing 2,4-D (0.5 mg/L) was optimal for callus formation. Chen et al. (2014) studied organogenesis and somatic embryogenesis in *V. jatamansi* using leaf as explant. Callogenesis was 100% on MS having NAA (1.0 mg/L), while MS + 2,4-D (0.1 mg/L) promoted globular embryoid differentiation. Adventitious shoots were obtained on MS + BAP (0.5 mg/L) + NAA (0.1 mg/L), achieving 90% regeneration. Purohit et al. (2015) attained direct shoot induction from shoot bud explants on MS fortified using BAP (0.3 mg/L) with NAA (0.1 mg/L) and GA_3 (0.03 mg/L). The medium supported 100% shoot regeneration with simultaneous root initiation. Regenerated shoots attained 3.20 ± 0.21 cm length and 15.33 ± 1.23 leaves. Saini et al. (2018) optimized direct rhizogenesis in *V. jatamansi* using leaf, petiole, nodal segment and shoot bud explants. Leaves produced the highest rooting efficiency (90%) on MS medium containing BAP (2.0 mg/L) and NAA (2.0 mg/L).

Sharma et al. (2021) developed a vitrification-based cryopreservation method for *V. jatamansi* shoot tips. Treatment with PVS2 at 0 °C for 110 min achieved the highest recovery (91.6%) after liquid nitrogen exposure, surpassing PVS3 (73.3%). Importantly, valepotriate contents remained unchanged, confirming biosynthetic stability of cryopreserved regenerants. Mathur et al. (1989) pioneered encapsulation studies in *V. wallichii* using Sodium alginate (6.0%) complexed with $CaCl_2 \cdot 2H_2O$ (75 mM) for 30 min produced optimal firm beads. Encapsulated explants grew efficiently in vitro, showing 96–100% plantlet formation within 20 days on 0.7% agar, with or without MS nutrients. Encapsulation significantly enhanced growth compared to non-encapsulated controls, which either failed to survive or showed retarded development. In conclusion, *Valeriana* tissue culture research has provided robust systems for plant regeneration, root induction and enhancement of bioactive compounds (Table 12.1 and Fig. 12.1). While these achievements support conservation and pharmaceutical applications, large-scale deployment and germplasm banking require continued efforts.

2 Future Direction and Challenges

Overcoming recalcitrance remains a critical challenge in the propagation of alpine plants, as many species exhibit resistance to in vitro growth; addressing this requires integrated approaches such as careful explant selection, standardized hormone regimes and optimized transformation protocols, as demonstrated in this chapter. Future directions may also emphasize synergistic approaches, combining tissue culture with plant growth-promoting microorganisms to strengthen conservation, particularly for Himalayan endemic species. Additionally, cryopreservation advancements and synthetic seed technologies could facilitate long-term germplasm banking, enabling reintroduction programs that restore populations in degrading alpine ecosystems. For plants not covered in this chapter, such as *Gentiana* spp., *Bergenia* spp., *Meconopsis* spp., etc., which are vital for their anti-inflammatory and ornamental values, similar protocols could be adapted to address their low seed viability and slow growth, potentially incorporating nanotechnology for better nutrient delivery in culture media. However, significant challenges persist, including physiological plasticity issues where alpine plants migrating to higher elevations due to warming encounter unpredictable stresses, complicating ex-situ propagation. Genetic instability during prolonged culture periods, high contamination risks in sterile setups and the need for species-specific media optimization remain barriers, especially for rare, endangered, threatened species (RET) which face endangerment from overharvesting and habitat loss in regions like India, Nepal, China and Tibet. Economic and regulatory hurdles, such as scaling laboratory successes to field applications and ensuring compliance with international trade agreements like CITES, further impede progress. Addressing these will require collaborative research, funding for high-altitude facilities and community involvement to balance conservation with pharmaceutical demands, ultimately safeguarding the biodiversity of alpine medicinal flora for future generations.

Data Applicability Not Applicable.

Author Contribution PK—conceived the idea and designed the article; SS, PK, Deepika—literature mining and wrote the draft manuscript; PK, HS, ARW & RS—reviewed and edited the manuscript; All the authors have read and approved the manuscript.

Conflicts of Interest/Competing Interests The authors declare that they have no known competing financial interests or personal relationships that could have appeared to influence the work reported in this paper.

References

Abdi, G. H., & Khosh-Khui, M. (2007). Shoot regeneration via direct organogenesis from leaf segments of valerian (Valeriana officinalis). *International Journal of Agricultural Research, 2*, 877–882.

Adhikari, P. P., & Paul, S. B. (2018). History of Indian traditional medicine: A medical inheritance. *History, 11*(1), 421–426.

Ambrish, K., & Srivastava, S. K. (2014). Taxonomic studies on the genus "Arnebia" Forssk. (Boraginaceae) in India. *Taiwania, 59*(4), 315–325.

Aneva, I., Zhelev, P., Sidjimova, B., Nikolova, M., & Savev, S. (2021). Population status and natural localities of Rhodiola rosea in Rila Mts., Bulgaria. *Ecologia Balkanica, 4*, 145–151.

Arora, R., & Bhojwani, S. S. (1989). In vitro propagation and low temperature storage of Saussurea lappa CB Clarke—An endangered, medicinal plant. *Plant Cell Reports, 8*(1), 44–47.

Ashraf, M. V., Khan, S., Misri, S., Gaira, K. S., Rawat, S., Rawat, B., Khan, M. A. H., Shah, A. A., Asgher, M., & Ahmad, S. (2024). High-altitude medicinal plants as promising source of phytochemical antioxidants to combat lifestyle-associated oxidative stress-induced disorders. *Pharmaceuticals (Basel), 17*(8), 975.

Bakane, P., Chincholkar, P., Badkhal, A., & Umare, S. (2023). Kutki: an important jadi buti in ayurveda. *Journal of Pharmacognosy and Phytochemistry, 12*(5), 157–161.

Balaraju, K., Agastian, P., & Ignacimuthu, S. (2009). Micro-propagation of Swertia chirata Buch. Hams. Ex Wall. a critically endangered medicinal herb. *Acta Physiologiae Plantarum, 31*, 487–494.

Banik, B., Das, S., & Das, M. K. (2020). Medicinal plants with potent anti-inflammatory and anti-arthritic properties found in eastern parts of the Himalaya: An ethnomedicinal review. *Pharmacognosy Reviews, 14*(28), 121–137.

Beaulieu, F., Cuerrier, A., Filion, V. J., Saleem, A., & Arnason, J. T. (2016). A gall mite, Aceria rhodiolae (Acari: Eriophyidae), altering the phytochemistry of a medicinal plant, Rhodiola rosea (Crassulaceae), in the Canadian Arctic. *Journal of Natural History, 50*(19–20), 1237–1263.

Belwal, N. S., Kamal, B., Sharma, V., Gupta, S., Kumar Dobriyal, A., & Jadon, V. S. (2016). Production of genetically uniform plants from shoot tips of Aconitum heterophyllum wall.–a critically endangered medicinal herb. *The Journal of Horticultural Science and Biotechnology, 91*(5), 529–535.

Bhandari, A. K., Bisht, V. K., Negi, J. S., & Baunthiyal, M. (2013). 1, 8-Cineole: A predominant component in the essential oil of large cardamom (Amomum subulatum Roxb.). *Journal of Medicinal Plant Research: Planta Medica, 7*(26), 1957–1960.

Bhardwaj, A. K., Naryal, A., Bhardwaj, P., Warghat, A. R., Arora, B., Dhiman, S., Saxena, S., Pati, P. K., & Chaurasia, O. P. (2018). High efficiency in vitro plant regeneration and secondary metabolite quantification from leaf explants of Rhodiola imbricata. *Pharmacognosy Journal, 10*(3), 470–475.

Bisht, A. K., Bhatt, A., Dhar, U., Bhatt, A., & Dhar, U. (2015). Note on somatic embryogenesis and synthetic seed production in Angelica glauca: A valuable medicinal plant of Himalaya. *Journal of Medicinal Plants Research, 9*(12), 419–425.

Butenkova, A. N., Prokopyev, A. S., & Yamburov, M. S. (2024). Anatomical and morphological features of leaves of Rhodiola rosea L. in natural conditions of the mountains in southern Siberia (Altai Mountains). *Acta Biologica Sibirica, 10*, 583–599.

Butola, J. S., & Samant, S. S. (2010). Saussurea species in Indian Himalayan region: Diversity, distribution and indigenous uses. *International Journal of Plant Biology, 1*(1), e9-e9.

Butt, T. I., & Amjad, M. S. (2015). Ethnopharmacology, phytochemistry and cytotoxicity of emerging biotechnological tool: Mayapple (Podophyllum hexandrum L.: Berberidaceae). *Journal of Coastal Life Medicine, 3*(8), 652–657.

Carasso, V., & Mucciarelli, M. (2014). *In vitro* bulblet production and plant regeneration from immature embryos of *Fritillaria tubiformis* gren. & godr. *Propagation of Ornamental Plants, 14*, 101–111.

Chaachouay, N., & Zidane, L. (2024). Plant-derived natural products: A source for drug discovery and development. *Drugs and Drug Candidates, 3*(1), 184–207.

Chakraborty, A., Bhattacharya, D., Ghanta, S., & Chattopadhyay, S. (2010). An efficient protocol for in vitro regeneration of Podophyllum hexandrum, a critically endangered medicinal plant. *Indian Journal of Biotechnology, 9*, 217–220.

Chandola, V., Chandra, S., Kumar, R. R., & Nautiyal, A. R. (2023). Repro-ductive behavior of an endangered Himalayan medicinal herb Trillium govanianum Wall. ex D. Don. *Vegetos, 36*, 1437–1445. https://doi.org/10.1007/s42535-022-00548-3

Chandra, B., Palni, L. M. S., & Nandi, S. K. (2006). Propagation and conservation of Picrorhiza kurrooa Royle ex Benth.: An endangered Himalayan medicinal herb of high commercial value. *Biodiversity and Conservation, 15*(7), 2325–2338.

Chandran, S., Raghu, A. V., & Mohanan, K. V. (2023). In vitro conservation of rare, endangered, and threatened plants. In S. T. Sukumaran & T. R. Keerthi (Eds.), *Conservation and sustainable utilization of bioresources. Sustainable development and biodiversity* (Vol. 30, pp. 391–408). Springer.

Chang, H. C., Xie, H. M., Lee, M. R., Lin, C. Y., Yip, M. K., Agrawal, D. C., & Tsay, H. S. (2020). *In vitro* propagation of bulblets and LC–MS/MS analysis of isosteroidal alkaloids in tissue culture derived materials of Chinese medicinal herb *Fritillaria cirrhosa* D. Don. *Botanical Studies, 61*(1), 1–9.

Chattopadhyay, S., Srivastava, A. K., Bhojwani, S. S., & Bisaria, V. S. (2002). Production of podophyllotoxin by plant cell cultures of Podophyllum hexandrum in bioreactor. *Journal of Bioscience and Bioengineering, 93*, 215–220.

Chaudhary, R. P., Bhattarai, S. H., Basnet, G., Bhatta, K. P., Uprety, Y., Bhatta, L. D., et al. (2017). *Traditional practice and knowledge of indigenous and local communities in Kailash sacred landscape, Nepal* (ICIMOD Working Paper 2017/1). ICIMOD.

Chaudhuri, R. K., Pal, A., & Jha, T. B. (2007). Production of genetically uniform plants from nodal explants of Swertia chirata Buch. Ham. Ex Wall: An endangered medicinal herb. *In Vitro Cellular and Developmental Biology-Plant, 43*, 467–472.

Chaudhuri, R. K., Pal, A., & Jha, T. B. (2008). Conservation of Swertia chirata through direct shoot multiplication from leaf explants. *Plant Biotechnology Reports, 2*(3), 213–218.

Chaudhuri, R. K., Pal, A., & Jha, T. B. (2009). Regeneration and characterization of Swertia chirata Buch.-Ham. ex Wall. plants from immature seed cultures. *Scientia Horticulturae, 120*(1), 107–114.

Chauhan, A., Shirkot, C. K., Kaushal, R., & Rao, D. L. N. (2015). Plant growth-promoting rhizobacteria of medicinal plants in NW Himalayas: Current status and future prospects. In D. Egamberdieva, S. Shrivastava, & A. Varma (Eds.), *Plant-growth-promoting rhizobacteria (PGPR) and medicinal plants* (pp. 381–412). Springer.

Chauhan, H. K., Bisht, A. K., Bhatt, I. D., & Bhatt, A. (2020). Protocol for vegeta-tive propagation of trillium govanianum Wall ex D. Don. *Journal of Applied Research on Medicinal and Aromatic Plants, 16*, 1–21. https://doi.org/10.1016/j.jarmap.2019.100233

Chen, Y., & Li, F. (2005). Micropropagation and callus culture of Saussurea laniceps- an alpine medicinal plant. *Forest Science, 7*(1), 16–19.

Chen, U. C., Tai, C. D., Chen, C. C., & Tsay, H. S. (2000). Studies on the tissue culture of *Fritillaria hupehensis* Hsiao et K. C. Hsia III. Influence of medium component and light treatment on *in vitro* rooting and acclimatization of bulblet. *Journal of Agricultural Research of China, 49*(4), 39–47.

Chen, R., Zhang, M., Lu, J., Zhang, X., Jaime, A., Silva, T. D., & Guohua, M. (2014). Shoot organogenesis and somatic embryogenesis from leaf explants of Valeriana jatamansi Jones. *Scientia Horticulturae, 165*, 392–397.

Das, J., Mao, A. A., & Handique, P. J. (2013). Callus-mediated organogenesis and effect of growth regulators on production of different valepotriates in Indian valerian (*Valeriana jatamansi* Jones). *Acta Physiologiae Plantarum, 35*(1), 55e63.

Dhiman, R., Sharma, M., Sharma, M., & Sharma, M. (2023). Reviewing the medicinal potential of Valeriana jatamansi Jones: Its traditional uses, phytochemistry, and pharmacological activities. *TMR Pharmacology Research, 3*(4), 23.

Dulta, K., Kaur, G., Kumar, K., Mazhar, M., Bhardwaj, G., & Sharma, A. (2023). Species of Arnebia genus found in the Western Himalayas: *Arnebia euchroma (Royle ex Benth.), Arnebia benthamii (Wall. Ex G Don) Johnston, Arnebia guttata Bunge*. In A. Sharma & G. A. Nayik (Eds.), *Immunity boosting medicinal plants of the Western Himalayas*. Springer.

Erst, A. A., & Yakubov, V. V. (2019). Regenerative in vitro capacity of rare species Rhodiola rosea L. from various habitats. *Contemporary Problems of Ecology, 12*(4), 368–376.

Gagliardo, R., Labarthe, M., Zaic, M., Cruse-Sanders, J., & Determann, R. O. (2012). In vitro propagation of trillium species with notes on root formation, cleaning protocols and media formulation. *Native Plants Journal, 13*, 56–63.
Gailite, A., Klavina, D., & Ievinsh, G. (2010). In vitro propagation of an endangered plant Saussurea esthonica. *Environmental and Experimental Botany, 8*, 43–48.
Gamborg, O. L., Miller, R., & Ojima, K. (1968). Nutrient requirements of suspension cultures of soybean root cells. *Experimental Cell Research, 50*(1), 151–158.
Ganie, A. H., Tali, B. A., Khuroo, A. A., Reshi, Z. A., & Nawchoo, I. A. (2019). Impact assessment of anthropogenic threats to high-valued medicinal plants of Kashmir Himalaya, India. *Journal for Nature Conservation, 50*, 125715.
Gao, S., & Bjork, L. (2000). Chemical characters, height and root weight in callus regenerated plants of Valeriana officinalis L. *Journal of Herbs, Spices and Medicine Plants, 7*, 31–39.
Gao, S. L., Zhu, D. N., Cai, Z. H., Jiang, Y., & Xu, D. R. (1999). Organ culture of a precious Chinese medical plant – *Fritillaria unibracteata. Plant Cell, Tissue and Organ Culture, 59*, 197–201.
Giri, A., Ahuja, P. S., & Ajay Kumar, P. V. (1993). Somatic embryogenesis and plant regeneration from callus cultures of Aconitum heterophyllum Wall. *Plant Cell, Tissue and Organ Culture, 32*(2), 213–218.
Giri, A., Banerjee, S., Ahuja, P. S., & Giri, C. C. (1997). Production of hairy roots in Aconitum heterophyllum Wall. using Agrobacterium rhizogenes. *In Vitro Cellular & Developmental Biology-Plant, 33*(4), 280–284.
Gochar, M., Goyal, P., Katoch, K., Wani, M., Gupta, A. P., Pandey, D. K., Lone, J. F., Gairola, S., & Gupta, S. (2024). Enhanced steroidal alkaloids accumulation under abiotic elicitation of in vitro cultures of Fritillaria roylei using response surface methodology. *In Vitro Cellular & Developmental Biology. Plant, 60*(5), 711–724.
Grabherr, G. (2009). Biodiversity in the high ranges of the Alps: Ethnobotanical and climate change perspectives. *Global Environmental Change, 19*(2), 167–172.
Guo, M., Gao, M., & Liu, C. Z. (2007). In vitro propagation of an endangered medicinal plant Saussurea involucrata Kar. et Kir. *Plant Cell Reports, 26*, 261–265.
Guo, B., Stiles, A. R., & Liu, C. Z. (2013). Changes in endogenous hormones and oxidative burst as the biochemical basis for enhanced shoot organogenesis in cold-treated Saussurea involucrata explants. *Acta Physiol Plant, 35*, 283–287.
Guo-feng, L., He-chun, Y., Jiao-wang, D., Ling-cheng, J., Long-hua, S., & Li-min, G. (1992). Cryopreservation of calli of Arnebia euchroma. *Journal of Integrative Plant Biology, 34*(12).
Gupta, S., Chaturvedi, P., & Joshi, K. (2016). Micropropagation and bioaugmentation of Picroside in leaves of in vitro cultured plants of Picrorhiza kurrooa Royle ex Benth. *National Academy Science Letters, 39*, 157–161.
Hadi, A., Rafiq, S., Singh, S., Ali, S., Nawchoo, I. A., & Wagay, N. A. (2023). Seed viability, seed germination, and micropropagation of Aconitum violaceum Jacq. ex Stapf—A rare and threatened medicinal plant of Ladakh Himalaya. *In Vitro Cellular & Developmental Biology-Plant, 59*(3), 315–328.
Hao, Y. R., Li, M. S., & Wu, Y. W. (1982). Callus induction and plant regeneration in tissue culture of *Fritillaria pallidiflora. Acta Botanica Boreali-Occidentalia Sinica, 2*, 38–43.
He, Y., Sun, H., Fan, H., He, C., Guo, Q., & Zhang, Y. (2023). Vitrification-cryopreservation of dormant buds of Fritillaria przewalskii Maxim. and evaluation of its histology injuries and genetic fidelity of cryo-derived plants. *Current Plant Biology, 1*, 1–22.
Huang, B., Han, L., Li, S., & Yan, C. (2015). Optimization of induction, subculture conditions, and growth kinetics of *Angelica sinensis* (Oliv.) Diels callus. *Pharmacognosy Magazine, 11*, 574–578.
Huang, T., Liu, D., Cui, X., Li, M., Jin, L., Paré, P. W., et al. (2023). In vitro bioactive metabolite production and plant regeneration of medicinal plant Angelica sinensis. *Industrial Crops and Products, 194*, 116276.
Hussain, A., Qarshi, I. A., Nazir, H., & Ullah, I. (2012). Plant tissue culture: current status and opportunities. *Recent Advances in Plant In Vitro Culture, 6*(10), 1–28.

Jabeen, N., Shawl, A. S., Dar, G. H., Jan, A., & Sultan, P. (2006). Callus induction and organogenesis from explants of Aconitum heterophyllum medicinal plant. *Biotechnology, 5*(3), 287–291.

Jagdish, S., Joginder, S., & Tewari, V. P. (2018). Screening and evaluation of superior Chemotypes of Podophyllum hexandrum Royle from different geographical locations of North-west Himalayas. *Journal of Plant Chemical and Ecophysiology, 3*(1), 1021.

Jan, S., & Abbas, N. (2018). Chemistry of Himalayan phytochemicals. In S. Jan & N. Abbas (Eds.), *Himalayan phytochemicals* (pp. 121–166). Elsevier.

Jan, A., Thomas, G., Shawl, A. S., Jabeen, N., & Kozgar, M. I. (2010). Improved micropropagation protocol of an endangered medicinal plant-Picrorhiza kurrooa Royle ex Benth. promptly through auxin treatments. *Chiang Mai Journal of Science, 37*(2), 304–313.

Jian-feng, L. I. U., Xiu-feng, Y. A. N., Yun-qing, C. H. E. N. G., & Xiao-mei, Z. H. O. U. (2007). Cryopreservation of calli by vitrification and plant regeneration of Rhodiola sachalinensis. *Journal of Beijing Forestry University, 29*(2), 147–151.

Jiang, B. O., Yang, Y. G., Guo, Y. M., Guo, Z. C., & Chen, Y. Z. (2005). Thidiazuron-induced in vitro shoot organogenesis of the medicinal plant Arnebia euchroma (Royle) Johnst. *In Vitro Cellular & Developmental Biology-Plant, 41*(5), 677–681.

Johnson, T. S., Narayan, S. B., & Narayana, D. B. (1997). Rapid in vitro propagation of Saussurea lappa-an endangered medicinal plant through multiple shoot cultures. *In Vitro Cellular & Developmental Biology, 33*, 128–130.

Joshi, M., & Dhar, U. (2003). In vitro propagation of Saussurea obvallata (DC.) Edgew-an endangered ethnoreligious medicinal herb of Himalaya. *Plant Cell Reports, 21*, 933–939.

Joshi, P., & Dhawan, V. (2007a). Axillary multiplication of Swertia chirayita (Roxb. Ex Fleming) H. Karst., a critically endangered medicinal herb of temperate Himalayas. *In Vitro Cellular & Developmental Biology – Plant, 43*, 631e638.

Joshi, P., & Dhawan, V. (2007b). Analysis of genetic diversity among Swertia chirayitagenotypes. *Biologia Plantarum, 51*(4), 764e768.

Joshi, S. K., Dhar, U., & Andola, H. C. (2007). *In vitro* bulblet regeneration and evaluation of *Fritillaria roylei* Hook. – a high value medicinal herb of the Himalaya. *Acta Horticulturae, 756*, 75–84.

Joshi, R. K., Satyal, P., & Setzer, W. N. (2016). Himalayan aromatic medicinal plants: A review of their ethnopharmacology, volatile phytochemistry, and biological activities. *Medicine, 3*(1), 6.

Jun, L. (1998). Research on the propagaton of Trillium tsehonoskii M. in vitro. *China Journal of Chinese Materla Medica, 23*(9), 522–524.

Kakkar, R. A., Haneen, M. A., Parida, A. C., & Sharma, G. (2023). The known, unknown, and the intriguing about members of a critically endangered traditional medicinal plant genus aconitum. *Frontiers in Plant Science, 14*, 1139215.

Kanaujia, K. A., Maurya, N., & Arya, D. K. (2023). Exploring the medicinal potential of trillium govanianum (a threatened plant): Current insights, challenges, and future prospects. *World Journal of Pharmacy and Pharmaceutical Sciences, 12*, 433–463.

Kaur, R., Sood, M., Chander, S., Mahajan, R., Kumar, V., & Sharma, D. R. (1999). In vitro propagation of Valeriana jatamansi. *Plant Cell, Tissue and Organ Culture, 59*, 227–229.

Kawoosa, T., Singh, H., Kumar, A., Sharma, S. K., Devi, K., Dutt, S., Vats, S. K., Sharma, M., Ahuja, P. S., & Kumar, S. (2010). Light and temperature regulated terpene biosynthesis: Hepatoprotective monoterpene picroside accumulation in Picrorhiza kurrooa. *Functional & Integrative Genomics, 10*(3), 393–404.

Khan, A., Shah, A. H., & Ali, N. (2021). In-vitro propagation and phytochemical profiling of a highly medicinal and endemic plant species of the Himalayan region (Saussurea costus). *Scientific Reports, 11*(1), 23575.

Khan, Z., Khan, B., Shah, S. T., Iqbal, J., Basit, A., Khan, M. S., Iqbal, W., Elsadek, M. F., Jamal, A., Ali, M. A., & Prisa, D. (2024). Preserving nature's treasure: A journey into the in vitro conservation and micropropagation of the endangered medicinal marvel—Podophyllum hexandrum Royle. *Horticulturae, 10*(8), 809.

Khan, Z., Khan, B., Shah, S. T., Iqbal, J., Basit, A., Ur Rehman, M., Ahmad, M. A., Fahim, M., Saeed, M. F., Székely, Á., & Jamal, A. (2025). Improved conservation of callus and rhizome microcuttings of Podophyllum hexandrum germplasm using the slow growth storage approach. *Scientific Reports, 15*(1), 28173.

Khoja, A. A., Hamid, M., Hamid, B., Waheed, M., Jameel, M. A., Bussmann, R. W., & Haq, S. M. (2024). Himalayan Apiaceae-A comprehensive ecological and ethnobotanical evaluation. *Ethnobotany Research and Applications, 27*, 1–18.

Kizil, S., Sogut, T., Sesiz, U., & Khawar, K. (2016). Accelerated micropropagation of endemic *Fritillaria aurea* Schott. *Scientific Bulletin Series F, 20*, 99–104.

Kukulezanka, K., Kromor, K., & Czastka, B. (1989). Propagation of *Fritillaria meleagris* L. through tissue culture. *Acta Horticulturae, 251*, 147–153.

Kumar, S., & Khanduri, V. P. (2024). Impact of climate change on the Himalayan alpine treeline vegetation. *Heliyon, 10*(23), e40797.

Kumar, V., Chauhan, R. S., & Tandon, C. (2017). Biosynthesis and therapeutic implications of iridoid glycosides from Picrorhiza genus: The road ahead. *Journal of Plant Biochemistry and Biotechnology, 26*(1), 1–13.

Kumar, P., Partap, M., Ashrita, Rana, D., Kumar, P., & Warghat, A. R. (2020). Metabolite and expression profiling of steroidal alkaloids in wild tissues compared to bulb derived *in vitro* cultures of *Fritillaria roylei* – High value critically endangered Himalayan medicinal herb. *Industrial Crops and Products, 145*, 111945.

Kumar, P., Ashrita, Partap, M., & Warghat, A. R. (2021). Fritillaria roylei. In *Himalayan medicinal plants* (pp. 57–66). Academic.

Kumar, V., Garima, Sharma, R., Thakur, A., Sharma, P., Sharma, Y. P., Thakur, A., & Sharma, S. (2024a). Conserving industrially important endangered medicinal herb trillium govanianum (Himalayan Trillium) through biotechnological based interventions. *Plant Cell, Tissue and Organ Culture, 159*(1), 3.

Kumar, V., Sharma, R. K., Singh, S. K., Arora, K., Joshi, M. C., Rawat, P., & Singh, M. (2024b). Morphology, chemical composition, and biological activities of Aconitum balfourii: A comprehensive review of important Himalayan and highly exploited medicinal plant. *Vegetos, 37*(5), 1690–1704.

Kumari, M., Kumar, P., Saini, V., Joshi, R., Shankar, R., & Kumar, R. (2025). Transcriptional landscape illustrates the diversified adaptation of medicinal plants to multifactorial stress combinations linked with high altitude. *Planta, 261*(5), 1–25.

Kuo, C. L., Agrawal, D. C., Chang, H. C., Chiu, Y. T., Huang, C. P., Chen, Y. L., & Tsay, H. S. (2015). In vitro culture and production of syringin and rutin in Saussurea involucrata (Kar. et Kir.)–an endangered medicinal plant. *Botanical Studies, 56*(3), 1–8.

Lal, N., & Ahuja, P. S. (1996). Plantlet regeneration from callus in *Picrorhiza kurrooa* Royle ex Benth.—An endangered medicinal plant. *Plant Tissue Culture, 6*(2), 127–134.

Lal, N., Ahuja, P. S., Kukreja, A. K., & Pandey, B. (1988). Clonal propagation of Picrorhiza kurroa Royle ex Benth. by shoot tip culture. *Plant Cell Reports, 7*(3), 202–205.

Lata, H., Moraes, R. M., Bertoni, B., & Pereira, A. M. (2010). *In vitro* germplasm conservation of *Podophyllum peltatum* L. under slow growth conditions. *In Vitro Cellular & Developmental Biology-Plant, 46*(1), 22–27.

Leng, R., Harrison, S., & Anderson, K. (2023). Himalayan alpine ecohydrology: An urgent scientific concern in a changing climate. *Ambio, 52*(2), 390–410.

Liang, E., Wang, Y., Piao, S., Lu, X., Camarero, J. J., Zhu, H., Zhu, L., Ellison, A. M., Ciais, P., & Peñuelas, J. (2016). Species interactions slow warming-induced upward shifts of treelines on the Tibetan plateau. *Proceedings of the National Academy of Sciences, 113*(16), 4380–4385.

Liu, Y., Wang, W., & Li, Z. (1996). Experiment on the rapid reproduction of *Fritillaria taipaiensis* P. Y. Li. *China Journal of Chinese Materia Médica, 21*(1), 15–17.

Mahajan, R., Kapoor, N., & Singh, I. (2015). Effect of growth regulators on in vitro cultures of Aconitum heterophyllum: An endangered medicinal plant. *International Journal of Pure and Applied Bioscience, 3*, 50e55.

Mahajan, R., Kapoor, N., & Singh, I. (2016). Somatic embryogenesis and callus proliferation in Picrorhiza kurroa Royle ex. Benth. *Journal of Experimental Biology and Agricultural Sciences, 4*, 201–209.

Majumder, A., Roychowdhury, D., & Ray, S. (2023). Biotechnological approaches for ex situ conservation of medicinal plants. In S. Jha & M. Halder (Eds.), *Medicinal plants: Biodiversity, Biotechnology and conservation* (Sustainable development and biodiversity, vol 33) (pp. 729–800). Springer.

Malik, S., Sharma, M., & Ahuja, P. S. (2016). An efficient and economic method for in vitro propagation of Arnebia euchroma using liquid culture system. *American Journal of Biotechnology and Medical Research, 1*(1), 19–25.

Manjkhola, S., Dhar, U., & Joshi, M. (2005). Organogenesis, embryogenesis, and synthetic seed production in Arnebia euchroma—A critically endangered medicinal plant of the Himalaya. *In Vitro Cellular & Developmental Biology-Plant, 41*(3), 244–248.

Mathur, J., Ahuja, P. S., Lal, N., & Mathur, A. K. (1989). Propagation of Valeriana wallichii DC. using encapsulated apical and axial shoot buds. *Plant Science, 60*(1), 111–116.

Matvieieva, N., Belokurova, V., Ratushniak, Y., Shcherbak, N., & Kuchuk, M. (2023). In vitro direct shoot regeneration from Rhodiola rosea L. leaf explants. *Biotechnologia Acta, 16*(3), 45–50.

Mehreen, H., Zafar, J., & Zishan, G. (2020). In vitro regeneration of Saussurea heteromalla (D. Don.) hand-Mazz using auxins and current status in Galiyat Abbottabad. *Ukrainian Journal of Ecology, 10*(5), 1–7.

Mishra, J., Singh, M., Palni, L. M. S., & Nandi, S. K. (2011). Assessment of genetic fidelity of encapsulated microshoots of Picrorhiza kurrooa. *Plant Cell, Tissue and Organ Culture, 104*, 181–186.

Mondal, T. K., Bantawa, P., Sarkar, B., Ghosh, P., & Chand, P. K. (2013). Cellular differentiation, regeneration, and secondary metabolite production in medicinal Picrorhiza spp. *Plant Cell, Tissue and Organ Culture, 112*(2), 143–158.

Murashige, T., & Skoog, F. (1962). A revised medium for rapid growth and bio assays with tobacco tissue cultures. *Physiologia Plantarum, 15*(3), 473–497.

Nadeem, M., Palni, L. M. S., Purohit, A. N., Pandey, H., & Nandi, S. K. (2000). Propagation and conservation of Podophyllum hexandrum Royle: An important medicinal herb. *Biological Conservation, 92*, 121–129.

Oseni, O. M., Pande, V., & Nailwal, T. K. (2018). A review on plant tissue culture, a technique for propagation and conservation of endangered plant species. *International Journal of Current Microbiology and Applied Sciences, 7*(7), 3778–3786.

Ozcan, S., Parmaksiz, I., Mirici, S., Cocu, S., Uranbey, S., Ipek, A., Sancak, C., Sarihan, E., Gurbuz, B., Sevimay, C. S., & Arslan, N. (2007). Efficient *in vitro* bulblet regeneration from immature embryos of endemic and endangered geophyte species in *Sternbergia, Muscari* and *Fritillaria* genera. In Z. Xu, J. Li, Y. Xue, & W. Yang (Eds.), *Biotechnology and sustainable agriculture 2006 and beyond* (pp. 381–383). Springer.

Pal, M., & Chaudhury, A. (2010). High frequency direct plant regeneration, micropropagation and shikonin induction in Arnebia hispidissima. *Journal of Crop Science and Biotechnology, 13*(1), 13–19.

Pan, S. Y., Litscher, G., Gao, S. H., Zhou, S. F., Yu, Z. L., Chen, H. Q., Zhang, S. F., Tang, M. K., Sun, J. N., & Ko, K. M. (2014). Historical perspective of traditional indigenous medical practices: The current renaissance and conservation of herbal resources. *Evidence-Based Complementary and Alternative Medicine, 2014*, 525340.

Pandey, H., Nandi, S. K., Kumar, A., Palni, U. T., Chandra, B., & Palni, L. (2004). In vitro propagation of Aconitum balfourii Stapf: An important aconite of the Himalayan alpines. *The Journal of Horticultural Science and Biotechnology, 79*, 34–41.

Pandey, M., Dhar, U., Samant, S. S., & Thengane, S. R. (2011). Recurrent somatic embryogenesis and plant regeneration in *Angelica glauca* Edgew., a critically endangered medicinal plant of the Western Himalaya. *The Journal of Horticultural Science and Biotechnology, 86*, 493–498.

Pandey, V., Bhatt, I. D., & Nandi, S. K. (2019). Environmental stresses in Himalayan medicinal plants: Research needs and future priorities. *Biodiversity and Conservation, 28*(8), 2431–2455.

Pence, V. C., & Soukup, V. G. (1993). Factors affecting the initiation of mini rhizomes from Trillium erectum and T. grandiflorumtissuesin vitro. *Plant Cell, Tissue and Organ Culture, 35*, 229–235.

Petric, M., Jevremovic, S., Trifunovic, M., Tadic, V., Milosevic, S., Dragicevic, M., & Subotic, A. (2013). The effect of low temperature and GA_3 treatments on dormancy breaking and activity of antioxidant enzymes in *Fritillaria meleagris* bulblets cultured *in vitro*. *Acta Physiologiae Plantarum, 35*(11), 3223–3236.

Pundir, A., Kad, A., & Sood, H. (2019). Elicitation of salidroside under tissue culture conditions in the trans- Himalayan plant Rhodiola imbricata. *WSEAS Transactions on Biology and Biomedicine, 16*, 75e89.

Purohit, S., Rawat, V., Jugran, A. K., Singh, R. V., Bhatt, I. D., & Nandi, S. K. (2015). Micropropagation and genetic fidelity analysis in Valeriana jatamansi Jones. *Journal of Applied Research on Medicinal and Aromatic Plants, 2*, 15–20.

Quadri, R. R., Kamili, A. N., Silva, J. A. T., & Shah, A. M. (2012). *In vitro* multiplication of Arnebia benthamii Wall., a critically endangered medicinal herb of the Western Himalayas. *Medicinal and Aromatic Plant Science and Biotechnology, 6*, 54–57.

Rajesh, M., Sivanandhan, G., Arun, M., Vasudevan, V., Theboral, J., Girija, S., & Ganapathi, A. (2014). Factors influencing podophyllotoxin production in adventitious root culture of Podophyllum hexandrum Royle. *Acta Physiologiae Plantarum, 36*, 1009–1021.

Rana, B., & Sreenivasulu, Y. (2013). Protein changes during ethanol induced seed germination in Aconitum heterophyllum. *Plant Science, 198*, 27–38.

Rodriguez, R. E. (1997). Micropropagation of *Valeriana edulis* ssp. Procera. *Planta Medica, 63*, 274–275.

Sadr, S. S., Zakizadeh, H., Naghavi, M. R., & Olfati, J. A. (2018). Regeneration of an endangered Fritillaria (*Fritillaria raddeana*) species via petal and leaf explants. *Iranian Journal of Rangelands and Forests Plant Breeding and Genetic Research, 26*(1), 71–82.

Saini, S., Dubey, A., Kumar, A., Taj, G., Budhori, S., & Kumar, V. A. (2018). In vitro direct rhizogenesis of an endangered medicinal herb Valeriana jatamansi Jones. *International Journal of Current Microbiology and Applied Sciences, 7*(4), 3108–3114.

Salaria, N., Kumar, R., & Kumar, A. (2025). Conservation and management of himalayan plant biodiversity. In *Biodiversity of the Himalaya* (pp. 71–90). Apple Academic Press.

Samaddar, T., Jha, S., & Jha, T. B. (2014). Indian Swertia from Eastern Himalaya: Strategies of conservation and biotechnological improvements. In J. Rybczyński, M. Davey, & A. Mikuła (Eds.), *The Gentianaceae – Volume 1: Characterization and ecology* (pp. 279–301). Springer.

Sarker, S. D., & Nahar, L. (2004). Natural medicine: The genus Angelica. *Current Medicinal Chemistry, 11*(11), 1479–1500.

Semwal, R. B., Joshi, K., Pandian, A., Badoni, P. P., & Semwal, D. K. (2020). Biological applications and secondary metabolites of Saussurea costus (Falc.) Lipsch. *Journal of Convergence Knowledge Holist Health, 4*(1), 1–8.

Seydi, S., Sedaghathoor, S., & Kaviani, B. (2020). Plant regeneration by organogenesis from bulbous explants in *Fritillaria imperialis* L., a wild rare ornamental species at the risk of extinction. *Advances in Horticultural Science., 33*(4), 503–510.

Shailja, KanwarK, Soni, M., & Singh, A. (2017). In vitro propagation and conservation of an endangered high value medicinal herb Swertia chirayita of temperate Himalayas. *Indian Journal of Plant Physiology, 22*, 247–257.

Sharma, S. (2016). Effect of temperature on in vitro organogenesis of Rhodiola imbricata Edgew. e a medicinal herb. *World Journal of Pharmacy and Pharmaceutical Sciences, 5*(12), 1228e1243.

Sharma, N., & Sharma, B. (2003). Cryopreservation of shoot tips of Picrorhiza kurroa Royle ex Benth, an indigenous endangered medicinal plant, through vitrification. *CryoLetters, 24*, 181–190.

Sharma, M., & Sharma, M. (2024). Assessment of current status and conservation strategies of some high valued medicinal plants from himalayan regions. *Annali di Botanica, 14*(1).

Sharma, S., & Thokchom, R. (2014). A review on endangered medicinal plants of India and their conservation. *Journal of Crop and Weed, 10*(2), 205–218.

Sharma, V., Kamal, B., Srivastava, N., Dobriyal, A. K., & Jadon, V. S. (2014). In vitro flower induction from shoots regenerated from cultured axillary buds of endangered medicinal herb Swertia chirayita H. Karst. *Biotechnology Research International, 2014*(1), 264690.

Sharma, V., Belwal, N., Kamal, B., Dobriyal, A. K., & Jadon, V. S. (2016a). Assessment of genetic fidelity of in vitro raised plants in Swertia chirayita through ISSR, RAPD analysis and peroxidase profiling during organogenesis. *Brazilian Archives of Biology and Technology, 59*, e16160389.

Sharma, N., Chauhan, R. S., & Sood, H. (2016b). Discerning picroside-I biosynthesis via molecular dissection of in vitro shoot regeneration in Picrorhiza kurroa. *Plant Cell Reports, 35*(8), 1601–1615.

Sharma, N., Gowthami, R., & Pandey, R. (2019). Synthetic seeds: A valuable adjunct for conservation of medicinal plants. In *Synthetic seeds: Germplasm regeneration, preservation and prospects* (pp. 181–216). Springer.

Sharma, S., Parasher, K., Mukherjee, P., & Sharma, Y. P. (2021). Cryopreservation of a threatened medicinal plant, Valeriana jatamansi Jones, using vitrification and assessment of biosynthetic stability of regenerants. *CryoLetters, 42*(5), 300–308.

Sharma, S., Kumar, P., Sharma, R., & Warghat, A. R. (2023). In vitro propagation and omics breakthroughs for understanding specialized metabolite production in high-value Himalayan Fritillaria species. *Industrial Crops and Products, 205*, 117541.

Sharma, S., Negi, S., Kumar, P., & Irfan, M. (2025a). Cellular strategies for surviving the alpine extremes: Methylerythritol phosphate pathway-driven isoprenoid biosynthesis and stress resilience. *Protoplasma, 262*, 1–20.

Sharma, S., Negi, S., Kumar, P., & Irfan, M. (2025b). Nitric oxide dynamics in high-altitude medicinal plants: Role in stress adaptation, signaling, and phytohormonal interactions. *Physiologia Plantarum, 177*(3), e70342.

Shiau, Y., Tai, C., Chen, C., & Tsay, H. (2000). Studies on the tissue culture of *Fritillaria hupehensis* Hsiao et KC Hsia. I. The induction of bulbscale and embryogenic callus from different source explants. *Journal of Agricultural Research of China., 49*(4), 29–38.

Singh, A., & Singh, K. S. (2022). Bioactive compounds from polar regions: An account of chemical ecology and biotechnological applications. *Current Organic Chemistry, 26*(11), 1055–1087.

Singh, C., Chauhan, N., Upadhyay, S. K., Singh, R., & Rani, A. (2021). The Himalayan natural resources: Challenges and conservation for sustainable development. *Journal of Pharmacognosy and Phytochemistry., 10*(1), 1643–1648.

Sood, H. (2020). Production of medicinal compounds from endangered and commercially important medicinal plants through cell and tissue culture technology for herbal industry, bioactive compounds. In *IntechOpen* (pp. 1–16).

Sood, H., & Chauhan, R. S. (2009). High frequency callus induction and plantlet regeneration from different explants of *Picrorhiza kurrooa*—A medicinal herb of Himalayas. *African Journal of Biotechnology, 8*(9), 1965–1972.

Sood, H., & Chauhan, R. S. (2010). Biosynthesis and accumulation of a medicinal compound, Picroside-I, in cultures of Picrorhiza kurroa Royle ex Benth. *Plant Cell, Tissue and Organ Culture (PCTOC), 100*(1), 113–117.

Sultan, P., Shawl, A. S., Ramteke, P. W., Jan, A., Chisti, N., Jabeen, N., & Shabir, S. (2006). *In vitro* propagation for mass multiplication of *Podophyllum hexandrum*: A high value medicinal herb. *Asian Journal of Plant Sciences, 5*, 179–184.

Sultan, P., Shawl, A. S., Fayaz, A. S., & Ramteke, P. W. (2009). *In vitro* embryogenesis and marker guided analysis of *Podophyllum hexandrum*: An endangered medicinal plant. *International Journal of Biodiversity and Conservation, 1*, 215–222.

Sun, C. S., Chu, C. C., & Wang, C. C. (1977). Callus formation and organ regeneration in the tissue culture of *Fritillaria thunbergii* Miq. *Acta Botanica Sinica, 19*, 161–162.

Talreja, S., & Tiwari, S. (2023). A comprehensive review of Aconitum heterophyllum. *Journal of Ayurveda and Integrated Medical Sciences, 8*(10), 195–201.

Tao, Q., Han, G., Ma, B., Jia, H., Zhao, C., Li, W., & Yan, Z. (2024). In vitro propagation and artificial seed production of Fritillaria cirrhosa D. Don, an Endangered Medicinal Plant. *Phyton (0031-9457), 93*(6).

Tasheva, K., & Kosturkova, G. (2010). Bulgarian golden root in vitro cultures for micropropagation and reintroduction. *Central European Journal of Biology, 5*(6), 853e863.

Thapa, B., Upadhyay, K., Chhetri, B., Gurung, N., & Meena, N. K. (2024). From wild to cultivated: Unraveling the potential of Swertia chirayita (Roxb.) H.Karst. for crop improvement and conservation. In A. P. Mishra, A. Kumar, N. Chandra, G. Singh, & C. B. Pande (Eds.), *Threatened medicinal plants in the Indian Himalayan region* (World Sustainability Series) (pp. 183–208). Springer.

Upadhyay, R., Arumugam, N., & Bhojwani, S. S. (1989). In vitro propagation of Picrorhiza kurroa Royle ex Benth. e An endangered species of medicinal importance. *Phytomorphology, 39*, 235e242.

Verma, R., Tapwal, A., Kumar, D., & Puri, S. (2021). Antimicrobial potential and phytochemical profiling of ethnomedicinal plant trillium govanianum Wall. ex D. Don in Western Himalaya. *Journal of Herbal Medicine, 29*, 100491.

Verma, D., Pundir, S., Goyal, R., Aljabali, A. A., Dua, K., Chellappan, D. K., et al. (2024). Versatile therapeutic potential of Swertia chirayita (roxb.) H. Karst: A review. *Chemistry Africa, 7*(6), 2941–2961.

Vishvamitera, S., Baghla, S., & Chauhan, R. (2024). Saussurea costus (Falc.) Lipsch: Botanical, biochemical, therapeutical aspects and conservation strategies. In *Ethnopharmacology and OMICS advances in medicinal plants volume 2: Revealing the secrets of medicinal plants* (pp. 75–92). Springer Nature Singapore.

Wang, L., An, L., Hu, Y., Wei, L., & Li, Y. (2009). Influence of phytohormonesand medium on the shoot regeneration from leaf of Swertia chirata Buch.-Ham. ex Wall. in vitro. *African Journal of Biotechnology, l8*(11), 2513e2517.

Wani, T. A., Kaloo, Z. A., & Dangroo, N. A. (2022). Aconitum heterophyllum Wall. ex Royle: A critically endangered medicinal herb with rich potential for use in medicine. *Journal of Integrative Medicine, 20*(2), 104–113.

Warghat, A. R., Bajpai, P. K., Rewang, S., Kapoor, S., Kumar, J., Chaurasia, O. P., & Srivastava, R. B. (2016). In vitro callus induction and plantlet regeneration of Saussurea lappa (Clarke.) from Ladakh region of India. *Proceedings of the National Academy of Sciences India Section B – Biological Sciences, 86*(2), 651–660.

Yang, Y., & Sun, H. (2009). The bracts of Saussurea velutina (Asteraceae) protect inflorescences from fluctuating weather at high elevations of the Hengduan Mountains, Southwestern China. *Arctic, Antarctic, and Alpine Research, 41*(4), 515–521.

Yang, M., Sun, L., Yu, Y., Zhang, H., Malik, I., Wistuba, M., & Yu, R. (2023). Predicting the potential geographical distribution of Rhodiola L. in China under climate change scenarios. *Plants, 12*(21), 3735.

Zaib-Un-Nisa, S. J., Anwar, M., Sajad, S. H., Farooq, G., & Ali, H. (2019). Micropropagation through apical shoot explants and morphogenic potential of different explants of Saussurea lappa: An endangered medicinal plant. *Pure and Applied Biology, 8*(1), 585–592.

Zhang, Y. H., Lu, Y. Y., He, C. Y., & Gao, S. F. (2019). A method for cell suspension culture and plant regeneration of Angelica sinensis (Oliv.) Diels. *Plant Cell, Tissue and Organ Culture, 136*, 313–322.

Zhang, Z., Xing, C., Su, H., Wang, J., Qi, Y., & Li, M. (2024). In vitro plant regeneration and bioactive metabolite production of endangered medicinal plant Fritillaria cirrhosa. *Current Plant Biology, 39*, 100363.

Zhao, Y., Sun, W., Wang, Y., Saxena, P. K., & Liu, C. Z. (2012). Improved mass multiplication Rhodiola crenulata shoots using temporary immersion bioreactor with forced ventilation. *Applied Biochemistry and Biotechnology, 166*(6), 1480e1490.

Zhou, B. J. (1983). A study on *in vitro* culture of bulb of *Fritillaria ussuriensis* Maxim. *Plant Physiology Communication, 1*, 30.

Zych, M., Furmanowa, M., Krajewska-Patan, A., Lowicka, A., Dreger, M., & Mendlewska, S. (2005). *Micropropagation of Rhodiola kirilowii plants using encapsulated axillary buds and callus* (Acta Biologica Cracoviensia Series Botanica, 2nd ed. 47), pp. 83e87.

[illegible]

Verma, R. [illegible]

[illegible] active metabolite production of endangered medicinal plant [illegible]

[illegible]

Zy[illegible] M. [illegible] [illegible] Series [illegible]

Microbial Interactions in Alpine Ecosystems

13

Suresh Kumar, Deepti Gupta, and Arti Devi

Abstract

Alpine ecosystem supports a wide variety of specialized microbial communities because of their extreme climatic conditions, which include low temperatures, nutritional shortages, intense UV radiations and varying moisture regimes. These microorganisms which include bacteria, viruses, fungi, archaea and algae are essential to several aspects of ecosystem functioning such as soil formation organic matter breakdown, cycling of nutrients and plantmicrobe symbioses. In alpine environments, microbial interactions are complex and constantly shifting. This chapter explores the nature of microbial interactions in alpine environments, highlighting how commensalism, competition, and mutualism influences the the makeup and operation of microbial communities. Different aspects of microbial interactions like relationships between microbes and plants, functions of microbes in nutrient cycling, breakdown of soils and organic matter stability are discussed in this chapter. Furthermore, chapter also emphasises the possible applications of alpine microorganisms in ecological restoration and emphasizes how climate change affects microbial activity in these challenging environments. Functions of microbial interactions in alpine ecosystems are also explored along with their wider implications for climate resilience, biogeochemical cycling and ecological balance.

Keywords

Alpine ecosystems · Microbial interactions · Nutrient cycling · Ecological restoration · Climate resilience

S. Kumar · D. Gupta (✉) · A. Devi
Department of Bio-Sciences, Himachal Pradesh University, Shimla, Himachal Pradesh, India

P. Kumar et al. (eds.), *Climate Resilience and Molecular Adaptation in Alpine Medicinal Plants*, https://doi.org/10.1007/978-981-95-6923-6_13

1 Introduction

Alpine ecosystems are characterized by harsh environment conditions and they are located in high-altitude regions above the treeline. They experience a variety of environmental factors, including low temperatures, decreased air pressure, strong ultraviolet (UV) radiations, limited nutrition supplies, and short growth seasons. The first colonizers are microbial communities, which play a crucial role in fundamental transformation, carbon and nitrogen fixation, soil fertility, and the development of pioneer plants (Ciccazzo et al., 2016). Diverse microbial populations including bacteria, viruses, fungi, algae and archaea have adapted to and flourished in these environments in spite of the adverse circumstances (Lazzaro et al., 2015). Climate, biological processes, parent material, surroundings, and time are the primary determinants of soil production in alpine ecosystems, just as they are in lowland ecosystems. On other side, biological activity as well as physical and chemical processes are significantly impacted by these topographically associated microclimatic conditions (Donhauser & Frey, 2018). Microbial communities' capacity to maintain stability is mainly determined by their wide range and complicated intercommunal relationships. These inter-microbiome interactions result in complexly connected microbial networks in alpine ecosystem (Li et al., 2025). Microbial communities show different type of interactions such as commensalism, parasitism, mutualism and competition. Mutualism, in which both microbial partners benefit from the relationship, is especially prevalent among microbial communities (Read & Haselwandter, 1981). On the other hand, parasitism and competition are relationships in which one creature gains an advantage over another, or both suffer as a result of competition for scarce resources like carbon or nitrogen molecules (Lazzaro et al., 2015). Soil microbes are crucial to the nutrient cycle of ecosystems because they are the primary driver of the breakdown of organic matter (OM). Changes in plant composition, microbial activity, and soil microbial structure may result from rising temperatures. In alpine regions, these changes may directly impact biogeochemical processes that impact ecosystem functioning and soil carbon storage (Djukic et al., 2010). The wider implications of microbial interactions for climate resilience are also interesting. Alpine microorganisms are essential to feedback loops in climate systems because they control the fluxes of greenhouse gases including CO_2, CH_4 and N_2O (Bardgett et al., 2008).

2 Microbial Community in Alpine Ecosystem

Extreme weather conditions including freezing temperatures, daily temperature swings and low oxygen levels are common in alpine ecosystem. These pressures favour bacteria that are psychrophilic or cold-loving and psychrotolerant (Margesin & Miteva, 2011). Different types of bacteria like Proteobacteria, Actinobacteria, and Acidobacteria are dominant taxa that are engaged in nitrogen cycling and the breakdown of organic materials in alpine ecosystem (Lazzaro et al., 2015). All mycorrhizal fungi have the trait of their extensive nutrient harvesting in the rhizosphere as well as in the surrounding soil. The structure and operation of the soil food web are maintained by their neutral, positive, or negative interactions with a diverse range of bacteria, saprotrophic fungi, protists, and soil animals. Additionally, the soil food web regulates the cycling of nutrients from soil organic matter (SOM), which in turn regulates the quantity of carbon dioxide (CO_2) that enters the soil and is either stored or breathed back into the atmosphere (Frey, 2019). Because of the various abiotic circumstances in alpine ecosystem, archaea are often found in the rhizosphere and endosphere and uncommon in endosphere. Apart from adaptation to long-term energy stress and abiotic factors, archaeal colonization is dependent on biotic factors such as bacterial competition, which may have led to microniche differentiation.They are vital to the plant as these microbiomes have the ability to change a plant's growth, production, adaptability, diversity, and health (Taffner et al., 2018).

Microalgae have garnered a lot of attention globally in recent decades due to their use in the pharmacological, nutraceutical, and renewable energy sectors. Microalgae allow heterotrophic microorganisms to coexist in their environment because of their significant function as primary producers and the release of dissolved organic nutrients (Krug et al., 2020). Plants of alpine ecosystem are suited to harsh climatic circumstances such as low temperatures, intense UV rays, severe winds, and brief growth seasons. These plants have developed adaptations including cushion growth patterns, hairy leaves, deep root systems, and dwarfism in order to live and flourish in shallow, nutrient-poor soils in alpine ecosystem (Korner, 2023). Over the last several decades, as summer temperatures have increased, alpine plants have spread to high elevations and colonized unvegetated soils. Soil bacterial populations exhibit patterns of co-occurrence with plants that can enhance plant dispersion models. Also, plant interactions with microbes are crucial for plant growth and survival as they colonize previously unvegetated soils as snow melts earlier and plant species have different effects on unvegetated soil microbial communities based on their litter chemistry and litter microbial communities (Mesquita, 2019).

Types of Microbial Interactions in Alpine Ecosystem

Microbial communities in alpine ecosystems are involved in a range of interactions that are essential to resilience, nutrient dynamics and ecosystem functioning regardless of extreme climatic conditions. These interactions are:

2.1 Competition

Microbial populations frequently compete strongly in alpine ecosystems because of the harsh surroundings and scarcity of essential resources including nutrients, space, moisture and energy sources. Microbial community structure and functions are greatly impacted by this competition, which takes place both inside and across populations of bacteria, fungi, archaea, and algae. While mycorrhizal fungi may outcompete saprotrophic fungi for access to plant roots, actinobacteria and protobacteria frequently vie for control of soil niches. Interference competition is a tactic used by certain bacteria to block rivals by secreting enzymes or antibiotics. This microbial competition influences ecosystem functions.

2.2 Commensalism

Commensalism is the type of interaction in which one plant or microbial species benefits from another without significantly altering it. This relationship is crucial in the harsh alpine conditions where species depend on one another for life support and resources of nutrients are limited (Read & Perez-Moreno, 2003; Schmidt et al., 2008). Microbes or non-mycorrhizal plants often benefit from the nutrient-rich rhizosphere of mycorrhizal plants without actively taking part in the mutualistic relationship (Read & Perez-Moreno, 2003).

2.3 Mutualism

It is an ecological relationship in which both the participating species gain benefits from each other. Mutualistic interactions are crucial for survival and ecological stability in alpine ecosystems. These interactions mostly take place inside microbial populations as well as between microorganisms and plants. Mycorrhizal association, lichen symbiosis, plant endophyte interaction and microbial syntropy are some mutualistic interactions seen among microbial communities of alpine ecosystem. Woody or non-woody herbaceous plants (such as Dryas octopetala, Kobresia myosuroides, and Polygonum viviparum) and dwarf shrubs (such as Betula nana and Salix spp.) can have ectomycorrhizal (ECM) fungi as symbionts. Although arbuscular mycorrhizal (AM) connections are common in alpine tundra, there is a wide range in the degree of root colonization. In cold-dominated environments, it is reasonable to assume that mycorrhizal associations will be a primary route for plants to receive nutrients due to the well-known capacity of some mycorrhizal fungi to mobilize both organic and inorganic forms of P and N (Gardes & Dahlberg, 1996). The symbiotic phenotype of nutritionally adapted fungi known as lichens receives fixed carbon from a colony of microscopic cyanobacterial or green algal cells called the photobiont. Macro-lichens, often known as reindeer lichens or even reindeer mosses, have the ability to take over the vegetation in alpine and boreal environments. Macro-lichens are crucial for the health and biodiversity of cold habitats in addition to providing reindeer with a significant winter food supply. Lichen species with podetia that may grow to a height of 12–14 cm dominates lichen heaths in alpine environments; the most significant species are found in the genera Cladonia and Flavocetraria. They are often seen in large numbers in Fennoscandia (the alpine and northern boreal zone) at high elevations and latitudes. Colonizing the root's endophytic fungi may have a variety of impacts on plant hosts, including varying degrees of virulence or even beneficial benefits. As a result, the endophyte-host interaction may be characterized as a continuum that includes parasitism, commensalism, and mutualism. In addition to being crucial to the structure, function, and health of plant communities, endophytes can change food webs, community composition, and ecosystem processes. Additionally, endophytic fungi have been shown to directly benefit certain plant species. They can improve the overall nutrition of the plant, lessen the host plant's water stress, and have been shown to assist certain plants cope with environmental stressors. The resources available for plant development and reproduction are nonetheless diminished since some carbon is transferred from the plant to the fungus (Molina-Montenegro et al., 2015).

3 Role of Microbial Interaction in Alpine Region

Alpine ecosystems are the most attractive and unique domains on Earth. Found in high mountainous regions, These habitats are distinguished by harsh circumstances, including low temperatures, high winds and low growing seasons. The alpine habitat can support a wide variety of specially adapted microbes, despite the severe environmental circumstances. They closely resemble the complexity of alpine soil in that their community structures are intimately linked to climate, vegetation, and soil characteristics. The alpine areas cover over 3% of the earth's geographical surface. Because soil microorganisms regulate ecosystem function and have an impact on several key ecosystem processes, including soil organic matter turnover and biogeochemical cycles, it is essential to comprehend the structure and function of the microbial population. Microbial interaction refers to the various ways in which microorganisms such as bacteria, fungi and protozoa relate to each other and to other organisms including plants, animals and humans. These interactions are fundamental to the structure, function and dynamics of ecosystems and plays crucial role in global biogeochemistry, health, agriculture and industry.

3.1 Key Roles

1. Soil fertility and nutrient cycling: Microbial communities in alpine soils are essential for decomposing organic matter and recycling nutrients, especially given the limited plant growth and organic input at high altitudes. Subalpine soils, for example, show higher microbial activity and nutrient cycling rates compared to higher, more extreme elevations.
2. Adaptation to Harsh Conditions: Microorganisms in alpine regions have developed a range of adaptations to survive and thrive under the harsh conditions characteristic of these environments. Key challenges include low temperatures (often below freezing), high UV radiations, limited nutrients, and variable moisture levels.
3. Community Structure and Dynamics: Microbial community composition and abundance are strongly influenced by seasonal changes and elevation. Bacterial populations, for example, are dominated by cold-adapted species under winter snowfall, but other groups grow more prevalent during seasons without snow.

4. Adaptation to Stress: Microbes in alpine regions form biofilms and biological soil crusts as protective strategies against environmental stresses like UV radiations and temperature fluctuations. These structures also facilitate nutrient acquisition and biological weathering.
5. Impact of Abiotic Factors: Environmental factors including temperature, moisture content, and nutrient availability influence the makeup of microbial communities. Microbial interactions mediate how communities respond to these changing conditions, influencing ecosystem processes like nutrient cycling and decomposition.

4 Impact of Climate Change on Microbial Interaction

Climate change is significantly altering microbial interactions in alpine ecosystems via a variety of channels, with significant implications for nitrogen cycle, soil biogeochemistry, and ecosystem stability. While the specific types of microbes present can vary depending on elevation, soil type and local environmental conditions, several groups are consistently found and are impacted by climatic shifts. Various effects of climate change on microbial communities are summarized as:

1. Seasonal and Climatic Shifts: Climate change is causing earlier snowmelt in alpine regions, which triggers abrupt transitions in the microbiological communities of soil.
2. These shifts are closely linked to rapid changes in arbon and nitrogen cycling. Communities that previously thrived under snow cover (such as BetaProteobacteria) may decline as snowpack duration decreases.
3. Temperature and Moisture: Rising temperatures and altered precipitation patterns affect soil moisture and nutrient availability which in turn influence microbial diversity and community structure. For example, enhanced rainfall has been shown to reduce microbial diversity in some alpine grasslands while warming can have variable effects depending on the ecosystem type.
4. Vegetation Changes: Climate change can lead to upward shifts in vegetation zones (e.g., shrub encroachment) which alters the composition and functional potential of soil microbiomes. Shrubland soils, for instance, have higher proportions of fungi and specific enzymes for decomposing recalcitrant plant biopolymers.
5. Modifications to Microbial Structure and Function: Alpine soil microbial communities exhibit pronounced annual cycles, with different groups dominating in winter (under snow) versus summer (snow-free seasons). Earlier snowmelt and warmer temperatures can disrupt these cycles, leading to changes in the abundance of key microbial groups.
6. Functional Shifts: The functional potential of soil microbiomes such as their ability to decompose organic matter and cycle nutrients, changes with climateinduced shifts in vegetation and soil properties. For example, a higher proportion

of fungi in shrubland soils is associated with increased decomposition of recalcitrant carbon, potentially reducing carbon storage and increasing CO_2 emissions.
7. Diversity Patterns: There is no universal altitudinal pattern for microbial diversity. Some studies show increased diversity with elevation, others show a decrease or a mid-elevation peak. However, environmental conditions related to altitude (e.g., temperature, moisture, nutrients) remain strong drivers of microbial community structure. Fungal communities are also significant, with notable genera such as Mortierella showing distinct adaptations to specific altitudinal zones.

Climate change is reshaping the types and activities of microbes in alpine regions. Predominant groups include Acidobacteria, Actinobacteria, Bacteroidetes, Chloroflexi, Cyanobacteria, Firmicutes, and Proteobacteria are playing important roles. Earlier snowmelt, rising temperatures, altered precipitation, and vegetation shifts are causing significant changes in microbial community structure, diversity, and function, with potential consequences for nutrient cycling and carbon storage in alpine ecosystems. These changes can disrupt the delicate balance of plant-microbe relationships and biogeochemical processes, with major implications for ecosystem function and resilience.

5 Ecological Significance of Microbial Interaction in Alpine Regions

The principal strategies to colonize and establish in various environments include microbe-microbe or microbe-host interactions. These interactions include all ecological aspects resulting in genotype selection like metabolite conversion, metabolite exchange, chemotaxis, signalling as well as genetic exchange. These relationships are linked to the coevolution process, which reduces biotic and abiotic stressors and leads to adaptation and specialization, allowing the occupancy of various niches. Microbial interactions are pivotal for successful establishment and maintenance of microbial population. The physiological flexibility and high adaptive capacity of microorganisms enable them to colonize different ecosytems. Microbial communities in soil adapt their species and functional composition to climatic and nutritional conditions in response to environmental changes (Grayston et al., 2001; Wallenstein & Hall, 2012). Microbial life depends on temperature, humidity, nutrient concentrations, and vegetation, all of which are significantly altered by the seasons. Therefore, it is anticipated that the microbial communities will exhibit similarly substantial changes in situations where such fluctuations are rather pronounced (Monson et al., 2006; McMahon et al., 2011).

In mountainous regions, microbes play a crucial role in soil formation and all of the main elemental cycles that affect the establishment, growth, and survival of plants and the global greenhouse gas balance. Temperature and other climatic factors, which are expected to change dramatically by the end of this century due to ongoing global change, have a considerable impact on microbial activity and diversity (Bardgett et al., 2008). As microbial activity rises over time, soil becomes enough to support plant colonization. By interacting with the soil microbiome, plant establishment influences subsequent ecosystem evolution (Schulz et al., 2013). Numerous climatic systems, latitudes, and altitudes exhibit notable variations in the pace of soil formation and, hence, primary succession in glacier forefields. Numerous climatic systems, latitudes, and altitudes exhibit notable variations in the pace of soil formation and, hence, primary succession in glacier forefields.

Both microbial and plant life depend on the soil. By altering the amount and quality of organic inputs, plant litter can change the carbon and nitrogen content of soil (Sitters et al., 2020). All of this might affect microbiological communities in soil and activity in the future. Microorganism also impacted by changes in the soil's environment. For instance, functional activities such as decreased soil water-holding capacity can lead to an increase in the diversity of anaerobic bacteria (Maestre et al., 2015; Xun et al., 2018). Thus, an essential component of ecological sciences is comprehending how soil microbes affect ecosystem activities, Different latitudes, elevations, and climatic systems have a significant impact on the rate of soil growth and, consequently, primary succession in glacier forefields. Both plants and microorganisms depend on the soil. Plant litter can alter soil carbon and nitrogen by changing the quantity and quality of organic inputs (Sitters et al., 2020). All of this could affect soil microbial populations and activity in the future. Changes in the environment of the soil itself can also affect microbial communities, for example, functional processes like reduced soil water holding capacity can increase the diversity of anaerobic bacteria (Maestre et al., 2015; Xun et al., 2018). Therefore, understanding the effects of soil microorganisms on ecosystem functions is an important aspect in ecological sciences (Cardinale et al., 2012; Tardy et al., 2014).

Organic matter declines significantly with altitude because plant life is more constrained by the climate at higher elevations (Egli et al., 2004, 2009). Microorganisms that live in highalpine soil must therefore adapt to oligotrophic, extremely nutrient-limited environments. Furthermore, there is little defense against environmental stressors like strong UV rays and temperature and moisture fluctuations that could affect the structures of microbial communities in coarse-textured, shallow soils with a low water retention capacity and little plant cover (Ciccazzo et al., 2016; Rime et al., 2015). Many microbes that have undergone unique adaptations can survive in mountain regions in spite of these harsh environmental constraints (Margesin, 2012; Rime et al., 2015). In alpine regions with sparse vegetation, microorganisms often aggregate in biofilms to produce biological soil crusts as a defense mechanism against various environmental stresses and to facilitate biological weathering and nutrient acquisition (Schulz et al., 2013).

Seasonal fluctuations are pronounced in alpine habitats. Seasonal snow and slopes in elevation cycles as well as varying precipitation pattern all have an impact on the species composition of plants and soil in these environments, which in turn affects the makeup of the microbial communities that are produced. Therefore, alpine ecosystems are perfect places to study how microbial populations react to changes in their surroundings. These investigations also shed light on how ecosystem functioning may be impacted by future climatic changes, which are predicted to have a significant impact on alpine plant and microbial communities (Djukic et al., 2010). By changing plant and soil characteristics, climate change can affect the soil microbial diversity and community composition which in turn impacts the ecosystem. In freezing alpine settings, where plant growth is severely constrained by climate constraints and nutrient scarcity, microbes have been demonstrated to have an unexpectedly high variety (Rime et al., 2015). Microbial organism therefore are primary ecosystem engineers controlling the cycling of carbon and nutrients in these environments. Warming has been demonstrated to have a sensitive effect on microbial diversity and function (Rui et al., 2015; Li et al., 2016a), which accelerates C- and N-cycling and may be linked to net losses from the system (Hagedornetal, 2010a; Dawes et al., 2017). On the contrary, Yue et al. (2015) found that certain alpine grassland soils contain a feedback mechanism that prevents increased mineralization in response to warming and eventually keeps the stock of carbon and nitrogen steady. Warming alters the characteristics of soil, vegetation, soil nutrient status, organic matter quality, and biotic interactions, all of which have an impact on microbial activity and diversity. The conflicting findings about C- and Nstock as a result of warming in alpine environments are likely highlighted by such simultaneous impacts and feedback processes. Furthermore, lowland settings are more likely to be hospitable due to ice, snow, and other unfavorable environmental conditions, which lowers the diversity of microbial niche. Thus, reduction of the cryosphere could signify a considerable loss of microbial diversity such as observed for plant diversity in alpine environments with warming (Pauli et al., 2003). Besides ecological consequences, melting of cryospheric components of alpine ecosystems also potentially entails the risk of releasing pathogens and indeed potentially pathogenic microbes have been recovered from cold environments (Butinar et al., 2007; Petrova et al., 2008; Legendre et al., 2014). This potential impact of global change is especially pertinent to alpine settings as they are closely related to communities of people that might be accessed via air circulation or by mechanisms related to hydrology.

6 Conclusion

Microbiology interactions in alpine regions are not only vital for local ecosystem health but also have far-reaching implications for global climate regulation. Understanding these interactions is crucial for forecasting how ecosystems will react to climate change and for formulating plans to reserve the ecological integrity of alpine environments. Microbial interactions in alpine areas are fundamental to ecosystem health and are being reshaped by climate change. Their adaptability is remarkable, but ongoing environmental shifts pose significant challenges with broad implications for alpine ecology and global biogeochemical cycles. Microorganisms rarely exist in isolation. Instead, they form complex and dynamic communities where they engage in a variety of interactions-mutualistic, commensal, competitive, antagonistic and parasitic that shape both their own populations and the broader environment. These interactions are crucial for processes such as nutrient cycling, disease resistance, enhancement of plant growth and sustainability of the environment. For example, mutualistic relationships like that of nitrogen-fixing bacteria with legumes or mycorrhizal fungi with plants directly enhance soil fertility and crop yields while competitive and antagonistic interactions (such as antibiotic production by Streptomyces) influence microbial diversity and pathogen suppression. In animals, especially humans, microbial communities are essential for digestion, immune function and protection against disease with disruptions often linked to health disorders. Understanding and harnessing microbial interactions is increasingly important for addressing global challenges such as sustainable agriculture, climate change mitigation and disease prevention. Advances in genomics and microbial ecology are opening new avenues for manipulating microbial communities to enhance crop production, restore degraded environments, and develop novel therapeutics. Thus, microbial interactions are indispensable for ecosystem health, human well-being and biotechnological innovations. Continued research is essential to fully unlock their potential and address the complex challenges posed by environmental change and human activities.

References

Bardgett, R. D., Freeman, C., & Ostle, N. J. (2008). Microbial contributions to climate change through carbon cycle feedbacks. *The ISME Journal, 2*(8), 805–814.

Butinar, L., Spencer-Martins, I., & Gunde-Cimerman, N. (2007). Yeasts in high Arctic glaciers: the discovery of a new habitat for eukaryotic microorganisms. *Antonie van Leeuwenhoek, 91*(3), 277–289.

Cardinale, B. J., Duffy, J. E., Gonzalez, A., Hooper, D. U., Perrings, C., Venail, P., Narwani, A., Mace, G. M., Tilman, D., Wardle, D. A., Kinzig, A. P., Daily, G. C., Loreau, M., Grace, J. B., Larigauderie, A., Srivastava, D. S., & Naeem, S. (2012). Biodiversity loss and its impact on humanity. *Nature, 486*(7401), 59–67.

Ciccazzo, S., Esposito, A., Borruso, L., & Brusetti, L. (2016). Microbial communities and primary succession in high altitude mountain environments. *Annals of Microbiology, 66*, 43–60.

Dawes, M. A., Schleppi, P., Hättenschwiler, S., Rixen, C., & Hagedorn, F. (2017). Soil warming opens the nitrogen cycle at the alpine treeline. *Global Change Biology, 23*(1), 421–434.

Djukic, I., Zehetner, F., Mentler, A., & Gerzabek, M. H. (2010). Microbial community composition and activity in different Alpine vegetation zones. *Soil Biology and Biochemistry, 42*(2), 155–161.

Donhauser, J., & Frey, B. (2018). Alpine soil microbial ecology in a changing world. *FEMS Microbiology Ecology, 94*(9), fiy099.

Egli, M., Mirabella, A., & Fitze, P. (2004). Clay mineral formation in soils of different ages in the alpine environment. *Geoderma, 124*(1–2), 43–62.

Egli, M., Sartori, G., Mirabella, A., & Favilli, F. (2009). Soil formation rates in alpine environments: Climatic and lithological controls. *Geoderma, 149*(3–4), 245–257.

Frey, S. D. (2019). Mycorrhizal fungi as mediators of soil organic matter dynamics. *Annual Review of Ecology, Evolution, and Systematics, 50*(1), 237–259.

Gardes, M., & Dahlberg, A. (1996). Mycorrhizal diversity in arctic and alpine tundra: An open question. *New Phytologist, 133*(1), 147–157.

Grayston, S. J., Griffith, G. S., Mawdsley, J. L., Campbell, C. D., & Bardgett, R. D. (2001). Accounting for variability in soil microbial communities of temperate upland grassland ecosystems. *Soil Biology and Biochemistry, 33*, 533–551.

Grayston, S. J., Wang, S., Campbell, C. D., & Edwards, A. C. (2001). Selective influence of plant species on microbial diversity in the rhizosphere. *Applied Soil Ecology, 18*(1), 1–12.

Hagedorn, F., Mulder, J., Jandl, R., & Richter, A. (2010). Mountain soils under a changing climate: Effects of soil warming on carbon and nitrogen dynamics. *Global Change Biology, 16*(7), 1931–1946.

Körner, C. (2023). Concepts in alpine plant ecology. *Plants, 12*(14), 2666.

Krug, L., Erlacher, A., Markut, K., Berg, G., & Cernava, T. (2020). The microbiome of alpine snow algae shows a specific inter-kingdom connectivity and algae-bacteria interactions with supportive capacities. *The ISME Journal, 14*(9), 2197–2210.

Lazzaro, A., Hilfiker, D., & Zeyer, J. (2015). Structures of microbial communities in alpine soils: Seasonal and elevational effects. *Frontiers in Microbiology, 6*, 1330.

Legendre, M., Bartoli, J., Shmakova, L., Jeudy, S., Labadie, K., Adrait, A., Lescot, M., Poirot, O., Bertaux, L., Bruley, C., Couté, Y., Rivkina, E., Abergel, C., & Claverie, J.-M. (2014). Thirty-thousand-year-old distant relative of giant icosahedral DNA viruses with a pandoravirus morphology. *Proceedings of the National Academy of Sciences of the United States of America (PNAS), 111*(11), 4274–4279.

Li, J., Delgado-Baquerizo, M., Wang, J., Hu, H., Cai, Z., Zhu, Y., & Singh, B. K. (2016). Fungal richness contributes to multifunctionality in boreal forest soil. *Soil Biology and Biochemistry, 102*, 30–39.

Li, J., Wang, X., Yuan, M., Duan, W., Xia, J., Zhang, X., & Wang, J. (2025). Effect of soil microbial community on ecosystem multifunctionality in an alpine grassland. *Catena, 249*, 108714.

Maestre, F. T., Delgado-Baquerizo, M., Jeffries, T. C., Eldridge, D. J., Ochoa, V., Gozalo, B., Quero, J. L., García-Gómez, M., Gallardo, A., Ulrich, W., Bowker, M. A., Arredondo, T., Barraza-Zepeda, C., Bran, D., Florentino, A., Gaitán, J., Gutiérrez, J. R., Huber-Sannwald, E., Jankju, M., … Singh, B. K. (2015). Increasing aridity reduces soil microbial diversity and abundance in global drylands. *Proceedings of the National Academy of Sciences of the United States of America (PNAS), 112*(51), 15684–15689.

Margesin, R. (2012). *Psychrophilic microorganisms in alpine soils* (pp. 187–198). Springer.

Margesin, R., & Miteva, V. (2011). Diversity and ecology of psychrophilic microorganisms. *Research in Microbiology, 162*(3), 346–361.

Mesquita, D. C. P. B. (2019). *Abiotic and biotic drivers of plant range shifts in the Alpine*. Doctoral dissertation, University of Colorado at Boulder.

McMahon, S. K., Williams, M. A., Bottomley, P. J., & Myrold, D. D. (2011). Dynamics of microbial communities during decomposition of carbon-13 labeled ryegrass fractions in soil. *Soil Biology and Biochemistry, 43*(9), 2041–2048.

Molina-Montenegro, M. A., Oses, R., Torres-Díaz, C., Atala, C., Núñez, M. A., & Armas, C. (2015). Fungal endophytes associated with roots of nurse cushion species have positive effects on native and invasive beneficiary plants in an alpine ecosystem. *Perspectives in Plant Ecology, Evolution and Systematics, 17*(3), 218–226.

Monson, R. K., Lipson, D. L., Burns, S. P., Turnipseed, A. A., Delany, A. C., Williams, M. W., et al. (2006). Winter forest soil respiration controlled by climate and microbial community composition. *Nature, 439*, 711–714.

Pauli, H., Gottfried, M., & Grabherr, G. (2003). Effects of climate change on the alpine and nival vegetation of the Alps. *Journal of Mountain Ecology, 7*(Supplement), 9–12.

Petrova, M. A., Gorlenko, V. M., & Ivanova, A. E. (2008). Psychrotrophic microorganisms from polar and alpine environments. *Microbiology, 77*(6), 642–650.

Read, D. J., & Haselwandter, K. (1981). Observations on the mycorrhizal status of some alpine plant communities. *New Phytologist, 88*(2), 341–352.

Read, D. J., & Perez-Moreno, J. (2003). Mycorrhizas and nutrient cycling in ecosystems–a journey towards relevance? *New Phytologist, 157*(3), 475–492.

Rime, T., Hartmann, M., Brunner, I., Widmer, F., Zeyer, J., & Frey, B. (2015). Vertical distribution of the soil microbiota along a successional gradient in a glacier forefield. *Molecular Ecology, 24*(5), 1091–1108.

Rui, J., Li, J., Wang, S., An, J., Liu, W., Lin, Q., Yang, Y., He, Z., & Li, X. (2015). Responses of bacterial communities to simulated climate changes in alpine meadow soil of the Qinghai–Tibet Plateau. *Applied Soil Ecology, 87*, 35–42.

Schmidt, S. K., Costello, E. K., Nemergut, D. R., et al. (2008). Microbial activity and diversity during early primary succession in recently deglaciated soils. *The ISME Journal, 2*(4), 352–363.

Schulz, S., Brankatschk, R., Dümig, A., Kögel-Knabner, I., & Schloter, M. (2013). The role of microorganisms at different stages of ecosystem development for soil formation. *Biogeosciences, 10*(6), 3983–3996.

Sitters, J., Bakker, E. S., Veldhuis, M. P., Veen, G. F., Olde Venterink, H., & Elzenga, J. T. M. (2020). Plant litter effects on carbon and nitrogen cycling depend on litter quality and soil nutrient availability. *Ecology, 101*(3), e02950.

Taffner, J., Erlacher, A., Bragina, A., Berg, C., Moissl-Eichinger, C., & Berg, G. (2018). What is the role of Archaea in plants? New insights from the vegetation of alpine bogs. *MSphere, 3*(3), 10–1128.

Tardy, V., Spor, A., Mathieu, O., Lévêque, J., Terrat, S., Plassart, P., Regnier, T., Bardgett, R. D., & van der Putten, W. H. (2014). Shifts in microbial diversity through land use intensity as drivers of carbon mineralization in soil. *Soil Biology and Biochemistry, 70*, 204–213.

Wallenstein, M. D., & Hall, E. K. (2012). A trait-based framework for predicting when and where microbial adaptation to climate change will affect ecosystem functioning. *Biogeochemistry, 109*(1–3), 35–47.

Xun, W., Zhao, J., Xue, C., Zhang, G., Ran, W., Wang, B., Shen, Q., & Zhang, R. (2018). Significant alteration of soil bacterial communities and functions by reduced soil moisture. *Soil Biology and Biochemistry, 121*, 74–82.

Yue, H., Wang, M., Wang, S., Gilbert, J. A., Sun, X., Wu, L., Lin, Q., Hu, Y., Li, X., & He, Z. (2015). The microbe-mediated mechanisms underlying soil carbon and nitrogen stabilization in alpine grassland under warming. *Global Change Biology, 21*(9), 3286–3298.

Pathway Elucidation in Medicinal Plant Metabolism

14

Anwesha Bandyopadhyay, Tarun Pal, and Ira Vashisht

Abstract

Plants produce a vast array of bioactive secondary metabolites, including phenylpropanoids, terpenoids and alkaloids that are essential for ecological adaptation and serve as critical resources for the pharmaceutical, cosmetic, and agrochemical industries. However, the low natural abundance of these compounds and the over-harvesting of medicinal species have created a significant supply-demand gap. This article provides a comprehensive overview of traditional and emerging strategies for the elucidation of plant biosynthetic pathways, a prerequisite for enhancing yields through genetic engineering and synthetic biology. Foundational biochemistry-based approaches was examined, such as stable isotope labelling and enzyme kinetics, and contrast them with modern "omics" technologies. The integration of genomics, transcriptomics, proteomics and metabolomics collectively termed "herbgenomics" has revolutionized the field by enabling the identification of biosynthetic gene clusters (BGCs), tissue-specific expression patterns and active protein landscapes. Furthermore, the review explores the role of computational tools, machine learning and synthetic biology in reconstructing complex metabolic networks in heterologous hosts like yeast and bacteria. Finally, the transformative potential of CRISPR/Cas9 genome editing was discussed for functional gene validation and metabolic flux optimization. By synthesizing these multi-disciplinary approaches, this article outlines

A. Bandyopadhyay
School of Biotechnology, Faculty of Applied Sciences & Biotechnology, Shoolini University, Bajhol, Himachal Pradesh, India

T. Pal
School of Bioengineering and Food Technology, Faculty of Applied Sciences & Biotechnology, Shoolini University, Bajhol, Himachal Pradesh, India

I. Vashisht (✉)
School of Biotechnology, Jawaharlal Nehru University, New Delhi, India

P. Kumar et al. (eds.), *Climate Resilience and Molecular Adaptation in Alpine Medicinal Plants*, https://doi.org/10.1007/978-981-95-6923-6_14

a holistic roadmap for deciphering specialized metabolism and ensuring the sustainable production of high-value phytoconstituents.

Keywords

Alkaloids · CRISPR/Cas9 · Functional genomics · Genetic engineering · Metabolomics · Multi-omics integration

1 Introduction

Plants, being sessile organisms employ a very unique strategy to interact with co-inhabitant animals and microbes while overcoming harsh terrestrial environments. Diversification of metabolite systems towards secondary metabolism have been observed which result in production of bioactive components. Chemically, these metabolites are classified based on the core structure and attached functional groups and broadly comprise of phenylpropanoids, terpenoids, alkaloids, benzoids, and N-encompassing molecules, phenolics, and other phytoconstituent (Garagounis et al., 2021). Approximately, more than two lacs phytoconstituents were discovered and many more are on the path of discovery (Kessler & Kalske, 2018). Medicinal plants provide a plethora of valuable bioactive compounds which form an integral component of traditional medicine. In addition to the ecological and pharmaceutical role, these compounds form a major ingredient of major commercial sectors like flavorings, fragrances, cosmetics, pigments, dyes, and agrochemicals. Their benefits extend to multiple areas including essential oil, drugs, industrial components and food additives. To harness the full potential of these compounds, an elaborate insight into the metabolomic network and associated regulatory and transport processes is required. Further, plants produce miniscule amounts of these metabolites which are often difficult to harvest, leading to a gap in demand vs supply chain. Furthermore, to fulfil industrial demand, these plants are over-harvested leading to their endangered status in natural habitat. Several efforts have been pushed towards tissue culture-based production of metabolites but the content remains low. The best approach for obtaining optimal yield would be to push conservation approaches and parallel efforts to understand the biosynthetic pathways and enhance yield via various genetic engineering or synthetic biology-based approaches. Biosynthetic pathway elucidation is an extremely complex process which involves understanding of specific secondary metabolite isolation, structural characterization (both qualitative and quantitative), biochemical and molecular characterization followed by in vivo assessments. In addition to traditional methods of elicitation, enzymatic approaches, biotechnological approaches have showed major potential in elucidation of biosynthetic pathways.

Recently, progressing research advances in 'omics' induced a deeper insight into the synthesis using various genomic, proteomics, transcriptomics and metabolomics-based approaches. After its inception in the 1970s, sequencing technology has led to development of several allied fields which are still in bloom and with advanced bioinformatics tools, has led to the design of various yield enhancement strategies.

Some good reviews have emphasized the role of sequencing approaches like (Chele et al., 2025)—reviewed medicinal plant genome application with review of sequencing technology developments, plant genome situation. Present book chapter provides a holistic overview of fundamental, advanced as well as progressing approaches which are instrumental in elucidation of plant secondary metabolism pathways.

2 Approaches for Pathway Discovery: Overview

Ensuring steady supply of medicinal plants requires streamlined approaches for large scale cultivation and management approaches. Accurate identification of superior varieties, optimum environment conditions, conventional breeding and culturing conditions and tissue culture approaches remain major strategies which explore the path for enhancement of metabolite content in raw materials. Classical genetics has been known to be a major method for elucidation of biosynthetic pathways as observed in case of model plant *Arabidopsis thaliana* where linkage of auxin homeostasis and glucosinolate biosynthesis was explored (Ruegger et al., 2001). It also caused major revolution in understanding of metabolic intermediates as well as enzyme activities pertaining to phenylpropanoid biosynthesis along with basis for understanding of transcriptional regulation of secondary metabolism. However, even after channelling resources, medicinal plant supply often falls short for quenching industrial demand. Thus, the best approach would be to extend approaches to enhancing metabolite content, which warrants identification of biosynthetic pathways as a starting point, which can further lead to the development of elite varieties with high precision capable of providing high-quality raw material, thus reducing overexploitation of habitat. The pathway elucidation studies have initially depended solely on traditional quantification, enzyme discovery, isotope labelling, and *in vitro* kinetic studies. Although these methods form the ultimate roadmap in functional analysis of pathway elucidation, they are extremely time-consuming and tedious. Herbgenomics explores molecular genetics via integration of genomics, transcriptomics, and metabolomics approaches (Hu et al., 2019; Sun et al., 2022). A combinatorial strategy involving reverse and forward genetics for gene positioning in order to place them in the right order in the biosynthetic pathway via creating and screening mutants, as has been depicted in *Artemisia annua* (Liao et al., 2022a, b).

Initially, when sequencing approaches were developed, major emphasis was placed on identifying whole-genome sequencing approaches exclusively for model plants. Recently, with increased affordability, the availability of reference genomes has increased for medicinal plants as well. This led to a push to lay a foundation for molecular genetics approaches. It has been observed that the complexity or size doesn't tend to hinder de novo assembly of genomes as represented by *Panax ginseng, Ginkgo biloba,* and *Taxus chinensis* (Sun et al., 2022). Advanced approaches based on metabolomics have integrated genomic as well as metabolomics of Perilla and Cannabis populations for deciphering association allelic variants with variation for generating bioactive molecules (Sun et al., 2022). A mineable genomic database has also been created for traditional medicines, which has been derived based on eight international pharmacopoeias (Liao et al., 2022a, b), which helps with genomic sequences along with identification of appropriate raw materials. Thus, it is safe to say that "omics" has significantly catapulted the approaches aimed at deciphering underlying biosynthetic pathways for a wide range of metabolites. However, among all, transcriptomics has emerged as a major contributor in these studies, resulting in pioneering achievements in the discovery of complete pathways, most notable being colchicine, glycyrrhizic acid, cannabidiol, diosgenin, and hyoscyamine (Sun et al., 2022). Owing that the metabolites are biosynthesized in specific cells/tissues at a certain developmental stage while responding to specific environmental cues, thus giving way to capturing the underlying mechanisms via transcriptome-based approaches. Transcriptome studies, however, base their inferences on genomic annotation where specific enzymes and intermediate metabolites allow identification of pathways (Polturak & Osbourn, 2021). In addition to biosynthetic pathway elements, transcriptomics collects important information on regulatory elements like transcription factors, microRNAs, molecular markers, and several other gene-regulatory networks, which allow a holistic understanding of metabolic pathways in vivo. Several transcription factors like R2-R3 MYB, AP2/ERF, basic helix-loop helix, WRKY, and bzip families have been well-characterized for their role in the regulation of biogenesis of *Catharanthus rose*us, *Artemisia annua,* and *Salvia miltiorrhiza*. Metabolic engineering, on the other hand, can give way to the reconstitution of metabolic pathways by heterologous expression. A major role has been played by the traditional approaches like molecular docking, identification of crystal structure, site-directed mutagenesis, molecular mechanics, and other biochemical approaches. Proteomics and mass spectroscopy-based approaches are also catching up in speed for elucidation of biosynthetic mechanisms, where investigation of epigenomic regulation, chromatin accessibility, and histone modification is being explored. Several computational tools and models are being developed to solve the problems associated with finding missing links in biosynthetic pathways (Fig. 14.1 and Table 14.1).

2.1 Biochemistry-Based Approaches

The traditional foundation for elucidating metabolic pathways in medicinal plants consists of metabolic pathway-focused approaches. By directly connecting enzyme activity to biochemical changes, these strategies enable researchers to determine the mechanistic specifics of biosynthetic pathways. The primary methods include time-course analysis, *in vitro* enzyme assays, substrate feeding experiments using labeled precursors, and enzyme separation and characterization. When combined, they offer the functional evidence required to support theories produced by transcriptomics or genomics research.

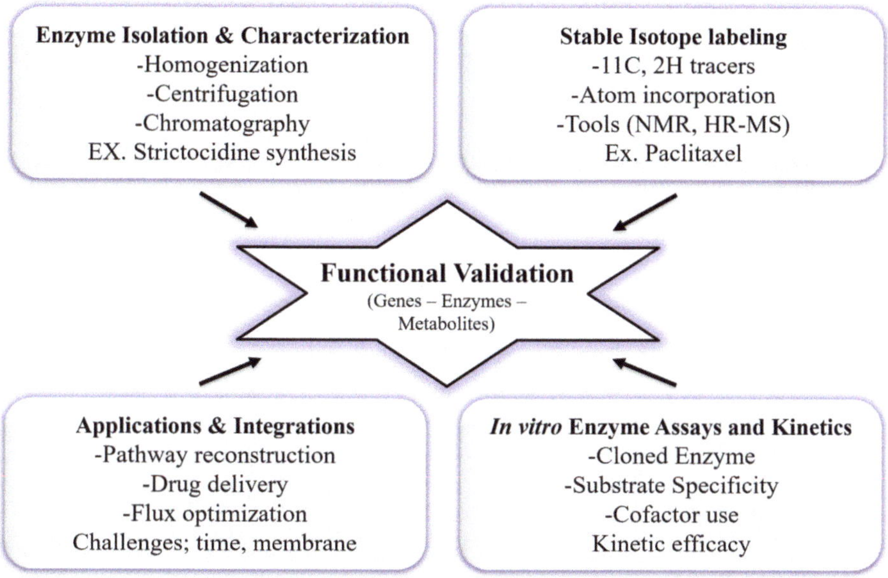

Fig. 14.1 Biochemistry-based approaches for functional validation of biosynthetic pathways in medicinal plants

2.1.1 Enzyme Isolation and Characterization

Enzyme isolation is a key component of biochemical cascade analysis. Plants frequently produce bioactive substances through multi-step enzymatic cascades that involve a variety of catalytic classes, such as terpene, glycosyltransferases, methyltransferases, and cytochrome P450 monooxygenases. To preserve enzymatic function, isolation often involves tissue homogenization, differential centrifugation, and chromatographic purification. To confirm the specific reaction catalyzed, activity experiments are subsequently carried out using synthetic or natural substrates. For instance, researchers successfully identified the first stage of monoterpene indole alkaloid anabolism after isolating strictosidine synthases from *Catharanthus roseus* (Lebaudy et al., 2008).

Recent advances in microfluidic enzyme screening tools now enable high-throughput characterization of plant enzymes at nanoliter scales. These techniques allow for quick testing of substrate libraries and reduce protein loss during purification (Elvira et al., 2022). Furthermore, by directly monitoring metabolite changes with gene knockout or overexpression, CRISPR-Cas9-mediated gene allows functional validation of isolated enzymes (Hsieh et al., 2023). This integration connects cutting-edge functional genomics and classical biochemistry.

2.1.2 Stable Isotope Labelling

Exogenous nutrient fortification with isotopically labelled molecules, such as ^13C- or ^2H-labeled precursors, enables tracing the incorporation of atoms into downstream metabolites. This technique can elucidate branching points, intermediate

Table 14.1 Overview of biochemistry, genomics, proteomics, and metabolomics approaches in medicinal plant research

Plants	Metabolite & pathway studied	Type of omics & biochemical method applied	Key findings	References
Catharanthus roseus	Monoterpene indole alkaloids (MIAs)	Enzyme isolation and characterization of strictosidine synthase (STR) and tryptophan decarboxylase (TDC)	Functional role of STR and TDC validated; enzyme kinetics and substrate specificity defined, aiding MIA biosynthesis reconstruction.	Lebaudy et al. (2008)
Catharanthus roseus	MIAs (Ajmalicine, Vindoline)	CRISPR-Cas9-mediated gene knockout and overexpression combined with biochemical assays	Functional validation of T16H and D4H genes confirmed their roles in alkaloid biosynthesis; pathway modulation achieved.	Hsieh et al. (2023)
Taxus spp.	Paclitaxel (Taxol)	Stable isotope labeling (^13C-mevalonate) + GC-MS/NMR flux analysis	Confirmed the mevalonate pathway as the main route for paclitaxel biosynthesis; isotope tracing revealed rate-limiting enzymatic steps.	Croteau et al. (2006)
Digitalis lanata	Cardiac glycosides	Enzyme purification and in vitro kinetics of 3β-hydroxysteroid dehydrogenase	Confirmed its catalytic function in digitoxigenin biosynthesis; enzyme characterized biochemically for substrate specificity.	Chacon (2024)
Papaver somniferum	Benzylisoquinoline alkaloids (Morphine pathway)	Substrate feeding and enzyme isolation (e.g., salutaridine synthase, codeinone reductase)	Clarified the enzymatic sequence in morphine biosynthesis; led to pathway engineering insights.	Wink (1999)

Taxus brevifolia	Paclitaxel biosynthesis	Enzyme reconstitution in vitro (Taxadiene synthase, DBAT) + protein expression in E. coli	Reconstructed core pathway steps; confirmed catalytic efficiency and cofactor requirements of paclitaxel enzymes.	Reshi et al. (2023)
Salvia miltiorrhiza	Diterpenoid tanshinones	Heterologous enzyme expression + kinetic assays	Characterized CYP76AH1 as a key monooxygenase; linked transcriptomics with biochemical function.	Chen (2024)
Saccharomyces cerevisiae (engineered)	Plant-derived alkaloids and flavonoids	Synthetic biology & metabolic engineering – reconstitution of complete plant biosynthetic pathways in microbial chassis	Enabled large-scale production of natural products (e.g., artemisinic acid, noscapine) via engineered microbes.	Deng et al. (2023)
Artemisia annua	Artemisinin	Feeding experiments with labeled precursors + enzyme assays of amorpha-4,11-diene synthase (ADS)	Identified ADS as the gateway enzyme in artemisinin biosynthesis; crucial for synthetic biology applications.	Ro et al. (2006)
Camptotheca acuminata	Camptothecin (anticancer alkaloid)	Enzyme assays + isotope labeling of tryptophan and secologanin pathways	Demonstrated branch-point enzymes linking terpenoid and indole pathways.	Liu et al. (2019)
Glycyrrhiza uralensis	Glycyrrhizin (triterpenoid saponin)	Enzyme isolation + LC-MS metabolic profiling	Characterized β-amyrin synthase and UDP-glycosyltransferase; clarified sequential glycosylation steps.	Wang et al. (2022)
Withania somnifera	Withanolides (steroidal lactones)	Protein expression + in vitro enzyme validation of cytochrome P450s and glycosyltransferases	Revealed oxidation and glycosylation steps leading to withaferin A formation.	Singh and Jayaraman (2024)

(continued)

Table 14.1 (continued)

Plants	Metabolite & pathway studied	Type of omics & biochemical method applied	Key findings	References
Panax ginseng	Ginsenosides	In vitro enzyme characterization + gene co-expression analysis	Identified dammarenediol synthase and key UGTs involved in protopanaxadiol biosynthesis.	Kim et al. (2023)
Stevia rebaudiana	Steviol glycosides	Enzyme kinetics and metabolite feeding studies	Validated role of ent-kaurene oxidase in steviol glycoside biosynthesis.	Brandle and Telmer (2007)
Curcuma longa	Curcuminoids	Enzyme isolation (Curcumin synthase 1 and 2) + metabolite profiling	Elucidated curcuminoid formation from p-coumaroyl-CoA and feruloyl-CoA precursors.	Katsuyama et al. (2009)
General medicinal plants	Terpenoids, alkaloids, phenolics	Integration of biochemistry with proteomics, metabolomics, and fluxomics	Provided comprehensive enzyme annotation and discovery of missing pathway steps.	Dejana (2023)
General medicinal plants	Terpenoids, alkaloids	Stable isotope-assisted metabolomics (SIAM) + machine learning-driven fluxomics	Automated reconstruction of pathway fluxes; enhanced precision in tracking carbon flow through biosynthetic networks.	Zhang et al. (2023a, b)
General secondary metabolites	Terpenoids, phenolics, alkaloids	Position-Specific Isotope Analysis (PSIA) using NMR/MS	Provided atom-level insight into precursor incorporation and enzyme reaction mechanisms.	Zhang et al. (2023a, b)
General medicinal plants	Multiple metabolites (enzymatic cascades)	High-throughput microfluidic enzyme activity screening using microdroplet-based assays	Enabled parallel screening of thousands of enzyme reactions at the nanoliter scale; efficient discovery of key biosynthetic enzymes.	Elvira et al. (2022)

formation, and pathway flux. For example, the biosynthesis of paclitaxel was outlined by ^13C-labeled mevalonate feeding experiments in *Taxus* species (Croteau et al., 2006). Unambiguous identification of labelled products is ensured by combining isotope labelling with nuclear magnetic resonance (NMR) or high-resolution mass spectrometry. Stable isotope-assisted metabolomics combined with machine learning is a cutting-edge discovery that enables automated reconstruction of pathway fluxes across many tissues and developmental stages (Zhang et al., 2023a, b). Furthermore, position-specific isotope analysis (PSIA) has surfaced, offering biosynthetic events at the atom-level, which is especially useful for deciphering intricate terpenoid and alkaloid pathways.

2.1.3 In Vitro Enzyme Assays and Kinetics

Cloned expression of putative enzymes, usually in yeast, *E. coli*, or alternatively in plant cell cultures, allows for controlled in vitro assays. Through presenting substrates under defined parameters and monitoring product formation, scientists can determine catalytic characteristics, substrate selectivity, cofactor utilizations, and stereochemical orientation. This data is essential for elucidation of kinetic efficacy and for rebuilding cellular metabolism circuits. Cell-free protein synthesis (CFPS) methods have become popular in the study of medicinal plants in recent years. The CFPS methods save time and prevent toxicity to host organisms by enabling quick prototyping of enzyme function without requiring full cellular expression (Chacon, 2024). Furthermore, transient enzyme–substrate interactions, which are undetectable in bulk assays, are being explored through single-molecule enzymology using fluorescence resonance energy transfer (FRET) (Okon et al., 2024).

Metabolic analysis strategies have been instrumental in deciphering metabolic pathways of clinically relevant biomolecules like vincristine, morphine, and paclitaxel. These studies have facilitated metabolite flux optimization strategies, allowing the production of biomolecules in exogenous host systems or tailored plant cellular systems. Furthermore, the integration of biochemical data with metabolomics facilitates the identification of novel phytoconstituents and their biosynthetic origins, enhancing natural product discovery (Wink, 1999). Enzymology-based techniques offer direct functional analysis and remain the validated procedure for confirming biocatalytic activity. They are especially helpful for transient or sparsely expressed intermediates that might not be seen in aggregated tissue. However, these techniques are time-consuming, and because of membrane attachment, instability, or cofactor dependence, certain plant enzymes are difficult to isolate or express under heterologous conditions. These obstacles are frequently removed when chemo-biology and omics data are combined, allowing for the focused discovery of active mediators. For conclusive evidence of an enzymatic phase, biochemical enzyme isolation, substrate specificity tests, and product identification are still necessary. These strategies are particularly effective when combined with targeted inhibitor studies or the reconstruction of minimal enzyme sets in heterologous systems to demonstrate sufficiency for product formation (Reshi et al., 2023). Maintaining enzyme activity during extraction, employing realistic substrate analogues, and utilizing high-resolution analytical readouts (LC-MS/MS, GC-MS, NMR) for clear product assignment are vital challenges. Activity experiments normally need microsomal preparation or heterologous expression platforms to produce functional protein when enzymes are membrane-associated or low-abundance, which is typical for plant cytochrome P450s and glycosyl-transferases. Thus, biochemistry provides causal evidence linking genes to metabolites and serves as the final layer of validation in pathway elucidation (Chen, 2024).

Looking ahead, the use of biochemical techniques is being revolutionized by metabolic engineering and synthetic biology. The restoration of entire biosynthetic pathways in microbial chassis (e.g., Saccharomyces cerevisiae, Streptomyces) enables scalable production of therapeutic plant compounds (Deng et al., 2023). To predict "missing enzymes" and experimentally validate them, biochemistry is also quickly integrating into multi-omics (proteomics, metabolomics, fluxomics) (Dejana, 2023). Faster route discovery and translation into medicinal applications are guaranteed by this convergence.

2.2 Computational Approaches to Pathway Elucidation

Biosynthetic pathway elucidation requires prior knowledge of potential chemical reaction scenarios that might lead to the formation of the bioactive compound. Metabolomics has been utilized for the identification of tissue-specific accumulation patterns and reaction intermediates. Molecular networking has also picked up speed by connecting mass spectra and separated compounds (Fox Ramos et al., 2019). On an ideal basis, connected compounds can indicate the whole biosynthetic branch, and once the reaction sequence has been deciphered, the intermediate enzymes and activity can be hypothesized from organic chemistry principles of catalytic reactions and literature data. Further, defining the first set of gene candidates can be derived from genomes or transcriptome-based sequences, following up with gene and transcript sequences to look for specific enzyme domains. Based on this combinatorial sequence information, pathway enzymes can be predicted and validated further.

Gene prioritization through homology-based search is reportedly among the most obvious and direct ways to identify missing links in the metabolite biosynthetic pathway. This requires inspiration from previous reports, and that leads to homology-based ortholog analysis derived from already characterized biosynthetic reactions in other plants. Homology search makes use of BLAST or phylogenetics and statistics as placed on evolutionary models and protein sequence information from diverse species (Salim et al., 2018).

Although this methodology is mostly successful, a word of caution is that it is not necessary that the closest orthologs would perform the same reaction (Farrow et al., 2018). In this respect, prioritization using random mutagenesis using techniques like ethyl methane sulfonate, where induced mutations are screened for any changes in metabolite production (Edge et al., 2018). Further, the mutated plants are subjected to sequencing, which results in linking phenotype with genotype and the steps of the pathway. Though this is a labor-intensive strategy, it turns out to be powerful and has been utilized in alkaloid profiling biosynthetic pathway (Qu et al., 2019).

2.3 Genomics-Based Approaches

Genomics provides a prediction for pathway discovery by revealing genomes, co-localized biosynthetic gene clusters (BGCs), and transcriptomics. The identification of specific enzyme families (e.g. terpene synthases, P450 monooxygenases, glycosyltransferases, and methyltransferases) that are frequently involved in secondary metabolism is made possible by high-quality genome assemblies that include precise gene models, repeat resolution, and functional annotation. Genes that co-evolve with canonical genes or have lineage-specific expansions can be given priority using comparative genomics and phylogenetic analysis. In certain plant lineages, physically clustered BGCs facilitate discovery via genomic neighborhoods, whereas in others, pathway genes are dispersed and must be identified through co-expression patterns or regulatory linkages. Integrating genome sequence with transcriptomes from inducible conditions (elicitation, developmental stages) is especially effective: genes that are co-expressed with metabolite accumulation patterns become strong candidates for functional testing. Genomics also enables the mining of natural variation (population resequencing, GWAS) to link allelic differences to chemical phenotypes, offering orthogonal evidence for pathway gene involvement (Singh et al., 2022; Malit et al., 2022). Wisecaver et al. (2017) presented an interesting research where they tested a hypothesis followed by an evaluation. The hypothesis was that since genes contributing to biosynthetic pathways exhibit environment-dependent co-regulation. However, BGCs predicted through bioinformatics (lacking any associated metabolite) were not found to be co-expressed more than the null distribution for the neighbouring genes. Thus, it could be inferred that co-expression analysis serves as a more trustworthy platform for decoding the genetic basis as well as structural blueprint of bioactive compounds.

2.4 Sequencing and Assembly of Genomes

The creation of smooth genomic contigs with accurate gene annotation is made possible by next-generation sequencing platforms, for e.g., long-read platforms (PacBio, Oxford Nanopore) in conjunction with short-read Illumina data. Long-read sequencing is essential for resolving co-localized genes and repetitive elements associated with specialized metabolic pathways in medicinal plants with large genomes or extensive tandem repeats, such as *Panax ginseng* and *Salvia miltiorrhiza* (Biedrzycka et al., 2017). Once put together, genome annotation algorithms identify gene families, predict open reading frames, and attribute potential activities to known enzymes based on sequence homology.

2.5 Identification of Biosynthetic Gene Clusters (BGCs)

Plant biosynthetic pathways can be encoded either by dispersed genes or by physically clustered biosynthetic gene clusters. BGCs often include genes for core enzymes, tailoring enzymes, and pathway-specific regulators. In plants, terpenoid, alkaloid, and polyketide clusters have been identified using computational tools like antiSMASH-P and PlantiSMASH, which scan for co-localized enzymatic domains. For example, the characterization of the noscapine alkaloid co-localized gene in Papaver somniferum showed a coordinated set of genes encoding cytochrome P450s, methyltransferases, and oxidoreductases, enabling pathway regeneration and heterologous production (Gorbunova et al., 2014).

2.6 Comparative Genomics and Phylogenetic Approaches

Intergenomic analyses across related species or cultivars can reveal lineage-specific expansions or contractions of gene families associated with adaptive metabolic pathways. Phylogenetic analysis helps assign predicted enzymatic activities by clustering uncharacterized genes with biochemically validated homologs. For instance, terpene synthase phylogenies in the Lamiaceae family have been used to predict enzyme products (Croteau et al., 2006). Genomic analyses of *Scutellaria baicalensis* and other Lamiaceae species revealed that the root-specific 4′-deoxyflavone metabolic pathway was formed by segmental gene duplications and subfunctionalization (Zhao et al., 2019). In a study on Nepeta species, Lichman et al. (2020) revealed that iridoid biosynthesis in Nepeta was lost and regained, highlighting the evolutionary flexibility of plant specialized metabolism. Another study looked at the gene families in *Papaver somniferum* that are involved in the manufacture of benzyl-iso-quinoline alkaloids, such as morphine and codeine (Winzer et al., 2012). An evolutionary benefit for diversified alkaloid manufacturing was indicated by the study's discovery of a notable expansion in enzyme-coding gene families, including cytochrome P450s and O-methyltransferases (Karalija et al., 2025).

2.7 Integration with Transcriptomics

Whether genes are actively regulated or under what circumstances they function cannot be determined just by genomics. Researchers can create gene co-regulation networks that connect potential genes to metabolite localization signatures by combining genomic-transcriptomic profiles (RNA-seq). Strong evidence for a functional link can be found in the fact that genes co-expressed in tissue-specific or inducible contexts frequently take part in the same metabolic pathway. For instance, potential cytochrome P450s implicated in tanshinone production were found in *Salvia miltiorrhiza* roots using transcriptome-guided mining (Quan et al., 2016). A cutting-edge transcriptomics method called single-cell RNA sequencing identifies unique metabolite production pathways that are particular to certain cell types or phases of development (Aldridge & Teichmann, 2020). It provides previously unheard-of resolution that gets around the drawbacks of traditional transcriptomic techniques and makes it possible to determine gene expression patterns at the cellular level. For instance, Sun et al. (2023) used single-cell RNA sequencing to study the spatial organisation of the metabolism of monoterpenoid indole alkaloids in *Catharanthus roseus* leaves, producing a biosynthetic model with localised transcripts of 20 monoterpenoid indole alkaloid (MIA) genes. By dividing the MIA route into three different cell types and identifying important transporters, the study provided information on how to improve the yields of monoterpenoid indole alkaloids in plants using metabolic engineering and synthetic biology. The leaf monoterpenoid indole alkaloid biosynthesis pathway is partitioned across multiple cell types (Li et al., 2023). In an investigation of *Hypericum perforatum*, Wu et al. (2024) demonstrated the importance of single-cell RNA sequencing by identifying specialised "Hyper cells" that are in charge of the manufacture of hyperforin in leaves and flowers. Candidate genes from genomic analyses require functional validation. Among the methods are CRISPR-Cas-mediated gene knockouts in the native plant, in vitro or in vivo enzyme tests, and heterologous expressions in *E. coli*, yeast, or *Nicotiana benthamiana*. The role of the enzyme in the pathway is confirmed by combining these functional investigations with metabolomic data. By concentrating experimental efforts on high-confidence candidates rather than thorough screening, this method speeds up discovery.

The biosynthetic pathways of important therapeutic chemicals, such as ginsenosides in Panax species, camptothecin in Camptotheca, and paclitaxel in Taxus, have been clarified thanks in large part to genomic techniques. In addition to finding structural genes, genomic data mining also reveals transcription factors and transporters that control metabolite flux, allowing metabolic engineering to produce more compounds. Additionally, by identifying hitherto unidentified pathways with possible medicinal relevance, genomics helps in the discovery of natural products. Large-scale metabolic potential forecasts are made possible by genomics, which also speeds up the identification of pathway genes in non-model plants and eliminates experimental redundancy. The inability to accurately anticipate enzyme specificity based only on sequence is one of its limitations, particularly for multifunctional enzymes with promiscuous catalytic activity. Furthermore, a gene's existence does not ensure expression or activity; for this reason, transcriptomic, proteomic, and metabolomic evidence must be added to genomic predictions.

3 Proteomics-Based Approaches

Proteomics enables the characterization and evaluation of proteins that actively participate in medicinal plant metabolic pathways. Unlike genomics, proteomics captures the actual expression of enzymes, subcellular localization, protein-protein interactions, and post-translational changes, offering direct evidence of pathway activity. Proteomic approaches help resolve compartmentalized steps, for example, enzymes localized to plastids, the ER, or vacuoles which is essential for high-fidelity modeling of pathway topology (Martínez-Esteso et al., 2024; Yin & Yang, 2024). Because proteomics requires high-quality reference databases (from genomes or transcriptomes) and sensitive instrumentation, integration with genomics and metabolomics is typically required to interpret proteomic hits confidently.

3.1 Protein Extraction and Sample Preparation

Plant tissues often contain substantial levels of phenolics, polysaccharides, and secondary metabolites, which can interfere with protein purification. Extraction protocols typically employ strong denaturants, detergents, and reducing agents along with protease inhibitors. Fractionation techniques, like subcellular fractionation, are used to isolate low-abundance biosynthetic proteins. For instance, microsomal fractions enriched for cytochrome P450s have been essential for studying terpenoid biosynthesis in *Artemisia annua* (Hayes et al., 2019).

3.2 Shotgun and Targeted Proteomics

Proteomics is dominated by two primary tactics: targeted and untargeted (shotgun). By identifying multiple proteins at once using LC-MS/MS, shotgun proteomics enables them to record the global enzymatic landscape in settings that elicit pathways. Higher sensitivity and quantitative accuracy are provided by targeted techniques that concentrate on predetermined peptides of specific enzymes in particular. When examining low-abundance enzymes engaged in specialized metabolite synthesis, these techniques are especially helpful (Singh et al., 2022).

3.3 Protein Profiling Based on Activity

Activity-based protein profiling (ABPP) represents a strong proteome profiling tool to mark functional enzymes. By this method, covalent probes are designed to bind the active sites of catalytic groups (e.g., oxidoreductases or glycosyltransferases). Subsequent enrichment and LC-MS/MS analysis identify active proteins *in situ*, providing tangible evidence of active involvement in the metabolic pathway. ABPP has been productively applied to identify key enzymes in alkaloid and flavonoid biosynthesis (Keller et al., 2005).

3.4 Subcellular and Temporal Proteomics

Specialized metabolism often involves spatially localized reactions, such as plastid-localized terpenoid synthesis or vacuolar storage of alkaloids. Subcellular proteomics isolates organelles or fractions of membranes, allowing subcellular enzyme localization. Temporal proteomics examines protein abundance changes across developmental stages or after elicitation, revealing enzymes that correlate with metabolite accumulation. This strategy has been critical in mapping enzymes responsible for the inducible production of defence-related bioactive constituents (Yue et al., 2023). Proteomics, by revealing functional enzymes and protein complexes, has supported metabolic engineering approaches to boost synthesis in both plant and heterologous systems. Integrating omics with metabolomics enables comprehensive reconstruction of biosynthetic pathways, thereby enhancing the efficiency of metabolite profiling and synthetic biology strategies. Using the latest mass spectrometry technology to improve precision and trueness, is essential to advancing proteomics for the study of specialised metabolism in medicinal plants. Proteogenomics can help make discoveries by identifying new enzymes engaged in biosynthetic processes. Proteomics with CRISPR-Cas9 integration speeds up enzyme characterisation by enabling functional confirmation of certain proteins. Furthermore, standardising proteomic procedures is essential to guaranteeing reproducibility and consistency, which will encourage the wider use and dependability of proteomic data in medicinal plant research (Wang et al., 2024). The existence and activity of enzymes are demonstrated by proteomics, which also identifies protein alterations that are essential for function and multi-enzyme complexes that might help metabolite channeling. Limitations, however, include the difficulty of identifying very low-abundance proteins and the need for reliable reference protein databases, which might not be complete in non-model medicinal plants. Furthermore, it is advised to combine proteomic data with metabolomics and enzyme assays because protein abundance does not always correspond with catalytic activity.

4 Metabolomics-Based Approaches

The most direct understanding of plant-specific metabolism can be gained through metabolomics, the in-depth analysis of tiny endogenous intracellular phytocompounds. Metabolomics can reveal pathway activity, uncover novel intermediates, and direct the activity profiling of genes and enzymes by comparing metabolites across different tissues, developmental stages, or environmental variables. In contrast to genomics or proteomics, metabolomics directly reflects the actual chemical outcomes of metabolic processes. One traditional method for bridging the gaps in biosynthetic pathways is to link transcriptome data with metabolomes. For instance, by comparing two Strychnos species with differing capacities for strychnine production, Hong et al. (2022) elucidated the complete strychnine biosynthetic cascade.

Metabolomics provides a first-order, systems-level readout of pathway activity by measuring small molecules, including substrates, intermediates, products, and side-products across tissues, time points, and treatments. The chemical diversity of an extract can be mapped using untargeted metabolomics (LC-MS, GC-MS, NMR), which can also identify co-accumulating metabolites whose interaction networks point to links with precursor products. MS/MS spectral networking and feature annotation expand identification beyond library matches by classifying structurally related metabolites (Durand et al., 2024).

4.1 Untargeted Metabolomics

Non-targeted metabolomics targets to detect and quantify as many compounds as feasible without any preceding selection. Mass spectroscopies are often employed techniques to find metabolites that change under various circumstances, data analysis entails peak detection, alignment, and normalization, then statistical and multivariate analyses. Finding previously unidentified metabolites or linking metabolic alterations to enzyme activity or gene expression are two applications where this method excels (Fiehn, 2002).

4.2 Targeted and Isotope-Resolved Metabolomics

Targeted metabolomics focuses on predefined sets of metabolites, allowing precise quantification and kinetic studies. When combined with isotopic labeling such as 13C-glucose, 15 N-amino acids, selective metabolite quantification can track flux through biochemical pathways, trace metabolic conversions, and identify branching points. For example, 13C-labeled mevalonate feeding in *Taxus* species has elucidated the flux toward paclitaxel biosynthesis (Croteau et al., 2006). Isotopologue-resolved metabolomics thus provides a dynamic picture of pathway activity, complementing static metabolite profiles.

4.3 MS/MS Spectral Networking and Structural Annotation

Researchers can infer biosynthetic linkages and cluster structurally similar compounds using advanced mass spectrometry techniques like tandem MS and spectral networking. Molecular networking techniques, like GNPS, make it easier to visualize chemical families and identify possible intermediates that library-based identification would overlook. In medicinal plants, where numerous secondary metabolites are rare, infrequent, or inadequately described in current databases, this method is quite helpful (Wang et al., 2016). The study includes using chemical logic to link transcript presence with metabolite measurement based on Liquid Chromatography-Mass Spectrometry. In a similar technique, genomics was used to initially distinguish the *Lophophora williamsii* from *Lophophora diffusa* (Watkins et al., 2023).

5 Multi-omics Approaches

Metabolomics is most powerful when integrated with genomics, transcriptomics, or proteomics. Correlating metabolite accumulation with gene expression (co-expression networks) or protein abundance can identify candidate enzymes and regulatory genes. In *Salvia miltiorrhiza*, such integrated analyses revealed candidate cytochrome P450s, and glycosyltransferases involved in tanshinone biosynthesis (Quan et al., 2016). Similarly, metabolomic profiling of mutant or RNAi-silenced lines can provide causal evidence linking specific genes to metabolite production. One prominent example is the identification of the colchicine route in *Gloriosa superba*, where tissue-resolved metabolomics and transcriptomics have been correlated using metabolic logic (Nett & Sattely, 2021). Likely, the metabolic cascade to paclitaxel biogenesis has recently been completely unravelled using a combination approach incorporating transcriptomics and metabolomics. Notably, this involved identifying hitherto mysterious enzymes including β-phenylalanine-CoA ligase, taxane 9ẓ-dioxygenase, taxane 1β-hydroxylase, and C4β-C20 epoxidase. The successful heterologous synthesis of paclitaxel in *Nicotiana benthamiana* was the result of this admirable endeavour (Zhang et al., 2023a, a).

Numerous therapeutic plant pathways, such as those involving phenolics in Salvia species, terpenoids in *Artemisia annua*, and alkaloids in *Catharanthus roseus*, have been studied using metabolomics. It makes it possible to map pathway fluxes, identify chemical markers for metabolic engineering, and find previously unidentified metabolites. Furthermore, in plant cell cultures and bioreactors, metabolomic insights are being increasingly utilized to optimize culture conditions for increased synthesis of phytocompounds. By connecting genotype alterations to metabolic effects and offering functional evidence for gene assignments, metabolomics also makes it possible to chemically phenotype mutant lines or gene perturbations (RNA interference, CRISPR). The detection of very low-abundance intermediates and incomplete spectral libraries for plant-specific chemicals is our challenge that is lessened by advanced sample preparation, derivatization, and orthogonal analytical techniques (Oh et al., 2023; Shen et al., 2023). Metabolomics can detect transitory intermediates, evaluate route outputs directly, and discover novel bioactive compounds. Among mass spectrometry's shortcomings are ionization biases, difficulties in identifying unknown metabolites, and insufficient coverage brought on by chemical variety. Furthermore, quantitative precision requires strict validation and standardization. However, when combined with isotopic tagging and computational methods, metabolomics provides essential evidence for pathway reconstruction.

5.1 Integration of Multi-omics Approaches

Multi-omics integrates omics approaches into a cohesive framework that leverages the strengths of each layer to assign pathway steps with higher confidence than any single approach. Typical workflows begin with genome and transcriptome mining to nominate candidates, use co-expression and co-accumulation networks to prioritize gene metabolite pairs, validate protein presence and activity via proteome-wide analysis, and confirm precursor derived relationship through metabolomics and isotopic feeding. Computational integration methods—correlation networks, machine-learning classifiers, Bayesian models, and causal inference frameworks—enable systematic linking of genes to metabolites and can predict regulatory nodes and transcription factors controlling pathway flux. Recent examples show that paired omics (transcriptome + metabolome) and tri-omics (adding proteome) dramatically accelerate discovery of both structural and regulatory genes of specialized metabolic pathways. Remaining challenges include harmonizing disparate data types (dynamic range, technical biases), designing matched sampling regimes, and developing standards for statistical confidence; nonetheless, multi-omics is now the most efficient route to comprehensive pathway reconstruction in understudied therapeutic plants (Zhu et al., 2023; Wolters et al., 2024).

Using a multi-omics approach, the MIA biosynthesis process in *Catharanthus roseus* was recently further elucidated. A secologanin transporter was discovered using Hi-C data, and a reductase that produces the bis-indole alkaloid anhydrovinblastine was identified by single-cell RNA sequencing (Li et al., 2023). The Ginseng Genome Database, for instance, provides a comprehensive range of multi-omics dataset for *Panax ginseng* (Jayakodi et al., 2018). The omics data for hundreds of medicinal plants are compiled by other repositories such HMOD, GPGP, MPOD, and BPGD (Wang et al., 2018; Liao et al., 2022a, b; He et al., 2022; Gupta et al., 2024).

5.2 Multi-omics: Find New Uses for Known Molecules

Some well-known compounds may find new uses if multi-omics analysis is incorporated into research aimed at determining the metabolic underpinnings of phenotypes or characteristics. By attracting beneficial soil bacteria, coumarin helps to overcome iron deficiency, according to the integration analysis and RNA-seq. Effects of coumarin on rhizosphere microbiome can be possibly methodically assessed by culture-free phylogenetic profiling using 16S rRNA gene amplicons (Harbort et al., 2020). Without the application of integrated analysis of transcriptomics and microbiomics, this emerging role of coumarins might not be identified. To uncover these hidden roles of other signaling molecules like flavones (Yu et al., 2021), benzoxazinoids (Kudjordjie et al., 2019), and strigolactones (Claassens & Hills, 2018) in governing plant-microbe associations, a fusion of transcriptomics, metabolomics, and microbiomics analysis was also used (Jiang et al., 2016).

5.3 Multi-omics: Finding Unknown Molecules with Known Functions

To find new compounds that may have contributed to specific biological processes, an untargeted metabolomics technique is necessary. The framework of research to extend the exploration of biological activities and structures of these compounds is made possible by the objective capture of chemical variations in samples with bioactive and non-bioactivities by comparative metabolic profiling. The identification of N-hydroxy-pipecolic acid as a phenom-mediated signalling molecule that causes Arabidopsis to develop systemic disease resistance is one such example (Chen et al., 2018; Rekhter et al., 2019; Rajniak et al., 2015; Jeon et al., 2020).

5.4 Multi-omics: Finding Unknown Molecules with Unknown Functions

In phytochemical research, finding new compounds with distinct biological functions has been a continuous endeavor. Unknown natural compounds are now found using genomic and transcriptomic exploration, thanks to the introduction and advancement of multi-omics technology, particularly genomics (Nützmann et al., 2016; Owen et al., 2017). When combined with effective heterologous expression platforms, the genomic and transcriptome characteristics that underline the production of natural phytocompounds which are bioactive can facilitate the rapid identification of hitherto unidentified plant metabolites. On the other hand, function guided investigations of genes thought to be involved in the production, regulation, or transport of metabolites can frequently uncover unidentified compounds with unique activities. Notable examples include Arabidopsis that is necessary for pathogen-triggered defense (Rajniak et al., 2015; Li et al., 2021; Huang et al., 2019). It is becoming more and more clear that gene clustering is a significant genomic trait that can be used to easily identify plant signaling metabolites (Polturak & Osbourn, 2021). Since several genes encoding biosynthetic enzymes operating in the same pathway may be readily recognized simultaneously, plant biosynthetic gene clusters offer an excellent starting point for identifying and clarifying hitherto unidentified biosynthetic pathways. To find new triterpene chemical structures, researchers start by doing untargeted metabolomics analysis and heterologous functional characterization of the cluster genes and their mutants (Luo et al., 2022).

Further developments in multi-layered dataset analysis techniques are essential for identifying plant bioactive pathways in the future (Li et al., 2023; Burlat et al., 2023; Sun et al., 2023). In the age of big data, the continuous democratisation of SML algorithms is also essential for outperforming cognitive inference. Lastly, focusing on data collection in conjunction with machine learning would surely speed up the process of identifying viable candidates, eventually opening up plant bioactive pathways for a variety of biotechnological uses in business and medicine (Chele et al., 2025).

6 Synthetic Biology-Based Approaches

Recent developments in the field of sustainable agriculture have highlighted the systems-synthetic biology integration, leading to the improvement of genetic engineering approaches aimed at boosting secondary metabolism. A collective approach involving expression network modelling, metabolite profiling, co-localized gene set, identification, neural network modelling, GWAS, and proteomic mapping. Gene circuit engineering is based on combining biology, computational science, and engineering approaches. This field works through the de novo biological part construction, systems, and devices. Synthetic biology offers an approach for validation as well as engineering of metabolic pathways. The tools allow systematic testing and pathway validation by reassembly in model organisms like bacteria or yeast, where analysis can be done outside a complex plant system. Paddon et al. (2013) reconstructed the artemisinin metabolic network in *S. cerevisiae* by the introduction of necessary genes, thereby validating the catalytic functional steps. Further, it also led to semi-synthetic artemisinin production, causing a significant impact on efforts against malaria globally. Another study by Galanie et al. (2015) used *S. cerevisiae* for the opioid biosynthesis pathway by introducing BIA biosynthetic genes from *Papaver somniferum*. This resulted in validation of the pathway and opened avenues for fermentation-based biosynthesis. Taking high growth efficiency and easy genetic manipulation of microbial hosts, complex pathway genes can be identified, validated, as well production of metabolites can be achieved (Thodey et al., 2014; Winzer et al., 2015). Similarly, synthetic biology has been utilized in pathway elucidation and engineering of several metabolites like colchicine, Vinblastine, Mescaline, Cocaine and Paclitaxel (Taxol) along with several other important bioactive compounds (Nett & Sattely, 2021; Zhang et al., 2022; Berman et al., 2024; Wang et al., 2022; Li et al., 2019; Yang et al., 2024). Integration of computational approaches with systems biology is on its way to pathway elucidation; further, artificial intelligence and machine learning are also picking up pace for accelerating pathway discovery and engineering (Carbonell, 2021). The identification and engineering of metabolic pathways is being accelerated by the combination of artificial intelligence (AI) and machine learning.

7 CRISPR/Cas9 Tool for Biosynthetic Pathway Elucidation

Newly developed CRISPR/Cas9-based editing has gathered a lot of attention due to its alluring features like precision, ease, specificity, multiplexing, and efficacy, thus leading to an advanced genome editing and synthetic biology era. Scientists worldwide have focused their attention on reaping the benefits of this tool, but the use of this tool is still in its infancy due to challenges like the limited genomic resources and suitable genetic transformation technologies in the majority of the plants. Most of the studies aim at amplify the bioactive yield as explored in plants like *Cannabis sativa, Atropa belladonna, Dendrobium officinale, Cichorium intybus L., Salvia miltiorrhiza, Dioscorea zingiberensis, Symphytum officinale, Rehmannia glutinosa, Monochasma savatieri,* and *Papaver somniferum L.,* as reviewed by Arun et al. (2023). A few significant reports have been instrumental in understanding of pathway and regulatory components. CRISPR/Cas9-based deletions induced in the MYB46 gene in *Fragaria vesca* revealed its function in controlling the phenylpropanoid biosynthesis and fruit set (Rai et al., 2025). The *F. vesca*-specific U6 promoter was used for the expression of two guideRNAs targeting the first exonic region of the *FvMYB*6 gene, causing in-frame 81-bp deletion or out-of-frame 82-bp deletion, leading to potential knockout of the gene. Along with transcriptome analysis, this study revealed that mutant flowers depicted reduced contents of flavonoids, mainly quercetin-glucoside, epicatechin, and kaempferol-3-coumaroylhexoside. Additionally, a localized decrease in lignin content was detected in the anthers.

CRISPR technology was used for the elimination of the biosynthesis of sesquiterpene lactones (STL) in *Cichorium intybus* L. plant (Cankar et al., 2021). Four genes encoding the first step of STL synthesis, *germacrene A synthase (CiGAS),* and null mutations carrying chicory lines in all four genes were developed. Inactivation of these genes led to an increase in phenolic compounds and showed that STLs are not essential for chicory growth, and inhibition of this pathway led to a reduction of STLs, thus facilitating inulin extraction. Another study by Newton et al. (2023) aimed at targeting control of supporting pathways involved in paclitaxel biosynthesis using CRISPR-guided methylation. Both chemical inhibitors and CRISPR-guided DNA methylation were used to regulate the flux of the phenylpropanoid pathway in *Taxus chinensis*. The study demonstrated the importance of metabolic flux control of supporting network in secondary metabolic pathways and projected the importance of CRISPR-guided methylation for *in-vitro* metabolic pathway optimization.

Presently, the use of CRISPR technology is mostly limited to strategies aimed at using the existing information on biosynthetic pathways regarding pathway intermediates or regulatory components. This is also because our current understanding of responsible gene networks and regulators remains limited. CRISPR-based studies can be put to use for gene function discovery through reverse genetic engineering. Advances in genetic transformation technologies are expected to further enhance the utilization of CRISPR-based technology for pathway elucidation in medicinal plants.

8 Conclusion

Plants are host to a plethora of specialized metabolites, but the mystery around biosynthetic pathways hinders efforts in tapping their full potential. Major efforts and progress have been made in this regard, but the medicinal plant genome sequencing needs to catch up in speed to design molecular biology approaches, either in pathway discovery or engineering. With progressing tools like genome editing, the pathway validation seems more feasible. The present chapter briefly discusses current procedures and methodologies for the characterization of missing genes from secondary metabolic pathways. Integrated approaches involving a mix of genome editing, "Omics", biochemistry, and various other in vitro approaches are expected to further allow streamlining the process of gene prioritization. Synthetic biology approaches also offer hope as a major tool for both functional validation and high-yield production of functional phytochemicals.

References

Aldridge, S., & Teichmann, S. A. (2020). Single-cell transcriptomics comes of age. *Nature Communications, 11*(1), 4307. https://doi.org/10.1038/s41467-020-18158-5

Arun, M., Halka, J., & Kowsalya, K. (2023). CRISPR/Cas9: A novel genetic tool to manipulate plant secondary metabolite pathways. In *Genetic manipulation of secondary metabolites in medicinal plants* (pp. 45–57). Springer Nature Singapore. https://doi.org/10.1007/978-981-99-1322-7_3

Berman, P., de Haro, L. A., Cavaco, A.-R., Panda, S., Dong, Y., Kuzmich, N., Lichtenstein, G., Peleg, Y., Harat, H., & Jozwiak, A. (2024). The biosynthetic pathway of the hallucinogen mescaline and its heterologous reconstruction. *Molecular Plant, 17*(6), 1129–1150. https://doi.org/10.1016/j.molp.2024.02.006

Brzycki Newton, C., Young, E. M., & Roberts, S. C. (2023). Targeted control of supporting pathways in paclitaxel biosynthesis with CRISPR-guided methylation. Frontiers in Bioengineering and Biotechnology, 11, 1272811

Brandle, J. E., & Telmer, P. G. (2007). Steviol glycoside biosynthesis. *Phytochemistry, 68*(14), 1855–1863.

Biedrzycka, A., Sebastian, A., Migalska, M., Westerdahl, H., & Radwan, J. (2017). Testing genotyping strategies for ultra-deep sequencing of a co-amplifying gene family: MHC class I in a passerine bird. *Molecular Ecology Resources, 17*(4), 642–655. https://doi.org/10.1111/1755-0998.12606

Burlat, V., Papon, N., & Courdavault, V. (2023). Medicinal plants enter the single-cell multi-omics era. *Trends in Plant Science, 28*(11), 1205–1207. https://doi.org/10.1016/j.tplants.2023.08.005

Cankar, K., Bundock, P., Sevenier, R., Häkkinen, S. T., Hakkert, J. C., Beekwilder, J., et al. (2021). Inactivation of the germacrene A synthase genes by CRISPR/Cas9 eliminates the biosynthesis of sesquiterpene lactones in *Cichorium intybus* L. *Plant Biotechnology Journal, 19*(12), 2442–2453. https://doi.org/10.1111/pbi.13668

Carbonell, P. (2021). Synthetic biology design tools for metabolic engineering. In V. Singh (Ed.), *Microbial cell factories engineering for production of biomolecules* (pp. 65–77). Elsevier. https://doi.org/10.1016/B978-0-12-823506-0.00005-1

Chacon, F. T. (2024). *The effects of cultivation and extraction conditions on the phytochemistry and bioactivity of Cannabis sativa L.* (Doctoral dissertation). The Pennsylvania State University.

Chele, K. H., Piater, L. A., van der Hooft, J. J., & Tugizimana, F. (2025). Bridging ethnobotanical knowledge and multi-omics approaches for plant-derived natural product discovery. *Metabolites, 15*(6), 362. https://doi.org/10.3390/metabo15060362

Chen, S. (2024). Biosynthesis of natural products from medicinal plants: Challenges, progress and prospects. *Chinese Herbal Medicines, 16*(1), 1–15. https://doi.org/10.1016/j.chmed.2023.100647

Chen, Y. C., Holmes, E. C., Rajniak, J., Kim, J. G., Tang, S., Fischer, C. R., et al. (2018). N-hydroxy-pipecolic acid is a mobile metabolite that induces systemic disease resistance in *Arabidopsis*. *Proceedings of the National Academy of Sciences of the United States of America, 115*(21), E4920–E4929. https://doi.org/10.1073/pnas.1805291115

Claassens, A. P., & Hills, P. N. (2018). Effects of strigolactones on plant roots. In *Root biology* (pp. 43–63). Springer. https://doi.org/10.1007/978-3-319-75910-4_3

Croteau, R., Ketchum, R. E., Long, R. M., Kaspera, R., & Wildung, M. R. (2006). Taxol biosynthesis and molecular genetics. *Phytochemistry Reviews, 5*(1), 75–97. https://doi.org/10.1007/s11101-005-3748-2

Dejana, L. (2023). *Signaling in induced resistance by arbuscular mycorrhizal fungi against insect pests in tomato* (Doctoral dissertation). Universidad de Granada.

Deng, L., Zhao, Z., Liu, L., Zhong, Z., Xie, W., Zhou, F., et al. (2023). Dissection of 3D chromosome organization in *Streptomyces coelicolor* A3(2) leads to biosynthetic gene cluster overexpression. *Proceedings of the National Academy of Sciences of the United States of America, 120*(11), e2222045120. https://doi.org/10.1073/pnas.2222045120

Durand, M., Besseau, S., Papon, N., & Courdavault, V. (2024). Unlocking plant bioactive pathways: Omics data harnessing and machine learning assisting. *Current Opinion in Biotechnology, 87*, 103135. https://doi.org/10.1016/j.copbio.2023.103135

Edge, A., et al. (2018). A *tabersonine 3-reductase Catharanthus roseus* mutant accumulates vindoline pathway intermediates. *Planta, 247*(1), 155–169. https://doi.org/10.1007/s00425-017-2778-3

Elvira, K. S., Gielen, F., Tsai, S. S., & Nightingale, A. M. (2022). Materials and methods for droplet microfluidic device fabrication. *Lab on a Chip, 22*(5), 859–875. https://doi.org/10.1039/D1LC00967C

Farrow, S. C., et al. (2018). Cytochrome P450 and O-methyltransferase catalyze the final steps in the biosynthesis of the anti-addictive alkaloid ibogaine from *Tabernanthe iboga*. *Journal of Biological Chemistry, 293*(35), 13821–13833. https://doi.org/10.1074/jbc.RA118.003282

Fiehn, O. (2002). Metabolomics—The link between genotypes and phenotypes. *Plant Molecular Biology, 48*(1), 155–171. https://doi.org/10.1023/A:1013713905833

Fox Ramos, A. E., et al. (2019). Natural products targeting strategies involving molecular networking: Different manners, one goal. *Natural Product Reports, 36*(7), 960–980. https://doi.org/10.1039/C9NP00006J

Galanie, S., Thodey, K., Trenchard, I. J., Filsinger Interrante, M., & Smolke, C. D. (2015). Complete biosynthesis of opioids in yeast. *Science, 349*(6252), 1095–1100. https://doi.org/10.1126/science.aac9373

Garagounis, C., Delkis, N., & Papadopoulou, K. K. (2021). Unraveling the roles of plant specialized metabolites: Using synthetic biology to design molecular biosensors. *New Phytologist, 231*(4), 1338–1352. https://doi.org/10.1111/nph.17406

Gorbunova, V., Boeke, J. D., Helfand, S. L., & Sedivy, J. M. (2014). Sleeping dogs of the genome. *Science, 346*(6214), 1187–1188. https://doi.org/10.1126/science.aaa2552

Gupta, P., Sharma, A., & Nagegowda, D. A. (2024). Multi-omics: Powerful accelerator for uncovering plant specialized metabolic pathways—The case of leonurine. *Molecular Plant, 17*(1), 13–15. https://doi.org/10.1016/j.molp.2023.10.012

Harbort, C. J., Hashimoto, M., Inoue, H., Niu, Y., Guan, R., Rombolà, A. D., et al. (2020). Root-secreted coumarins and the microbiota interact to improve iron nutrition in *Arabidopsis*. *Cell Host & Microbe, 28*(6), 825–837. https://doi.org/10.1016/j.chom.2020.09.006

Hayes, P. E., Clode, P. L., Guilherme Pereira, C., & Lambers, H. (2019). Calcium modulates leaf cell-specific phosphorus allocation in Proteaceae from south-western Australia. *Journal of Experimental Botany, 70*(15), 3995–4009. https://doi.org/10.1093/jxb/erz169

He, S., Yang, L., Ye, S., Lin, Y., Li, X., Wang, Y., et al. (2022). MPOD: Applications of integrated multi-omics database for medicinal plants. *Plant Biotechnology Journal, 20*(5), 797–799. https://doi.org/10.1111/pbi.13787

Hong, B., Grzech, D., Caputi, L., Sonawane, P., López, C. E. R., Kamileen, M. O., et al. (2022). Biosynthesis of strychnine. *Nature, 607*(7919), 617–622. https://doi.org/10.1038/s41586-022-04913-8

Hsieh, Y. F., Suslov, D., Espen, L., Schiavone, M., Rautengarten, C., Griess-Osowski, A., et al. (2023). *cis*-Golgi phosphate transporters harboring an EXS domain are essential for plant growth and development. *Plant Physiology, 192*(2), 1000–1015. https://doi.org/10.1093/plphys/kiad188

Hu, H., Shen, X., Liao, B., et al. (2019). Herbgenomics: A stepping stone for research into herbal medicine. *Science China Life Sciences, 62*(7), 913–920. https://doi.org/10.1007/s11427-018-9472-y

Huang, A. C., Jiang, T., Liu, Y. X., Bai, Y. C., Reed, J., Qu, B., et al. (2019). A specialized metabolic network selectively modulates *Arabidopsis* root microbiota. *Science, 364*(6440), eaau6389. https://doi.org/10.1126/science.aau6389

Jayakodi, M., Choi, B. S., Lee, S. C., Kim, N. H., Park, J. Y., Jang, W., et al. (2018). Ginseng Genome Database: An open-access platform for genomics of *Panax ginseng*. *BMC Plant Biology, 18*(1), 62. https://doi.org/10.1186/s12870-018-1276-8

Jeon, J. E., Kim, J. G., Fischer, C. R., Mehta, N., Dufour-Schroif, C., Wemmer, K., et al. (2020). A pathogen-responsive gene cluster for highly modified fatty acids in tomato. *Cell, 180*(1), 176–187. https://doi.org/10.1016/j.cell.2019.12.003

Jiang, N., Doseff, A. I., & Grotewold, E. (2016). Flavones: From biosynthesis to health benefits. *Plants, 5*(2), 27. https://doi.org/10.3390/plants5020027

Karalija, E., Macanović, A., & Ibragić, S. (2025). Revisiting traditional medicinal plants: Integrating multiomics, *in vitro* culture, and elicitation to unlock bioactive potential. *Plants, 14*(13), 2029. https://doi.org/10.3390/plants14132029

Katsuyama, Y., Kita, T., Funa, N., & Horinouchi, S. (2009). Curcuminoid biosynthesis by two type III polyketide synthases in the herb Curcuma longa. *Journal of Biological Chemistry, 284*(17), 11160–11170.

Keller, N. P., Turner, G., & Bennett, J. W. (2005). Fungal secondary metabolism—From biochemistry to genomics. *Nature Reviews Microbiology, 3*(12), 937–947. https://doi.org/10.1038/nrmicro1286

Kessler, A., & Kalske, A. (2018). Plant secondary metabolite diversity and species interactions. *Annual Review of Ecology, Evolution, and Systematics, 49*(1), 115–138. https://doi.org/10.1146/annurev-ecolsys-110617-062406

Kim, A., Park, S. M., Kim, N. S., & Lee, H. (2023). Ginsenoside Rc, an active component of Panax ginseng, alleviates oxidative stress-induced muscle atrophy via improvement of mitochondrial biogenesis. *Antioxidants, 12*(8), 1576

Kudjordjie, E. N., Sapkota, R., Steffensen, S. K., Fomsgaard, I. S., & Nicolaisen, M. (2019). Maize-synthesized benzoxazinoids affect the host-associated microbiome. *Microbiome, 7*(1), 59. https://doi.org/10.1186/s40168-019-0677-7

Lebaudy, A., Hosy, E., Simonneau, T., Sentenac, H., Thibaud, J. B., & Dreyer, I. (2008). Heteromeric K^+ channels in plants. *The Plant Journal, 54*(6), 1076–1082. https://doi.org/10.1111/j.1365-313X.2008.03476.x

Li, J., Mutanda, I., Wang, K., Yang, L., Wang, J., & Wang, Y. (2019). Chloroplastic metabolic engineering coupled with isoprenoid pool enhancement for committed taxanes biosynthesis in *Nicotiana benthamiana*. *Nature Communications, 10*, 4850. https://doi.org/10.1038/s41467-019-12772-0

Li, J., Halitschke, R., Li, D., Paetz, C., Su, H., Heiling, S., et al. (2021). Controlled hydroxylations of diterpenoids allow for plant chemical defense without autotoxicity. *Science, 371*(6526), 255–260. https://doi.org/10.1126/science.abe3369

Li, C., Wood, J. C., Vu, A. H., Hamilton, J. P., Rodriguez Lopez, C. E., Payne, R. M., et al. (2023). Single-cell multi-omics in the medicinal plant *Catharanthus roseus*. *Nature Chemical Biology, 19*(8), 1031–1041. https://doi.org/10.1038/s41589-023-01352-z

Liao, B., Hu, H., Xiao, S., Zhou, G., Sun, W., Chu, Y., Meng, X., Wei, J., Zhang, H., Xu, J., & Chen, S. (2022a). Global Pharmacopoeia Genome Database is an integrated and mineable genomic database for traditional medicines derived from eight international pharmacopoeias. *Science China Life Sciences, 65*(4), 809–817. https://doi.org/10.1007/s11427-021-1968-7

Liao, B., Shen, X., Xiang, L., et al. (2022b). Allele-aware chromosome-level genome assembly of *Artemisia annua* reveals the correlation between ADS expansion and artemisinin yield. *Molecular Plant, 15*, 1310–1328. https://doi.org/10.1016/j.molp.2022.05.013

Liu, Z., Wu, Q., Li, Y., Liu, S., Li, L., Chen, Q., ... & Yu, F. (2019). Effects of exogenous salicylic acid on accumulation of camptothecin and gene expression in *Camptotheca acuminata*. *Canadian Journal of Forest Research, 49*(1), 104–110

Lichman, B. R., Godden, G. T., Hamilton, J. P., Palmer, L., Kamileen, M. O., Zhao, D., et al. (2020). The evolutionary origins of the cat attractant nepetalactone in catnip. *Science Advances, 6*(20), eaba0721. https://doi.org/10.1126/sciadv.aba0721

Luo, F., Yu, Z., Zhou, Q., & Huang, A. (2022). Multi-omics-based discovery of plant signaling molecules. *Metabolites, 12*(1), 76. https://doi.org/10.3390/metabo12010076

Malit, J. J. L., Leung, H. Y. C., & Qian, P. Y. (2022). Targeted large-scale genome mining and candidate prioritization for natural product discovery. *Marine Drugs, 20*(6), 398. https://doi.org/10.3390/md20060398

Martínez-Esteso, M. J., Morante-Carriel, J., Samper-Herrero, A., Martínez-Márquez, A., Sellés-Marchart, S., Nájera, H., & Bru-Martínez, R. (2024). Proteomics: An essential tool to study plant-specialized metabolism. *Biomolecules, 14*(12), 1539. https://doi.org/10.3390/biom14121539

Nett, R. S., & Sattely, E. S. (2021). Total biosynthesis of the tubulin-binding alkaloid colchicine. *Journal of the American Chemical Society, 143*(46), 19454–19465. https://doi.org/10.1021/jacs.1c09410

Nützmann, H. W., Huang, A., & Osbourn, A. (2016). Plant metabolic clusters—From genetics to genomics. *New Phytologist, 211*(3), 771–789. https://doi.org/10.1111/nph.13981

Oh, S. W., Imran, M., Kim, E. H., Park, S. Y., Lee, S. G., Park, H. M., et al. (2023). Approach strategies and application of metabolomics to biotechnology in plants. *Frontiers in Plant Science, 14*, 1192235. https://doi.org/10.3389/fpls.2023.1192235

Okon, S., Zimowska, B., & Rai, M. (2024). *Microbial genetics*. CRC Press.

Owen, C., Patron, N. J., Huang, A., & Osbourn, A. (2017). Harnessing plant metabolic diversity. *Current Opinion in Chemical Biology, 40*, 24–30. https://doi.org/10.1016/j.cbpa.2017.05.004
Paddon, C. J., Westfall, P. J., Pitera, D. J., Benjamin, K., Fisher, K., McPhee, D., et al. (2013). High-level semi-synthetic production of the potent antimalarial artemisinin. *Nature, 496*(7446), 528–532. https://doi.org/10.1038/nature12051
Polturak, G., & Osbourn, A. (2021). The emerging role of biosynthetic gene clusters in plant defense and plant interactions. *PLoS Pathogens, 17*(7), e1009698. https://doi.org/10.1371/journal.ppat.1009698
Qu, Y., et al. (2019). Completion of the canonical pathway for assembly of anticancer drugs vincristine/vinblastine in *Catharanthus roseus*. *The Plant Journal, 97*(2), 257–266. https://doi.org/10.1111/tpj.14114
Quan, Q., Zhu, H., Liu, S. Y., Fei, S. M., Fan, H., & Yang, W. L. (2016). Steering Bell-diagonal states. *Scientific Reports, 6*(1), 22025. https://doi.org/10.1038/srep22025
Rai, A., Skårn, M. N., Elameen, A., Tengs, T., Amundsen, M. R., Bjorå, O. S., et al. (2025). CRISPR-Cas9-mediated deletions of *FvMYB46* in *Fragaria vesca* reveal its role in regulation of fruit set and phenylpropanoid biosynthesis. *BMC Plant Biology, 25*(1), 256. https://doi.org/10.1186/s12870-025-04935-y
Rajniak, J., Barco, B., Clay, N. K., & Sattely, E. S. (2015). A new cyanogenic metabolite in *Arabidopsis* required for inducible pathogen defence. *Nature, 525*(7569), 376–379. https://doi.org/10.1038/nature14907
Rekhter, D., Lüdke, D., Ding, Y., Feussner, K., Zienkiewicz, K., Lipka, V., et al. (2019). Isochorismate-derived biosynthesis of the plant stress hormone salicylic acid. *Science, 365*(6452), 498–502. https://doi.org/10.1126/science.aaw1720
Reshi, Z. A., Ahmad, W., Lukatkin, A. S., & Javed, S. B. (2023). From nature to lab: A review of secondary metabolite biosynthetic pathways, environmental influences, and in vitro approaches. *Metabolites, 13*(8), 895. https://doi.org/10.3390/metabo13080895
Ro, D. K., Paradise, E. M., Ouellet, M., Fisher, K. J., Newman, K. L., Ndungu, J. M., et al. (2006). Production of the antimalarial drug precursor artemisinic acid in engineered yeast. *Nature, 440*(7086), 940–943. https://doi.org/10.1038/nature04640
Ruegger, M., et al. (2001). Mutations that reduce sinapoylmalate accumulation in *Arabidopsis thaliana* define loci with diverse roles in phenylpropanoid metabolism. *Genetics, 159*(4), 1741–1749.
Salim, V., et al. (2018). *Camptotheca acuminata* 10-hydroxycamptothecin O-methyltransferase: An alkaloid biosynthetic enzyme co-opted from flavonoid metabolism. *The Plant Journal, 95*(1), 112–125. https://doi.org/10.1111/tpj.13940
Shen, S., Zhan, C., Yang, C., Fernie, A. R., & Luo, J. (2023). Metabolomics-centered mining of plant metabolic diversity and function: Past decade and future perspectives. *Molecular Plant, 16*(1), 43–63. https://doi.org/10.1016/j.molp.2022.10.007
Singh, P. K., & Jayaraman, N. K. (2024). Unlocking ethnomedicinal plant potentials: Advancing drug discovery in the high-throughput omics era. In *Ethnomedicinal plants for drug discovery: Current developments* (p. 443). Springer.
Singh, K. S., van der Hooft, J. J., van Wees, S. C., & Medema, M. H. (2022). Integrative omics approaches for biosynthetic pathway discovery in plants. *Natural Product Reports, 39*(9), 1876–1896. https://doi.org/10.1039/d2np00038h
Sun, W., Xu, Z., Song, C., & Chen, S. (2022). Herbgenomics: Decipher molecular genetics of medicinal plants. *The Innovation, 3*(6), 100322. https://doi.org/10.1016/j.xinn.2022.100322
Sun, S., Shen, X., Li, Y., Li, Y., Wang, S., Li, R., et al. (2023). Single-cell RNA sequencing provides a high-resolution roadmap for understanding the multicellular compartmentation of specialized metabolism. *Nature Plants, 9*(1), 179–190. https://doi.org/10.1038/s41477-022-01307-9
Thodey, K., Galanie, S., & Smolke, C. D. (2014). A microbial biomanufacturing platform for natural and semisynthetic opioids. *Nature Chemical Biology, 10*(10), 837–844. https://doi.org/10.1038/nchembio.1613
Wang, M., Carver, J. J., Phelan, V. V., Sanchez, L. M., Garg, N., Peng, Y., et al. (2016). Sharing and community curation of mass spectrometry data with Global Natural Products Social Molecular Networking. *Nature Biotechnology, 34*(8), 828–837. https://doi.org/10.1038/nbt.3597

Wang, X., Zhang, J., He, S., Gao, Y., Ma, X., Gao, Y., et al. (2018). HMOD: An omics database for herbal medicine plants. *Molecular Plant, 11*(5), 757–759.

Wang, J. Y., Huang, Z. Y., Wu, Q. Y., Pan, J., Li, C. X., & Xu, J. H. (2022). Facile biosynthesis of taxadiene by a newly constructed *Escherichia coli* strain fusing enzymes taxadiene synthase and geranylgeranyl pyrophosphate synthase. *Process Biochemistry, 122*, 129–136. https://doi.org/10.1016/j.procbio.2022.08.016

Wang, M., Zhang, S., Li, R., & Zhao, Q. (2024). Unraveling the specialized metabolic pathways in medicinal plant genomes: A review. *Frontiers in Plant Science, 15*, 1459533. https://doi.org/10.3389/fpls.2024.1459533

Watkins, J. L., Li, Q., Yeaman, S., & Facchini, P. J. (2023). Elucidation of the mescaline biosynthetic pathway in peyote (*Lophophora williamsii*). *The Plant Journal, 116*(3), 635–649.

Wink, M. (1999). *Biochemistry of plant secondary metabolism* (Vol. 2). CRC Press.

Winzer, T., Gazda, V., He, Z., Kaminski, F., Kern, M., Larson, T. R., et al. (2012). A *Papaver somniferum* 10-gene cluster for synthesis of the anticancer alkaloid noscapine. *Science, 336*(6089), 1704–1708.

Winzer, T., Kern, M., King, A. J., Larson, T. R., Teodor, R. I., Donninger, S. L., Li, Y., Dowle, A. A., Cartwright, J., Bates, R., et al. (2015). Morphinan biosynthesis in opium poppy requires a P450-oxidoreductase fusion protein. *Science, 349*(6249), 309–312.

Wisecaver, J. H., Borowsky, A. T., Tzin, V., Jander, G., Kliebenstein, D. J., & Rokas, A. (2017). A global coexpression network approach for connecting genes to specialized metabolic pathways in plants. *The Plant Cell, 29*(5), 944–959. https://doi.org/10.1105/tpc.17.00009

Wolters, F. C., Del Pup, E., Singh, K. S., Bouwmeester, K., Schranz, M. E., van der Hooft, J. J., & Medema, M. H. (2024). Pairing omics to decode the diversity of plant specialized metabolism. *Current Opinion in Plant Biology, 82*, 102657.

Wu, S., Morotti, A. L. M., Yang, J., Wang, E., & Tatsis, E. C. (2024). Single-cell RNA sequencing facilitates the elucidation of the complete biosynthesis of the antidepressant hyperforin in St. John's wort. *Molecular Plant, 17*(9), 1439–1457.

Yang, C., Wang, Y., Su, Z., Xiong, L., Wang, P., Lei, W., Yan, X., Ma, D., Zhao, G., & Zhou, Z. (2024). Biosynthesis of the highly oxygenated tetracyclic core skeleton of Taxol. *Nature Communications, 15*, 2339.

Yin, Q., & Yang, M. (2024). Chemoproteomics approach to elucidating biosynthetic pathway of plant natural products. *Frontiers in Plant Science, 15*, 1506569.

Yu, P., He, X., Baer, M., Beirinckx, S., Tian, T., Moya, Y. A., et al. (2021). Plant flavones enrich rhizosphere *Oxalobacteraceae* to improve maize performance under nitrogen deprivation. *Nature Plants, 7*(4), 481–499.

Yue, T., Wang, Y., Zhang, L., Gu, C., Xue, H., Wang, W., et al. (2023). Deep learning for genomics: From early neural nets to modern large language models. *International Journal of Molecular Sciences, 24*(21), 15858.

Zhang, J., Hansen, L. G., Gudich, O., Viehrig, K., Lassen, L. M. M., Schrubbers, L., Adhikari, K. B., Rubaszka, P., Carrasquer-Alvarez, E., Chen, L., et al. (2022). A microbial supply chain for production of the anti-cancer drug vinblastine. *Nature, 609*(7927), 341–347.

Zhang, R., Chen, B., Zhang, H., Tu, L., & Luan, T. (2023a). Stable isotope-based metabolic flux analysis: A robust tool for revealing toxicity pathways of emerging contaminants. *TrAC Trends in Analytical Chemistry, 159*, 116909.

Zhang, Y., Wiese, L., Fang, H., Alseekh, S., de Souza, L. P., Scossa, F., et al. (2023b). Synthetic biology identifies the minimal gene set required for paclitaxel biosynthesis in a plant chassis. *Molecular Plant, 16*(12), 1951–1961.

Zhao, Q., Yang, J., Cui, M. Y., Liu, J., Fang, Y., Yan, M., et al. (2019). The reference genome sequence of *Scutellaria baicalensis* provides insights into the evolution of wogonin biosynthesis. *Molecular Plant, 12*(7), 935–950.

Zhu, F., Wen, W., Cheng, Y., Alseekh, S., & Fernie, A. R. (2023). Integrating multiomics data accelerates elucidation of plant primary and secondary metabolic pathways. *aBIOTECH, 4*(1), 47–56.

Validation of Traditional Knowledge Related to the Use of Alpine Medicinal Plants

15

Shriyanshu Thakur, Hemant Sood, and Rolika Gupta

Abstract

Traditional medical practices in mountainous areas have relied heavily on alpine medicinal plants, which are cultivated in the harsh environments of high-altitude habitats. In addition to being essential for curing regional illnesses, these plants have great cultural and historical value for indigenous groups. Generations have passed down traditional knowledge about their usage, which is a vital component of health systems and cultural identity. The history, significance, and incorporation of alpine medicinal plants into traditional knowledge systems are examined in this chapter. It explores the cultural significance, historical background, and applications of common alpine plants, including *Picrorhiza kurroa* and *Arnica montana*. Scientific validation by ethnobotanical surveys, phytochemical analysis, and socio-cultural documentation is crucial to guaranteeing their sustained effectiveness and safe usage. These techniques promote the sustainable use and protection of alpine flora and the complex knowledge systems that underpin it by bridging traditional practices with contemporary research.

Keywords

Phytochemical composition · Cultural significance · Ethnomedicine · Herbal medicine · Biodiversity conservation

S. Thakur · R. Gupta (✉)
Division of Microbiology, Career Point University, Hamirpur, Himachal Pradesh, India

H. Sood
Department of Biotechnology & Bioinformatics, Jaypee University of Information Technology, Solan, Himachal Pradesh, India

P. Kumar et al. (eds.), *Climate Resilience and Molecular Adaptation in Alpine Medicinal Plants*, https://doi.org/10.1007/978-981-95-6923-6_15

1 Introduction

1.1 Traditional Knowledge and Alpine Medicinal Plants

Alpine medicinal plants are a distinct group of plants found worldwide that are valued for their extensive pharmacological potential. They thrive in high-altitude environments above the treeline. These plants are suited to endure harsh climatic factors like strong UV rays, frigid temperatures, and poor soil nutrient levels. Their strong medicinal effects are a result of the greater amounts of secondary metabolites they frequently generate, such as terpenoids, flavonoids, and alkaloids. These plants have long been utilized in traditional medical systems, particularly those in the Himalayas, Andes, and other alpine regions, to treat various illnesses, ranging from inflammation and infections to long-term conditions like diabetes and asthma. The ethnomedical value of species like *Saussurea costus*, *Rheum emodi*, *Swertia chirayita* and *Nardostachys jatamansi* has been well established (Sharma & Sharma, 2024).

Even while alpine plants have therapeutic uses, habitat loss, overharvesting, and climate change are posing a growing threat to them. Conservation initiatives and sustainable use plans are essential because the delicate alpine ecosystems are especially vulnerable to changes in the environment. In order to preserve these priceless resources, recent studies have placed a strong emphasis on combining ethnobotanical expertise with conservation biology and pharmaceutical science (Dangwal et al., 2025).

Despite their therapeutic benefits, alpine plants are becoming threatened by climate change, excessive harvesting, and habitat destruction. Due to changes in plant distribution and the reduction of the native habitat range of some high-value species, alpine ecosystems are especially vulnerable to the effects of global warming. Furthermore, several endemic species have seen population losses due to overexploitation brought on by growing economic demand; some of these species are now severely endangered. The traditional applications of alpine plants have been validated by recent ethnobotanical and pharmacological investigations, which have shown their medicinal value and identified possible avenues for pharmaceutical development. As a result, there is currently more interest in recording indigenous knowledge and creating conservation plans that combine scientific methods with conventional ecological understanding (Khakurel et al., 2024).

In conclusion, alpine medicinal plants have promise as resources for new drug discovery in addition to being essential for conventional healthcare systems. Innovation in global health, cultural heritage, and environmental services all depend on their preservation and sustainable use.

1.2 Importance of Alpine Medicinal Plants

Alpine medicinal plants are particularly significant because of their ecological distinctiveness, pharmacological potential, and longstanding ties to conventional medical systems. These plants flourish under harsh climatic circumstances, such as strong UV radiation, low oxygen levels, high wind exposure, and inadequate soil nutrients, and they grow in high-altitude areas over 3000 m. Alpine plants have adapted to these severe environments by producing a wide range of secondary metabolites, including alkaloids, flavonoids, and glycosides, many of which have strong medicinal qualities. These plants have been used for generations in traditional medicine by indigenous groups in areas such as the Tibetan Plateau, the Andes, and the Himalayas. Commonly used to treat fever, inflammation, liver problems, digestive issues, and neurological illnesses include species like *Swertia chirayita*, *Nardostachys jatamansi*, *Saussurea costus*, and *Aconitum heterophyllum*. In addition to being an essential component of cultural identity, these ethnobotanical activities are vital to the continuation of basic healthcare in isolated, underdeveloped areas (Sharma & Sharma, 2024).

Apart from their cultural and therapeutic significance, alpine plants play a vital role in maintaining ecological equilibrium. They play an important role in maintaining a distinctive biodiversity, controlling hydrological cycles, and halting soil erosion in delicate alpine environments. However, overharvesting, habitat degradation, and climate change are putting this ecological balance in jeopardy, leading to altitudinal migration and population declines in a number of alpine species (Singh et al., 2024).

Alpine plants have been shown to have pharmacological value in recent research, with bioassays verifying their cytoprotective, antibacterial, anti-inflammatory, and antioxidant properties. Bioprospecting and pharmacological validation of alpine flora have gained attention as a result of this mounting data (Pandey et al., 2023). The necessity for conservation plans and sustainable management techniques that combine ecological science and traditional knowledge is also brought up. Alpine medicinal plants are also an important source of new bioactive chemicals for the pharmaceutical industry across the world, in addition to being essential for the health and cultural preservation of highland populations. The protection of biodiversity, traditional knowledge systems, and the future of natural medicine all depend on maintaining this botanical legacy (Dangwal et al., 2025).

1.3 Importance of Conventional Knowledge in the Healthcare System

In order to meet health requirements in isolated and frequently neglected areas, traditional knowledge systems that are based on millennia of empirical observation and cultural practice include the identification, gathering, preparation, and use of alpine medicinal herbs. With little access to contemporary medical facilities, populations living in the Tibetan Plateau, Andes, and Himalayas have been able to live and prosper in biologically demanding regions because of this expertise. The traditional applications of alpine medicinal plants have caught the interest of contemporary scientific investigation, in addition to their local value. Since pharmacological data now backs up many ethnomedical claims, alpine flora's biochemical diversity and significance for worldwide drug discovery are highlighted. The convergence of contemporary science and traditional wisdom creates opportunities for advancement in integrative healthcare and complementary medicine. However, cultural uniformity, generational loss, and environmental degradation are posing a growing challenge to the preservation of traditional knowledge. While the sale of medicinal plants without sufficient benefit-sharing systems raises ethical problems, modernization and migration frequently impair the transfer of traditional knowledge (Khakurel et al., 2024).

However, environmental deterioration, generational loss, and cultural uniformity pose a growing challenge to the preservation of traditional knowledge. Migration and modernization frequently interfere with the transfer of traditional knowledge, and the commercialization of medicinal plants without sufficient benefit-sharing arrangements raises ethical questions. Therefore, biodiversity conservation, community health resilience, and equitable development need to acknowledge and safeguard traditional knowledge as a valid and significant source of medical information.

In conclusion, traditional knowledge systems about alpine medicinal plants constitute a living legacy of ecological awareness, cultural continuity, and workable medical treatments. To guarantee their ongoing relevance and sustainability, their incorporation into national and international healthcare initiatives needs documentation, validation, and ethical assistance (Khakurel et al., 2024).

2 Traditional Knowledge and Alpine Medicinal Plants

2.1 Historical Concept of Alpine Medicinal Plant Use

The ethnobotanical history of alpine populations in the European Alps and Himalayan areas, as well as traditional treatment methods, are intricately linked to the past usage of Alpine medicinal plants. Numerous plant species with strong bioactive chemicals have grown in these high-altitude habitats, which are distinguished by their harsh climate and distinctive biological niches. According to empirical information that has been verbally transmitted through the centuries, these plants have historically been essential to the survival and well-being of alpine communities (Scherrer et al., 2023).

Alpine plants were widely used by ancient civilizations who lived in the Alps, including the Celts and subsequently the Romans, to cure illnesses, including fevers, infections, and stomach issues, according to ethnobotanical research. The usage of medicinal plants in prehistoric times was demonstrated by notable archaeological finds such as the 5300-year-old Tyrolean Iceman "Ötzi," whose belongings contained traces of the fungus *Piptoporus betulinus*, which is known to have antibacterial characteristics. Many of the distinctive traditional knowledge systems that flourished in the Alpine area predate contemporary pharmacology because of its isolation (Chizzola & Gansinger, 2023).

Alpine plant knowledge was also greatly conserved and organized during the Middle Ages by monastic medicine. Modern herbal therapy still makes use of plants like *Arnica montana*, *Gentiana lutea*, and *Valeriana officinalis*, whose medicinal

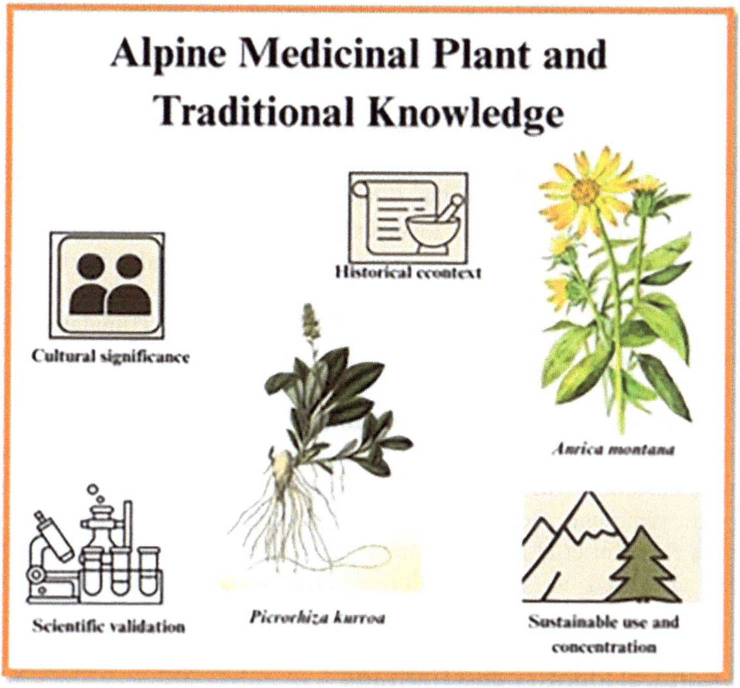

Fig. 15.1 Shows the alpine medicinal plant and traditional knowledge

qualities were recorded in herbal literature like Hildegard of Bingen's Physica. By watching the effects of plants over centuries of trial and error, local shepherds, midwives, and healers further enhanced this knowledge. Traditional usage was validated in the nineteenth and early twentieth centuries as a result of growing scientific interest in phytochemistry and pharmacognosy. The introduction of biomedicine and industrialization, however, caused a lot of this traditional knowledge to be lost. The anti-inflammatory, antioxidant, and antibacterial qualities of Alpine botanicals have, however, sparked renewed interest in them due to current ethnopharmacological studies. These days, conserving this historical information is essential for modern medicine development, biodiversity preservation, and cultural heritage. The historic uses of the alpine flora continue to inspire contemporary scientific research, and it is still a largely unexplored resource (Milani et al., 2023) (Fig. 15.1).

2.2 Cultural Significance of Alpine Flora

The cultural, spiritual, and social life of mountain communities worldwide is profoundly influenced by alpine flora. Native American and rural communities' customs, culture, and sense of self are intricately entwined with high-altitude flora, from the European Alps to the Himalayas and Andes. A vital component of biocultural history, these plants are prized not just for their ecological and therapeutic properties but also as potent cultural icons. The following subheadings provide a clearer understanding of the cultural relevance of Alpine flora:

2.2.1 Symbolism and Spirituality

Some Alpine plants have spiritual and metaphorical implications that make them highly valued. For example, in the European Alps, *Leontopodium alpinum*, or Edelweiss, is a symbol of courage, love, and innocence. In addition to being utilized historically in charms thought to ward off evil, it may be found in traditional songs, jewelry, and folklore. Symbolizing purity and spiritual elevation, *Nardostachys jatamansi* is utilized in religious rites and incense in the Himalayas (Rüdisser et al. 2019).

2.2.2 Traditional Festivals and Rituals

Alpine plants are frequently essential to regional celebrations, medical practices, and farming customs. *Arnica montana* and *Gentiana lutea* are among the herbs used in therapeutic ointments and in solstice celebrations in the alpine areas of Switzerland and Austria (Vitalini et al., 2023). Alpine herbs are vital to regional healing practices and are used as incense during Buddhist ceremonies in Himalayan civilizations.

2.2.3 Cultural Identity and Heritage

Ancestral knowledge and cultural identity are alive and well in the alpine plants. A cultural continuity that has endured despite industrialization is seen in the elders', midwives', and shepherds' sustainable usage and oral transfer of plant knowledge. In addition to being survival tools, alpine plants serve as symbols of location, memory, and community for many mountain communities (Petelka et al., 2022).

3 Methods of Validation

3.1 Ethnobotanical Survey and Documentation

Traditional knowledge on the use of alpine medicinal herbs is documented and validated mostly through ethnobotanical surveys. A scientific basis for comprehending plant-based healing systems that have evolved over generations in high-altitude locations is provided by these surveys, which methodically document indigenous and local traditions. The delicate ecosystems of the Himalayas, Alps, and other mountainous areas are home to a variety of plants with distinctive phytochemical profiles, many of which have been used in traditional medicine to treat rheumatic, digestive, dermatological, and respiratory conditions. The goal of recent field research has been to document these ethnomedicinal practices and evaluate their scientific viability.

According to Sharma et al. (2021), Himalayan populations have a wealth of traditional knowledge about plants, including *Rheum australe*, *Picrorhiza kurroa*, and

Nardostachys jatamansi, whose anti-inflammatory, hepatoprotective, and neuroprotective qualities have been scientifically shown. In addition to confirming traditional knowledge, the combination of ethnobotanical documentation and phytochemical and pharmacological research supports conservation and sustainable bioprospecting. These multidisciplinary methods highlight the critical need to preserve this intangible cultural legacy while reaffirming the importance of local knowledge systems in contemporary healthcare.

3.2 Phytochemical Composition

The ancient therapeutic uses of alpine medicinal plants are supported by their exceptional phytochemical richness, which is formed by the severe circumstances of high-altitude habitats. These plants have long been utilized in traditional medical systems in mountainous areas like the Himalayas, the Alps, and the Andes to treat a range of illnesses, from neurological and inflammatory problems to digestive and respiratory issues. Their distinct secondary metabolites, which include alkaloids, flavonoids, phenolic compounds, terpenoids, tannins, glycosides, and essential oils, are primarily responsible for the biological activity ascribed to these plants. These phytochemicals provide pharmacological advantages supported by contemporary science in addition to ecological roles, including UV protection and insect resistance (Rautela et al., 2021). The existence and bioactivity of these chemicals have been verified by recent phytochemical studies employing gas chromatography-mass–mass spectrometry (GC-MS), high-performance liquid chromatography (HPLC), and other cutting-edge analytical methods. These studies demonstrate the medicinal potential of alpine flora while offering a scientific basis for traditional applications. In order to validate indigenous knowledge, find novel medications, and support the conservation of both species and cultural legacy, it is crucial to comprehend the phytochemical profiles of alpine medicinal plants.

3.3 Socio-cultural Uses of Alpine Medicinal Plants

In high-altitude areas like the Himalayas, Alps, Carpathians, and Andes, indigenous and local cultures place a great deal of sociocultural value on alpine medicinal plants. These plants have long been essential to spiritual, ceremonial, and commercial life in addition to being the main source of healing. Cultural identity, local belief systems, and community knowledge systems are intricately linked to traditional alpine plant-based healing techniques. Information on how to utilize, harvest, preserve, and apply these plants is frequently passed down orally through the generations, creating a living legacy that promotes both cultural continuity and physical well-being.

For instance, *Saussurea obvallata*, often referred to locally as Brahma Kamal, is used in traditional treatments for heart and liver conditions and is revered in Indian Himalayan tribes. It is administered during significant spiritual events. *Arnica*

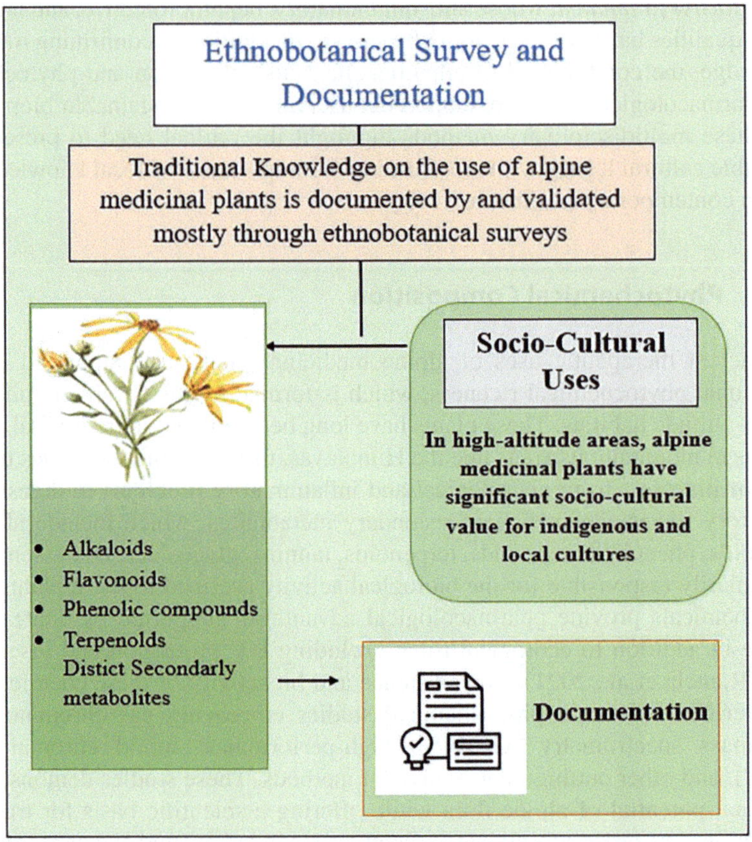

Fig. 15.2 Shows the ethnobotanical survey and documentation

montana is also valued culturally in folk festivals and symbolic healing practices in the European Alps, in addition to being used for medical salves. These plants' sociocultural usage transcends medical applications and is a reflection of holistic worldviews that incorporate spirituality, the environment, and health (Vitalini et al., 2023) (Fig. 15.2).

Recent ethnobotanical research has highlighted how crucial it is to preserve and record this traditional knowledge in light of growing risks from habitat degradation, climate change, and globalization. These kinds of recording initiatives are essential for preserving intangible cultural assets and assisting with community-based conservation plans.

4 Conclusion

Alpine medicinal plants represent a valuable intersection of biodiversity and traditional knowledge that has sustained indigenous communities in harsh mountain environments for centuries. Their significance extends beyond therapeutic use, encompassing cultural, spiritual, and ecological dimensions deeply embedded in local heritage. The historical and cultural contexts of alpine flora underscore the richness of traditional knowledge systems and their vital role in primary healthcare, particularly in remote regions. However, to ensure the continued relevance and safe application of these medicinal resources, scientific validation through ethnobotanical documentation, phytochemical analysis, and socio-cultural studies is essential. Such interdisciplinary approaches not only authenticate traditional claims but also promote sustainable harvesting and conservation efforts. Integrating traditional wisdom with modern research can foster innovative healthcare solutions while preserving both the biological and cultural diversity of alpine ecosystems. Ultimately, safeguarding this knowledge and its botanical sources is crucial for enhancing health security, cultural identity, and environmental stewardship in alpine regions worldwide.

References

Chizzola, R., & Gansinger, D. (2023). Chemical composition and antioxidant activity of Artemisia argyi essential oil and hydrolate. *Compounds, 3*(3), 521–531.

Dangwal, B., Arya, D., Bhatt, I. D., & Rana, S. K. (2025). Research on medicinal plants in the Himalaya: Current knowledge and future directions. In *Climate change impact on Himalayan biodiversity* (pp. 1–20).

Khakurel, D., Uprety, Y., Karki, S., Khadka, B., Poudel, B. D., Ahn, G., Cha, J. Y., Kim, W. Y., Lee, S. H., & Rajbhandary, S. (2024). Assessing the risks to valuable medicinal plants in Nepal from human activities and environmental factors. *Global Ecology and Conservation, 51*, e02860.

Milani, F., Bottoni, M., Bardelli, L., Colombo, L., Colombo, P. S., Bruschi, P., Giuliani, C., & Fico, G. (2023). Remnants from the past: From an 18th century manuscript to 21st century ethnobotany in Valle Imagna (Bergamo, Italy). *Plants, 12*(14), 2748.

Pandey, I., Bhattarai, K., Bhattarai, R., Kunwar, R., & Baral, B. (2023). Indigenous knowledge and bioactive compounds of Berberis aristata confirm its therapeutic potential: An ethnopharmacological appraisal in Nepal. *Ethnobotany Research and Applications, 26*, 1–21.

Petelka, J., Bonari, G., Säumel, I., Plagg, B., & Zerbe, S. (2022). Conservation with local people: Medicinal plants as cultural keystone species in the Southern Alps. *Ecology and Society, 27*(4).

Rautela, I., Rani, A., & Negi, J. S. (2021). Phytochemistry and pharmacology of *Nardostachysjatamansi*: A review. *Journal of Ethnopharmacology, 265*, 113308.

Rüdisser, J., Schirpke, U., & Tappeiner, U. (2019). Symbolic entities in the European Alps: Perception and use of a cultural ecosystem service. *Ecosystem Services, 39*, 100980.

Scherrer, M. M., Zerbe, S., Petelka, J., & Säumel, I. (2023). Understanding old herbal secrets: The renaissance of traditional medicinal plants beyond the twenty classic species. *Frontiers in Pharmacology, 14*, 1141044.

Sharma, M., & Sharma, M. (2024). Assessment of current status and conservation strategies of some high valued medicinal plants from Himalayan regions. *Annali Di Botanica, 14*(1).

Sharma, R., Manhas, R. K., & Magotra, R. (2021). Traditional knowledge and cultural significance of medicinal plants used by indigenous communities in the Indian Himalayas. *Journal of Ethnopharmacology, 276*, 114192.
Singh, G., Chandra, N., Arya, D., & Tewari, L. M. (2024). Extent, abundance, threats and information gap on alpine medicinal and aromatic plants of Western Himalaya, India: A review. In *Threatened medicinal plants in the Indian Himalayan region: Sustainability challenges and conservation strategies* (pp. 67–87).
Vitalini, S., Iriti, M., Puricelli, C., & Fico, G. (2023). Ethnobotanical survey on alpine plants: Cultural importance and traditional uses. *Ethnobotany Research and Applications, 25*, 1–17.

Intellectual Property Rights and Plant Commercialization: Legal and Economic Perspectives

16

Shriyanshu Thakur, Hemant Sood, and Rolika Gupta

Abstract

In agriculture, Intellectual Property Rights (IPRs) are essential for promoting investment and innovation, especially when it comes to creating and marketing plant varieties and biotechnological breakthroughs. The introduction of legal protections such as patents, plant breeders' rights, trademarks, and geographical indications has transformed the agriculture sector into a knowledge-driven enterprise. This research examines the economic and legal aspects of IPR in plant commercialization. It begins with an overview of international frameworks such as the TRIPS agreement, the UPOV Convention, and the Convention on Biological Diversity, in addition to national regulations like the Farmers' Rights Act and India's protection of plant varieties.

The economic analysis investigates how IPR impacts research and development, stimulates innovation in high-value crops and genetically modified organisms (GMOs), and modifies market dynamics through seed industry consolidation, licensing arrangements, and royalty structures. While IPR benefits breeders and agribusiness by stimulating innovation and commercialization, it also presents issues for farmers, such as increasing reliance on proprietary seeds and increased input costs.

Critical difficulties such as biopiracy, access and benefit sharing, market monopolization, and ethical concerns underline the necessity for a balanced strategy. Emerging problems such as digital sequence information and climate-

S. Thakur · R. Gupta (✉)
Division of Microbiology, Career Point University, Hamirpur, Himachal Pradesh, India

H. Sood
Department of Biotechnology & Bioinformatics, Jaypee University of Information Technology, Solan, Himachal Pradesh, India

P. Kumar et al. (eds.), *Climate Resilience and Molecular Adaptation in Alpine Medicinal Plants*, https://doi.org/10.1007/978-981-95-6923-6_16

resilient agriculture are complicating the IPR landscape; climate change and globalization are linked.Graphical Abstract

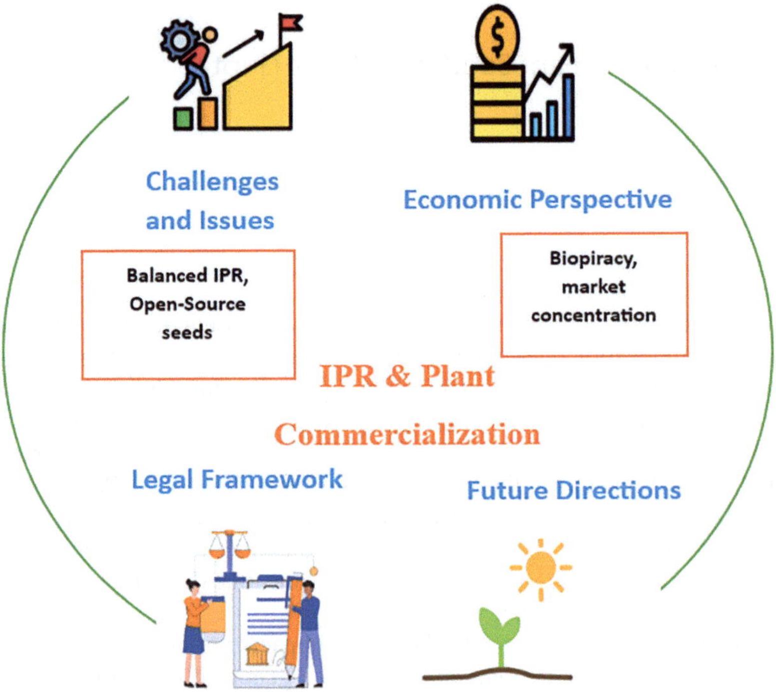

Keywords

Intellectual Property Rights · Plant Breeders' Rights · Agriculture · Plant commercialization · Biotechnology · Farmers' Rights · TRIPS Agreement · PPV&FR Act

1 Introduction

1.1 Overview

The term IPRs refers to the set of unique legal rights that provide authors or inventors of original works, concepts, or discoveries control over how their products are used and sold for a predetermined amount of time. Safeguarding intangible assets, fostering innovation, and encouraging creativity across a variety of industries are the main goals of intellectual property rights (Patel, 2021). Intellectual property, which includes names, symbols, designs, inventions, literary and artistic works, and distinguishing marks, is intangible in nature in contrast to physical property. The World Intellectual Property Organization (WIPO) defines intellectual property rights (IPR) as the legal protection resulting from creative labor in the fields of industry, science, literature, or the arts (Citaristi, 2022). Since creativity and innovation are now key factors in economic growth and competitiveness in today's knowledge-based economy, these rights are crucial.PRs can be broadly categorized into two primary categories: copyright, which covers literary and creative works, and industrial property, which includes patents, trademarks, industrial designs, and geographical indications. Trademarks protect distinctive signs, logos, and symbols used in commerce; copyrights protect the rights of authors, artists, and musicians over their original works; geographical indications identify goods associated with specific regions known for unique qualities; and patents protect new inventions that demonstrate novelty, inventive steps, and industrial applicability.

The World Trade Organization (WTO)'s Agreement on Trade-Related Aspects of Intellectual Property Rights (TRIPS) and treaties run by the WIPO serve as the main governing bodies for the worldwide framework for IPR protection (Taubman et al., 2020). Rapid technological advancements in industries like digital platforms, biotechnology, artificial intelligence, and pharmaceuticals have broadened the definition of intellectual property in recent years. This has brought about both new opportunities and challenges, including the need to protect traditional knowledge, protect data, and own AI-generated content. In addition, IPRs are crucial for maintaining equitable access to innovations and fair competition while striking a balance between inventors' rights and society's larger interests (Mahala & Chauhan, 2025). Since intellectual property has become a strategic asset for governments, corporations, and research institutions in today's globalized information economy, effective IPR protection is essential for long-term growth, innovation, and international commerce.

1.2 Significance of IPR in the Global Economy

The global economy is significantly shaped by intellectual property rights (IPRs), which encourage innovation, fair competition, and sustainable economic growth. Trade secrets, patents, trademarks, and copyrights are examples of intangible assets that are now crucial in determining a company's value and competitiveness in a knowledge-based economy. The growing dependence on intellectual property as a strategic resource for economic development is reflected in the record-breaking global registrations for patents, trademarks, and industrial designs, according to the World Intellectual Property Organization. Businesses make significant investments in research and development (R&D) with the knowledge that intellectual property rights (IPRs) would offer their inventions legal protection and financial gains, fostering an atmosphere that encourages innovation and technical progress (Beltrán-Urvina et al., 2025).The importance of IPRs is demonstrated by their capacity to draw in foreign direct investment (FDI) and ease international trade. Countries with robust intellectual property rights (IPR) policies typically attract more investment, especially in high-tech industries like biotechnology, computer technology, and pharmaceuticals (Meghwal et al., 2023). To promote industrial progress and job creation, multinational businesses, for example, frequently form partnerships, license agreements, and technology transfers in nations with strong intellectual property protection (Sun et al., 2024). Furthermore, IPRs support the digital economy by protecting advancements in software development, big data analytics, and artificial intelligence, which together generate trillions of dollars in global GDP each year.

In addition to promoting economic progress, IPRs are essential for the development of the cultural and creative sectors, which generate substantial global revenue streams and include media, entertainment, and design. According to the, the creative industries account for almost 6% of the world's GDP, illustrating how copyrights and associated rights enable artists and innovators in the increasingly international digital marketplace. IPR's importance in the global economy does, however, come with drawbacks, such as IP infringement, piracy, and the moral dilemma of access to necessary technologies and medications in developing nations. For international institutions like the WIPO and WTO, striking a balance between robust IPR protection and fair access continues to be a major policy concern.

1.3 IPR'S Significance in Agriculture and Plant Marketing

For many countries, especially developing nations, agriculture is the foundation of both economic growth and food security. The protection of intellectual discoveries has become essential for promoting research, investment, and commercialization in the agricultural biotechnology, seed production, and plant breeding industries due to their rapid improvements. To foster innovation and guarantee equitable financial rewards for breeders and inventors, intellectual property rights (IPRs) offer a legislative framework for safeguarding inventions about plants, seeds, genetic resources, and agricultural practices.

1.4 Agricultural Innovation and IPR

With the development of genetically modified (GM) crops, hybrid seeds, biofertilizers, and precision farming equipment, the agricultural industry has experienced a technological revolution. Large sums of money must be spent on R&D for such advancements. IPRs encourage innovators by giving them exclusive rights to their inventions through mechanisms like patents, plant breeders' rights (PBR), and geographical indications (GIs). For example, PBRs promote the creation of new plant varieties with increased yield and resilience to biotic and abiotic challenges, whereas patents protect biotechnological innovations such as pest-resistant crops (Blakeney & Siddique, 2020). Private businesses and academic organizations would have little incentive to spend money on pricey agricultural research if intellectual property rights were not protected.

1.5 Farmers' Privileges and Plant Breeders' Rights

The International Union for the Protection of New Varieties of Plants (UPOV) recognizes the rights of plant breeders, which give them sole authority over the propagation materials of their plant varieties. These rights promote innovation in the production of seeds, resulting in higher-quality seeds with increased yields. However, the majority of national laws allow farmers to conserve and reuse seeds for non-commercial uses, striking a balance between these rights and their privileges. The Plant Varieties and Farmers' Rights Protection Act of 2001 in India is a noteworthy example that ensures fair benefit-sharing and agro-biodiversity conservation by acknowledging the interests of both farmers and breeders (Government of India, 2023).

1.6 IPR in Plant Commercialization

Protection of intellectual property rights has a major impact on the commercialization of plant varieties and seeds. Agribusinesses and private seed companies depend on intellectual property rights (IPR) to protect their discoveries and guarantee a profit. Agreements for licensing, royalty sharing, and intellectual transfer are typical in the seed sector, where local businesses and international enterprises work together to distribute patented seeds on a large scale. For instance, biotechnology companies have created seeds that can withstand drought and pests, which necessitate stringent IPR enforcement to ensure their commercial viability (FAO, 2023). In addition, geographical indications (GIs) aid in the branding of distinctive agricultural goods associated with a particular area, like Darjeeling tea or Basmati rice, allowing rural communities to command higher prices in both native and foreign markets (Fig. 16.1).

2 Concept of IPR in Agriculture

Agriculture is an essential industry that supports human life by giving millions of people worldwide access to food, raw resources, and jobs. As agricultural technology, biotechnology, and plant breeding techniques have advanced, the idea of ownership of agricultural innovations has become more significant. Agrochemicals, farming equipment, genetically modified crops, and new plant varieties are among the inventions that are intended to be protected under agricultural intellectual property rights, or IPRs. Research and development is encouraged by these rights, which provide exclusive ownership and commercialization advantages for a predetermined amount of time.

Fig. 16.1 IPR in agriculture & plant marketing

2.1 Definition and Scope of IPR

A collection of unique legal rights known as intellectual property rights (IPRs) safeguards the innovations and creative works of people or organizations. To provide inventors control over the use, commercialization, and distribution of their creations for a predetermined amount of time, the government grants these rights. By giving creators legal ownership and financial rewards, intellectual property rights (IPR) seek to reward invention, foster creativity, and stimulate investment in R&D. Technological developments like precision farming, digital agriculture, genetic engineering, and biotechnology have greatly expanded the application of intellectual property rights (IPR) in the agricultural sector(Meghwal et al., 2023). Specialized agricultural machinery, enhanced agronomic practices, and sophisticated genetic modification techniques are examples of agricultural advancements that have expanded beyond traditional plant breeding.

2.2 The Scope of IPR in Agriculture Covers Multiple Areas, Including

2.2.1 The Development of New Plant Varieties by Means of Genetic Engineering or Breeding

To create plant types with enhanced traits, including increased yield, resilience to disease and drought, and improved nutritional value, breeding programs and genetic engineering techniques are frequently employed. By preserving these types, we can make sure that innovators and breeders get the recognition they deserve and profit from their work (Smith, 2019).

2.2.2 Agricultural Biotechnology Processes

Biotechnology advancements have transformed agriculture by making it possible to modify plant genetics to produce desired characteristics. Biotechnological procedures that are protected under IPR frameworks include tissue culture, gene editing (CRISPR/Cas9), and molecular marker-assisted selection. Significant research expenditures are needed for these procedures, and IPR protection encourages further innovation.

2.2.3 Innovation Farming Tool and Machinery

Precision seeders, computerized irrigation systems, drones for crop monitoring, and robotic harvesters are just a few examples of the specialized equipment and instruments that are essential to modern agriculture. The ability to patent these innovations grants the creators the sole authority to produce and sell their goods. In addition to helping the inventors, protecting such ideas speeds up agricultural mechanization (Pandey, 2025).

2.2.4 Agrochemical and Formulation

Extensive study, testing, and regulatory approvals are necessary for the production of novel fertilizers, insecticides, herbicides, and other agrochemicals. By preventing copying and unapproved commercialization by rivals, these compositions, when patented, guarantee equitable rewards on research and development expenditures.

2.2.5 Unique Geographical Produce with Distinct Qualities

Some agricultural products have distinctive qualities related to their climate, place of origin, and customs. Geographical Indications (GIs) protect these items by maintaining their validity and preventing others from abusing the label (Savale & Savale, 2016).

3 Types of IPR Relevant to Plant

3.1 Patent

Under national patent laws, patents grant inventors the only right to create, utilize, and market their ideas for a certain period of time, usually 20 years from the date of filing (Fig. 16.2). Patents are granted for a variety of advancements in agriculture, including:

- **Genetically Modified Organisms (GMOs)** such as Bt Cotton and herbicide-tolerant crops, which enhance yield and resistance to pests and diseases.
- **Biotechnological processes** like CRISPR-Cas9-based gene editing, tissue culture, and micropropagation methods.
- **Innovative agrochemicals**, biofertilizers, and pest-resistant traits that improve agricultural productivity and sustainability.

However, patents on living organisms remain controversial due to ethical concerns, biodiversity conservation, and farmers' traditional seed-saving practices. Under the Indian Patents Act, 1970 (as amended in 2005), patents cannot be granted on plants and animals as a whole or on essentially biological processes. However, biotechnological processes and microorganisms are patentable. Globally, countries adhering to TRIPS must provide patent protection for biotechnological innovations, though national variations exist (Meghwal et al., 2023).

3.2 Plant Breeders' Rights

A particular class of intellectual property rights referred to as Plant Variety Protection (PVP) or Plant Breeders' Rights (PBR), gives plant breeders legal protection for the new kinds they create (Vasudevan et al., 2024) (Fig. 16.2). The breeder is given complete control over the protected plant variety's production, reproduction, sale, marketing, distribution, and even import or export. This implies that throughout the protected period, which varies from 15 to 18 years, depending on the plant type, no one may economically utilize the protected variation without the breeder's permission.

3.2.1 PBR Eligibility Requirements

A plant variety must meet the globally accepted NDUS requirements to be eligible for protection:

- **Novelty**: The variety must be brand-new and not marketed or sold before the protection application.
- **Distinctness**: Based on one or more key attributes, such as plant height, leaf form, flower color, or yield features, it must be easily distinguished from any other recognized variety.

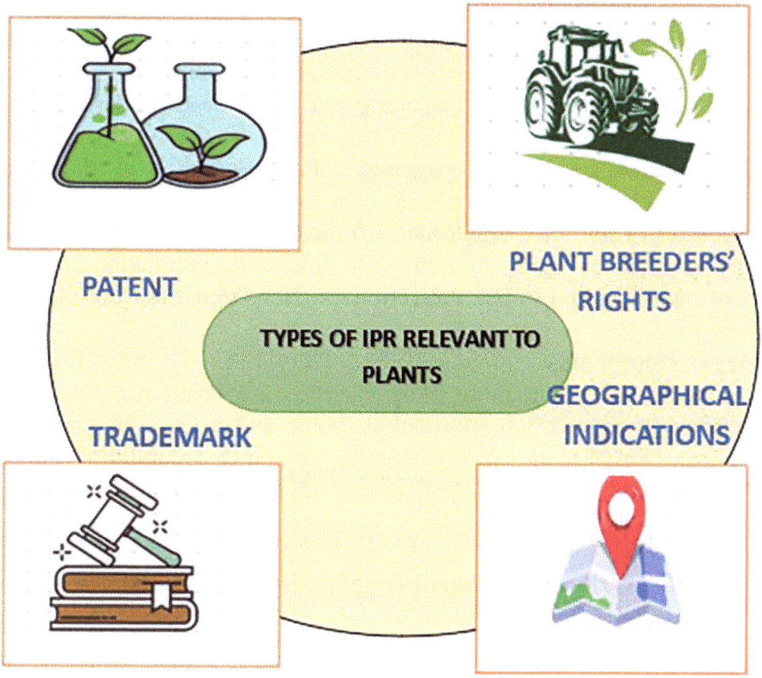

Fig. 16.2 Type of IPR relevant to plants

- **Uniformity**: The variety must have few differences between individual plants and be consistent in its key characteristics.
- **Stability**: To ensure a variety's stability, its traits should not alter over the course of multiple generations.

3.2.2 Indian Legal Framework

PBR in India is governed by the Protection of Plant Varieties and Farmers' Rights (PPV&FR) Act, 2001. This law is distinctive because it aims to reconcile the customary rights of farmers with those of breeders:

- **Breeders' Rights**: For as long as the variety is protected, breeders have the sole right to use it for commercial purposes
- **Farmers' Rights**: For personal use or non-commercial purposes, farmers are legally permitted to conserve utilize, plant, replant, trade, and distribute seeds of a protected cultivar.
- They are unable to sell protected variety seeds under a brand name, nevertheless.

3.2.3 PBR'S Importance and Benefits

Promotes Innovation: Offers rewards to breeders that create better cultivars with increased yields, resistance to pests, or climate adaptability.

Assures Quality: Assists in keeping market-available seeds under quality control by avoiding unapproved or inferior replication.

Protects Farmers' Interests: It guarantees the preservation of traditional seed-saving methods by enacting provisions about farmers' rights.

International Commerce Compliance: Promotes agricultural commerce and technology transfer by bringing India into line with international such as the International Union for the Protection of New Varieties of Plants (UPOV) framework.

Current Developments

Recent patterns show that more plant varieties are being registered under the PPV&FR Act, especially for horticulture plants and crops, including rice, wheat, and maize. Digital platforms are being developed to streamline the registration process, ensuring wider participation by small breeders and farmers.

3.3 Trademark and Geographical Indications (GIs)

Two significant categories of intellectual property rights (IPRs) that support the branding, marketing, and protection of plant-based products, seeds, and agricultural products are trademarks and geographical indications (GIs) (Fig. 16.2). Although they both function as identifiers, their goals and purview are very different (Tripathi, 2024).

3.3.1 Agriculture Trademarks

A distinctive sign, symbol, term, or combination of these that sets one company's products or services apart from another is called a trademark (Meghwal et al., 2023). Trademarks are mostly used in agriculture to brand and market organic products, seeds, fertilizers, and agrochemicals.

Characteristics of Trademarks

- Preserve the brand identity of agricultural products, such as fertilizers like IFFCO and seed firms like Mahyco.
- Words, logos, forms, colors, or even sounds can all be used.

In India, protection is given upon registration under the Trademarks Act, 1999, and has a ten-year expiration date with the potential for perpetual renewal.

Example of Agriculture:

- **Mahyco Seeds** (trademark of the seed brand)
- **Annapurna** (products of agrochemicals and fertilizer)
- **Organic India** (brand of organic products)

3.3.2 Geographical Indications (GIs) in Agriculture

Products with a particular place of origin and qualities, reputations, or characteristics that are inherently connected with that area are designated with a Geographical Indication (GI) (Johnson, 2017). GIs stand for a region or community, as opposed to trademarks, which identify a business or brand (Fig. 16.2).

Legal Framework:

- Controlled in India by the 1999 act on Geographical Indications of Goods (Registration and Protection).
- Ten years of validity, with an endless renewal option.
- Gives all authorized users in the area group rights.

Qualifications:

- The product must come from a certain area.
- Must possess special traits or a reputation brought about by the climate, land, or customs of that area.

India's Agricultural GI Examples:

- Tea from Darjeeling (West Bengal)
- Indo-Gangetic Plains Basmati Rice
- The orange of Nagpur, Maharashtra
- Keralan Malabar pepper
- Green Cardamom Coorg

4 Future Perspective

In agriculture and plant commercialization, intellectual property rights (IPR) are changing due to powerful legal, technological, and economic influences (Meghwal et al., 2023). Hybrid IP strategy, digital integration, and compliance-driven innovation are the directions of the future.

4.1 Outlook for Law and Regulation

- **Required Disclosure of Genetic Resource Origin**: The 2024 WIPO convention mandates provenance tracking and benefit-sharing procedures, additionally patent applicants disclose the origin of traditional knowledge and genetic resources.
- **Digital Sequence Information (DSI)**: New international procedures force businesses to preserve data traceability by connecting the use of sequence data to benefit-sharing commitments.
- **Genome Editing Regulations**: The U.S. provides exemptions under APHIS SECURE, while the EU suggests loosening regulations for specific NGTs. This has an effect on IP strategy for crops

- **Plant Variety Protection (PVP)**: India's PPVFR Act continues to emphasize farmers' rights and benefit-sharing, while UPOV's e-PVP platforms will streamline cross-border registrations.

4.2 Market and Economic Perspectives

- **Hybrid IP Portfolios**: It will soon be common practice to combine trade secrets (data, models), PVP/PBR (varieties), patents (traits, enabling tech), and brands/GIs.
- **Data as an IP Asset**: Data lineage and contractual protections are essential in breeding driven by AI and big data.
- **GI and Brand Value**: As a competitive advantage, geographical indications will be strengthened by consumer demand for products linked to origin, such as Darjeeling tea and Basmati rice.
- **Compliance as a Differentiator**: Investment, trade, and market trust will be impacted by demonstrated adherence to ABS, DSI, and traditional knowledge frameworks.

5 Conclusion

The future of intellectual property rights (IPR) in agriculture depends on implementing hybrid protection measures, including digital compliance tools, and tackling new problems like climate-resilient agriculture and digital sequence information (DSI). It will continue to be crucial to have legal frameworks such as TRIPS, UPOV, and the PPV&FR Act in addition to economic models that encourage benefit-sharing and sustainable commercialization. Ultimately, in a world that is changing quickly, a fair and inclusive IPR system—one that protects innovation while preserving farmers' rights and biodiversity—will be essential to attaining food security, agricultural sustainability, and equitable economic growth.

Plant breeding, biotechnology, and commercialization processes are all made legally and financially viable by IPRs, which are necessary to promote agricultural innovation. IPR frameworks promote international trade and competitiveness while encouraging research and development by providing exclusive rights through patents, plant breeders' rights, trademarks, and geographical indications. They do, however, pose particular difficulties, such as preventing biopiracy, guaranteeing equitable access to genetic resources, and striking a balance between the incentives of breeders and farmers' ancient methods.

References

Beltrán-Urvina, L. I., Acosta-Andino, B. F., Gallegos-Varela, M. C., & Vallejos-Orbe, H. M. (2025). Intellectual property as a strategy for business development. *Laws, 14*(2), 18.

Blakeney, M., & Siddique, K. H. (2020). *Local knowledge, intellectual property and agricultural innovation*. Springer.

Citaristi, I. (2022). World Intellectual Property Organization—WIPO. In *The Europa Directory of International Organizations 2022* (pp. 395–398). Routledge.

FAO. (2023). *Intellectual property rights and agricultural development: Opportunities and challenges*. Food and Agriculture Organization.

Government of India. (2023). *Protection of Plant Varieties and Farmers' Rights Act, 2001: Recent updates*. Ministry of Agriculture.

Johnson, R. (2017). Geographical indications (GIs) in US food and agricultural trade. *Congressional Research Service Report, 21*.

Mahala, A., & Chauhan, B. (2025). AI-generated innovations: Developing intellectual property (IP) protection framework for the digital age. *International Cybersecurity Law Review*, 1–17.

Meghwal, M. L., Dhurve, L., Raj, S., Maurya, D. K., & Kumar, A. (2023). A comprehensive review on the impacts of intellectual property rights on the global agricultural economy. *Asian Journal of Agriculture and Extension Economics & Sociology, 41*(12), 160–173.

Pandey, A. (2025). Technological innovation and intellectual property in agriculture: A critical study with reference to farmers' rights. In *Precision and intelligence in agriculture: Advanced technologies for sustainable farming* (pp. 29–50). IGI Global Scientific Publishing.

Patel, H. K. (2021). *The challenging relationship between contemporary art and intellectual property rights* (Doctoral dissertation).

Savale, S. K., & Savale, V. K. (2016). Intellectual property rights (IPR). *World Journal of Pharmaceutical Sciences, 5*(6), 2529–2559.

Smith, S. (2019). The foundations, continuing evolution, and outcomes from the application of intellectual property protection in plant breeding and agriculture. *Plant Breeding Reviews, 43*, 121–213.

Sun, H., Liu, J., Chen, B., & Yang, L. (2024). Exploring intellectual property in the digital realm: A bibliometric study on research on the management and protection of data-based intellectual property. *Information (Basel), 15*(12), 780.

Taubman, A., Wager, H., & Watal, J. (Eds.). (2020). *A handbook on the WTO TRIPS agreement*. Cambridge University Press.

Tripathi, V. (2024). Analyzing the role of legal protection for trademarks and geographical indications in preserving cultural heritage and enhancing global trade. *Journal of Law and Intellectual Property Rights, 1*(1), 50–63.

Vasudevan, S. N., Raju, T. J., Pooja, S. K., Krishnaprasad, B. T., & Govind, G. (2024). Regulatory framework of plant variety protection for modernized plant breeding approaches. In *Climate-smart rice breeding* (pp. 351–371). Springer Nature Singapore.

Morphological and Anatomical Adaptations of Alpine Plants to Extreme Environment

17

Jerin James, Poonam Meena, Ashiya Naaz,
Nishesh Sharma, and Manisha Nanda

Abstract

Alpine medicinal plants thrive in harsh environments characterized by low temperatures, intense ultraviolet radiation, high winds, and limited water availability. To survive these extreme conditions, these plants exhibit specialized morphological and anatomical adaptations. Leaf modifications are prominent, with many species developing small, narrow, and inrolled leaves to minimize water loss. These leaves often feature thick cuticles, multiple epidermal cell layers, and sunken stomata, reducing transpiration and shielding against cold and radiation. Leaf pubescence enhances solar reflectance, moderating leaf temperature, while heterophylly enables seasonal leaf size variation. Some plants relies on photosynthetic stems with chlorenchyma tissue, replacing reduced or absent leaves. Root adaptations include deep taproots or wide-spreading lateral roots for water access, and rhizosheaths that retain moisture and support nitrogen fixation. Cushion plants, with their dense, hemispherical growth, minimize wind exposure and insulate against temperature extremes. Succulence in leaves or stems allows water storage to endure freeze-thaw cycles and short growing seasons. These integrated adaptations enable alpine medicinal plants to maintain metabolic activity and resilience in variable climates. This study relies on these structural adaptations, highlighting their role in physiological function and survival, with implications for conservation and potential medicinal applications.

J. James · P. Meena · A. Naaz · N. Sharma (✉)
Department of Biotechnology, School of Applied and Life Sciences, Uttaranchal University, Dehradun, Uttarakhand, India

M. Nanda
Department of Microbiology, Graphic Era deemed to be University, Dehradun, Uttarakhand, India

P. Kumar et al. (eds.), *Climate Resilience and Molecular Adaptation in Alpine Medicinal Plants*, https://doi.org/10.1007/978-981-95-6923-6_17

Keywords

Adaptations · Conservation · Environmental factors · Metabolic activity · Oxidative stress · ROS

1 Introduction

Alpine ecosystems found above 2, 000 meters elevation with extreme climatic conditions and distinguished by brief, cold and erratic growing seasons. Alpine conditions are mountains rise above the local timberline through the high middle-latitude mountains of North America and Eurasia, the tropical mountains of the Old and New Worlds, the subantarctic mountains of Tierra del Fuego, New Zealand, and even Australia (Amen, 1966). They were also found in the Arctic and in places like Lappland's mountains and Alaska's Brooks Range. The majority of alpine zones nowadays are distinguished from one another by sizable tracts of lowlands, despite the fact that some mountain systems, like the Himalayas, are extremely massive. There is almost an uninterrupted chain from the Arctic to the Subantarctic only in North and South America. The vast Cordillera extends through the Andes and along the Rocky Mountains from the Brooks and Alaska Ranges in the north (Chabot & Billings, 1972). High altitude is defined as 1500–3500 m, ultrahigh altitude as 3500–5500 m, and extremely high altitude as more than 5500 m, according to commonly accepted criteria for altitude classification. This variability driven by latitudinal and altitudinal macrogradients, as well as mesotopographic gradients shaped by local topography, creates a complex mosaic of conditions that challenge plant survival. High-altitude vegetation has been impacted by climate change for many generations. Alpine plants have evolved a diverse array of adaptive mechanisms to thrive in these dynamic habitats, balancing resource scarcity, climatic extremes, and emerging pressures from climate change. Species with wide attitudinal adaptations variation have sophisticated, complex adaptive characteristics to ensure their survival. Climate change poses significant new challenges to alpine ecosystems, amplifying existing stressors and introducing novel pressures. These shifts were observed across various mountain systems that reflect the attempt of plants to track suitable climatic conditions as lower elevations become warmer (Pollock & Eagles, 1987; Xiong et al., 1999). However, the fragmented nature of alpine habitats that are separated by vast lowland expanses limits seed dispersal and gene flow which constrains the ability of species to colonize new areas. In tropical mountains, where altitudinal gradients are longer, plants may have more region to shift upward but in arctic and middle-latitude regions, rocky peaks and permanent snowlines often cap this movement, leading to potential habitat loss. The upward migration of subalpine species into alpine zones further intensifies competition threatening the ecological stability of these ecosystems. The studies are to be applied in evaluating phenotypic adaptations, phenology synchronizations and secondary metabolites bio-synthesis all critical survival strategies to acclimatize to environments (Moore et al., 2014). Alpine winds are robust, and the plants adopt low, cushion-forming growth patterns to minimize exposure and desiccation. Alpine plants possess a short growth season. The species opt to flower and develop seed early in the growth season to achieve reproductive success prior to the harsh conditions (Landi et al. (2015). Alpine environments and the adaptive mechanisms used by these plants are affected by any environmental conditions. In the high-altitude area, plant growth is limited by temperature fluctuations, intense UV radiation, salinity, low oxygen levels, and strong winds (Alyas et al., 2023; Zhao et al., 2023). This chapter comprises the morphological and anatomical adaptations of alpine medicinal plants that enable their survival in extreme high-altitude environments, with their responses to climatic stressors and implications for conservation and medicinal value.

2 Physiological Characteristics of Alpine Medicinal Plants

Plant physiology plays an important role in determining the ability of plant to tolerate/ respond to biotic and abiotic stress (Gimenez et al., 2018). Water logging, drought, salinity, temperature and anaerobiosis are the major abiotic stressors and light intensity, whereas biotic stresses are of insect pest attack, herbivory and pathogen infection (Magaña Ugarte et al. 2019). The Himalayan region among other aspects has shown evidence of varied climate conditions and CO_2 levels. Survey studies pertaining to season and diurnal atmospheric CO_2 variation report the morning to be the most variable and the afternoon the least (Rathore et al. 2015). Moreover, the findings showed that CO_2 was variable from winter to summer, its maximum in March and minimum in July and August. Photosynthesis of the plant may be impacted due to variation in CO_2 (Candra & Joshi, 2002). Northwest Himalayan air temperature rose by 1.6 °C during winter, warming at a higher rate (Bhutiyani et al., 2010). An annual elevation of about 0.15 °C in annual temperature was reported during 1960 to 2000 in the region of Western Himalayan region (Kumar et al., 2018). There have been findings indicating elevation dependent warming in mountain ecosystems with high mountain environments undergoing change at a higher rate (Pepin et al., 2015).

Alpine plants have evolved adaptations like small, compact morphs, hairy or waxy leaves to prevent heat loss, and low-temperature photosynthesis. Atmospheric oxygen levels are reduced at higher altitudes, and less atmospheric pressure (García-Plazaola et al., 2015). Plants have adaptations like higher density of stomata to facilitate better gas exchange, changed respiratory pathways, and increased efficiency of energy (Rawat et al., 2014). Alpine plants possess special pigments, including anthocyanins, which provide protection from potential damage from UV radiation (Bashyal, 2023). These plants adaptations share many physical and physiological traits with Arctic plant adaptations. But compared to arctic ecotypes, they seem to adapt metabolically to temperature fluctuations more readily (Billings, 1974). Alpine medicinal plants are those that occur naturally at high latitudes such as in mountain ranges such as the Himalayas, the Alps, and the Andes. Alpine plants have become accustomed to growing in poor environmental conditions such as low temperatures, intense sunlight, and poor soils (Billings et al., 1973). As a result, they produce high levels of secondary metabolites, which are accountable for their active medicinal properties. These alpine medicinal plants are utilized in Ayurveda, Tibetan, and traditional Chinese medicinal system. This widespread medicinal utilization is due to antimicrobial, adaptogenic, antioxidant, anti-inflammatory, and immunomodulatory properties of these plants attributes to different types of secondary metabolites produced in such plants (Billings, 1973; Niazian, 2019; Kumar et al., 2021; Kapoor et al., 2021).

A prominent stress against which alpine plants exhibit characteristic adaptation is freezing temperatures. In accordance to the immediate risk of organ and tissue damage, high-mountain plants have adapted a variety of mechanisms (physiology, biochemistry, etc) response to low temperatures. Figure 17.1 depicts the pathway mechanisms that influence in the anti-freezing mechanism (Shi et al. 2004); Sklenář 2017). These mechanisms include synthesis of compatible metabolites, solutes like proline, carbohydrates (low-molecular weight), or modification in photosynthetic machinery (Ladinig et al., 2013). The capability of alpine species to tolerate freezing temperatures can also be influenced by changes in the environment, snow cover, and favourable interactions between plants (Briceño et al., 2014; Venn et al., 2013). Timing of snowmelt, early spring stopping events, and low temperatures all have a major impact on the proliferation of high-mountain plants. Although substantial studies have been conducted in context to alpine flora (Pescador et al., 2016; Sierra-Almeida et al., 2009, 2016), there are relatively few research pertaining to impact of these factors on high-mountain taxa of Mediterranean region. Avoidance and tolerance are the two strategies for dealing with below-freezing temperatures. "Freezing avoidance" refers to the process by which many plants prevent ice formation (nucleation) inside their cells by temporarily "supercooling" or adjusting their osmotic balance.

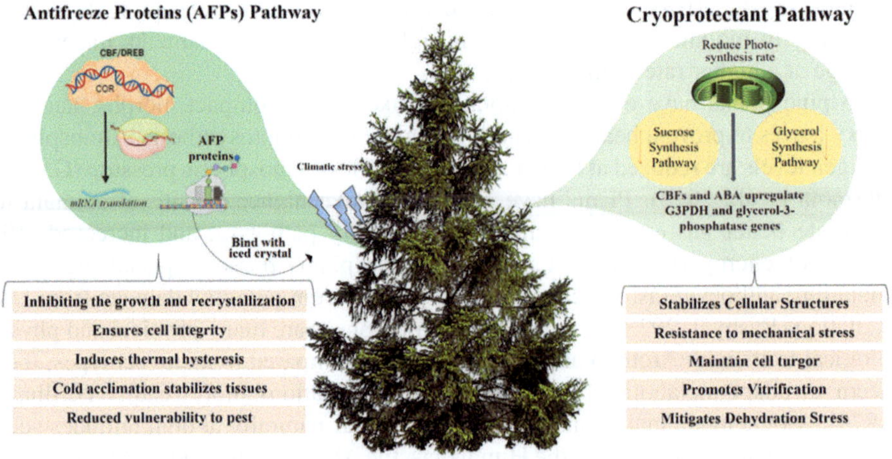

Fig. 17.1 The anti-freezing pathway mechanism that influence in alpine adaptation

Numerous studies, highlighted that freezing tolerance serves as a key survival strategy in the leaves of alpine plant species when exposed to such conditions (Ladinig et al., 2013; Taschler & Neuner, 2004). This tolerance is closely influenced by elevation, indicating that plants inhabiting higher altitudes are subject to distinct environmental pressures that drive the evolution of cool resistance (Pescador et al., 2016; Sierra-Almeida et al., 2009). Structural and functional adjustments such as reduced chloroplast and antenna sizes for optimised light capture and dynamic regulation of photo system II (PS-II) enable alpine plans to efficiently manage excess light energy. This regulation particularly through non photochemical quenching (NPQ), ensures the safe dissipation of surplus light energy (Germino & Smith, 2000; Loik et al., 2004). Together, underline the robust and multifaceted adaptation strategies alpine plants have developed to persist in extreme mountainous environments (Briceño et al., 2014; Loik et al., 2004; Germino & Smith, 2000).

2.1 Morphological and Anatomical Adaptations

Alpine medicinal plants experience numerous abiotic stresses such as low temperatures, intense ultraviolet radiation, high winds, and limited water availability. Alpine ecosystem have developed a variety of morphological and anatomical adaptations to survive and maintain physiological function in such habitats (Fig. 17.2). One of the most notable features is the modification of leaves. Many species possess small, narrow, and inrolled leaves that reduce surface area and water loss. These leaves often have thick cuticles, multiple layers of epidermal cells, and sunken stomata, which together minimize transpiration and offer protection from cold and radiation (Wickens, 1998; Körner, 2003). Leaf pubescence, or hairiness, is another common

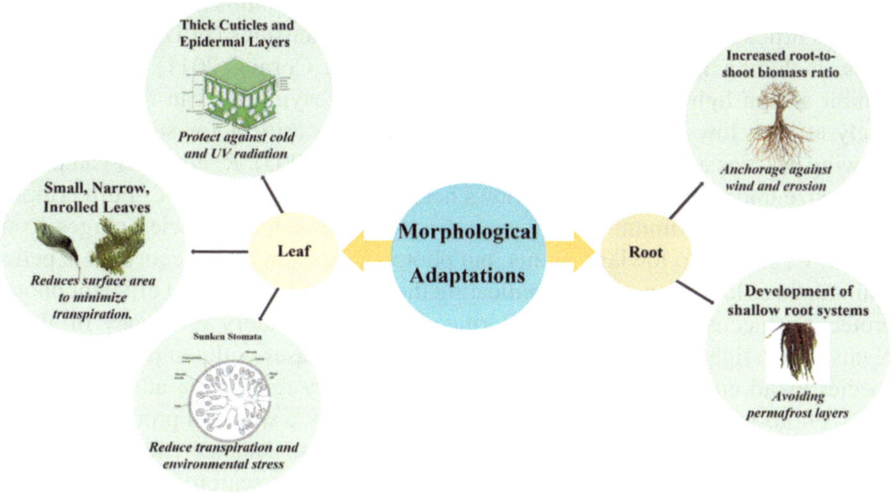

Fig. 17.2 Morphological adaptation of leaves and root that influence in alpine tree

trait that enhances solar reflectance and reduces leaf temperature (Larcher, 2003). In several species, heterophylly allows the production of larger leaves during the wet season and smaller, more pubescent leaves during dry periods. Some alpine plants shed their leaves entirely during drought, conserving moisture through deciduous behavior (Wickens, 1998). In extreme cases, the leaves are reduced or absent, and photosynthesis is taken over by modified stems. These stems, often ridged or furrowed, contain embedded chlorenchyma tissue and are structurally reinforced to withstand environmental stressors (Wickens, 1998; Mauseth, 2006).

Root adaptations are equally important for alpine survival. Some plants develop deep taproots or wide-spreading lateral roots to access limited water sources, while others form rhizosheaths soil-binding root structures that help retain moisture and facilitate nitrogen fixation (Wickens, 1998; Körner, 2003). Cushion plants are dense, low-growing, and hemispherical, are another key adaptation in alpine environments. Their compact form minimizes wind exposure, conserves internal moisture, and insulates the plant body against temperature extremes (Körner, 2003). Although less common, some alpine species also show signs of succulence in their leaves or stems, storing water in specialized parenchyma tissues to endure freeze-thaw cycles and short growing seasons (Larcher, 2003; Mauseth, 2006). These integrated structural adaptations help alpine medicinal plants maintain metabolic activity and survive in harsh, variable climates.

2.2 Photosynthetic and Respiratory Efficiency

Alpine medicinal plants, adapted to high elevations and extreme climates, show remarkable physiological strategies to maintain photosynthetic efficiency despite fluctuations in light intensity, temperature, and CO_2 availability. Photosynthetic carbon assimilation in alpine species like *Ranunculus glacialis* and *Geum reptans* remains efficient across a wide range of photon flux densities (PFDs) and temperatures, including near-freezing conditions (Streb & Cornic, 2011). These plants exhibit a high light-saturation point, and their photosynthesis can function effectively at both low and moderately high temperature critical for survival in short growing seasons and unpredictable weather (Körner, 2003). *R. glacialis* can photosynthesize under snow cover and shows net carbon gain even at 0 °C. Furthermore, the temperature optimum for photosynthesis in most alpine species ranges from 16–25 °C, similar to lowland plants, but photosynthesis in alpine ecotypes is better maintained at low temperatures, indicating thermal acclimation and efficient photoprotection mechanisms (Streb & Cornic, 2011). Respiratory efficiency in alpine plants is also tightly regulated. While respiration increases with temperature, alpine species avoid energy loss by minimizing unnecessary respiration at low temperatures. Photo-respiration in alpine species is not merely a wasteful process it plays a critical role as an alternative electron sink, helping to balance excess excitation energy when carbon assimilation is limited by cold or CO_2 scarcity (Streb & Cornic, 2011). This protective role of photo-respiration is particularly important under stress conditions like high light and low temperature. Inhibition of photorespiration leads to a higher reduction state of PSII and greater photodamage, especially in species like *R. glacialis*, which rely on this mechanism to maintain redox balance. The high efficiency and flexibility of photosynthetic and respiratory responses in alpine medicinal plants possess resilience and capacity to synthesize valuable secondary metabolites even in challenging mountain environments.

2.3 Water-Use Efficiency and Transpiration Regulation

Alpine medicinal plants, particularly those inhabiting permafrost regions such as the Qinghai-Tibet Plateau, exhibit finely tuned physiological adaptations to regulate water use and transpiration under conditions of climatic variability. Remarkably, there were not any substantial variations in water-use efficiency WUE between the dry, wet, and normal years of precipitation patterns. This indicates that alpine plants have evolved throughout time to protect their populations from hydrological extremes. This stability is maintained by prioritizing carbon assimilation primarily driven by plant growth status (NDVI) over absolute water availability. NDVI, a proxy for green biomass, emerged as the strongest predictor of WUE, underscoring the role of active vegetation in modulating ecosystem gas exchange. Air temperature (Tair) also had a positive influence on WUE, especially when moderate increases enhanced stomatal conductance and photosynthetic activity without proportionally increasing transpiration. On the other hand, factors like radiation and soil moisture showed complex or limited impacts, particularly during mid-season when transpiration is dominated by evaporation rather than direct plant water loss. Overall, alpine medicinal plants maintain relatively stable WUE by synchronizing photosynthetic demand with available water resources, an evolutionary strategy that enables them to thrive in high-altitude, climate-sensitive environments (Hu et al., 2024).

3 Environmental Factors and Their Effects

3.1 Nutrients

Growth of alpine plant species are strongly influenced on the availability of nutrients and at high altitude due to the limited nutrient availability which lead to the adverse effect on their development and growth. In alpine areas, nutrient availability is restricted by underdeveloped soils. Alpine plants frequently demonstrate slowly nutrient cycling to preserve internal resources and develop symbiotic connections with mycorrhizal fungus to improve nitrogen uptake. Better growth of *Aconitum carmichaeli* was established when treated with NPK and a lower biomass was established on exposure to limited nutrient levels (Hou et al., 2006). When *Aconitum heterophyllum* and *Aconitum balfourii* cultivated in nutrient enriched condition, height of the plant, tuber length, number of leaves, aconitine and pseudoaconitine content were found to be enhanced (Bahuguna et al., 2013). Like wise, the factors influences on the root development including root length, root diameter and root fresh weight were enhanced with increase in the nutrient level when macronutirents like phosphorus and potassium applied on the growth of *Rheum tanguticum* (Shen et al., 2023). Similar type of the nutrient proportion changes applied in alpine conditions for growth of species *Rhododendron anthopogon* procure adaption mechanisms in adapting to limited availability of nutrition (Rathore et al., 2018). Mooneyh and Billings (1960) documented the storage and use of lipids and carbohydrates in the overall course of the carbohydrate cycle for alpine plants, while Russellr (1940) described it for Arctic plants. Low temperatures and dormancy bring on winter; such subterranean stores are rarely used. The stored carbohydrates are needed in the early rapid shoot growth and are mostly depleted when released during early summer dormancy. At this stage, growth is so quick that respiration greatly outpaces photosynthesis and food is consumed before it can be started (Hadleye & Bliss, 1964). Up to the completion of 75–90 years of shoot growth, this is maintained. After then, the respiration rate decreases, shoot growth slows, and the stocks of carbohydrates are depleted. Subterranean storage of carbohydrates is at its highest point of the year once dormancy commences.

3.2 Temperature

Alpine plants have evolved to adapt in low temperature conditions. The physiology of alpine plants has a strong influence on different levels of growth, photosynthesis, respiration, and metabolic processes in diverse temperature conditions. The specialized root structure and leaf morphology enhance adaptation to alpine environmental stress in *Potentilla saundersiana* by low temperature and UV radiation. Alpine plants such as *Soldanella alpina* and *Ranunculus glacialis* were reported to exhibit minimal effect on exposure to low temperature (Streb et al., 2003). Low temperature conditions lead to a short growing season of the alpine plants, which may limit the growth. Thus, the high temperatures can affect the growth of alpine plants (such as *R. glacialis, T. farfara, R. alpinus, P. alpina)* when there is sufficient availability of nutrients (Nagelmüller et al., 2017; Jiang et al., 2021). Net photosynthetic has been reported to increase in *Abies koreana* (Woo et al., 2008) and decrease in *Abies faxoniana* (Yu et al., 2022) under low temperature conditions. Low-growing alpine plant such as *P. euryphylla* exhibit net photosynthesis as well as respiration rate with increase in temperature (Atkin et al., 2006).

Low temperatures limits the growth phase and biomass production of plant cover in arctic-alpine communities (Bliss, 1956; Pollock & Eagles, 1987; Körner, 2003). Similar growth inhibition has also been reported in grassland plant species and winter crops (Gallagher, 1979; Wingler & Hennessy, 2016). Alpine plants, though cold-acclimated, face growth limitations, and their survival is challenged once temperatures reach a critical tipping point (Körner, 2008). Zero soil temperatures must take place over a minimum 6 weeks for the survival of angiosperms (Körner, 2011). Previous research has demonstrated that tissue growth reduces at temperatures as low as 5 °C and ceases entirely at or below 0 °C (Alvarez-Uria & Körner,

2007; Körner, 2008; Nagelmüller et al., 2016). Notably, at this temperature plants are still capable of CO_2 uptake at approximately 30% of their maximum photosynthetic capacity, indicating that growth limitation under such conditions is not carbon-restricted (Körner, 2003, 2011). Comparable low-temperature thresholds have been reported for leaf expansion, root elongation and and radial growth of xylem (Körner & Woodward, 1987; Schenker et al., 2014, Rossi et al., 2007),

3.3 UV- Radiation

Long term global climate change, as well as growing anthropogenic stress, natural ecosystem vegetation has to withstand extreme environmental stress conditions with a high rate of fluctuation in humidity, UV radiation, temperature, precipitation ionizing radiation (IR). While UV radiation is an essential component of numerous physiological processes such as plant growth and development, and photosynthesis. UV radiation consists of three broad bands: UV-A (315–400 nm), UV-B (280–315 nm) and UV-C (100–280 nm). The exposure to UV-A photons is considerably larger, UV-B photons are more energetic and can inflict extensive biological damage (Moan, 2001). UV-C radiation undergoes significant attenuation by stratospheric ozone layer and does not reach terrestrial habitats. Overexposure to UV-B in plants is associated with increased ROS formation, resulting in oxidative stress in the cells (Czégény et al., 2016). This stress can disruption of DNA, membranes and photosynthetic machinery, inhibition of growth and division, biochemical and physiological imbalance and cell death (Hollosy, 2002; Frohnmeyer & Staiger, 2003; Kumari et al., 2010; Pristov et al., 2013; Aphalo et al., 2015; Takahashi et al., 2015) Stress disrupts the cellular equilibrium of vegetation (Kapoor et al., 2021; Kumari et al., 2020).

UV stress, long daylight hours, low temperatures, a large day-night temperature differential, low pressure, low concentrations of carbon dioxide and oxygen, poor soils, and inconsistent humidity and precipitation are all characteristics of high-altitude plateau climates (Ahmadi-Lahijani & Moori, 2022; Shrestha et al., 2022; Hadi et al., 2022). Physiological and metabolic processes, are influenced by these attributes. Transpiration, respiration, chlorophyll content, cytochrome, size of leaves long with different metabolites produced in metabolic processes aid plants in adapting to abiotic stress under high altitude conditions (Kumar et al., 2020, 2023; Kapoor et al., 2021; Satyakam & Kumar, 2023). Glutathione reductase, superoxide dismutase, ascorbate oxidase, catalase along with other enzymatic and non-enzymatic antioxidants can be activated by the vast majority of alpine plants to scavenge ROS. Plants can adapt to environmental stress through the use of Ca burst and kinase cascade signals (Niazian, 2019; Ahmadi-Lahijani & Moori, 2022). ROS and oxidative stress induced by high UV-B radiation and inhibit deleterious effects on the fundamental cellular structures (Bornman et al., 2019; Kumari et al., 2010; Kurdziel et al., 2017). UV radiation exposure is greater with greater altitude because of greater atmospheric transparency. Consequently, plant communities at higher altitudes receive proportionally greater doses of UV-A and UV-B radiation.

Numerous other factors influence surface UV radiation exposure, including solar angle, season, time of day, latitude, cloudiness, air pollution and local weather seizure cloudiness local weather at airports (Wuttke et al., 2007). Alpine plants possess a range of adaptation mechanisms to cope with these stresses. These include the synthesis of photoprotective compounds, stress hormone regulation specialized morphological changes and activation of antioxidants systems (Wright et al., 2005; Tusevski et al., 2010; Huang et al., 2014; Cui et al., 2018; Agati et al., 2013).

4 Environmental Stresses in Alpine Zones

Alpine medicinal plants of the Himalayan region are increasingly affected by a complex interplay of environmental stresses, driven largely by climate change. These stresses, both abiotic and biotic, have profound impacts on growth, survival, and the production of pharmacological important bioactive metabolites. Alpine environments often undergo abiotic stresses such as temperature fluctuations, drought, high light and low soil nutrients. Northwest Himalayas, for example deep warming trends, like elevation-driven warming effects and an increase in air temperature of 1.6 °C. Plant-pollinator interaction and recolonization of medicinal species are directly influenced by these changes, thereby inducing changes in plant penology such as alterations in the duration of flowering and reproduction (Pandey et al., 2019). Physiological and biological responses of alpine plants are influenced by flood and drought cycles, which are exacerbated by increasing and altered precipitation patterns. The climate change effects has also escalated biotic stresses like invasive weed competition, infections due to pathogens and intensifying insect-pest infestation. In addition to lowering biomass and reproductive success they also compromised the production and quality of bio-active chemicals that are critical for medical use. Under water and heat stress, plants such as *Veleriana jatamansi* and *Podophyllum hexandrum* have shown significant shifts in photosynthetic efficiency, antioxidant activity and metabolite accumulation, Long term experiments linking in these physiological alterations with stressors caused by environment is still few though. Several alpine species have been observed shift in their ranges upward in response to the changes in their habitat and possible loss of habitat. The ecological stability and medicinal value of plant species are typically at risk from environmental stress in alpine habitats which require immediate conservation efforts and further psychological studies to understand tolerance mechanisms (Pandey et al., 2019).

5 Climate Change and Alpine Ecosystems

The particular climatic conditions and geographical constraints among the alpine ecosystems cause them to be one of the most susceptible to climate change. Narrow ecological niches, low average temperatures (~6.7 °C), and short growth seasons distinguish these high-altitude ecosystems, which occur above the climatic treeline. As global warming intensifies, these zones face significant ecological transformations, including shifts in species distributions, phenology, and vegetation dynamics. The limited elevation space in mountains restricts the upward migration of alpine species, leading to habitat compression or loss, especially for cryophilic and endemic species that have no refuge beyond the current alpine belt (Verrall & Pickering, 2020). Recent decades have seen a surge in research interest surrounding alpine vegetation and its response to climate change, particularly in regions like the Tibetan Plateau, which has become a focal point for global alpine research. Key ecological concerns include the advancement of tree lines, increased shrub encroachment, altered community composition, and biodiversity loss. The warming is also more intense at higher altitudes (elevation-dependent warming), compounding stress on these ecosystems. Despite technological advancements in monitoring such as remote sensing (e.g., NDVI, MODIS) large geographic gaps remain in the literature, especially in the Andes, Central Asia, and parts of North America. Long-term, geographically varied ecological monitoring is vital because alpine zones support a variety of biodata and provide vital ecosystem services like soil stabilization, carbon storage, and water provisioning (Verrall & Pickering, 2020).

6 Impact of Climate Change on Alpine Medicinal Plants

Climate change and the effect of climate change on vegetation have been a natural process over Earth's five billion years (Anonymous, 2007). The high mountain ecosystem has been most affected by global climate change during the last few decades. Snow, glaciers, and permafrost are the most sensitive to atmospheric changes; thus, plant habit problems have been generated due to their proximity. Furthermore, these effects negatively impacted therapeutic plants (Anonymous, 2007). Now, the climate of Nepal has also seen a drastic change directly impacting the natural vegetation at the Himalayan, tropical, and subtropical levels (Shrestha et al., 2001). Fuel consumption and deforestation, which release carbonic substances into the environment and raise temperatures, are termed as the major causes for climatic imbalance. Greenery and the forest are declining as a result of excessive population and uneven exploitation of natural resources. Climatic change has caused alterations in chemical constitution in a number of medicinal plants owing to alterations in growth and ripening periods (Zobayed et al., 2005).

The aggressive, lower-elevation species that are increasing their range, slow-growing, endemic, or range-restricted medicinal plants face particular challenges due to this geographical constraint. Furthermore, changes in the length and timing of snow cover have an impact on parameters that are essential to plant physiology and the synthesis of pharmacologically significant chemicals, such as soil moisture and growth season length (Pandey et al., 2019).

Plant species upward migration into cooler elevations is one of the most obvious effects of warming in alpine environments. However, many species face "mountaintop extinction," where inadequate habitat remains above their existing range, because alpine zones are constrained by mountain summits (Verrall & Pickering, 2020). Because they cannot compete with aggressive, lower-elevation species that are increasing their range, slow-growing, endemic, or range-restricted medicinal plants face particular challenges due to this geographical constraint. Furthermore, changes in the length and timing of snow cover have an impact on parameters that are essential to plant physiology and the synthesis of pharmacologically significant chemicals, such as soil moisture and growth season length (Pandey et al., 2019).

7 Evolutionary Responses and Adaptations

Alpine medicinal plants, especially graminoids found in the northwest Himalayas exhibit a variety of evolutionary adaptations influenced by water supply and elevation-driven cooling. To maintain bigger measurements and withstand mechanical stress, species in low-elevation dry steppes, on the other hand, evolve large-diameter capillaries and vast sclerenchymatic tissues. Additionally, moisture gradients influence clonal growth forms: wetter alpine meadows promote long-rhizomatous plants with reduced mechanical tissues, while arid habitats select for short-rhizomatous graminoids with hollow stems and thick epidermis for structural stability and water conservation. Despite strong environmental filtering, the overall phylogenetic signal was weak (Kmult = 0.0888), reinforcing the view that alpine plant form and function have been shaped more by environment than evolutionary lineage. These anatomical and physiological traits reflect evolutionary adjustments to the cold, arid, and variable climates of alpine zones, offering key insights into how medicinal plants may continue to evolve under ongoing climatic shifts (Dolezal et al., 2019).

8 Role of Genetic Adaptation in Alpine Plants

The concept of convergent adaptation- where unrelated species evolve similar traits in response to comparable environmental challenges (Endler, 1986; Losos, 2011; Stern, 2013). Historically, instances of such adaptations have been identified across various disciplines, including experimental evolution, phylogenetic studies and quantitative genetic studies (Arendt & Reznick, 2008; Wood et al., 2005). Rellstab et al. (2020) investigated the genomic basis of convergent adaptation across three alpine Brassicaceae species, *Arabidopsis helleri, Arabis alpina,, Cardamine resedifolia,* along environmental gradients in the Swiss Alps. Based upon whole genome sequencing the study reported extensive overlaps between the genes under selection but particularly the ones associated with water-related variables like precipitation and soil moisture. While most adaptive signals were species-specific, a significant majority of shared genes had non synonymous mutations, which suggest functionally relevant significance and proof of convergent evolution. Convergence was highest between the closest relatives, showing that shared ancestry as well as comparable environmental pressures induce parallel genetic responses. The study captures the complex interplay between shared environmental pressures and lineage-specific evolutionary routes in alpine plant adaptation.

9 Conservation Strategies and Sustainable Utilization

The Himalayan region is home to a rich and distinctive plant diversity that is unparalleled globally. Within the Indian Himalayas, which represent a highly diverse and evolving ecological zone, nearly 8, 644 plant species have been documented (Samant et al., 1998). A significant portion of this biodiversity comprises medicinal plants, with approximately 1, 748 species traditionally used for healing purposes (Kala, 2005). Due to their extensive therapeutic applications, these plants are in high demand both nationally and internationally. However, in the absence of well-regulated cultivation systems, they are often subjected to widespread, unregulated, and unsustainable wild harvesting. Around 90% of the herbal raw materials used in pharmaceutical production are gathered from natural habitats, with nearly 70% of such collection directly causes damage to diodiversity (Kala, 2010).

Despite their limited availability, the market demand for high-value Himalayan medicinal and aromatic plants (MAPs) continues to rise. These plants not only support the local economy and provide livelihoods for mountain communities but also play a vital role in traditional healthcare systems (Samant et al., 2007). Unfortunately, numerous Himalayan medicinal plants are now at risk, primarily due to excessive and unsystematic harvesting, habitat degradation, a lack of ecological understanding, and insufficient research and measures for not only conservation but also to support sustained utilization of plants as raw material for diverse applications. IUCN has classified around 120 Himalayan medicinal species under various threat categories (IUCN, 2023). Safeguarding these valuable plants and ensuring their sustainable use requires urgent, coordinated actions involving both scientific and governmental bodies. Strategies must include enforcing harvest restrictions, reintroducing species into their native environments, and enhancing both in situ and ex situ conservation efforts (Kala, 2006). Micropropagation is among the approach which can be utilized for conservation studies for endangered medicinal plants. Table 17.1 summarizes some of the micropropagation studies conducted in reference to Alpine plants. Substantial efforts are essential to protect these medicinal resources for future generations and to maintain the ecological integrity of the Himalayan region (Kala, 2010; Samant et al., 2007).

10 Future Research Directions

Prioritizing integrative approaches that leverage the power of omics technologies, such as genomics, transcriptomics, metabolomics, and proteomics, will help future research directions for alpine medicinal plants completely understand the molecular mechanisms underlying the biosynthesis of bioactive compounds and stress adaptation in these high-altitude species (Sarheed et al., 2021; Zhang et al., 2020). In order to determine how the distribution, phytochemical profiles, and therapeutic efficacy of alpine medicinal plants are impacted by ongoing climate change, climate modeling in conjunction with plant response predictions will be essential. The development of methods for preservation and sustainable harvesting procedures will be

Table 17.1 Summary of prominent studies conducted to conserve and mass propagate Himalayan plants through micropropagation

Species	Explant	Hormones	Activities	References
Adenostyles alpina	Leaf	NAA and 2, 4-D	Showed highest percentage of callus and shoot regeneration.	Gianguzzi et al. (2023)
Aconitum balfourii Stapf	Tubers	Benzylaminopurine (BAP) and α-Naphthaleneacetic acid (NAA).	Callus formation	Pandey et al. (2002)
Podophyllum hexandrum Royle	Fruits	Benzylaminopurine (BAP) and Basal media	Somatic embryogenesis and multiple shoot regeneration	Pandey et al. (2002)
Picrorhiza kurrooa Royle	Seeds	Benzylaminopurine (BAP) and Indole-acetic acid (IAA)	Shoot multiplication	Pandey et al. (2002)
Aconitum napellus	Shoot tip/ Stem nodes	IAA	Plantlets	Watad et al. (1995)
Aconitum heterophyllum	Leaf	BAP + IAA	Embryo, plantlets	Giri et al. (1993)
Aconitum carmichaeli	Anther/ Shoot	NAA	Embryo, Multiple shoots, Microtubers	Hatano et al. (1988)
Nardostachys jatamansi	Root, Petiole	BAP + IAA+ NAA	Shoot, bud regeneration	Mathur (1992)
P. kurrooa	Shoot tip and single node cutting	IAA+ NAA	Multiple shoots, Plantlets	Lal et al. (1988)
Rheum emodi	Shoot tip, leaf	IAA	Multiple shoots, Plantlets	Upadhyay et al. (1989)
Saussurea lappa	Shoot tip	BAP	Shoot, bud regeneration	Arora and Bhojwani (1989)
Scutellaria alpina	Seeds	NAA and Thidiazuron (TDZ)	Callus and multiple shoot formation	Grzegorczyk-Karolak et al. (2016)
Rungia klossii	Seeds	BAP or Kinetin	Multiple shoot development	Ermayanti and Mukhsia (2012)
Chryanthemum	Shoot/ Root	2, 4-D kinetin and BAP	Somatic embryo and plantlet development	Shinoyama et al. (2004)

made possible by this (Chaudhary et al., 2022). Additionally, another essential frontier that may identify new medicinal compounds while preserving indigenous knowledge and cultural heritage is the integration of ethnobotanical insights with state-of-the-art analytical tools. This can be achieved by integrating contemporary scientific techniques with traditional ecological knowledge. These collaborative efforts will not only advance our understanding of alpine medicinal plants but also ensure their preservation and continued usage for future generations in spite of environmental and human-caused stressors.

11 Conclusion

Alpine medicinal plants are able to survive at very high elevations because of special physiological and morphological adaptations. Despite their adaptability, these plants are struggle to survive due to rising pressures from habitat loss, deforestation and climate change. Habitat loss, altered precipitation patterns and higher temperatures are obstructing their growth, distributions and medicinal properties. To survive, conservation on a large scale in the forms of conservation of their habitats, sustainable harvesting and studies on their adaptation processes must be undertaken. An integrated approach is presently required to address the detoriating biodiversity in Alpine region. Habitat restoration, implementation of regulations to restrict human activity in these regions, scientific studies to analyze the growth pattern and identify the factors having director indirect impact on Alpine plants can significantly contribute towards restoration and sustaining the plant biodiversity. Scientists, policymakers, and local communities must come together to preserve these vital plants for their ecological worth, medicinal worth and role in mountain livelihoods.

References

Agati, G., Brunetti, C., di Ferdinando, M., Ferrini, F., Pollastri, S., et al. (2013). Functional roles of flavonoids in photoprotection: New evidence, lessons from the past. *Plant Physiology and Biochemistry, 72*, 35–45.

Ahmadi-Lahijani, M. J., & Moori, S. (2022). Current status of medicinal plants in perspective of environmental challenges and global climate changes. In T. Aftab (Ed.), *Environmental challenges and medicinal plants. Environmental challenges and solutions* (pp. 1–28). Springer.

Alvarez-Uria, P., & Körner, C. (2007). Low temperature limits of root growth in deciduous and evergreen temperate tree species. *Functional Ecology, 21*, 211–218.

Alyas, J., Khalid, N., Ishaque, S., Fatima, H., Hashim, M., Hassan, S., Bukhari, S., Hano, C., Abbasi, B. H., & Anjum, S. (2023). Light (high light/UV radiation) modulates adaptation mechanisms and secondary metabolite production in medicinal plants. In *Medicinal plants: Their response to abiotic stress* (pp. 363–390). Springer Nature Singapore.

Amen, R. D. (1966). The extent and role of seed dormancy in alpine plants. *The Quarterly Review of Biology, 41*(3), 271–281.

Anonymous. (2007). IPCC (Intergovernmental Panel on Climate Change). Climate change 2007: The physical science basis..

Aphalo, P. J., Jansen, M. A. K., McLeod, A. R., & Urban, O. (2015). Ultraviolet radiation research: From the field to the lab-oratory and back. *Plant Cell and Environment, 38*, 853–855.

Arendt, J., & Reznick, D. (2008). Convergence and parallelism reconsidered: What have we learned about the genetics of adaptation? *Trends in Ecology & Evolution, 23*, 26–32.

Arora, R., & Bhojwani, S. S. (1989). In vitro propagation and low temperature storage of Saussurea lappa C.B. Clarke—An endangered medicinal plant. *Plant Cell Reports, 8*, 44–47.

Atkin, O. K., Scheurwater, I., & Pons, T. L. (2006). High thermal acclimation potential of both photosynthesis and respiration in two lowland Plantago species in contrast to an alpine congeneric. *Global Change Biology, 12*(3), 500–515.

Bahuguna, R., Prakash, V., & Bisht, H. (2013). Quantitative enhancement of active content and biomass of two aconitum species through suitable cultivation technology. *International Journal of Conservation Science, 4*(1).

Bashyal, S. (2023). *Plant functional traits of Aconitum Spicatum and Aconitum naviculare along the elevation gradients in central Nepal* (Doctoral dissertation). Department of Botany.

Bhutiyani, M. R., Kale, V. S., & Pawar, N. J. (2010). Climate change and the precipitation variations in the northwestern Himalaya: 1866-2006. *International Journal of Climatology, 30*(4), 535–548.

Billings, W. D. (1973). Arctic and alpine vegetations: Similarities, differences, and susceptibility to disturbance. *Bioscience, 23*(12), 697–704.

Billings, W. D. (1974). Adaptations and origins of alpine plants. *Arctic and Alpine Research, 6*(2), 129–142.

Billings, W. D., Shaver, G. R., & Trent, A. W. (1973). Temperature effects of growth and respiration of roots and rhizomes in tundra graminoids.

Bliss, L. C. (1956). A comparison of plant development in microenvironments of arctic and alpine tundras. *Ecological Monographs, 26*, 303–337.

Bornman, J. F., Barnes, P. W., Robson, T. M., Robinson, S. A., Jansen, M. A. K., et al. (2019). Linkages between stratospheric ozone, UV radiation and climate change and their implications for terrestrial ecosystems. *Photochemical and Photobiological Sciences, 18*(3), 681–716.

Briceño, V. F., Harris-Pascal, D., Nicotra, A. B., Williams, E., & Ball, M. C. (2014). Variation in snow cover drives differences in frost resistance in seedlings of the alpine herb Aciphylla glacialis. *Environmental and Experimental Botany, 106*, 174–181.

Candra, S., & Joshi, S. C. (2002). Diurnal and seasonal variation in CO2 levels in the surface air of Garhwal Himalaya. *Indian Journal of Forestry, 25*(2), 205–208.

Chabot, B. F., & Billings, W. D. (1972). Origins and ecology of the Sierran alpine flora and vegetation. *Ecological Monographs, 42*(2), 163–199.

Chaudhary, V. B., Liu, Y., Luo, T., & Bardgett, R. D. (2022). Complementary trade-offs between plant roots and mycorrhizal fungi traits across four alpine grasslands along a precipitation gradient in the northern Xizang Plateau. *Journal of Plant Ecology*, rtaf058. https://doi.org/10.1093/jpe/rtaf058

Cui, G., Li, B., He, W., Yin, H., Liu, S., et al. (2018). Physiological analysis of the effect of altitudinal gradients on Leymus secalinus on the Qinghai-Tibetan plateau. *PLoS One, 13*(9), e0202881.

Czégény, G., Mátai, A., & Hideg, É. (2016). UV-B effects on leaves–oxidative stress and acclimation in controlled environments. *Plant Science, 248*, 57–63.

Dolezal, J., Klimes, A., Dvorsky, M., Riha, P., Klimesova, J., & Schweingruber, F. (2019). Disentangling evolutionary, environmental and morphological drivers of plant anatomical adaptations to drought and cold in Himalayan graminoids. *Oikos, 128*(11), 1657–1672.

Endler, J. A. (1986). *Natural selection in the wild.* Princeton University Press.

Ermayanti, T. M., & Mukhsia, A. (2012, September). Applications of tissue culture for the production of native plant species to accelerate the alpine reclamation program at PT Freeport Indonesia. In *Proceedings of the seventh international conference on mine closure*. Australian Centre for Geomechanics, Perth (pp. 1–10).

Frohnmeyer, H., & Staiger, D. (2003). Ultraviolet-b radiation-mediated responses in plants. Balancing damage and protection. *Plant Physiology, 133*, 1420–1428.

Gallagher, J. N. (1979). Field studies of cereal leaf growth. I. Initiation and expansion in relation to temperature and ontogeny. *Journal of Experimental Botany, 30*, 625–636.

García-Plazaola, J. I., Rojas, R., Christie, D. A., & Coopman, R. E. (2015). Photosynthetic responses of trees in high-elevation forests: Comparing evergreen species along an elevation gradient in the Central Andes. *AoB Plants, 7*, plv058.

Germino, M. J., & Smith, W. K. (2000). High resistance to low-temperature photoinhibition in two alpine, snowbank species. *Physiologia Plantarum, 110*(1), 89–95.

Gianguzzi, V., Barone, G., Di Gristina, E., Sottile, F., & Domina, G. (2023). Micropropagation of endemic endangered taxa of the Italian flora: Adenostyles alpina subsp. macrocephala (Asteraceae), as a case study. *Plants, 12*(7), 1530.

Gimenez, E., Salinas, M., & Manzano-Agugliaro, F. (2018). Worldwide research on plant defense against biotic stresses as improvement for sustainable agriculture. *Sustainability, 10*(2), 391.

Giri, A., Ahuja, P. S., & Kumar, P. V. A. (1993). Somatic embryogenesis and plant regeneration from callus culture of Aconitum heterophyllum Wall. *Plant Cell, Tissue and Organ Culture, 32*, 213–218.

Grzegorczyk-Karolak, I., Kuźma, Ł., & Wysokińska, H. (2016). In vitro cultures of Scutellaria alpina as a source of pharmacologically active metabolites. *Acta Physiologiae Plantarum, 38*, 1–9.

Hadi, A., Singh, S., Nawchoo, I. A., Rafiq, S., & Ali, S. (2022). Impacts of habitat variability on the phenotypic traits of Aconitum violaceum Jacq. Ex Stapf. at different altitudes and environmental conditions in the Ladakh Himalaya India. *Plant Science Today, 9*, 546–554.

Hadleye, B., & Bliss, L. C. (1964). Energy relationships of alpine plants on Mt Washington, New Hampshire. *Ecological Monographs, 34*, 331–357.

Hatano, K., Kamura, K., Shoyama, Y., & Nishioka, I. (1988). Clonal multiplication of Aconitum carmichaeli Debx. by tip tissue culture and alkaloid contents of clonally propagated plants. *Planta Medica, 54*, 152–154.

Hollosy, F. (2002). Effects of ultraviolet radiation on plant cells. *Micron, 33*, 179–197.

Hou, D., Shu, G., & Ren, Z. (2006). Effect of nitrogen, phosphorus, and potassium on several nitrogen metabolism enzymes of Aconitum carmichaeli Dexb. (Ranunculaceae). *Southwest China Journal of Agricultural Sciences, 19*(5), 857–862.

Hu, Z., Wang, G., Sun, X., Huang, K., Song, C., Li, Y., Sun, S., Sun, J., & Lin, S. (2024). Biological factor controls the variations in water use efficiency of an alpine meadow during the growing season in a permafrost region of the Qinghai-Tibet Plateau. *Agricultural Water Management, 296*, 108811. https://doi.org/10.1016/j.agwat.2024.108811

Huang, B., da Costa, M., & Jiang, Y. (2014). Research advances in mechanisms of turfgrass tolerance to abiotic stresses: From physiology to molecular biology. *Critical Reviews in Plant Sciences, 33*(2–3), 141–189.

IUCN. (2023). IUCN World Heritage Outlook 4: A conservation assessment of all natural World Heritage sites. *International Union for Conservation of Nature*

Jiang, Y., Cheng, X. I., Zi, H., & Huang, X. (2021). Physiological and ecological responses of an alpine plant picea likiangensis at different altitudinal gradient Sn. *Pakistan Journal of Botany, 53*(6), 2233–2240.

Kala, C. P. (2005). Indigenous uses, population density, and conservation of threatened medicinal plants in protected areas of the Indian Himalayas. *Conservation Biology, 19*(2), 368–378.

Kala, C. P. (2006). Medicinal plants: Future source of new drugs. *International Journal of Herbal Medicine, 4*(1), 1–7.

Kala, C. P. (2010). Status of medicinal and aromatic plants and their conservation strategies in Uttaranchal State of India. *International Journal of Biodiversity Science, Ecosystem Services & Management, 6*(1–2), 43–52.

Kapoor, B., Kumar, A., & Kumar, P. (2021). Transcriptome repository of North-Western Himalayan endangered medicinal herbs: A paramount approach illuminating molecular perspective of phytoactive molecules and secondary metabolism. *Molecular Genetics and Genomics, 296*, 1177–1202.

Körner, C., & Woodward, F. I. (1987). The dynamics of leaf extension in plants with diverse altitudinal ranges: II. Field studies in Poa species between 600 and 3200 m altitude. *Oecologia, 72*(2), 279–283. https://doi.org/10.1007/BF00379279

Körner, C. (2003). *Alpine plant life: Functional plant ecology of high mountain ecosystems* (2nd ed.). Springer.

Körner, C. (2008). Winter crop growth at low temperature may hold the answer for alpine treeline formation. *Plant Ecology and Diversity, 1*, 3–11.

Körner, C. (2011). Coldest places on earth with angiosperm plant life. *Alpine Botany, 121*, 11–22.

Kumar, K., Joshi, S., & Joshi, V. (2018). Climate variability, vulnerability, and coping mechanism in Alaknanda catchment, Central Himalaya, India. *Ambio*, 286–291.

Kumar, R., Joshi, R., Kumari, M., et al. (2020). Elevated CO2 and temperature influence key proteins and metabolites associated with photosynthesis, antioxidant and carbon metabolism in Picrorhiza kurroa. *Journal of Proteomics, 219*, 103755.

Kumar, P., Acharya, V., & Warghat, A. R. (2021). Comparative transcriptome analysis infers bulb derived in vitro cultures as a promising source for sipeimine biosynthesis in Fritillaria cirrhosa D. Don (Liliaceae, syn. Fritillaria roylei Hook.)-High value Himalayan medicinal herb. *Phytochemistry, 183*, 112631.

Kumar, V., Kumar, P., Bhargava, B., Sharma, R., Irfan, M., & Chandora, R. (2023). Transcriptomic and metabolomic reprogramming to explore the high-altitude adaptation of medicinal plants: A review. *Journal of Plant Growth Regulation, 42*(12), 7315–7329.

Kumari, R., Singh, S., & Agrawal, S. B. (2010). Response of ultraviolet-b induced antioxidant defense system in a medicinal plant. Acoruscalamus. *Journal of Environmental Biology, 31*(6), 907–911.

Kumari, M., Joshi, R., & Kumar, R. (2020). Metabolic signatures provide novel insights to Picrorhiza kurroa adaptation along the altitude in Himalayan region. *Metabolomics, 16*, 1–17.

Kurdziel, M., Filek, M., & Łabanowska, M. (2017). The impact of short-term UV irradiation on grains of sensitive and tolerant cereal genotypes studied by EPR. *Journal of the Science of Food and Agriculture, 98*(7), 2607–2616.

Ladinig, U., Hacker, J., Neuner, G., & Wagner, J. (2013). How endangered is sexual reproduction of high-mountain plants by summer frosts? Frost resistance, frequency of frost events and risk assessment. *Oecologia, 171*, 743–760.

Lal, N., Ahuja, P. S., Kukreja, A. K., & Pandey, B. (1988). Clonal propagation of Picrorhiza kurrooa Royle ex Benth. by shoot tip culture. *Plant Cell Reports, 7*, 202–205.

Landi, M., Tattini, M., & Gould, K. S. (2015). Multiple functional roles of anthocyanins in plant-environment interactions. *Environmental and Experimental Botany, 119*, 4–17.

Larcher, W. (2003). *Physiological plant ecology: Ecophysiology and stress physiology of functional groups* (4th ed.). Springer.

Loik, M. E., Still, C. J., Huxman, T. E., & Harte, J. (2004). In situ photosynthetic freezing tolerance for plants exposed to a global warming manipulation in the Rocky Mountains, Colorado, USA. *New Phytologist, 162*(2), 331–341.

Losos, J. B. (2011). Convergence, adaptation, and constraint. *Evolution, 65*, 1827–1840.

Magaña Ugarte, R., Escudero, A., & Gavilán, R. G. (2019). Metabolic and physiological responses of Mediterranean high-mountain and alpine plants to combined abiotic stresses. *Physiologia Plantarum, 165*(2), 403–412.

Mathur, J. (1992). In vitro morphogenesis in Nardostachys jatamansi DC: Shoot regeneration from callus derived roots. *Annals of Botany, 70*, 419–422.

Mauseth, J. D. (2006). Structure–function relationships in highly modified shoots of Cactaceae. *Annals of Botany, 98*(5), 901–926. https://doi.org/10.1093/aob/mcl177

Moan, J. (2001). Non-ionizing Radiation. 7. Visible light and UV radiation. In A. Brune, R. Hellborg, B. R. R. Persson, & R. Pääkkönen (Eds.), *Radiation at home, outdoors and in the workplace* (pp. 69–85). Scandinavian Science Publisher.

Mooneyh, A., & Billings, W. D. (1960). The annual carbohydrate cycle of alpine plants as related to growth. *American Journal of Botany, 47*, 594–598.

Moore, M. J., Mota, J. F., Douglas, N. A., & Flores Olvera, H. (2014). The ecology, assembly, and evolution of gypsophile floras. In A. Rajkaran (Ed.), *Gypsum Ecosystems* (pp. 97–128). Nova Science Publishers

Nagelmüller, S., Hiltbrunner, E., & Körner, C. (2016). Critically low soil temperatures for root growth and root morphology in three alpine plant species. *Alpine Botany, 126*, 11–21.

Nagelmüller, S., Hiltbrunner, E., & Körner, C. (2017). Low temperature limits for root growth in alpine species are set by cell differentiation. *AoB Plants, 9*(6), plx054.

Niazian, M. (2019). Application of genetics and biotechnology for improving medicinal plants. *Planta, 249*(4), 953–973.

Pepin, N., Bradley, R. S., Diaz, H. F., Baraer, M., Caceres, E. B., Forsythe, N., Fowler, H., Greenwood, G., Hashmi, M. Z., Liu, X. D., Miller, J. R., Ning, L., Ohmura, A., Palazzi, E., Rangwala, I., Schöner, W., Severskiy, I., Shahgedanova, M., Wang, M. B.,... Zawar-Reza, P. (2015). Elevation-dependent warming in mountain regions of the world. *Nature Climate Change, 5*(5), 424–430. https://doi.org/10.1038/nclimate2563

Pandey, H., Chandra, B., Nadeem, M., Kumar, A., Nandi, S. K., & Palni, L. M. S. (2002). In vitro propagation of some alpine medicinal herbs of the Himalayan region. In *Role of plant tissue culture in biodiversity conservation and economic development* (pp. 297–323).

Pandey, V., Bhatt, I. D., & Nandi, S. K. (2019). Environmental stresses in Himalayan medicinal plants: Research needs and future priorities. *Biodiversity and Conservation, 28*(9), 2431–2455.

Pescador, D. S., Sierra-Almeida, Á., Torres, P. J., & Escudero, A. (2016). Summer freezing resistance: A critical filter for plant community assemblies in Mediterranean high mountains. *Frontiers in Plant Science, 7*, 194.

Pollock, C. J., & Eagles, C. F. (1987). Low temperature and the growth of plants. In F. I. Woodward & S. P. Long (Eds.), *Plants and temperature* (pp. 157–180). Cambridge University Press.

Pristov, J. B., Jovanović, S. V., Mitrović, A., & Spasojević, I. (2013). UV-irradiation provokes generation of superoxide on cell wall polygalacturonic acid. *Physiologia Plantarum, 148*, 574–581.

Rathore, M. S., Paliwal, N., Anand, K. G. V., & Agarwal, P. K. (2015). Somatic embryogenesis and in vitro plantlet regeneration in Salicornia brachiata Roxb. *Plant Cell, Tissue and Organ Culture, 120*, 355–360.

Rathore, N., Thakur, D., & Chawla, A. (2018). Seasonal variations coupled with elevation gradient drives significant changes in eco-physiological and biogeochemical traits of a high-altitude evergreen broadleaf shrub, Rhododendron anthopogon. *Plant Physiology and Biochemistry, 132*, 708–719.

Rawat, J. M., Rawat, B., Mishra, S., Negi, R., Mishra, S. N., Chandra, A., & Nautiyal, S. 2014. Altitudinal and seasonal variation in bioactive compound aconitine in Aconitum violaceum, a threatened medicinal plant of Indian Himalayan region. *International Journal of Advanced Research, 2*, 981–988.

Rellstab, C., Zoller, S., Sailer, C., Tedder, A., Gugerli, F., Shimizu, K. K., et al. (2020). Genomic signatures of convergent adaptation to alpine environments in three Brassicaceae species. *Molecular Ecology, 29*(22), 4350–4365.

Russellr, S. (1940). Physiological and ecological studies on an arctic vegetation. 111. Observations of carbon assimilation, carbohydrate storage and stomatal movement in relation to the growth of plants on Jan Mayen Island. *Journal of Ecology, 28*, 289–309.

Rossi, S., Deslauriers, A., Anfodillo, T., & Carraro, V. (2007). Evidence of threshold temperatures for xylogenesis in conifers at high altitudes. Oecologia, 152(1), 1–12. https://doi.org/10.1007/s00442-006-0625-7

Samant, S. S., Dhar, U., & Palni, L. M. S. (1998). *Medicinal plants of Indian Himalaya: Diversity, distribution and potential values*. Gyanodaya Prakashan.

Samant, S. S., Pant, S., & Arya, S. C. (2007). Medicinal plants of the Indian Himalayan Region: Diversity, distribution and potential values. *Current Science, 92*(7), 889–898.

Sarheed, M. M., Rajabi, S., Kunová, V., & Kučera, L. (2021). Plant metabolomics: An overview of the role of primary and secondary metabolites against different environmental stress factors. *Life, 11*(7), 706.

Satyakam, J. R., & Kumar, R. (2023). Unique metabolic shift reveals potential mechanism of cold and freezing acclimatization. *Journal of Plant Growth Regulation, 16*, 1–7.

Shen, N., Xu, W., Guo, F., Cui, Y., Chen, Y., & Su, X. (2023). Effect of fertilizer and harvesting stage on growth and bioactive compounds content of rheum tanguticum. *Journal of Plant Nutrition, 46*(18), 4462–4476.

Shi, S. B., Zhu, W. Y., Li, H. M., Zhou, D. W., Han, F., Zhao, X. Q., & Tang, Y. H. (2004). Photosynthesis of Saussurea superba and Gentiana straminea is not reduced after long-term enhancement of UV-B radiation. *Environmental and Experimental Botany, 51*(1), 75–83.

Shinoyama, H., Nomura, Y., Tsuchiya, T., & Kazuma, T. (2004). A simple and efficient method for somatic embryogenesis and plant regeneration from leaves of chrysanthemum [Dendranthema× grandiflorum (Ramat.) Kitamura]. *Plant Biotechnology, 21*(1), 25–33.

Shrestha, K.K., Tiwari, N. N., & Ghimire S. K.. (2001). MAPDON—Medicinal and Aromatic Plant Database of Nepal. In *Proceedings of Nepal-Japan Joint Symposium 2000*. Department of Plant Resources, Thapathali, Kathmandu, Nepal, pp. 53–74.

Shrestha, U. B., Lamsal, P., Ghimire, S. K., Shrestha, B. B., Dhaka, S., Shrestha, S., & Atreya, K. (2022). Climate change- induced distributional change of medicinal and aromatic plants in the Nepal Himalaya. *Ecology and Evolution, 12*(8), 9204.

Sierra-Almeida, A., Cavieres, L. A., & Bravo, L. A. (2009). Freezing resistance varies within the growing season and with elevation in high-Andean species of central Chile. *New Phytologist, 182*(2), 461–469.

Sierra-Almeida, A., Reyes-Bahamonde, C., & Cavieres, L. A. (2016). Drought increases the freezing resistance of high-elevation plants of the Central Chilean Andes. *Oecologia, 181*, 1011–1023.

Sklenář, P. (2017). Seasonal variation of freezing resistance mechanisms in north-temperate alpine plants. *Alpine Botany, 127*, 31–39.

Streb, P., & Cornic, G. (2011). Photosynthesis and antioxidative protection in alpine herbs. In C. Lütz (Ed.), *Plants in Alpine Regions* (pp. 75–89). Springer. https://doi.org/10.1007/978-3-7091-0136-0_7

Streb, P., Aubert, S., Gout, E., & Bligny, R. (2003). Reversibility of cold-and light-stress tolerance and accompanying changes of metabolite and antioxidant levels in the two high mountain plant species Soldanella alpina and Ranunculus glacialis. *Journal of Experimental Botany, 54*(381), 405–418.

Schenker, G., Lenz, A., Körner, C., & Hoch, G. (2014). Physiological minimum temperatures for root growth in seven common European broad-leaved tree species. *Tree Physiology, 34*(3), 302–313. https://doi.org/10.1093/treephys/tpu003

Stern, D. L. (2013). The genetic causes of convergent evolution. *Nature Reviews Genetics, 14*(11), 751–764. https://doi.org/10.1038/nrg3483

Takahashi, S., Kojo, K. H., Kutsuna, N., Endo, M., Toki, S., et al. (2015). Differential responses to high- and lowdose ultraviolet-b stress in tobacco bright yellow-2 cells. *Frontiers in Plant Science, 6*, 254.

Taschler, D., & Neuner, G. (2004). Summer frost resistance and freezing patterns measured in situ in leaves of major alpine plant growth forms in relation to their upper distribution boundary. *Plant, Cell & Environment, 27*(6), 737–746.

Tusevski, O., Trpevski, M., Lozanovska, I., Talevska, A., Ugurovsk, D., et al. (2010). Antioxidant activity of phenolic compounds in methanolic extracts of some Macedonian medicinal plants collected on Jablanica Mountain. *Biology Students' Research Society-Skopje, 4*, 103–111.

Upadhyay, R., Arumugam, N., & Bhojwani, S. S. (1989). In vitro propagation of Picrorhiza kurrooa Royel ex Benth. - An endangered species of medicinal importance. *Phytomorphology, 39*(2, 3), 235–242.

Venn, S. E., Morgan, J. W., & Lord, J. M. (2013). Foliar freezing resistance of Australian alpine plants over the growing season. *Austral Ecology, 38*(2), 152–161.

Verrall, B., & Pickering, C. M. (2020). Alpine vegetation in the context of climate change: A global review of past research and future directions. *Science of the Total Environment, 748*, 141344.

Watad, A. A., Kochba, M., Nissim, A., & Gaba, V. (1995). Improvement of Aconitum napellus micropropagation by liquid culture on floating membrane rafts. *Plant Cell Reports, 14*, 345–348.

Wickens, G. E. (1998). *Ecophysiology of economic plants in arid and semi-arid lands*. Springer Berlin Heidelberg.

Wingler, A., & Hennessy, D. (2016). Limitation of grassland productivity by low temperature and seasonality of growth. *Frontiers in Plant Science, 7*, 1130.

Woo, L. S., Hoon, Y. B., Don, H. S., Ho, S. J., & Joo, L. J. (2008). Genetic variation in natural populations of Abies nephrolepis Max. in South Korea. *Annals of Forest Science, 65*(3), 1.

Wood, T. E., Burke, J. M., & Rieseberg, L. H. (2005). Parallel genotypic adaptation: When evolution repeats itself. *Genetica, 123*, 157–170.

Wright, I. J., Reich, P. B., Cornelissen, J. H. C., Falster, D., Groom, P., et al. (2005). Modulation of leaf economic traits and trait relationships by climate. *Global Ecology and Biogeography, 14*, 411–421.

Wuttke, S., Naggar, S. E., Bluszcz, T., & Schrems, O. (2007). Ship-borne measurements of erythemal UV irradiance and ozone content in various climate zones. *Photochemical & Photobiological Sciences, 6*, 1081–1088.

Xiong, F. S., Ruhland, C. T., & Day, T. A. (1999). Photosynthetic temperature response of the Antarctic vascular plants Colobanthus quitensis and Deschampsia antarctica. *Physiologia Plantarum, 106*(3), 276–286. https://doi.org/10.1034/j.1399-3054.1999.106304.x13.1

Yu, J., Bing, H., Chang, R., Cui, Y., Shen, G., Wang, X., et al. (2022). Microbial metabolic limitation response to experimental warming along an altitudinal gradient in alpine grasslands, eastern Tibetan Plateau. *Catena, 214*, 106243.

Zhang, J., Liu, Y. X., Zhang, N., Hu, B., Jin, T., Xu, H., et al. (2020). NRT1.1B is associated with root microbiota composition and nitrogen use in field-grown rice. *Nature Biotechnology, 37*(6), 676–684.

Zhao, Q., Dong, M., Li, M., Jin, L., & Pare, P. W. (2023). Light-induced flavonoid biosynthesis in Sinopodophyllum hexandrum with high-altitude adaptation. *Plants, 12*(3), 575.

Zobayed, S. M. A., Afreen, F., & Kozai, T. (2005). Temperature stress can alter the photosynthetic efficiency and secondary metabolite concentrations in St. John's wort. *Plant Physiology and Biochemistry, 43*, 977–984.

Correction to : Functional Diversity and Ecophysiological Traits Along Elevational Gradients

Shiv Rattan

Correction to:
Chapter 3 in : P. Kumar et al. (eds.), *Climate Resilience and Molecular Adaptation in Alpine Medicinal Plants*,
https://doi.org/10.1007/978-981-95-6923-6_3

The original version of the chapter "Functional Diversity and Ecophysiological Traits Along Elevational Gradients" was inadvertently published with incorrect abstract. The complete abstract has now been updated in the chapter.

The updated version of this chapter can be found at
https://doi.org/10.1007/978-981-95-6923-6_3

P. Kumar et al. (eds.), *Climate Resilience and Molecular Adaptation in Alpine Medicinal Plants*, https://doi.org/10.1007/978-981-95-6923-6_18

MIX
Papier aus verantwortungsvollen Quellen
Paper from responsible sources
FSC® C105338

If you have any concerns about our products,
you can contact us on
ProductSafety@springernature.com

In case Publisher is established outside the EU,
the EU authorized representative is:
Springer Nature Customer Service Center GmbH
Europaplatz 3, 69115 Heidelberg, Germany

Printed by Libri Plureos GmbH
in Hamburg, Germany